Tumor-Suppressing Viruses, Genes, and Drugs

Tumor-Suppressing Viruses, Genes, and Drugs

Innovative Cancer Therapy Approaches

Edited by
Hiroshi Maruta
Ludwig Institute for Cancer Research
Royal Melbourne Hospital
Parkville/Melbourne, Australia

ACADEMIC PRESS

A Harcourt Science and Technology Company

San Diego   San Francisco   New York
Boston   London   Sydney   Tokyo

Academic Press
A Harcourt Science and Technology Company
525 B Street, Suite 1900, San Diego, California 92101-4495, USA
http://www.academicpress.com

Academic Press
Harcourt Place, 32 Jamestown Road, London NW1 7BY, UK
http://www.academicpress.com

Library of Congress Catalog Card Number: 2001095847

International Standard Book Number: 0-12-476249-2

Printed and bound in the United Kingdom
Transferred to Digital Printing, 2011

Contents

Contributors

Numbers in parentheses indicate the pages on which the authors' contributions begin.

Marc G. Achen (261), Ludwig Institute for Cancer Research, Royal Melbourne Hospital, Parkville/Melbourne 3050, Australia.

Yukinobu Arata (123), Graduate School of Biological Sciences, Nara Institute of Science and Technology, Ikoma, Nara 630-0101, Japan.

Lydia Armstrong (379), Department of Tumor Biology, Schering-Plough Research Institute, Kenilworth, New Jersey 07033.

Antony W. Burgess (341), Ludwig Institute for Cancer Research/CRC for Cellular Growth Factors, Parkville/Melbourne 3050, Australia; and Institute for Advanced Studies, Hebrew University, Jerusalem 91904, Israel.

Eve Damiens (145), CNRS, Station Biologique, 29682 Roscoff Cedex, Bretagne, France.

Michael I. Ebright (45), Department of Surgery, Memorial Sloan-Kettering Cancer Center, New York, New York 10021.

Wafik S. El-Deiry (77), Laboratory of Molecular Oncology and Cell Cycle Regulation, Howard Hughes Medical Institute, Department of Medicine, Genetics, and Pharmacology, University of Pennsylvania School of Medicine, Philadelphia, Pennsylvania 19104.

Yuman Fong (45), Department of Surgery, Memorial Sloan-Kettering Cancer Center, New York, New York 10021.

Lynda K. Hawkins (1), Imperial Cancer Research Fund, Program for Viral and Genetic Therapy of Cancer, Imperial College School of Medicine, Hammersmith Hospital, London W12 0NN, England.

Hong He (361), Ludwig Institute for Cancer Research, Royal Melbourne Hospital, Parkville/Melbourne 3050, Australia.

Jun-ya Kato (123), Graduate School of Biological Sciences, Nara Institute of Science and Technology, Ikoma, Nara 630-0101, Japan.

Masahiro Kawabata (199), Department of Biochemistry, The Cancer Institute of the Japanese Foundation for Cancer Research (JFCR), Toshima-Ku, Tokyo 170-8455, Japan.

Hiroaki Kawasaki (233), Department of Chemistry and Biotechnology, Graduate School of Engineering, The University of Tokyo, Hongo, Tokyo 113-8656, Japan; and Gene Discovery Research Center, National Institute of Advanced Industrial Science and Technology (AIST), Tsukuba Science City 305-8562, Japan.

David Kirn (1), Imperial Cancer Research Fund, Program for Viral and Genetic Therapy of Cancer, Imperial College School of Medicine, Hammersmith Hospital, London W12 0NN, England.

Chandra C. Kumar (379), Department of Tumor Biology, Schering-Plough Research Institute, Kenilworth, New Jersey 07033.

Tomoko Kuwabara (233), Department of Chemistry and Biotechnology, Graduate School of Engineering, The University of Tokyo, Hongo, Tokyo 113-8656, Japan; and Gene Discovery Research Center, National Institute of Advanced Industrial Science and Technology (AIST), Tsukuba Science City 305-8562, Japan.

Noboru Kuzumaki (177), Institute for Genetic Medicine, Hokkaido University, Sapporo 060-0815, Japan.

Patrick W. K. Lee (31), Cancer Biology Research Group and Department of Microbiology and Infectious Diseases, Faculty of Medicine, University of Calgary, Calgary, Alberta, Canada T2N 4N1.

Wen-Hwa Lee (97), Department of Medicine and Molecular Medicine, The University of Texas Health Science Center at San Antonio, San Antonio, Texas 78245.

Hiroshi Maruta (169, 177, 329, 361), Ludwig Institute for Cancer Research, Royal Melbourne Hospital, Parkville/Melbourne 3050, Australia.

Laurent Meijer (145), CNRS, Station Biologique, 29682 Roscoff Cedex, Bretagne, France.

Oliver Müller (311), Department of Structural Biology, Max-Planck-Institut für Moleculare Physiologie, 44227 Dortmund, Germany.

Thao Nheu (361), Ludwig Institute for Cancer Research, Royal Melbourne Hospital, Parkville/Melbourne 3050, Australia.

George C. Prendergast (293), The DuPont Pharmaceuticals Company, Glenolden Laboratory, Glenolden, Pennsylvania 19036; and The Wistar Institute, Tumor Biology Group, Philadelphia, Pennsylvania 19104.

Samuel D. Rabkin (45), Molecular Neurosurgery Laboratory, Massachusetts General Hospital, Harvard Medical School, Charlestown, Massachusetts 02129.

Daniel J. Riley (97), Departments of Medicine and Molecular Medicine, The University of Texas Health Science Center at San Antonio, San Antonio, Texas 78245.

Shigeru Sakiyama (221), Chiba Cancer Center Research Institute, Chuoh-ku, Chiba 260-8717, Japan.

Galina Selivanova (397), Department of Oncology–Pathology, Karolinska Institute, Cancer Center Karolinska, Karolinska Hospital, SE-171 76 Stockholm, Sweden.

Steven A. Stacker (261), Ludwig Institute for Cancer Research, Royal Melbourne Hospital, Parkville/Melbourne 3050, Australia.

Kazunari Taira (233), Department of Chemistry and Biotechnology, Graduate School of Engineering, The University of Tokyo, Hongo, Tokyo 113-8656, Japan; and Gene Discovery Research Center, National Institute of Advanced Industrial Science and Technology (AIST), Tsukuba Science City 305-8562, Japan.

Rishu Takimoto (77), Laboratory of Molecular Oncology and Cell Cycle Regulation, Howard Hughes Medical Institute, Department of Medicine, Genetics, and Pharmacology, University of Pennsylvania School of Medicine, Philadelphia, Pennsylvania 19104.

Toshiaki Tanaka (123), Graduate School of Biological Sciences, Nara Institute of Science and Technology, Ikoma, Nara 630-0101, Japan.

Tomoki Todo (45), Molecular Neurosurgery Laboratory, Massachusetts General Hospital, Harvard Medical School, Charlestown, Massachusetts 02129.

Kiichiro Tomoda (123), Graduate School of Biological Sciences, Nara Institute of Science and Technology, Ikoma, Nara 630-0101, Japan.

Klas G. Wiman (397), Department of Oncology–Pathology, Karolinska Institute, Cancer Center Karolinska, Karolinska Hospital, SE-171 76 Stockholm, Sweden.

Alfred Wittinghofer (311), Department of Structural Biology, Max-Planck-Institut für Moleculare Physiologie, 44227 Dortmund, Germany.

Noriko Yoneda-Kato (123), Graduate School of Biological Sciences, Nara Institute of Science and Technology, Ikoma, Nara 630-0101, Japan.

Preface

Owing to the discovery during the past decade or so of oncogenes such as RAS and SRC and tumor suppressor genes such as p53 and RB and to intensive studies of the signal transduction pathways leading to mitogenesis of cells or viral replication, new generations of potential cancer therapeutics have been rationally designed or developed and are about to replace or complement the conventional approaches to cancer therapy such as surgery, radiation, and DNA-damaging drugs in this new century. These innovative approaches include unique viral therapy using specific viral mutants such as Onyx-015, gene therapy using a variety of tumor suppressor genes, and chemotherapy using ribozymes or chemical signal therapeutics that selectively block one of these oncogenic kinase/G protein-mediated transduction cascades.

Although these new therapies are much more selective than conventional therapies, their efficacy can still be improved. In this unique book, world-leading molecular oncologists introduce these new cancer therapeutics and discuss their future uses. This book not only will serve as an advanced textbook for undergraduate and graduate students in biomedical fields, but also will stimulate the productive "lateral thinking" of many professional scientists in both academia and industry who are dedicated to the development or improvement of cancer therapies. This book is the first to handle these matters so comprehensively in a single volume.

I express my gratitude to all the authors, who contributed their excellent chapters to this book, and to the editorial staff of Academic Press, who made this publication possible.

1

Oncolytic Viruses:
Virotherapy for Cancer

Lynda K. Hawkins and David Kirn[1]

Imperial Cancer Research Fund
Program for Viral and Genetic Therapy of Cancer
Imperial College School of Medicine
Hammersmith Hospital
London W12 0NN, England

[1] To whom correspondence should be addressed.

Tumor-Suppressing Viruses, Genes, and Drugs
Innovative Cancer Therapy Approaches

I. INTRODUCTION

The clinical utility of any cancer treatment is defined by both its antitumoral potency and its therapeutic index between cancerous and normal cells. Most currently available therapies for metastatic solid tumors fail on one or both of these counts. Countless changes in dose, frequency, and/or combinations of standard cytotoxic chemotherapies or radiotherapy have not overcome the inherent resistance to these approaches. Although standard chemotherapies and radiotherapy target a variety of different structures within cancer cells, almost all of them kill cancer cells through the induction of apoptosis. Not surprisingly, therefore, apoptosis-resistant clones almost universally develop following standard therapy for metastatic solid epithelial cancers (e.g., non-small cell lung, colon, breast, prostate, pancreatic), even if numerous high-dose chemotherapeutic agents are used in combination. The overall survival rates for most metastatic solid tumors have changed relatively little despite decades of work with this approach (adjuvant therapy in contrast, applied by definition when tumor burden is low, has resulted in clinically significant improvements in mortality). Novel therapeutic approaches must therefore have not only greater potency and greater selectivity than currently available treatments, they should also have novel mechanisms of action that will not lead to cross-resistance with existing approaches (i.e., do not rely exclusively on apoptosis induction in cancer cells).

Tumor-targeted oncolytic viruses (virotherapy with replication-selective viruses) have the potential to fulfill these criteria. Viruses have evolved over millions of years to infect target cells, multiply, cause cell death and release of viral particles, and finally to spread in human tissues. Their ability to replicate in tumor tissue allows for amplification of the input dose (e.g., 1000- to 10,000-fold increases) at the tumor site, while their lack of replication in normal tissues results in efficient clearance and reduced toxicity (Figure 1). This selective replication within tumor tissue can theoretically increase the therapeutic index of these agents dramatically over standard replication-incompetent approaches. In addition, viruses lead to infected cell death through a number of unique and distinct mechanisms. In addition to direct lysis at the conclusion of the replicative cycle, viruses can kill cells through expression of toxic proteins, induction of both inflammatory cytokines and T-cell-mediated immunity, and enhancement of cellular sensitivity to their effects. Therefore, since activation of classic apoptosis pathways in the cancer cell is not the exclusive mode of killing, cross-resistance with standard chemotherapeutics or radiotherapy is much less likely to occur.

Revolutionary advances in molecular biology and genetics have led to a fundamental understanding of both (1) the replication and pathogenicity of viruses and (2) carcinogenesis. These advances have allowed novel agents to be engineered to enhance their safety and/or their antitumoral potency.

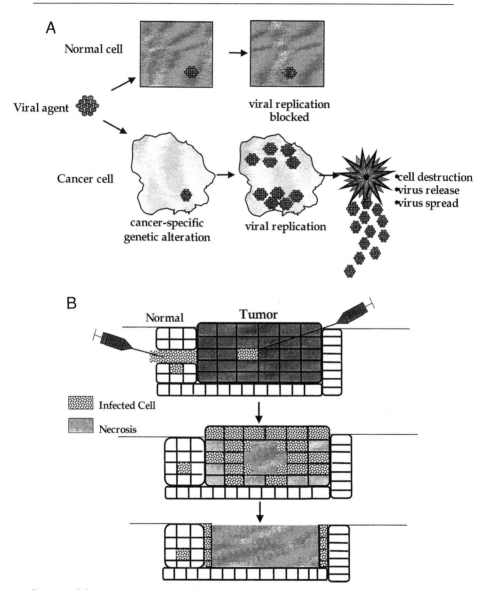

Figure 1 Schematic representation of tumor-selective viral replication and cell killing (A) and tumor-selective tissue necrosis (B).

Genetically engineered viruses in development have included adenoviruses, herpesviruses (HSV), and vaccinia. Inherently tumor-selective viruses such as reovirus, autonomous parvoviruses, Newcastle disease virus, measles virus strains, and vesicular stomatitis virus (VSV) have each been characterized. Each of these agents has shown tumor selectivity *in vitro* and/or *in vivo,* and efficacy has been demonstrated in murine tumor models with many of these agents following intratumoral (i.t.), intraperitoneal (i.p.), and/or intravenous (i.v.) routes of administration.

This chapter will review the discovery and development of replication-selective oncolytic adenoviruses, with an emphasis on recently acquired data from phase I and II clinical trials. The goal will be to summarize (1) the genetic targets and mechanisms of selectivity for these agents; (2) clinical trial data and what it has taught us to date about the promise but also the potential hurdles to be overcome with this approach; (3) future approaches to overcome these hurdles.

II. ATTRIBUTES OF REPLICATION-SELECTIVE VIRUSES FOR CANCER TREATMENT

A number of efficacy, safety, and manufacturing issues need to be assessed when considering a virus species for development as an oncolytic therapy. First, by definition the virus must replicate in and destroy human tumor cells. An understanding of the genes modulating infection, replication, or pathogenesis is necessary if rational engineering of the virus is to be possible. Because most solid human tumors have relatively low growth fractions, the virus should ideally infect noncycling cells. In addition, receptors for viral entry must be expressed on the target tumor(s) in patients (Wickham *et al.,* 1996). From a safety standpoint, the parental wild-type virus should ideally cause only mild, well-characterized human disease(s). Nonintegrating viruses have potential safety advantages, as well. A genetically stable virus is desirable from both safety and manufacturing standpoints. Finally, the virus must be amenable to high titer production and purification under Good Manufacturing Practices (GMP) guidelines for clinical studies.

III. APPROACHES TO OPTIMIZING TUMOR-SELECTIVE VIRAL REPLICATION

Four broad approaches are currently being used for tumor-selective virus replication. First, viruses such as reovirus and autonomous parvoviruses are inherently tumor-selective (Norman and Lee, 2000; Coffey *et al.,* 1998; Dupressoir *et al.,* 1989). A second approach is to limit the expression of the E1A gene product to tumor tissues through the use of tumor- and/or tissue-specific promoters. For example, the prostate-specific antigen (PSA) promoter/enhancer element has been inserted upstream of the *E1A* gene; the result is that viral replication correlates with the level of PSA expression in a given cell (Rodriguez *et al.*, 1997); similar results are obtained with the alphafetoprotein promoter targeting hepatocellular carcinoma cell lines (Hallenbeck, 1999). A third broad approach to tumor selectivity is to delete gene functions that are critical for efficient viral replication in normal cells but are expendable in tumor cells. The deletion approach was first described by Martuza *et al.* (1991) with herpesviruses deleted in the thymidine kinase gene (*dlsptk*). Subsequently, two adenovirus deletion mutants have been described (Bischoff *et al.*, 1996; Heise *et al.*, 1997, 2000). Finally, a fourth approach is to limit the uptake of viruses to tumor cells while ablating the uptake into normal cell populations (Kim *et al.*, 1997; Roelvink *et al.*, 1999).

IV. ADENOVIRUSES

Human adenovirus is a nonenveloped, double-stranded DNA virus of approximately 38 kb. The virus infects both proliferating and quiescent cells, and it can be manufactured to high titers. Adenoviruses can kill tumor cells by many mechanisms (Table 1). The first general approach to engineering tumor selectivity into adenovirus was to complement loss-of-function mutations in cancers with loss-of-function mutations within the adenovirus genome. Many of the

Table 1 Potential Mechanisms of Antitumoral Efficacy with Replication-Selective Adenoviruses

Mechanism	Examples of adenoviral genes modulating effect
I. Direct cytotoxicity due to viral proteins	E3 11.6kD
	E4ORF4
II. Augmentation of antitumoral immunity	
CTL[a] infiltration, killing	E3 gp19kD[b]
Tumor cell death, antigen release	E3 11.6kD
Immunostimulatory cytokine induction	E3 10.4/14.5, 14.7kD[b]
Antitumoral cytokine induction (e.g., TNF[c])	E3 10.4/14.5, 14.7kD[b]
Enhanced sensitivity to cytokines (e.g., TNF)	E1A
III. Sensitization to chemotherapy	Unknown (? E1A, others)
IV. Expression of exogenous therapeutic genes	NA[d]

[a] CTL, cytotoxic T-lymphocyte.
[b] Viral protein inhibits antitumoral mechanism.
[c] TNF, tumor necrosis factor.
[d] NA, not applicable.

same critical regulatory proteins that are inactivated by viral gene products during adenovirus replication are also inactivated during carcinogenesis (Barker and Berk, 1987; Nielsch et al., 1991; Sherr, 1996; Olson and Levine, 1994). Because of this convergence, the deletion of viral genes that inactivate these cellular regulatory proteins can be complemented by genetic inactivation of these proteins within cancer cells (Heise et al., 1997; Kirn et al., 1998a). This pioneering approach was first described with herpesvirus. Martuza et al. (1991) deleted the thymidine kinase gene (dlsptk). Subsequently, Bischoff et al. (1996) hypothesized that an adenovirus with a deletion of the gene encoding a p53-inhibitory protein E1B-55kD (dl1520, or Onyx-015) would be selective for tumors that already had inhibited or lost p53 function. p53 function is lost in the majority of human cancers through mechanisms including gene mutation, overexpression of p53-binding inhibitors (e.g., mdm2, human papillomavirus E6), and loss of the p53-inhibitory pathway modulated by p14ARF (Scheffner et al., 1991; Zhang et al., 1998; Hollstein et al., 1991). However, the precise role of p53 in the inhibition of adenoviral replication has not been defined to date. In addition, other adenoviral proteins also have direct or indirect effects on p53 function (e.g., E4orf6, E1B 19kD, E1A) (Dobner et al., 1996). Finally, E1B-55kD itself has important viral functions that are unrelated to p53 inhibition (e.g., viral mRNA transport, host cell protein synthesis shutoff) (Yew et al., 1994) (Figure 2). Not surprisingly, therefore, the role of p53 in the replication selectivity of dl1520 has been difficult to confirm despite extensive in vitro experimentation by many groups. Clinical trials were ultimately necessary to determine the selectivity and clinical utility of dl1520.

A second class of deletion mutants have now been described in E1A. Mutants in the E1A conserved region 2 are defective in retinoblastoma protein (pRB) binding. These viruses are being evaluated for use against tumors with pRB pathway abnormalities (Kirn et al., 1998a; Fueyo et al., 2000). With dl922/947, for example, S-phase induction and viral replication are reduced in quiescent normal cells, whereas replication and cytopathic effects are not reduced in tumor cells; interestingly, dl922/947 demonstrates significantly greater potency than dl1520 both in vitro and in vivo (Kirn et al., 1998a), and in a nude mouse–human tumor xenograft model, intravenously administered dl922/947 had significantly superior efficacy to even wild-type adenovirus (Heise et al., 2000). Unlike the complete deletion of E1B-55kD in dl1520, these mutations in E1A are targeted to a single conserved region and may therefore leave intact other important functions of the gene product.

The expression of the E1A gene product, and therefore replication, has been limited to tumor tissues through the use of tumor- and/or tissue-specific promoters. For example, the prostate-specific antigen (PSA), MUC-1, and αfetoprotein promoter/enhancer elements have been utilized to target prostate, breast, and hepatocellular carcinomas, respectively (Rodriguez et al.,

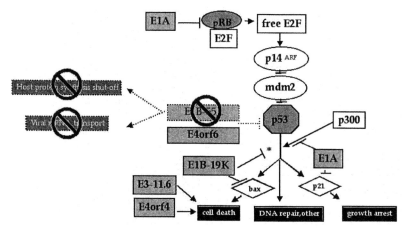

Figure 2 Diagram of both p53 pathway interactions with adenoviral gene products and functions of E1B-55kD: complexity of cancer cell and adenoviral biology. Note that adenoviral proteins target multiple components of this pathway at sites upstream of p53, downstream of p53, and at the level of p53 itself. Examples of p53-regulated cell functions are shown. In addition, the known functions of E1B-55kD are shown. In addition to the loss of p53 binding when E1B-55kD was deleted in *dl*1520 (Onyx-015), these functions are also lost.

1997; Kurihara *et al.*, 2000; Hallenbeck, 1999). Phase I clinical trials of the prostate-targeted adenovirus are underway.

V. POLIOVIRUS

Attenuated poliovirus is one of the most recently described oncolytic viruses for the treatment of cancer (Gromeier *et al.*, 2000). Poliovirus is a nonenveloped positive-strand RNA virus and the agent that causes paralytic poliomyelitis. Cell tropism is to lower motor neurons within the spinal cord and brainstem, and infection results in the clinical syndrome dominated by flaccid paralysis. The cellular receptor for this virus is CD155, a member of the immunoglobulin (Ig) superfamily. Additional cell-type-specific factors may play a role in the productive infection. The neuropathogenicity of poliovirus has been mapped to the 5′ IRES (internal ribosome entry site) (Gromeier *et al.*, 1996). Deletion of this sequence and replacement with an IRES from HRV2 (human rhinovirus type 2), as in PV1 (RIPO), eliminated the neurovirulent phenotype and allowed the virus to replicate in other cell lines (e.g., HeLa cells) (Gromeier *et al.*, 1999). This mutant was no longer pathogenic in the mouse model for poliomyelitis and after intraspinal inoculation into *Cynomolgus* monkeys (Gromeier *et al.*, 1996, 1999). In addition to demonstrating poor growth in neuronal-derived cell lines in tissue culture and lack of virulence in a mouse model, PV1(RIPO) has been shown to infect and propagate in cell lines derived from malignant gliomas (Gromeier *et al.*, 2000). Additionally, treatment of subcutaneous (s.c.) or intracerebral tumors in nude mice inhibited tumor growth and resulted in tumor elimination (Gromeier *et al.*, 2000). This research points to the potential use of attenuated poliovirus as an anticancer therapeutic.

VI. VESICULAR STOMATITIS VIRUS

Vesicular stomatitis virus (VSV) is an enveloped, negative-sense RNA virus that infects a wide variety of mammalian and insect cells (Dietzschold *et al.*, 1996). Infections in humans are asymptomatic or associated with a mild fibrile illness. This virus is also exquisitely sensitive to interferons (IFN). Its advantages as an oncolytic virus include that it can be grown to high titers (up to 10^9 pfu/ml in some tumor cell lines), has a short replication cycle (1 to 2 hr in tumor cells), and is not endemic to North America, implying that preexisting antibodies in patients would not interfere with its infection process. Its sensitivity to IFN might potentially be used as a safety measure.

In *in vitro* experiments, VSV was shown to replicate better (resulting in higher virus yield) in tumor cells versus normal cells (Stojdl *et al.*, 2000); this difference was even more striking in the presence of IFNα. Treatment of xenografts with VSV (intratumoral injection of VSV or VSV-infected cells) in nude mice resulted in regression or growth inhibition. When VSV administration was coupled with exogenous administration of IFNα (i.p.), all of the mice survived, whereas in the absence of IFN, the mice died after about 10 days. Whether the amount of IFN administered is similar to physiological levels is unclear from this report. Although the recently discovered oncolytic potential of this agent is intriguing, much more research is required, including research into its selectivity mechanism, toxicity, mechanism(s) of selective cell killing, optimal route of administration, and manufacturing issues.

VII. REOVIRUS

Reovirus is a highly stable, nonenveloped double-stranded RNA virus that results in either an asymptomatic, nonpathogenic infection or mild gastroenteritis in humans. Approximately 50% of the population possess antibodies to reovirus. A wide variety of mammalian cells can be infected via its receptor, sialic acid (Tyler and Fields, 1996). Early work (1977) had shown that reovirus could productively infect transformed cells but not normal cells (such as primary human and primate, rat kidney, and baby hamster kidney cells). Recent research has shown that it replicates selectively in cells with an activated Ras signaling pathway (Coffey *et al.*, 1998; Norman and Lee, 2000). This is an attractive feature since approximately 30% of all cancers have Ras mutations, and most other tumors have Ras pathway activation due to growth factor receptor overexpression (e.g., EGF-R).

Intratumoral injection of xenografts in immunodeficient mice has resulted in tumor regressions in several models (Coffey *et al.*, 1998). Importantly, however, there was a high mortality rate among the SCID mice; 50–60% of the reovirus-treated SCID mice eventually developed hind limb necrosis adjacent to the site of injection and died. This toxicity was not seen in the immunocompetent C3H mouse model. In immunocompetent models, significant tumor regressions resulted following multiple intratumoral injections. Therefore, the selectivity of reovirus for tumors *in vivo* is questionable, and selectivity may be dependent on T-cell-mediated clearance from normal tissues.

Given the selectivity for activated Ras pathway and the relative nonpathogenicity of the virus, reovirus may be an attractive candidate for an oncolytic agent. Future study of this virus is necessary to (1) further address selectivity *in vivo*, (2) determine the feasibility of genome manipulation to enhance potency, and (3) identify manufacturing issues.

VIII. BACTERIA

Bacteria have been explored as oncolytic agents over many decades. They have a number of desirable attributes which make them potentially suitable including (1) motility (facilitates intratumoral spread), (2) very large genome size capable of accommodating large inserts, (3) their spread can be controlled by antibiotics, and (4) some species are anaerobic so they can grow in the hypoxic areas of tumors. Notably, unlike viruses, most of the intratumoral bacteria are in the extracellular milieu and are not intracellular.

One example of an obligate anaerobe that has made it to clinical trials is *Clostridium beijerinckii* (Lemmon *et al.*, 1997). In clinical trials using *Clostridium* with an i.v. dose of up to 10^{10} spores, there was partial lysis of the malignancies with no evidence of spore germination elsewhere in normal tissues. This trial was stopped due to a lack of tumor responses. Since then, transgenes have been added to the bacterial genome of *C. beijerinckii*, namely the prodrug converting enzymes cytosine deaminase (CD) and nitroreductase (NR). With i.v. administration of this agent carrying the nitroreductase gene into BALB/c mice bearing EMT6 tumors, tumor regressions were documented, but only under hypoxic conditions (Lemmon *et al.*, 1997). The lack of toxicity and the highly selective growth imply that this bacteria may be a viable clinical candidate.

It was subsequently discovered that *Salmonella typhimurium*, a gram-negative bacterium, could infect and preferentially accumulate within implanted tumors in mice (Sznol *et al.*, 2000). As a facultative anaerobe, replication is possible both within well-oxygenated and hypoxic regions of the tumor. Because *S. typhimurium* infects both human and mouse cell lines, modeling is possible in both immunodeficient and immunocompetent mice. This strain was able to retard tumor growth in a broad range of both syngeneic and xenograft mouse models.

The prodrug converting enzyme, tk (thymidine kinase), has been added to an attenuated salmonella (Pawelek *et al.*, 1997). This agent was still shown to be hyperinvasive in tumors and show a high tumor:normal cell ratio. In an immunocompetent animal model, the coadministration of gancyclovir (GCV) resulted in a dose-dependent suppression of tumor growth and prolonged survival, compared to the bacterium without gancyclovir.

Because the preliminary data looked encouraging for this agent, steps were taken to increase the safety through further attenuation of the bacteria. The *msbB* gene, whose product is involved in the myristoylation of lipid A, was deleted (Low *et al.*, 1999). This deletion led to reduced TNFα (tumor necrosis factor) induction in mice and pigs, which lowered the possibility of septic shock. This attenuated strain still selectively colonized the tumor, had a high tumor:normal cell ratio, and showed efficacy in mice melanoma models (Low *et al.*, 1999). As expected, this mutant retained sensitivity to antibiotics.

To further attenuate this agent, another mutation was made in the *purI* gene (Clairmont *et al.*, 1999). This gene deletion resulted in a requirement for an external source of purine; this agent is called VNP20009. Following i.v. administration into tumor-bearing mice, the agent is rapidly cleared from the bloodstream (<24 hr), and resulted in both a 1000-fold decrease in virulence compared with wild-type bacteria as well as a very high tumor:normal tissue infection ratios (e.g., 1000:1 tumor:liver). There was an inhibition in tumor growth in both murine and human tumor xenograft models. VNP20009 was also tested in *Cynomolgus* monkeys by i.v. administration for preclinical toxicology studies. This trial showed that toxicity was mild (confined predominantly to changes in liver enzymes) and confirmed its safety at doses of 10^9 colony-forming units (cfu)/monkey or $1-2 \times 10^6$ cfu/mouse in several tumor models. Therefore, this agent is proven to be (1) attenuated compared to the parental strain, (2) efficacious, and (3) tumor specific. Phase I clinical trials are underway to determine the maximally tolerated dose following either intratumoral or intravenous administration.

IX. VACCINIA VIRUS

Vaccinia virus (VV) is an enveloped, double-stranded DNA virus, spanning almost 200 kb (Moss, 1996a). This virus replicates in the cytoplasm, bringing with its infectious particle the enzymes required for DNA replication and transcription of its genes. VV has a broad host range. Its advantages as an oncolytic virus are its large genome size with the capacity to accommodate multiple foreign genes, its wide tropism, rapid infection, and ease of recombination for making viral mutants (insertions and/or deletions). Its safety record has been well established with its use as a smallpox vaccine. Only in rare cases can wild-type VV be lethal in severely immunocompromised patients (Moss, 1996b). Most VV recombinants have inserted transgenes into the VV's own *tk* gene; this gene deletion may help give the virus selectivity in dividing cells, although it does remove a potential safety valve.

VV has been used both as a tool to induce an antitumoral immune response and as a means of lysing tumor cells directly following virus replication. A number of genes have been inserted into VV, including the costimulatory molecules ICAM-1 (Uzendoski *et al.*, 1997) and CD70 (Lorenz *et al.*, 1999); EMAP-II (endothelial monocyte-activating factor) (Gnant *et al.*, 1999); cytokines such as IL-2 (Mukherjee *et al.*, 2000) and GM-CSF (Mastrangelo *et al.*, 1999); viral antigens such as HPV-16 viral proteins (He *et al.*, 2000) and HIV-1 envelope protein (Moss *et al.*, 1996b); tumor associated antigens (TAA) such as Mart-1 (Zajac *et al.*, 1997); tyrosinase (Colella *et al.*, 2000; Drexler *et al.*, 1999) CEA (Lorenz *et al.*, 1999; Conry *et al.*, 2000); PSA (Eder *et al.*, 2000), and Melan A peptide gene (Valmori *et al.*, 2000); markers such as lacZ or luciferase (Puhlmann *et al.*, 2000); and finally the prodrug converting enzyme cytosine deaminase (CD) (McCart *et al.*, 2000). These viruses generally are made by homologous recombination replacing the *tk* gene. Many of these constructs have been tested in nude and immunocompetent mouse models, and the results have been encouraging. Injected tumors usually undergo growth inhibition or regression, survival of the animal is prolonged, and, in many cases, a protective immune response against rechallenge is elicited.

Wild-type VV was used in clinical studies for patients with recurrent superficial melanoma and intravesicle treatment of bladder cancer ($n = 4$) (Mastrangelo *et al.*, 2000 and references therein). In both cases, an immune response to VV developed. However, even with multiple injections, virus replication occurred, indicating that the immune response did not totally inhibit virus replication. No significant toxicity occurred.

VV containing the GM-CSF gene was subsequently used in a phase I dose escalation study of patients with surgically incurable cutaneous melanoma ($n = 7$) (Mastrangelo *et al.*, 1999); GM-CSF is a potent immune enhancing agent in murine tumor models. The agent was injected into superficial lesions twice weekly for 6 weeks. All patients developed an anti-VV response. GM-CSF expression was detectable for as long as 31 weeks following treatment initiation. Toxicity was minimal, consisting primarily of flulike symptoms of ≤ 24 hr duration. All patients had local inflammation at higher doses, dense infiltration by CD4$^+$ and CD8$^+$ lymphocytes, and anti-VV humoral immune responses above baseline levels (all patients had been vaccinated against smallpox as children). The local responses for the seven patients in this study were as follows: one complete response, one partial response, three mixed responses, and two had no response. Unlike studies with autologous tumor cell preparations infected with retroviral GM-CSF, responses were not demonstrated at distant sites. This study showed that VV could be administered safely with potential for a local antitumoral response, but the ability to induce effective systemic antitumoral immunity was not demonstrated.

VV encoding the *IL-2* gene was administered i.t. weekly for 12 weeks to 6 patients with refractory malignant mesothelioma (Mukherjee *et al.*, 2000). A T lymphocyte infiltrate was observed, and VV IgG rose in all patients. Nevertheless, expression of IL-12 continued up to 3 weeks postinjection. There was no significant toxicity; however, no tumor regressions were observed.

In another phase I study, VV carrying the *CEA* gene was administered to 20 patients with metastatic adenocarcinoma, either intradermally or subcutaneously (Conry *et al.*, 1999). CEA is a TAA expressed in a majority of colorectal, gastric, and pancreatic cancers, among others. Both routes of injections were equivalent for development of cellular and humoral immune responses to the vector. Although there was only minimal toxicity in less than 20% of the patients, there were no objective responses.

A phase I trial was also performed with VV encoding PSA via intradermal inoculation in 33 patients with prostate cancer (who had elevated PSA levels despite previous treatment) (Eder *et al.*, 2000). There were three doses administered per patient at three escalating dose levels. This treatment was in the presence or absence of exogenously administered GM-CSF. A specific T-cell response to PSA was induced in some patients, although no objective responses occurred. Nine patients had no apparent clinical progression for 11–25 months. Thus, repeated vaccination with VV-PSA was safe (grade 1 toxicity only) and resulted in a specific tumor antigen response.

In summary, VV-expressing TAAs, IL-2, or GM-CSF were well-tolerated in a number of phase I trials using subcutaneous, intradermal, or intratumoral inoculation. Not surprisingly, no objective systemic antitumoral responses were seen in these patients with advanced disease on phase I trials. Phase II testing in patients with minimal disease will be necessary to assess the usefulness of VVs as cancer vaccines.

X. HERPESVIRUS

Herpesvirus (HSV) is an enveloped double-stranded DNA virus of approximately 152 kbp (Martuza, 2000). It contains a number of genes with identified essential and nonessential functions (*in vitro*) whose expression is regulated in a complex manner during the early and late phases of infection; DNA replication occurs between the early and late phases. These genes are grouped as immediate early (IE or α), early (E or β), and late (L or γ) genes, depending on the timing of their expression. HSV has a natural tropism for neuronal and mucosal cells. Techniques have been developed for genetic engineering, and it is possible to replace some genes, allowing space for approximately 30 kb of foreign DNA by replacing nonessential sequences. It has a built-in safety mechanism through its *tk* gene; antiviral agents such as acyclovir and gancyclovir can be administered to halt the spread of the virus. Conversely, the deletion of this gene leads to attenuation of the virus in nondividing cells, and this may give it some selectivity for replication in cycling cancer cells. When a *tk⁻* HSV was tested in murine models, neurotoxicity still occurred at higher doses. Loss of its sensitivity to these drugs results in a lack of a method to control the infection if need be.

HSV replication has also been controlled through transcriptional mechanisms. The ICP4 gene was put under the control of the albumin enhancer/promoter with the objective of getting expression specifically in hepatocellular carcinoma cells (G92A) (Miyatake *et al.*, 1997), while the *tk* gene was also replaced with the *lacZ* marker gene. The ICP4 protein is the main transactivator of HSV transcription and hence, regulation of this gene should affect the entire infection process. A productive infection occurred in the albumin-expressing hepatocellular carcinoma, but unfortunately this was not tumor-specific because normal liver cells also express albumin. However, it did show proof of concept for control using a tissue-specific promoter for HSV therapy.

It was then discovered that deletion of the gene for ICP34.5 attenuates HSV in nondividing cells (such as neurons) but allows replication in cycling cells (such as tumor cells) (Martuza, 2000; Roizman, 1996; Kucharczuk *et al.*, 1997). This approach leaves the *tk* gene intact, thereby retaining a potential safety mechanism. ICP34.5 gives the virus the ability to replicate in central nervous system (CNS) cells for unknown reasons (Roizman, 1996). Therefore the deletion abrogates this ability. ICP34.5 also blocks the phosphorylation of eIF-2 (translation initiation factor) by PKR (protein kinase RNA-dependent kinase) to prevent premature protein synthesis shutoff. Additionally, it promotes infection in nondividing cells and inhibits apoptosis of infected cells. Deletion of both copies of this gene attenuates the propagation in normal cells. One version of this ICP34.5-deleted virus is known as HSV-1716 (originally made by M. Brown, Glasgow, Scotland) (Kucharczuk *et al.*, 1997). In malignant mesothelioma tumors grown i.p. in SCID mice, i.p. virus injections reduced tumor burden and prolonged survival. Although there did not appear to be any cytopathic changes in normal tissues, there were still some unexplained animal deaths (Kucharczuk *et al.*, 1997). This virus is now in phase I clinical trials for patients with recurrent glioblastoma (Markert *et al.*, 2000a; Rampling *et al.*, 2000; Kirn, 2000a).

Other HSV mutants were created also deleting the *ICP34.5* gene. In one study, both copies were either deleted (R3616) or there was an insertion of stop codons (R4009) (Andreansky *et al.*, 1996). These viruses were used to treat intracranial or glioma tumors in SCID mice. Both mutants significantly prolonged survival, and tumor eradication was documented in some cases. For intracerebral injections, these deletion mutants have an LD_{50} of over 10^7 pfu, whereas wild-type HSV-1 has an LD_{50} of less than $10^{2.5}$, demonstrating quantitatively the enhanced safety associated with the deletion of *ICP34.5* genes. In addition, intracranial injections of up to 2.5×10^6 pfu with R3616 or R4009 into SCID mice had no effect over a 90-day observation period, whereas 10^3 pfu of the wild-type HSV-1 resulted in 100% mortality in 7 days.

In addition to *ICP34.5*, another *HSV-1* gene has been deleted to enhance safety and selectivity. This gene encodes the protein ICP6, the large subunit of ribonucleotide reductase (RR), and this doubly deleted virus is known as G207 (Mineta *et al.*, 1995). In this instance the *ICP6* gene was disrupted by insertion of *lacZ*, the marker gene. Loss of the RR subunit further decreases the ability of G207 to replicate in nondividing cells and therefore increases specificity for tumor cells. The favorable attributes of G207 include: (1) its replication-competence in dividing cells, (2) attenuated neurovirulence, (3) temperature sensitivity, (4) enhanced sensitivity to gancyclovir or acyclovir, (5) multiple deletions (making reversion to wild-type virus less likely), and (6) the presence of the *lacZ* marker. Animal models in which G207 has been tested for antitumoral efficacy and toxicity include head and neck cancer (Carew *et al.*, 1999), colorectal cancer with liver metastasis (Kooby *et al.*, 1999), neuroblastoma (Todo *et al.*, 1999), and epithelial ovarian cancer (Coukos *et al.*, 2000); safety has been tested with intracerebral inoculations in *Aotus nancymae* Owl Monkeys (Hunter *et al.*, 1999). In all of these examples, the virus was nontoxic at therapeutic doses, and tumor-selectivity and efficacy were observed. Nude mice with s.c. or intracerabral gliomas (xenografts) showed decreased tumor growth and increased survival after i.t. injection with G207 (Mineta *et al.*, 1995). To demonstrate the lack of neurovirulence, nude mice and Owl monkeys (*Aotus nancymae*) were given intracerebral injections of 10^7 pfu of G207. All of the mice survived; in contrast, the LD_{50} for the HSV-1-injected animals was 10^3 (established LD_{50}). *Aotus nancymae* are extremely susceptible to HSV infection, and all of the G207-injected animals ($n = 3$) survived when given a dose of 10^7 pfu, whereas as the HSV-injected animal ($n = 1$) at a dose of 10^3 pfu did not. Therefore, this vector was demonstrated to have significantly reduced toxicity in two animal models, while being efficacious in nude mice with gliomas. G207 has now moved forward into phase I trials (Kirn, 2000a; Markert *et al.*, 2000b; Rampling *et al.*, 2000).

To boost the immune response against the tumor, the cytokine gene for IL-12 was added to the replication-defective genome of HSV. This was used in combination with G207 (Toda *et al.*, 1998). IL-12 has multifunctional roles, including promotion of a TH1 cellular immune response. In an immunocompetent syngeneic model, efficacy was associated with tumor-specific CTL activity, and there was a significant reduction in tumor growth in inoculated and noninoculated contralateral tumors. However, there was no significant growth inhibition of murine s.c. tumors in athymic mice. (G207 is more restricted and unable to grow in most rodent cell lines. The cells used here, CT26, have only limited susceptibility to G207.) Therefore in this system, G207 is not functioning by direct tumoricidal action but more as a vaccine. The IL-12 is acting as an adjuvant for the development of antitumor immunity. A similar, more recent study used the HSV-1 mutant R3659 engineered to carry the *IL-12* gene (Parker *et al.*, 2000). This virus, known as M002, was used to treat experimental murine brain tumors in syngeneic models. It was demonstrated that (1) physiological levels of IL-12 were produced, (2) there was an increase in median survival as compared to R3659, and (3) there was infiltration of inflammatory mediators into the tumor. M002 had an enhanced oncolytic effect. Therefore, local cytokine expression resulted in antitumoral activity in animals with an intact immune system.

The role of preexisting immunity to HSV-1 has been assessed. Fischer rats were preimmunized by i.p. injection in one study, and the titers of neutralizing antibodies were determined (Herrlinger *et al.*, 1998). After 21 days, D74 gliomas were implanted. An i.t. injection of the virus hrR3 (RR-LacZ$^+$) was given, and there was a further increase in titer of neutralizing antibodies. These results showed that although there was a substantial decrease in *lacZ* expression from the virus (down to 15% of nonimmunized animals), it was not completely abolished.

Since cancer management usually combines several therapies, HSV-1716 plus various chemotherapeutic drugs were used together *in vitro* to look at the effect on viral replication. Most seem to augment oncolytic activity at relevant concentrations, implying that use of these agents together could lead to increased efficacy (Toyoizumi *et al.*, 1999). The combination of HSV plus radiation was also tested (Advani *et al.*, 1998). Tumor-bearing mice were injected with 2×10^7 pfu of R3616 on day zero and then subjected to ionizing radiation on days 1 and 2. In this regimen, 13 of 23 animals showed complete tumor regression by day 60. When R3616 was administered on days 0, 1, and 2 along with radiation, there were complete responses in 9 out of 10 animals. R3616 was shown to have enhanced replication as a result of radiation in U-87 malignant glioma xenografts, and this resulted in greater tumor reduction. This same phenomenon was also true when using HSV mutants with other deletions and thus does not appear to be specific for the ICP34.5 mutation.

The *tk* gene is still intact in most of these HSV mutants, including G207. To see if there could be an enhancement of antitumor activity using the prodrug gancyclovir (GCV), one group tested administration of both together

(Todo *et al.*, 2000). They found that GCV had no significant effect on survival or tumor growth inhibition and concluded that GCV may not be beneficial. However, retention of *tk* does allow the virus infection to be controlled, and therefore can be an important safety valve.

Two phase I clinical trials employing selectively replicating HSV mutants have been completed in patients with glioblastoma multiforme (GBM) or anaplastic astrocytomas (AA). In one trial, the vector used was 1716, containing a deletion in the *ICP34.5* gene (Rampling *et al.*, 2000). Nine patients were injected with up to 10^5 pfu of virus and, although there was no incidence of encephalitis, there was no clinical benefit associated with the virus. Similar results were found in the other phase I study where the HSV mutant G207 (described previously) was injected at doses up to 3×10^9 pfu at five sites (Markert *et al.*, 2000a). This study enrolled 21 patients and again, although no toxicity was found, no benefit was attributed to the therapy. These studies demonstrate that these HSV mutants can be safely administered, but more studies need to be conducted to determine the maximum tolerated doses that are needed for clinical benefit (Kirn, 2000a).

The use of HSV-1 viral mutants is continuing in preclinical models and clinical trials. Other mutations that have improved selectivity for tumor cells versus normal tissue are being explored. The inclusion of additional therapeutic genes, including multiple genes from different classes, should further enhance the oncolytic function of this replicating agent.

XI. CLINICAL TRIAL RESULTS WITH REPLICATION-COMPETENT ADENOVIRUSES IN CANCER PATIENTS

A. Clinical trial results with wild-type adenovirus: Flawed study design

During the twentieth century a diverse array of viruses were injected into cancer patients by various routes, including adenovirus, Bunyamwara, coxsackie, dengue, feline panleukemia, Ilheus, mumps, Newcastle disease virus, vaccinia, and West Nile (Kirn, 2000b; Southam and Moore, 1952; Asada, 1974; Smith *et al.*, 1956). These studies illustrated both the promise and the hurdles to overcome with oncolytic viral therapy. Unfortunately, these previous clinical studies were not performed to current clinical research standards, and therefore none give interpretable and definitive results. At best, these studies are useful in generating hypotheses that can be tested in future trials.

Although suffering from many of the trial design flaws listed below, a trial with wild-type adenovirus is one of the most useful for hypothesis generation and also for illustrating how clinical trial design flaws severely curtail the utility of the study results. The knowledge that adenoviruses could eradicate a variety of tumor cells *in vitro* led to a clinical trial in the 1950s with wild-type adenovirus. Ten different serotypes were used to treat 30 cervical cancer patients (Smith *et al.*, 1956). Forty total treatments were administered by either direct intratumoral injection ($n = 23$), injection into the artery perfusing the tumor ($n = 10$), treatment by both routes ($n = 6$), or intravenous administration ($n = 1$). Characterization of the material injected into patients was minimal. The volume of viral supernatant injected is reported, but actual viral titers/doses are not; injection volumes (and by extension doses) varied greatly. When possible, the patients were treated with a serotype to which they had no neutralizing antibodies present. Corticosteroids were administered as nonspecific immunosuppressive agents in roughly half of the cases. Therefore, no two patients were treated in identical fashion.

Nevertheless, the results are intriguing. No significant local or systemic toxicity was reported. This relative safety is notable given the lack of pre-existing immunity to the serotype used and concomitant corticosteroid use in many patients. Some patients reported a relatively mild viral syndrome lasting 2–7 days (severity not defined); this viral syndrome resolved spontaneously. Infectious adenovirus was recovered from the tumor in two-thirds of the patients for up to 17 days postinoculation.

Two-thirds of the patients had a "marked to moderate local tumor response" with necrosis and ulceration of the tumor (definition of "response" not reported). None of the seven control patients treated with either virus-free tissue culture fluid or heat-inactivated virus had a local tumor response (statistical significance not reported). Therefore, clinically evident tumor necrosis was only reported with viable virus. Neutralizing antibodies increased within 7 days after administration. Although the clinical benefit to these patients is unclear, and all patients eventually had tumor progression and died, this study did demonstrate that wild-type adenoviruses can be safely administered to patients and that these viruses can replicate and cause necrosis in solid tumors despite a humoral immune response. The maximally tolerated dose, dose-limiting toxicity, objective response rate, and time to tumor progression, however, remain unknown for any of these serotypes by any route of administration.

B. A novel staged approach to clinical research with replication-selective viruses: The example of *dl*1520 (Onyx-015)

For the first time since viruses were first conceived as agents to treat cancer over a century ago, we now have definitive data from numerous phase I and II clinical trials with a well-characterized and well-quantitated virus. *dl*1520 (Onyx-015, now CI-1042, Pfizer, Groton, CT) is a novel agent with a novel mechanism of action. This virus was to become the first virus to be used in humans that had been genetically engineered for replication selectivity. We predicted that both toxicity and efficacy would be dependent on multiple factors including (1) the inherent ability of a given tumor to replicate and shed the virus (2) the location of the tumor to be treated (e.g., intracranial vs. peripheral), and (3) the route of administration of the virus. In addition, we felt it would be critical to obtain biological data on viral replication, antiviral immune responses, and their relationship to antitumoral efficacy in the earliest phases of clinical research.

We therefore designed and implemented a novel staged clinical research and development approach with this virus (Figure 3). The goal of this approach was to sequentially increase systemic exposure to the virus only after safety with more localized delivery had been demonstrated. Following demonstration of safety and biological activity by the intratumoral route, trials were sequentially initiated to study intracavitary instillation (initially intraperitoneal), intra-arterial infusion (initially hepatic artery), and eventually intravenous administration. In addition, only patients with advanced and incurable cancers were initially enrolled on trials. Only after safety had been demonstrated in terminal cancer patients were trials initiated for patients with premalignant conditions. Finally, clinical trials of combinations with chemotherapy were initiated only after the safety of dl1520 as a single agent had been documented by the relevant route of administration.

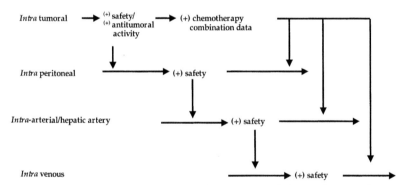

Figure 3 A staged clinical research and development approach for a replication-selective agent in cancer patients. Following demonstration of safety and biological activity by the intratumoral route, trials were sequentially initiated to study intracavitary instillation (initially intraperitoneal), intra-arterial infusion (initially hepatic artery), and eventually intravenous administration. In addition, only patients with advanced and incurable cancers were initially enrolled on trials. Only after safety had been demonstrated in terminal cancer patients were trials initiated for patients with premalignant conditions. Finally, clinical trials of combinations with chemotherapy were initiated only after the safety of dl1520 as a single agent had been documented by the relevant route of administration.

XII. RESULTS FROM CLINICAL TRIALS WITH *dl*1520 (Onyx-015, or CI-1042)

A. Toxicity

No maximally tolerated dose or dose-limiting toxicities were identified at doses up to 2×10^{12} particles administered by intratumoral injection. Flulike symptoms and injection-site pain were the most common associated toxicities (Ganly *et al.*, 2000). This safety is remarkable given the daily or even twice-daily dosing that was repeated every 1–3 weeks in the head and neck region or pancreas (Nemunaitis *et al.*, 2000a).

Intraperitoneal, intra-arterial, and intravenous administration were also remarkably well-tolerated, in general. Intraperitoneal administration was feasible at doses up to 10^{13} particles divided over 5 days (Vasey *et al.*, 2000). The most common toxicites included fever, abdominal pain, nausea/vomiting, and bowel motility changes (diarrhea, constipation). The severity of the symptoms appeared to correlate with tumor burden. Patients with heavy tumor burdens reached a maximally tolerated dose at 10^{12} particles (dose-limiting toxicities were abdominal pain and diarrhea), whereas patients with a low tumor burden tolerated 10^{13} without significant toxicity.

No dose-limiting toxicities were reported following repeated intravascular injection at doses up to 2×10^{12} particles (hepatic artery) (Reid *et al.*, 2000) or 2×10^{13} particles (intravenous) (Nemunaitis *et al.*, 2000b). Fever, chills, and asthenia following intravascular injection were more common and more severe than after intratumoral injections (grade 2–3 fever and chills vs. grade 1). Dose-related transaminitis was reported infrequently. The transaminitis was typically transient (<10 days), low-grade (grade 1–2), and not clinically relevant. Further dose escalation was limited by supply of the virus.

B. Viral replication

Viral replication has been documented at early time points after intratumoral injection in head and neck cancer patients (Figure 4) (Nemunaitis *et al.*, 2000a,c). Roughly 70% of patients had evidence of replication on days 1–3 after their last treatment. In contrast, day 14–17 samples were uniformly negative. Intratumoral injection of liver metastases (primarily colorectal) led to similar results at the highest doses of a phase I trial. Patients with injected pancreatic tumors, in contrast, showed no evidence of viral replication by plasma PCR or fine needle aspiration. Similarly, intraperitoneal *dl*1520 could not be shown to reproducibly infect ovarian carcinoma cells within the peritoneum. Therefore, different tumor types can vary dramatically in their permissiveness for viral infection and replication.

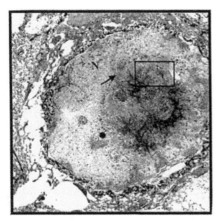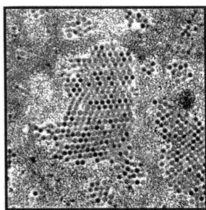

Figure 4 Replication of adenoviral agent (*dl*1520) in nucleus of squamous carcinoma cell in the head and neck region of a patient 3 days after intratumoral virus injection. The dark staining individual particles (arrow) and clusters within the nucleus (box) are adenoviral particles.

Proof-of-concept for tumor infection following intra-arterial (Reid *et al.*, 2000) or intravenous (Nemunaitis *et al.*, 2000c) administration with human adenovirus has also been achieved. Approximately half of the roughly 25 patients receiving hepatic artery infusions of 2×10^{12} particles were positive by PCR 3–5 days following treatment. Patients with elevated neutralizing antibody titers prior to treatment were substantially less likely to have evidence of viral replication 3–5 days posttreatment. Three of four patients with metastatic carcinoma to the lung treated intravenously with $\geq 2 \times 10^{12}$ particles were positive for replication by PCR on day 3 (± 1). Therefore, it is feasible to infect distant tumor nodules following intravenous or intra-arterial administration.

C. Immune response

Neutralizing antibody titers to the coat (Ad5) of *dl*1520 were positive but relatively low in roughly 50–60% of all clinical trial patients at baseline (Nemunaitis *et al.*, 2000a). Antibody titers increased uniformly following administration of *dl*1520 by any of the routes tested, in some cases to levels >1:80,000. Antibody increases occurred regardless of evidence for replication or shedding into the bloodstream (Nemunaitis *et al.*, 2000a). Acute inflammatory cytokine levels were determined prior to treatment (by hepatic artery infusion), 3 hr post- and 18 hr posttreatment: IL-1, IL-6, IL-10, γ-interferon (INFγ), tumor necrosis factor.

Significant increases were demonstrated within 3 hr for IL-1, IL-6, tumor necrosis factor, and to a lesser extent INFγ; all cytokines were back down to pretreatment levels by 18 hr (Reid *et al.*, 2000). In contrast, IL-10 did not increase until 18 hr.

D. Efficacy with *dl*1520 (Onyx-015) as a single agent

Two phase II trials enrolled a total of 40 patients with recurrent head and neck cancer (Nemunaitis *et al.*, 2000a,c). Tumors were treated very aggressively with 6–8 daily needle passes for 5 consecutive days (30–40 needle passes per 5-day cycle; $n = 30$) and 10–15 per day on a second trial (50–75 needle passes per cycle; $n = 10$). The median tumor volume on these studies was approximately 25 cm^3; an average volume of 10 cm^3 of tumor therefore received an estimated 4–5 needle passes per cycle. Despite the intensity of this local treatment, the unconfirmed response rate at the injected site was only 14%, and the confirmed local response rate was <10%. Interestingly, there was no correlation between evidence of antitumoral activity and neutralizing antibody levels at baseline or posttreatment (Nemunaitis *et al.*, 2000a). No objective responses were demonstrated in patients with tumor types that could not be so aggressively injected (due to their deep locations), although some evidence of minor shrinkage or necrosis was obtained. In summary, single agent responses across all studies were rare, and therefore combinations with chemotherapy were explored.

E. Efficacy in combination with chemotherapy: Potential synergy discovered

Evidence for a potentially synergistic interaction between adenoviral therapy and chemotherapy has been obtained on multiple trials. Encouraging clinical data has been obtained in patients with recurrent head and neck cancer treated with intratumoral *dl*1520 in combination with intravenous cispaltin and 5-fluorouracil (Khuri *et al.*, 2000). Thirty-seven patients were treated and 19 responded (54%, intent-to-treat; 63%, evaluable); this compares favorably with response rates to chemotherapy alone in previous trials (30–40%, generally). The time-to-tumor progression was also superior to previously reported studies. However, comparisons to historical controls are unreliable. We therefore used patients as their own controls whenever possible ($n = 11$ patients). Patients with more than one tumor mass had a single tumor injected with *dl*1520 while the other mass(es) was left uninjected. Since both masses were exposed to chemotherapy, the effect of the addition of viral therapy to chemotherapy could be assessed. The *dl*1520-injected tumors were significantly more likely to respond ($p = 0.017$) and less likely to progress ($p = 0.06$) than were noninjected tumors. Noninjected control tumors that progressed on chemotherapy alone were subsequently treated with

Onyx-015 in some cases; two of the four injected tumors underwent complete regressions. These data illustrate the potential of viral and chemotherapy combinations. The clinical utility of *dl*1520 in this indication will be definitively determined in a phase III randomized trial.

A phase I/II trial of *dl*1520 administered by hepatic artery infusion in combination with intravenous 5-fluorouracil and leukovorin was carried out (*n* = 33 total) (Reid *et al.*, 2000). Following phase I dose escalation, 15 patients with colorectal carcinoma who had previously failed the same chemotherapy were treated with combination therapy after failing to respond to *dl*1520 alone; one patient underwent a partial response and 10 had stable disease ($2-7^+$ months). Chemosensitization of colorectal liver metastases is therefore possible via hepatic artery infusions, although the magnitude and frequency of this effect remains to be determined. In contrast, data from a phase I/II trial studying the combination of *dl*1520 and gemcitabine chemotherapy were disappointing (*n* = 21); the combination resulted in only two responses, and these patients had not received prior gemcitabine (Hecht *et al.*, 2000). Therefore, potential synergy was demonstrated with *dl*1520 and chemotherapy in two tumor types that supported viral replication (head and neck, colorectal), but not in a tumor type that was resistant to viral replication (pancreatic).

F. Results from clinical trials with *dl*1520 (Onyx-015): Summary

*dl*1520 has been well tolerated at the highest practical doses that could be administered ($2 \times 10^{12} - 2 \times 10^{13}$ particles) by intratumoral, intraperitoneal, intra-arterial, and intravenous routes. The lack of clinically significant toxicity in the liver or other organs was notable. Flulike symptoms (fever, rigors, asthenia) were the most common toxicities and were increased in patients receiving intravascular treatment. Acute inflammatory cytokines (especially IL-1 and IL-6) increased within 3 hr following intra-arterial infusion. Neutralizing antibodies increased in all patients, regardless of dose, route, or tumor type. Viral replication was documented in head and neck and colorectal tumors following intratumoral or intra-arterial administration. Neutralizing antibodies did not block antitumoral activity in head and neck cancer trials of intratumoral injection. However, viral replication/shedding into the blood was inhibited by neutralizing antibodies; intra- arterial virus was more sensitive to antibody inhibition than was intratumorally injected virus. Single agent antitumoral activity was minimal ($\cong 15\%$) in head and neck cancers that could be repeatedly and aggressively injected. No objective responses were documented with single agent therapy in phase I or I/II trials in patients with pancreatic, colorectal, or ovarian carcinomas. A favorable and potentially synergistic interaction with chemotherapy was discovered in some tumor types and by different routes of administration.

G. Future directions: Why has *dl*1520 (Onyx-015) failed to date as a single agent for refractory solid tumors?

Future improvements with this approach will be possible if the reasons for *dl*1520 failure as a single agent, and success in combination with chemotherapy, are uncovered. Factors that are specific to this adenoviral mutant, as well as factors that may be generalizable to other viruses, should be considered. Regarding this particular adenoviral mutant, it is important to remember that this virus is attenuated relative to wild-type adenovirus in most tumor cell lines *in vitro* and *in vivo*, including even p53 mutant tumors (Harada and Berk, 1999; Goodrum and Ornelles, 1997; Kirn *et al.*, 1998b; Rothmann *et al.*, 1998; Heise *et al.*, 2000). This is not an unexpected phenotype because this virus has lost critical E1B-55kD functions that are unrelated to p53, including viral mRNA transport. This attenuated potency is not apparent with other adenovirus mutants such as *dl*922/947 (Heise *et al.*, 2000).

In addition, a second deletion in the E3 gene region (10.4/14.5 complex) may make this virus more sensitive to the antiviral effects of tumor necrosis factor; an immunocompetent animal model will need to be identified in order to resolve this issue. Factors likely to be an issue with any virus include barriers to intratumoral spread, antiviral immune responses, and inadequate viral receptor expression (e.g., CAR, integrins). Viral coat modifications may be beneficial if inadequate CAR expression plays a role in the resistance of particular tumor types (Roelvink *et al.*, 1999; Douglas *et al.*, 1996).

XIII. FUTURE DIRECTIONS: APPROACHES TO IMPROVING THE EFFICACY OF REPLICATION-SELECTIVE VIRAL AGENTS

Second generation viruses are being engineered for greater potency while maintaining safety. For example, a promising adenoviral E1A CR-2 mutant (*dl*922/947) has been described that demonstrates not only tumor-selectivity (based on the G1-S checkpoint status of the cell) but also significantly greater antitumoral efficacy *in vivo* compared to the first-generation virus *dl*1520, and even wild-type adenovirus in one model (Kirn *et al.*, 1998a; Heise *et al.*, 2000). Deletion of antiapoptotic genes (Chinnadurai, 1983; Sauthoff *et al.*, 2000; Medina *et al.*, 1999) or overexpression of death proteins (Doronin *et al.*, 2000) may result in more rapid tumor cell killing *in vitro*; it remains to be seen whether these *in vitro* observations are followed by evidence for improved efficacy *in vivo* over wild-type adenovirus (Table 1).

Potency can also be improved by arming viruses with therapeutic genes (e.g., prodrug-activating enzymes and cytokines) (Hermiston, 2000; Hawkins *et al.*, 1999; Freytag *et al.*, 1998; Wildner *et al.*, 1999). Viral coat modifications may be beneficial if inadequate CAR expression plays a role in the resistance of particular tumor types (Roelvink *et al.*, 1999; Kim *et al.*, 1997). Improved systemic delivery may require novel formulations or coat modifications, as well as suppression of the humoral immune response. Finally, the identification of mechanisms leading to the favorable interaction between replicating viral therapy and chemotherapy may allow augmentation of this interaction.

XIV. SUMMARY

Replication-selective oncolytic viruses have a number of attractive features as novel cancer treatments. Clinical studies have demonstrated that viral treatment can be well tolerated and that tumor necrosis can result. The feasibility of virus delivery to tumors through the bloodstream has also been demonstrated (Heise *et al.*, 1999; Reid *et al.*, 1999; Nemunaitis *et al.*, 1999). The inherent ability of replication-competent adenoviruses to sensitize tumor cells to chemotherapy was a novel discovery that has led to chemosensitization strategies. These data will support the further development of viral agents, including second-generation constructs containing exogenous therapeutic genes to enhance both local and systemic antitumoral activity (Heise and Kirn, 2000; Hermiston, 2000; Agha-Mohammadi and Lotze 2000). Given the limited ability of *in vitro* cell-based assays and murine tumor model systems to accurately predict the efficacy and therapeutic index of replication-selective viruses in patients, we believe that the timely translation of encouraging viral agents into well-designed clinical trials with relevant biological end points is critical. Only then can the true therapeutic potential of these agents be realized.

Acknowledgments

The following individuals have been instrumental in making this chapter possible: John Nemunaitis, Stan Kaye, Tony Reid, Fadlo Khuri, James Abruzzesse, Eva Galanis, Joseph Rubin, Antonio Grillo-Lopez, Carla Heise, Larry Romel, Chris Maack, Sherry Toney, Nick LeMoine, Britta Randlev, Patrick Trown, Fran Kahane, Frank McCormick, and Margaret Uprichard.

References

Advani, S. J., Sibley, G. S., Song, P. Y., Hallahan, D. E., Kataoka, Y., Roizman, B., and Weichselbaum, R. R. (1998). Enhancement of replication of genetically engineered herpes simplex viruses by ionizing radiation: A new paradigm for destruction of therapeutically intractable tumors. *Gene Ther.* **5,** 160–165.

Agha-Mohammadi, S., and Lotze, M. (2000). Immunomodulation of cancer: Potential use of replication-selective agents. *J. Clin. Invest.* **105,** 1173–1176.

Andreansky, S. S., He, B., Gillespie, G. Y., Soroceanu, L., Markert, J., Chou, J., Roizman, B., and Whitley, R. J. (1996). The application of genetically engineered herpes simplex viruses to the treatment of experimental brain tumors. *Proc. Natl. Acad. Sci. U.S.A.* **93,** 11313–11318.

Asada, T. (1974). Treatment of human cancer with mumps virus. *Cancer* **34,** 1907–1928.

Barker, D. D., and Berk, A. J. (1987). Adenovirus proteins from both E1B reading frames are required for transformation of rodent cells by viral infection and DNA transfection. *Virology* **156,** 107–121.

Bischoff, J. R., Kirn, D. H., Williams, A., Heise, C., Horn, S., Muna, M., Ng, L., Nye, J. A., Sampson-Johannes, A., Fattaey, A., and McCormick, F. (1996). An adenovirus mutant that replicates selectively in p53-deficient human tumor cells. *Science* **274,** 373–376.

Carew, J. F., Kooby, D. A., Halterman, M. W., Federoff, H. J., and Fong, Y. (1999). Selective infection and cytolysis of human head and neck squamous cell carcinoma with sparing of normal mucosa by a cytotoxic herpes simplex virus type 1 (G207). *Hum. Gene Ther.* **10,** 1599–1606.

Chinnadurai, G. (1983). Adenovirus 2 Ip+ locus codes for a 19 kd tumor antigen that plays an essential role in cell transformation. *Cell* **33,** 759–766.

Clairmont, C., Bermudes, D., Low, K. B., Pawelek, J., Pike, J., Ittensohn, M., Li, Z., Lou, X., Margitich, D., Lee, K., Zheng, L. M., and King, I. (1999). VNP20009, a genetically modified *Salmonella typhimurium:* Anti-tumor efficacy, toxicology, and biodistribution in preclinical models. *Clin. Cancer Res.* **5,** 3830s.

Coffey, M., Strong, J., Forsyth, P., and Lee, P. (1998). Reovirus therapy of tumors with activated ras pathway. *Science* **282,** 1332–1334.

Colella, T. A., Bullock, T. N., Russell, L. B., Mullins, D. W., Overwijk, W. W., Luckey, C. J., Pierce, R. A., Restifo, N. P., and Engelhard, V. H. (2000). Self-tolerance to the murine homologue of a tyrosinase-derived melanoma antigen: Implications for tumor immunotherapy. *J. Exp. Med.* **191,** 1221–1232.

Conry, R. M., Khazaeli, M. B., Saleh, M. N., Allen, K. O., Barlow, D. L., Moore, S. E., Craig, D., Arani, R. B., Schlom, J., and LoBuglio, A. F. (1999). Phase I trial of a recombinant vaccinia virus encoding carcinoembryonic antigen in metastatic adenocarcinoma: Comparison of intradermal versus subcutaneous administration. *Clin. Cancer Res.* **5,** 2330–2337.

Coukos, G., Makrigiannakis, A., Montas, S., Kaiser, L. R., Toyozumi, T., Benjamin, I., Albelda, S. M., Rubin, S. C., and Molnar-Kimber, K. L. (2000). Multi-attenuated herpes simplex virus-1 mutant G207 exerts cytotoxicity against epithelial ovarian cancer but not normal mesothelium and is suitable for intraperitoneal oncolytic therapy. *Cancer Gene Ther.* **7,** 275–283.

Dietzschold, B., Rupprecht, C. E., Fu, Z. F., and Koprowski, H. (1996). Rhabdoviruses. *In* "Fields Virology" (D. M. K. B. N. Fields, P. M. Howley, *et al.*, eds.), Lippincott-Raven, Philadelphia.

Dobner, T., Horikoshi, N., Rubenwolf, S., and Shenk, T. (1996). Blockage by adenovirus E4orf6 of transcriptional activation by the p53 tumor suppressor. *Science* **272,** 1470–1473.

Doronin, K., Toth, K., Kuppuswamy, M., Ward, P., Tollefson, A., and Wold, W. (2000). Tumor-specific, replication-competent adenovirus vectors overexpressing the adenovirus death protein. *J. Virol.* **74,** 6147–6155.

Douglas, J. T., Rogers, B. E., Rosenfeld, M. E., Michael, S. I., Feng, M., and Curiel, D. T. (1996). Targeted gene delivery by tropism-modified adenoviral vectors. *Nat. Biotechnol.* **14,** 1574–1578.

Drexler, I., Antunes, E., Schmitz, M., Wolfel, T., Huber, C., Erfle, V., Rieber, P., Theobald, M., and Sutter, G. (1999). Modified vaccinia virus Ankara for delivery of human tyrosinase as melanoma-associated antigen: Induction of tyrosinase- and melanoma-specific human leukocyte antigen A*0201-restricted cytotoxic T cells *in vitro* and *in vivo*. *Cancer Res.* **59,** 4955–4963.

Dupressoir, T., Vanacker, J. M., Cornelis, J. J., Duponchel, N., and Rommelaere, J. (1989). Inhibition by parvovirus H-1 of the formation of tumors in nude mice and colonies *in vitro* by transformed human mammary epithelial cells. *Cancer Res.* **49,** 3203–3208.

Eder, J. P., Kantoff, P. W., Roper, K., Xu, G. X., Bubley, G. J., Boyden, J., Gritz, L., Mazzara, G., Oh, W. K., Arlen, P., Tsang, K. Y., Panicali, D., Schlom, J., and Kufe, D. W. (2000). A phase I trial of a recombinant vaccinia virus expressing prostate-specific antigen in advanced prostate cancer. *Clin. Cancer Res.* **6,** 1632–1638.

Freytag, S. O., Rogulski, K. R., Paielli, D. L., Gilbert, J. D., and Kim, J. H. (1998). A novel three-pronged approach to kill cancer cells selectively: Concomitant viral, double suicide gene, and radiotherapy [see comments]. *Hum Gene Ther.* **9,** 1323–1333.

Fueyo, J., Gomez-Manzano, C., Alemany, R., Lee, P., McDonnell, T., Mitlianga, P., Shi, Y., Levin, V., Yung, W., and Kyritsis, A. (2000). A mutant oncolytic adenovirus targeting the Rb pathway produces anti-glioma effect *in vivo. Oncogene* **19,** 2–12.

Ganly, I., Kirn, D., Eckhardt, S., Rodriguez, G., Souter, D., Von Hoff, D., and Kaye, S. (2000). A phase I study of Onyx-015, an E1B attenuated adenovirus, administered intratumorally to patients with recurrent head and neck cancer. *Clin. Cancer Res.* **6,** 798–806.

Gnant, M. F., Berger, A. C., Huang, J., Puhlmann, M., Wu, P. C., Merino, M. J., Bartlett, D. L., Alexander, H. R., Jr., and Libutti, S. K. (1999). Sensitization of tumor necrosis factor α-resistant human melanoma by tumor-specific *in vivo* transfer of the gene encoding endothelial monocyte-activating polypeptide II using recombinant vaccinia virus. *Cancer Res.* **59,** 4668–4674.

Goodrum, F. D., and Ornelles, D. A. (1997). The early region 1B 55-kilodalton oncoprotein of adenovirus relieves growth restrictions imposed on viral replication by the cell cycle. *J. Virol.* **71,** 548–561.

Gromeier, M., Alexander, L., and Wimmer, E. (1996). Internal ribosomal entry site substitution eliminates neurovirulence in intergeneric poliovirus recombinants. *Proc. Natl. Acad. Sci. U.S.A.* **93,** 2370–2375.

Gromeier, M., Bossert, B., Arita, M., Nomoto, A., and Wimmer, E. (1999). Dual stem loops within the poliovirus internal ribosomal entry site control neurovirulence. *J. Virol.* **73,** 958–964.

Gromeier, M., Lachmann, S., Rosenfeld, M. R., Gutin, P. H., and Wimmer, E. (2000). Intergeneric poliovirus recombinants for the treatment of malignant glioma [see comments]. *Proc. Natl. Acad. Sci. U.S.A.* **97,** 6803–6808.

Hallenbeck, P. (1999). Oncolytic adenovirus driven by the alpha-fetoprotein promoter. *Hum. Gene Ther.* **10,** 1721–1733.

Harada, J., and Berk, A. (1999). p53-independent and -dependent requirements for E1B-55kD in adenovirus type 5 replication. *J. Virol.* **73,** 5333–5344.

Hawkins, L., Nye, J., Castro, D., Johnson, L., Kirn, D., and Hermiston, T. (1999). Replicating adenoviral gene therapy. *Proc. Am. Assoc. Cancer Res.* **40,** 476.

He, Z., Wlazlo, A. P., Kowalczyk, D. W., Cheng, J., Xiang, Z. Q., Giles-Davis, W., and Ertl, H. C. (2000). Viral recombinant vaccines to the E6 and E7 antigens of HPV-16. *Virology* **270,** 146–161.

Hecht, R., Abbruzzese, J., Bedford, R., Randlev, B., Romel, L., Lahodi, S., and Kirn, D. (2000). Endoscopic ultrasound-guided intratumoral injection of pancreatic carcinomas with a replication-selective adenovirus: A phase I/II clinical trial. *Proc. Am. Soc. Clin. Oncol.* **19,** 1039.

Heise, C., and Kirn, D. (2000). Replication-selective adenoviruses as oncolytic agents. *J. Clin. Invest.* **105,** 847–851.

Heise, C., Sampson, J. A., Williams, A., McCormick, F., Von, H. D., and Kirn, D. H. (1997). ONYX-015, an E1B gene-attenuated adenovirus, causes tumor-specific cytolysis and antitumoral efficacy that can be augmented by standard chemotherapeutic agents [see comments]. *Nat. Med.* **3,** 639–645.

Heise, C., Williams, A., Xue, S., Propst, M., and Kirn, D. (1999). Intravenous administration of ONYX-015, a selectively-replicating adenovirus, induces antitumoral efficacy. *Cancer Res.* **59,** 2623–2628.

Heise, C., Hermiston, T., Johnson, L., Brooks, G., Sampson-Johannes, A., Williams, A., Hawkins, L., and Kirn, D. (2000). An adenovirus E1A mutant that demonstrates potent and selective antitumoral efficacy. *Nat. Med.* **6,** 1134–1139.

Hermiston, T. (2000). Gene delivery from replication-selective viruses: Arming guided missiles in the war against cancer. *J. Clin. Invest.* **105,** 1169–1172.

Herrlinger, U., Kramm, C. M., Aboody-Guterman, K. S., Silver, J. S., Ikeda, K., Johnston, K. M., Pechan, P. A., Barth, R. F., Finkelstein, D., Chiocca, E. A., Louis, D. N., and Breakefield, X. O. (1998). Pre-existing herpes simplex virus 1 (HSV-1) immunity decreases, but does not abolish, gene transfer to experimental brain tumors by a HSV-1 vector. *Gene Ther.* **5**, 809–819.

Hollstein, M., Sidransky, D., Vogelstein, B., and Harris, C. C. (1991). p53 mutations in human cancers. *Science* **253**, 49–53.

Hunter, W. D., Martuza, R. L., Feigenbaum, F., Todo, T., Mineta, T., Yazaki, T., Toda, M., Newsome, J. T., Platenberg, R. C., Manz, H. J., and Rabkin, S. D. (1999). Attenuated, replication-competent herpes simplex virus type 1 mutant G207: Safety evaluation of intracerebral injection in nonhuman primates. *J. Virol.* **73**, 6319–6326.

Khuri, F., Nemunaitis, J., Ganly, I., Gore, M., MacDougal, M., Tannock, I., Kaye, S., Hong, W., and Kirn, D. (2000). A controlled trial of Onyx-015, an E1B gene-deleted adenovirus, in combination with chemotherapy in patients with recurrent head and neck cancer. *Nat. Med.* **6**, 879–885.

Kim, M., Wright, M., Deshane, J., Accavitti, M. A., Tilden, A., Saleh, M., Vaughan, W. P., Carabasi, M. H., Rogers, M. D., Hockett, R. J., Grizzle, W. E., and Curiel, D. T. (1997). A novel gene therapy strategy for elimination of prostate carcinoma cells from human bone marrow. *Hum. Gene Ther.* **8**, 157–170.

Kirn, D. (2000a). A tale of two trials: Selectively replicating herpesviruses for brain tumors. *Gene Ther.* **7**, 815–816.

Kirn, D. (2000b). Replication-selective micro-organisms: Fighting cancer with targeted germ warfare. *J. Clin. Invest.* **105**, 836–838.

Kirn, D., Heise, C., Williams, M., Propst, M., and Hermiston, T. (1998a). Adenovirus E1A CR2 mutants as selectively-replicating agents for cancer. *In* "Cancer Gene Therapy". Academic Press, San Diego.

Kirn, D., Hermiston, T., and McCormick, F. (1998b). ONYX-015: Clinical data are encouraging [letter; comment]. *Nat. Med.* **4**, 1341–1342.

Kooby, D. A., Carew, J. F., Halterman, M. W., Mack, J. E., Bertino, J. R., Blumgart, L. H., Federoff, H. J., and Fong, Y. (1999). Oncolytic viral therapy for human colorectal cancer and liver metastases using a multi-mutated herpes simplex virus type-1 (G207). *FASEB J.* **13**, 1325–1334.

Kucharczuk, J. C., Randazzo, B., Chang, M. Y., Amin, K. M., Elshami, A. A., Sterman, D. H., Rizk, N. P., Molnar-Kimber, K. L., Brown, S. M., MacLean, A. R., Litzky, L. A., Fraser, N. W., Albelda, S. M., and Kaiser, L. R. (1997). Use of a "replication-restricted" herpes virus to treat experimental human malignant mesothelioma. *Cancer Res.* **57**, 466–471.

Kurihara, T., Brough, D. E., Kovesdi, I., and Kufe, D. W. (2000). Selectivity of a replication-competent adenovirus for human breast carcinoma cells expressing the MUC1 antigen. *J. Clin. Invest.* **106**, 763–771.

Lemmon, M. J., van Zijl, P., Fox, M. E., Mauchline, M. L., Giaccia, A. J., Minton, N. P., and Brown, J. M. (1997). Anaerobic bacteria as a gene delivery system that is controlled by the tumor microenvironment. *Gene Ther.* **4**, 791–796.

Lorenz, M. G., Kantor, J. A., Schlom, J., and Hodge, J. W. (1999). Anti-tumor immunity elicited by a recombinant vaccinia virus expressing CD70 (CD27L). *Hum. Gene Ther.* **10**, 1095–1103.

Low, K., Ittensohn, M., Le, T., Platt, J., Sodi, S., Amoss, M., Ash, O., Carmichael, E., Chakraborty, A., Fischer, J., Lin, S., Luo, X., Miller, S., Zheng, L., King, I., Pawelek, J., and Bermudes, D. (1999). Lipid A mutant salmonella with suppressed virulence and TNF-α induction retain tumor-targeting *in vivo*. *Nat. Biotechnol.* **17**, 37–41.

McCart, J. A., Puhlmann, M., Lee, J., Hu, Y., Libutti, S. K., Alexander, H. R., and Bartlett, D. L. (2000). Complex interactions between the replicating oncolytic effect and the enzyme/prodrug effect of vaccinia-mediated tumor regression. *Gene Ther.* **7**, 1217–1223.

Markert, J., Medlock, M., Rabkin, S., Gillespie, G., Todo, T., Hunter, W., Palmer, C., Feigenbaum, F., Tornatore, C., Tufaro, F., and Martuza, R. (2000a). Conditionally replicating herpes simplex virus mutant, G207 for the treatment of malignant glioma: Results of a phase I trial. *Gene Ther.* **7,** 867–874.

Martuza, R. (2000). Conditionally replicating herpes viruses for cancer therapy. *J. Clin. Invest.* **105,** 841–846.

Martuza, R. L., Malick, A., Markert, J. M., Ruffner, K. L., and Coen, D. M. (1991). Experimental therapy of human glioma by means of a genetically engineered virus mutant. *Science* **252,** 854–856.

Mastrangelo, M. J., Maguire, H. C., Jr., Eisenlohr, L. C., Laughlin, C. E., Monken, C. E., McCue, P. A., Kovatich, A. J., and Lattime, E. C. (1999). Intratumoral recombinant GM-CSF-encoding virus as gene therapy in patients with cutaneous melanoma. *Cancer Gene Ther.* **6,** 409–422.

Mastrangelo, M., Eisenlohr, L., Gomella, L., and Lattime, E. (2000). Poxvirus vectors: Orphaned and underappreciated. *J. Clin. Invest.* **105,** 1031–1034.

Medina, D. J., Sheay, W., Goodell, L., Kidd, P., White, E., Rabson, A. B., and Strair, R. K. (1999). Adenovirus-mediated cytotoxicity of chronic lymphocytic leukemia cells. *Blood* **94,** 3499–3508.

Mineta, T., Rabkin, S. D., Yazaki, T., Hunter, W. D., and Martuza, R. L. (1995). Attenuated multi-mutated herpes simplex virus-1 for the treatment of malignant gliomas. *Nat. Med.* **1,** 938–943.

Miyatake, S., Iyer, A., Martuza, R. L., and Rabkin, S. D. (1997). Transcriptional targeting of herpes simplex virus for cell-specific replication. *J. Virol.* **71,** 5124–5132.

Moss, B. (1996a). Poxiviridae: The viruses and their replication. *In:* "Fields Virology" (B. N. Fields, ed.), 3rd Ed. Lippincott-Raven, Philadelphia.

Moss, B. (1996b). Genetically engineered poxviruses for recombinant gene expression, vaccination, and safety. *Proc. Natl. Acad. Sci. U.S.A.* **93,** 11341–11348.

Mukherjee, S., Haenel, T., Himbeck, R., Scott, B., Ramshaw, I., Lake, R. A., Harnett, G., Phillips, P., Morey, S., Smith, D., Davidson, J. A., Musk, A. W., and Robinson, B. (2000). Replication-restricted vaccinia as a cytokine gene therapy vector in cancer: Persistent transgene expression despite antibody generation. *Cancer Gene Ther.* **7,** 663–670.

Nemunaitis, J., Cunningham, C., Edelman, G., Berman, B., and Kirn, D. (1999). Phase I dose-escalation trial of intravenous ONYX-015 in patients with refractory cancer. Proceedings of the Cancer Gene Therapy Meeting.

Nemunaitis, J., Ganly, I., Khuri, F., Arsenau, J., Kuhn, J., McCarty, T., Landers, S., Maples, P., Romel, L., Randlev, B., Reid, T., Kaye, S., and Kirn, D. (2000a). Selective replication and oncolysis in p53 mutant tumors with Onyx-015, an E1B-55kD gene-deleted adenovirus, in patients with advanced head and neck cancer: A phase II trial. *Cancer Res.* **60,** 6359–6366.

Nemunaitis, J., Cunningham, C., Randlev, B., and Kirn, D. (2000b). A phase I trial of intravenous administration with a replication-selective adenovirus, dl 1520. *Proc. Am. Soc. Clin. Oncol.* **19,** 724.

Nemunaitis, J., Khuri, F., Posner, M., Vokes, E., Romel, L., and Kirn, D. (2000c). Phase II trials with an E1B-55kD gene-deleted adenovirus in patients with recurrent head and neck carcinomas. *Cancer Res.* **60,** 6359–6366.

Nielsch, U., Fognani, C., and Babiss, L. E. (1991). Adenovirus E1A–p105(Rb) protein interactions play a direct role in the initiation but not the maintenance of the rodent cell transformed phenotype. *Oncogene* **6,** 1031–1036.

Norman, K, and Lee, P. (2000). Reovirus as a novel oncolytic agent. *J. Clin. Invest.* **105,** 1035–1038.

Olson, D. C., and Levine, A. J. (1994). The properties of p53 proteins selected for the loss of suppression of transformation. *Cell Growth Differ.* **5,** 61–71.

Parker, J., Gillespie, G., Love, C., Randall, S., Whitley, R., and Markert, J. (2000). Engineered herpes simplex virus expressing IL-12 in the treatment of experimental murine brain tumors. *Proc. Natl. Acad. Sci. U.S.A.* **97,** 2208–2213.

Pawelek, J., Low, K., and Bermudes, D. (1997). Tumor-targeted Salmonella as a novel anticancer vector. *Cancer Res.* **57,** 4537–4544.

Puhlmann, M., Brown, C. K., Gnant, M., Huang, J., Libutti, S. K., Alexander, H. R., and Bartlett, D. L. (2000). Vaccinia as a vector for tumor-directed gene therapy: Biodistribution of a thymidine kinase-deleted mutant. *Cancer Gene Ther.* **7,** 66–73.

Rampling, R., Cruickchank, G., Papanastassiou, V., Nicoll, J., Hadley, D., Brennan, D., Petty, R., Maclean, A., Harland, J., McKie, E., Mabbs, R., and Brown, M. (2000). Toxicity evaluation of replication-competent herpes simplex virus (ICP 34.5 null mutant 1716) in patients with recurrent malignant glioma. *Gene Ther.* **7,** 1–4.

Reid, A., Galanis, E., Abbruzzese, J., Romel, L., Rubin, J., and Kirn, D. (1999). A phase I/II trial of ONYX-015 administered by hepatic artery infusion to patients with colorectal carcinoma. EORTC-NCI-AACR Meeting on Molecular Therapeutics of Cancer.

Reid, T., Galanis, E., Abbruzzese, J., Randlev, B., Romel, L., Rubin, J., and Kirn, D. (2000). Hepatic arterial infusion of a replication-selective adenovirus, Onyx-015: A phase I/II clinical trial. *Proc. Am. Soc. Clin. Oncol.* **19,** 953.

Rodriguez, R., Schuur, E. R., Lim, H. Y., Henderson, G. A., Simons, J. W., and Henderson, D. R. (1997). Prostate attenuated replication competent adenovirus (ARCA) CN706: A selective cytotoxic for prostate-specific antigen-positive prostate cancer cells. *Cancer Res.* **57,** 2559–2563.

Roelvink, P., Mi, G., Einfeld, D., Kovesdi, I., and Wickham, T. (1999). Identification of a conserved receptor-binding site on the fiber proteins of CAR-recognizing adenoviridae. *Science* **286,** 1568–1571.

Roizman, B. (1996). The function of herpes simplex virus genes: A primer for genetic engineering of novel vectors. *Proc. Natl. Acad. Sci. U.S.A.* **93,** 11307–11312.

Rothmann, T., Hengstermann, A., Whitaker, N. J., Scheffner, M., and zur Hausen, H. (1998). Replication of ONYX-015, a potential anticancer adenovirus, is independent of p53 status in tumor cells. *J. Virol.* **72,** 9470–9478.

Sauthoff, H., Heitner, S., Rom, W., and Hay, J. (2000). Deletion of the adenoviral E1B-19kD gene enhances tumor cell killing of a replicating adenoviral vector. *Hum. Gene Ther.* **11,** 379–388.

Scheffner, M., Munger, K., Byrne, J. C., and Howley, P. M. (1991). The state of the p53 and retinoblastoma genes in human cervical carcinoma cell lines. *Proc. Natl. Acad. Sci. U.S.A.* **88,** 5523–5527.

Sherr, C. J. (1996). Cancer cell cycles. *Science* **274,** 1672–1677.

Smith, R., Huebner, R. J., Rowe, W. P., Schatten, W. E., and Thomas, L. B. (1956). Studies on the use of viruses in the treatment of carcinoma of the cervix. *Cancer* **9,** 1211–1218.

Southam, C. M., and Moore, A. E. (1952). Clinical studies of viruses as antineoplastic agents, with particular reference to Egypt 101 virus. *Cancer* **5,** 1025–1034.

Stojdl, D. F., Lichty, B., Knowles, S., Marius, R., Atkins, H., Sonenberg, N., and Bell, J. C. (2000). Exploiting tumor-specific defects in the interferon pathway with a previously unknown oncolytic virus. *Nat. Med.* **6,** 821–825.

Sznol, M., Lin, S., Bermudes, D., Zheng, L., and King, I. (2000). Use of preferentially replicating bacteria for the treatment of cancer. *J. Clin. Invest.* **105,** 1027–1030.

Toda, M., Martuza, R. L., Kojima, H., and Rabkin, S. D. (1998). *In situ* cancer vaccination: An IL-12 defective vector/replication-competent herpes simplex virus combination induces local and systemic antitumor activity. *J. Immunol.* **160,** 4457–4464.

Todo, T., Rabkin, S. D., Sundaresan, P., Wu, A., Meehan, K. R., Herscowitz, H. B., and Martuza, R. L. (1999). Systemic antitumor immunity in experimental brain tumor therapy using a multimutated, replication-competent herpes simplex virus. *Hum. Gene Ther.* **10,** 2741–2755.

Todo, T., Rabkin, S. D., and Martuza, R. L. (2000). Evaluation of ganciclovir-mediated enhancement of the antitumoral effect in oncolytic, multimutated herpes simplex virus type 1 (G207) therapy of brain tumors [in process citation]. *Cancer Gene Ther.* **7,** 939–946.

Toyoizumi, T., Mick, R., Abbas, A. E., Kang, E. H., Kaiser, L. R., and Molnar-Kimber, K. L. (1999). Combined therapy with chemotherapeutic agents and herpes simplex virus type 1 ICP34.5 mutant (HSV-1716) in human non-small cell lung cancer. *Hum. Gene Ther.* **10,** 3013–3029.

Tyler, K. L., and Fields, B. N. (1996). Reoviruses. *In* "Fields Virology" (B. N. Fields, ed.), 3rd Ed. Lippincott-Raven, Philadelphia.

Uzendoski, K., Kantor, J. A., Abrams, S. I., Schlom, J., and Hodge, J. W. (1997). Construction and characterization of a recombinant vaccinia virus expressing murine intercellular adhesion molecule-1: Induction and potentiation of antitumor responses. *Hum. Gene Ther.* **8,** 851–860.

Valmori, D., Levy, F., Miconnet, I., Zajac, P., Spagnoli, G. C., Rimoldi, D., Lienard, D., Cerundolo, V., Cerottini, J. C., and Romero, P. (2000). Induction of potent antitumor CTL responses by recombinant vaccinia encoding a melan-A peptide analogue. *J. Immunol.* **164,** 1125–1131.

Vasey, P., Shulman, L., Gore, M., Kirn, D., and Kaye, S. (2000). A phase I trial of an E1B-55kD gene-deleted adenovirus administered by intraperitoneal injection into patients with advanced, refractory ovarian carcinoma. *Proc. Am. Soc. Clin. Oncol.* **19,** 1512.

Wickham, T. J., Segal, D. M., Roelvink, P. W., Carrion, M. E., Lizonova, A., Lee, G. M., and Kovesdi, I. (1996). Targeted adenovirus gene transfer to endothelial and smooth muscle cells by using bispecific antibodies. *J. Virol.* **70,** 6831–6838.

Wildner, O., Blaese, R. M., and Morris, J. M. (1999). Therapy of colon cancer with oncolytic adenovirus is enhanced by the addition of herpes simplex virus-thymidine kinase. *Cancer Res.* **59,** 410–413.

Yew, P. R., Liu, X., and Berk, A. J. (1994). Adenovirus E1B oncoprotein tethers a transcriptional repression domain to p53. *Genes Dev* **8,** 190–202.

Zajac, P., Oertli, D., Spagnoli, G. C., Noppen, C., Schaefer, C., Heberer, M., and Marti, W. R. (1997). Generation of tumoricidal cytotoxic T lymphocytes from healthy donors after *in vitro* stimulation with a replication-incompetent vaccinia virus encoding MART-1/Melan-A 27-35 epitope. *Int. J. Cancer* **71,** 491–496.

Zhang, Y., Xiong, Y., and Yarbrough, W. G. (1998). ARF promotes MDM2 degradation and stabilizes p53: ARF-INK4a locus deletion impairs both the Rb and p53 tumor suppression pathways. *Cell* **92,** 725–734.

2

Reovirus Therapy of Ras-Associated Cancers

Patrick W. K. Lee

Cancer Biology Research Group
and Department of Microbiology and Infectious Diseases
Faculty of Medicine, University of Calgary
Calgary, Alberta, Canada T2N 4N1

I. INTRODUCTION

A. Reovirus

Reovirus is a member of the family Reoviridae which is characterized by the presence of a segmented, double-stranded (ds) RNA genome as well as certain common features of their replication strategies (Tyler and Fields, 1996). Members of this family include many mammalian, bird, plant, and insect viruses, the most notorious of which is the important human pathogen, rotavirus. Unlike rotavirus, however, reovirus is considered to be benign and relatively nonpathogenic. It can normally be isolated from the respiratory and digestive tracts of humans but is not associated with any disease state; hence the acronym reo (respiratory, enteric, orphan) that was coined in 1959 (Sabin, 1959). Based on neutralization and hemagglutination-inhibition tests, three prototypical reovirus serotypes have been isolated from the respiratory and enteric tracts of children. These are designated Type 1 Lang, Type 2 Jones, Type 3 Abney, and Type 3

Dearing. Since most people have detectable antireovirus antibodies by the time they reach adolescence, the virus most likely causes mild upper respiratory infections and/or mild diarrheal illnesses in children with no serious consequences. The benign nature of this virus was further confirmed from studies on adult human volunteers, in which intranasal inoculation of reovirus produced only mild symptoms (Rosen *et al.*, 1963).

Despite its relatively benign nature, reovirus has been used extensively as a model for viral pathogenesis in neonatal mice (Tyler and Fields, 1996). The three reovirus serotypes manifest marked differences in pathogenicity in these animals. For example, type 1 reovirus grows well in the ileum after oral inoculation, whereas type 3 reovirus does not. Regardless of the route of introduction, both serotypes eventually spread to the central nervous system, including the brain. Here again, the two serotypes differ significantly in their tropism. Type 3 reovirus preferentially infects neuronal cells, resulting in encephalitis, whereas type 1 reovirus infects ependymal cells lining the ventricles, resulting in hydrocephalus. Based on genetic reassortment studies (made possible due to the segmented nature of the genome), different pathogenicity manifestations have been assigned to distinct genomic segments. For example, the *S1* gene encoding the reovirus cell attachment protein σ1 has been shown to be responsible for distinct tropism in the brain (Weiner *et al.*, 1977, 1980); and the M1 gene, encoding a viral core protein, is associated with the myocarditic phenotype of a reovirus variant (Sherry and Fields, 1989). Overall, the use of reovirus reassortants for the study of viral gene function related to pathogenesis, however rudimentary, has shed considerable light on viral pathogenesis in general, and has made reovirus an important model for the study of this branch of virology to this day.

Presently, studies on reovirus have focused on detailed structural and functional analysis of the reovirus proteins encoded by the 10 gene segments. Of all the reovirus proteins, the S1 gene product (protein σ1) is the most thoroughly studied, probably because it is the cell attachment protein and therefore plays a major role in tissue tropism. It is through the study of the σ1 protein [and the receptor(s) it interacts with] that the story of the Ras (and hence, cancer) connection came about.

B. Ras-associated cancers and current anti-Ras strategies

Ras proteins are small membrane-associated guanine nucleotide binding proteins involved in the transduction of signals from the cell surface to appropriate downstream effectors inside the cell (for reviews, see Bos, 1997, Olson and Marais, 2000; Reuther and Der, 2000). These proteins therefore play important roles in cell growth, differentiation, survival, and migration. Following ligand binding, receptor tyrosine kinases [such as epidermal growth factor receptor

(EGFR)] undergo autophosphorylation which provides appropriate docking sites for different elements including adaptor molecules such as Grb2 and Shc. These in turn recruit guanine-nucleotide exchange factors (GEFs such as Sos) which promote GDP/GTP exchange, resulting in Ras activation (in the GTP-bound form). Activated Ras in turn stimulates a group of downstream pathways including the family of mitogen activated protein kinases (MAPKs) which are involved in many biological responses varying from cell proliferation and differentiation to apoptosis and cell migration.

In view of the central role of Ras in cell signaling, it is not surprising that Ras is one of the most commonly mutated oncogenes in human neoplasia, accounting for at least 30% of human cancer, ranging from 95% in pancreatic cancer to 5% in breast cancer (Bos, 1989). Ras mutations are found in 50% of sporadic colorectal cancer, 25% of lung cancer, and 30% of myeloid leukemia. Mutations are often found at codons 12, 13, and 61, and result in Ras being constitutively active due to the inability of the Ras-bound GTP to be hydrolyzed to GDP. It is important to note, however, that deregulation of Ras can also occur in the absence of mutations in Ras itself (Lowe and Skinner, 1994; Levitzki, 1994). For example, overexpression of upstream elements such as EGFR, platelet-derived growth factor receptor (PDGFR) (found in 40–50% of gliomas and glioblastomas) (Guha et al., 1995, 1997; Nister et al., 1991), or related receptor tyrosine kinases (RTKs) such as HER2/Neu/ErbB2 (found in 25–30% of breast cancer) (Slamon et al., 1987, 1989) could lead to Ras activation. Overall, it is likely that at least 50% of all cancers manifest an activated Ras pathway.

The high incidence of Ras activation in human cancer has led to intensive research into the mechanism of Ras signaling, in the hope that appropriate molecular modulators could be designed to attenuate its activity. To this end, a number of approaches have been attempted (for a review, see Kloog and Cox, 2000), the most promising of which are those directed at the inhibition of Ras anchoring to the cell membrane. Association of Ras with the plasma membrane requires a series of modification to the Ras protein, one of which is farnesylation of the cysteine in the carboxy terminal CAAX motif. This process is carried out by the cytosolic enzyme farnesyltransferase (FTase). Based on this information, farnesyltransferase inhibitors (FTIs) have been designed that block the activity of this enzyme (Kohl et al., 1993; James et al., 1993), and have shown to be effective in causing tumor regression in experimental animals (Kohl et al., 1995; Liu et al., 1998). However, some studies showed that the efficacy of FTIs does not correlate with the mutation status of Ras, suggesting that inhibition of Ras farnesylation is but one of the effects of these drugs, and that the anticancer activity of FTIs is probably due in large part to their effects on other targets (Sepp-Lorenzino et al., 1995; Whyte et al., 1997). Another drug that prevents Ras membrane anchorage is trans-farnesylthiosalicylic acid (FTS) which resembles the farnesylcysteine of Ras, and therefore directly competes with Ras for membrane anchorage sites (Marom *et al.*, 1995; Haklai *et al.*, 1998). Unlike FTI, however, FTS appears to be relatively Ras-specific, and has been shown to inhibit Ras transforming activity *in vitro* and to inhibit tumor growth in animals (Jansen *et al.*, 1999). It remains to be seen whether FTI and/or FTS are effective against cancers in humans. Another approach for downregulating the Ras signaling pathway is through the use of inhibitors that block downstream effectors. These have included sulindac sulfide which blocks the interaction of Ras with Raf-1 (Herrmann *et al.*, 1998), and MEK inhibitors such as PD98059 which blocks the activation of ERK1/2 (Dudley *et al.*, 1995).

It is important to note that, for Ras-associated cancer, essentially all the approaches attempted thus far focus on the use of drugs to tone down the Ras signaling pathway. Although these approaches are by and large conceptually sound, the major problem pertains to specificity; it is simply not easy to tone down activated Ras without affecting the function of normal Ras. As a result, problems with toxicity are likely to be encountered, which could in turn limit the scope of application.

II. REOVIRUS ONCOLYSIS

A. Discovery of the exploitation of the Ras signaling pathway by reovirus

1. The first experiments: The "good" red herring

The discovery that reovirus exploits the host cell Ras signaling pathway for infection actually came from studies that were unrelated to cancer research at all. For many years my laboratory has been involved in the study of the reovirus cell attachment protein (protein $\sigma 1$) and its interaction with the host cell receptor (Lee and Gilmore, 1998). Protein $\sigma 1$ is a lollipop-shaped structure located at the 12 vertices of the icosahedral virus. It is a trimeric protein with a rodlike fibrous tail that is anchored to the body of the virion and a globular head most distal from it. Thus, protein $\sigma 1$ morphologically resembles the adenovirus fiber protein that is also responsible for the attachment of the virus to cellular receptors.

The receptor for reovirus has also been extensively probed. It is now generally accepted that the initial interaction between reovirus and the host cell involves the binding of the $\sigma 1$ protein to sialic acid moieties on the cell surface (Gentsch and Pacitti, 1985, 1987; Paul *et al.*, 1989). This binding presumably leads to conformational changes in $\sigma 1$ as well as other viral capsid proteins, processes that may be necessary for a second stage binding step (which likely involves a second receptor), as well as subsequent viral entry and programmed disassembly of viral capsids inside the host cell. Because most cell surface proteins contain sialic acid moieties, it stands to reason that reovirus should bind to multiple sialoglycoproteins, rather than a single homogeneous species. That was in fact found to be the case based on ligand-blot analysis (Choi et al., 1990). Indeed, reovirus is able to bind to most, if not all, of the cell lines that we have tested. Interestingly, not all the cell lines that allow reovirus to bind, or even to internalize, can support virus growth. This observation represents the first indication that factors inside the cell probably play a major role in dictating host cell permissiveness to reovirus infection.

In the fall of 1992, one of my students was investigating the possible existence of a "second" receptor for reovirus binding. He was using the human epidermoid carcinoma cell line A431 for his study because this cell line was highly susceptible to reovirus infection and because neuraminidase treatment (which removes sialic acid) completely abrogated reovirus binding as well as infection in these cells. At that time we felt that these cells would be suitable for the identification of a presumptive "second" reovirus receptor. In one study, he attempted to release cell surface factors (using proteases and glycosidases) that could interact with the virus. During the course of this study, he accidentally found that spent A431 medium (as control) alone was capable of blocking both virus binding and infection. Our first reaction to this observation was a mixed one. On one hand this could complicate our receptor release experiment mentioned above; on the other hand, whatever blocked reovirus binding could be a potential reovirus receptor secreted by A431 cells. A trip to the library revealed that A431 cells do secrete a 105 kDa epidermal growth factor receptor-related protein (ERRP) which corresponds to the N-terminal extracellular domain of the EGF receptor protein (Weber et al., 1984). This revelation generated considerable excitement in my laboratory because it really would be a great story if the second receptor for reovirus turned out to be none other than the EGFR!

Follow-up experiments quickly established that indeed, secreted ERRP was able to bind reovirus (Tang et al., 1993). Furthermore, EGFR on A431 plasma membrane was also found to be recognizable by the virus. That EGFR confers enhanced host cell permissiveness to reovirus infection was further confirmed by the observation that NIH-3T3 cells (or their derivatives), which are nonpermissive to reovirus infection, became permissive on transfection with the EGFR gene (Strong et al., 1993). Permissiveness to reovirus infection was characterized by drastically enhanced viral protein synthesis, progeny virus output, and cytopathic effects. This enhancement of infection efficiency requires a functional EGFR, since it was not observed in cells expressing a mutated (kinase-inactive) EGFR. Both parental and transfected cells nonetheless manifest similar levels of virus binding and internalization. Based on all these observations, our initial suspicion was that the binding of reovirus to EGFR (or a related receptor) is a necessary step for productive reovirus infection. In view of the essential role of EGFR in signal transduction, it was speculated that binding of reovirus activates EGFRs tyrosine kinase activity, triggering the cell signaling cascade, thereby priming the cell for subsequent virus infection. Support for this interpretation also came from the additional observation that reovirus binding can sometimes trigger limited enhancement of EGF receptor autophosphorylation. Attractive as this hypothesis may seem, there are major theoretical inconsistencies. First of all, the ubiquitous nature of reovirus binding (to sialic acid) suggests that reovirus binding to EGFR is likely via the sialic acid moieties of the latter, making EGFR but one of the many cell surface structures recognized by reovirus. The improbability of the reovirus to find an EGFR (or an EGFR-like receptor) in the presence of a milieu of surface sialoglycoproteins is inconsistent with the observation that a single infectious virus particle has little problem initiating an infection. Second, even if a reovirus particle was fortunate enough to find an EGFR, it is highly doubtful that the signal triggered by the binding of a single infectious virion would be strong enough to generate an intracellular environment that is now more conducive to the infection process. It is interesting to note that had this hypothesis (i.e., the priming of the host cell by reovirus) turned out to be correct, then the premise of reovirus as a potential anticancer therapeutic would not have come about.

2. The Ras–PKR connection

The above considerations made us entertain a more likely scenario in which reovirus plays a relatively passive role in dictating the outcome of an infection. In this scenario, reovirus takes advantage of an already activated signal transduction pathway conferred by the presence of functional EGFR on the host cell. To test this possibility, we used the v-erbB oncogene to see if it could confer permissiveness to cells that are normally nonpermissive to reovirus infection (Strong and Lee, 1996). The v-erbB gene product is essentially a mutated EGFR whose main features include a truncated extracellular binding domain (lacking the N-terminal 555 amino acids compared with EGFR, and therefore would not bind reovirus) as well as point mutations and deletions within the cytoplasmic domain. As a result of these mutations, v-erbB possesses ligand-independent, constitutive tyrosine kinase activity and is highly transforming. The subsequent observation that v-erbB confers enhanced permissiveness to reovirus infection in NIH-3T3 cells clearly indicates that reovirus utilizes an already activated signaling pathway downstream of EGFR for infection. Our earlier demonstration that reovirus can bind EGFR therefore represents a fortuitous event that is unrelated to the ensuing infection.

These results led us to entertain the possibility of the Ras pathway, a major pathway downstream of EGFR, as a major determinant of host cell permissiveness to reovirus infection. This was indeed confirmed by subsequent experiments which showed that the relatively nonpermissive NIH-3T3 cells became highly permissive after transformation by either activated Sos, or activated Ras (Strong et al., 1998). The important role of Ras in reovirus infection was further confirmed through the use of cells containing zinc-inducible activated Ras, shown to be infectible only in the presence of zinc.

How is activated Ras involved in reovirus infection? We felt that the only way to find out was to examine both nonpermissive and permissive cells in terms of the infection process. It seemed reasonable to suggest that the step that is blocked in the nonpermissive cells is where Ras is required for the infection to proceed in permissive cells. Since we already demonstrated earlier that virus binding and entry are comparable for both cell types, we proceeded to examine postentry events, the most crucial of which was early transcription of viral genes. Interestingly, early transcription proceeded normally in nonpermissive cells; however, they were not translated (Strong et al., 1998). This would explain the lack of viral proteins in these cells. We therefore concluded that translation of viral genes is the step that requires an activated Ras pathway of the host cell.

A key element involved in viral gene translation is the double-stranded RNA-activated protein kinase (PKR). This protein is normally activated (phosphorylated) during viral infection (possibly by double-stranded structures on viral replication intermediates such as viral transcripts), and could lead to inhibition of viral gene translation if left unchecked [through phosphorylation of the eukaryotic initiation factor (eIF)2-α] (for a review, see Williams, 1999). It has been suggested that PKR activation is a major antiviral strategy employed by mammalian cells. To overcome PKR activation, many viruses have developed anti-PKR measures. In the case of reovirus, it has been shown that the major outer capsid protein, σ3, is capable of binding double-stranded RNA, thereby preventing PKR from being activated (Imani and Jacobs, 1988; Lloyd and Shatkin, 1992). However, based on our experiments described above, we felt that it is more likely that the anti-PKR capacity of the host cell is a major determinant of cellular permissiveness to virus infection. It is known that reovirus transcripts (particularly the s1 transcript) can activate PKR (Bischoff and Samuel, 1989). The inability of these transcripts to be translated in nonpermissive cells could be due to the lack of intrinsic anti-PKR activity in these cells, whereas permissive cells such as Ras-transformed NIH-3T3 would be able to downregulate PKR, thereby allowing viral protein synthesis to proceed. This was indeed found to be the case (Strong et al., 1998). PKR phosphorylation was demonstrable only in NIH-3T3 cells, and only after exposure to reovirus. In transformed cells that supported reovirus growth, PKR phosphorylation did not occur. When PKR phosphorylation was artificially suppressed using the drug 2-aminopurine, untransformed NIH-3T3 cells became more permissive to reovirus infection. Furthermore, primary embryo fibroblasts from PKR$^{-/-}$ mice were found to support reovirus growth significantly better than PKR$^{+/+}$ cells. Taken together, these experiments demonstrate that PKR inactivation or ablation enhances host cell permissiveness to reovirus infection in the same way as does transformation by Ras or elements of the Ras pathway. Together with the phosphorylation data, these results provide strong support for the notion that elements of the Ras signaling pathway negatively regulate PKR, leading to enhanced infectibility of cells transformed by them. The PKR–Ras connection has been noted previously by Mundschau and Faller (1992), but not in the context of viral infection. It is interesting that reovirus exploits this particular mechanism for its own replication, thereby making it oncolytic. The overall scheme of this exploitation by reovirus is depicted in Figure 1.

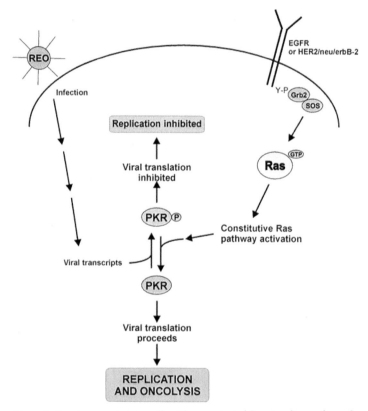

Figure 1 Reovirus oncolysis in cells with an activated Ras signaling pathway. In untransformed cells, infection by reovirus and production of viral transcripts stimulates the cellular double-stranded RNA-activated protein kinase, PKR. PKR activation (phosphorylation) leads to inhibition of viral protein synthesis and thus viral replication. In cancer cells with an activated Ras signaling pathway, PKR activation is inhibited or reversed, thereby allowing viral protein translation to proceed and resulting in the oncolytic effect of reovirus.

How is PKR downregulated by an activated Ras pathway? At present it is unclear as to whether double-stranded RNA-activated PKR phosphorylation is blocked in transformed cells or if phosphorylated PKR becomes dephosphorylated in these cells. Using an *in vitro* approach, we have more recently found that phosphorylated PKR can be dephosphorylated by lysates from transformed cells, but not by those from untransformed cells (P. W. K. Lee, unpublished data, 2001). This suggests that a phosphatase downstream of Ras is likely responsible for PKR inactivation. The exact nature of this putative phosphatase is presently under investigation.

B. Animal models

The discovery that reovirus exploits an activated Ras signaling pathway quickly led us to determine the tumor cell killing potential of reovirus. Of all the human cancer cell lines tested (including those from the brain, breast, pancreas, colorectal, prostate, ovary, etc.), over 80% were effectively lysed by the virus. The first animal experiments were carried out using the human glioblastoma cell line U87 implanted subcutaneously in immune-compromised animals (SCID mice) as xenografts (Coffey *et al.*, 1998). Typically a single injection of reovirus (1×10^7 plaque forming units) directly into the tumor was enough to result in complete tumor regression within 4 weeks. However, SCID mice are known to be highly susceptible to reovirus infection because they are severely immune compromised, and most of them died shortly after complete tumor remission. We have since substituted SCID with nude mice for our studies; tumor regression rate was found to be comparable, but now the survival rate is significantly higher (over 90%).

A most relevant question pertains to the potential use of reovirus in treating cancers that have metastasized. We have found that, at least in immune compromised animals (SCID or nude mice), there is evidence of systemic spread of the virus. For example, in a bilateral breast tumor model in which reovirus was injected unilaterally, dramatic regression was observed for both the injected and remote tumors (P. W. K. Lee, unpublished data). Intravenous administration of the virus has also been shown to be equally effective. Overall these results raise the therapeutic possibility of systemic delivery of the virus.

Another important issue relates to the efficacy of reovirus tumor treatment in immune competent animals. To address this, we have used murine tumors implanted in syngeneic mice (including those with prior exposure to reovirus and thus harbor anti-reovirus antibodies), and have found that direct and repeated intratumoral injections of reovirus were very effective against such tumors. No side effects were noted, and no tumor regrowth occurred for a period of 6 months after eradication of the tumor mass. Intravenous administration of reovirus in these animals, however, is not as effective, although significant reduction in tumor growth rate is often observed (P. W. K. Lee, unpublished data). On the other hand, intravenous administration of reovirus combined with chemotherepeutic drugs (such as cisplatin) or immune suppressing drugs (such as cyclosporin) has yielded very promising preliminary results (P. W. K. Lee, unpublished). Further research is needed to evaluate the applicability of such combined therapy in the treatment of cancers in humans, particularly in view of the fact that most individuals have had previous exposure to the virus and thus carry neutralizing anti-reovirus antibodies.

C. Primary human biopsy tissues

Although reovirus is effective against a variety of human cancer cell lines *in vitro* and *in vivo* as xenografts, the ultimate test will rest on its efficacy in the treatment of cancers in humans. In the absence of such data, it was of interest to determine if human tumor biopsy tissues would manifest infectibility by reovirus. To this end, human biopsy samples of ovarian cancer, breast cancer, and lung cancer were obtained and exposed to reovirus as single cell suspensions. Viral protein synthesis was clearly evident in these cell preparations (P. W. K. Lee, unpublished data). Additionally, thin slices of some of these biopsy samples were prepared and exposed to reovirus. Subsequent immunohistochemical staining using antibodies against total reovirus proteins revealed distinctive positive staining for reovirus proteins in tumor cell clusters but not in the surrounding normal tissue (P. W. K. Lee, unpublished data). Our demonstration that reovirus is capable of replicating in primary tumor tissues *ex vivo* represents a crucial positive step toward the potential use of reovirus as an anticancer therapeutic in humans.

III. CONCLUDING REMARKS

While the results from all of our studies [*in vitro, in vivo* (mice), and *ex vivo* (human)] thus far have by and large been positive in terms of specific tumor cell killing by reovirus, a most crucial question remains: will it work in humans? We hope it will, at least on some cancers, particularly early stage cancers that are localized and have not yet metastasized. It is difficult to predict how well reovirus will work on those cancers that have metastasized, although results from our metastatic animal models look very encouraging. Since most individuals have previously been exposed to the ubiquitous reovirus, and thus harbor neutralizing antibodies, it may be necessary to combine reovirus therapy with immunosuppressive drugs. On the other hand, a competent immune system on the part of the patient could be beneficial and could complement reovirus therapy in some cancers. Another strategy would be sequential administration of the various reovirus serotypes that are known to invoke different immune responses. Combined therapy with chemotherapies and/or with other oncolytic viruses such as the adenovirus mutant ONYX-015 (Heise and Kirn, 2000) and the herpes simplex virus-1 mutant G207 (Martuza, 2000) are other viable options. Regardless of what the future may hold, it is worth mentioning that reovirus phase I clinical studies (focusing on toxicity and safety issues) are already underway at the Tom Baker Cancer Centre in Calgary. The outcome of these studies should provide valuable information on short-term as well as long-term approaches toward the ultimate goal of using reovirus as an anticancer therapeutic.

Acknowledgments

The author thanks the following individuals who have contributed to this work: J. E. Strong, M. C. Coffey, P. Forsyth, K. L. Norman, K. Hirasawa, S. Nishikawa, E. Wilcox, and D. Tang. This research was supported by a grant from the Canadian Institutes of Health Research.

References

Bischoff, J. R., and Samuel, C. E. (1989). Mechanism of interferon action. Activation of the human P1/eIF-2α protein kinase by individual reovirus s-class mRNAs: s1 mRNA is a potent activator relative to s4 mRNA. *Virology* **172,** 106–115.

Bos, J. L. (1989). Ras oncogenes in human cancer: A review. *Cancer Res.* **49,** 4682–4689.

Bos, J. L. (1997). Ras-like GTPases. *Biochim. Biophys. Acta* **1333,** M19–M31.

Choi, A. H., Paul, R. W., and Lee, P. W. K. (1990). Reovirus binds to multiple plasma membrane proteins of mouse L fibroblasts. *Virology* **178,** 316–320.

Coffey, M. C., Strong, J. E., Forsyth, P. A., and Lee, P. W. K. (1998). Reovirus therapy of tumors with activated Ras pathway [see comments]. *Science* **282,** 1332–1334.

Dudley, D. T., Pang, L., Decker, S. J., Bridges, A. J., and Saltiel, A. R. (1995). A synthetic inhibitor of the mitogen-activated protein kinase cascade. *Proc. Natl. Acad. Sci. U.S.A.* **92,** 7686–7689.

Gentsch, J. R., and Pacitti, A. F. (1985). Effect of neuraminidase treatment of cells and effect of soluble glycoproteins on type 3 reovirus attachment to murine L-cells. *J. Virol.* **56,** 356–364.

Gentsch, J. R., and Pacitti, A. F. (1987). Differential interaction of reovirus type 3 with sialylated receptor components on animal cells. *Virology* **161,** 245–248.

Guha, A., Dashner, K., Black, P. M., Wagner, J. A., and Stiles, C. D. (1995). Expression of PDGF and PDGF receptors in human astrocytoma operation specimens supports the existence of an autocrine loop. *Int. J. Cancer* **60,** 168–173.

Guha, A., Feldkamp, M. M., Lau, N., Boss, G., and Pawson, A. (1997). Proliferation of human malignant astrocytomas is dependent on Ras activation. *Oncogene* **15,** 2755–2765.

Haklai, R., Weisz, M. G., Elad, G., Paz, A., Marciano, D., Egozi, Y., Ben-Baruch, G., and Kloog, Y. (1998). Dislodgment and accelerated degradation of Ras. *Biochemistry* **37,** 1306–1314.

Hashiro, G., Loh, P. C., and Yau, J. T. (1977). The preferential cytotoxicity of reovirus for certain transformed cell lines. *Arch. Virol.* **54,** 307–315.

Heise, C., and Kirn, D. H. (2000). Replication-selective adenoviruses as oncolytic agents. *J. Clin. Invest.* **105,** 847–851.

Herrmann, C., Block, C., Geisen, C., Haas, K., Weber, C., Winde, G., Moroy, T., and Muller, O. (1998). Sulindac sulfide inhibits Ras signalling. *Oncogene* **17,** 1769–1776.

Imani, F., and Jacobs, B. L. (1988). Inhibitory activity for the interferon-induced protein kinase is associated with the reovirus serotype 1 σ3 protein. *Proc. Natl. Acad. Sci. U.S.A.* **85,** 7887–7891.

James, G. L., Goldstein, J. L., Brown, M. S., Rawson, T. E., Somers, T. C., McDowell, R. S., Crowley, C. W., Lucas, B. K., Levinson, A. D., and Marsters, J. C., Jr. (1993). Benzodiazepine peptidomimetics: Potent inhibitors of Ras farnesylation in animal cells. *Science* **260,** 1937–1942.

Jansen, B., Schlagbauer-Wadl, H., Kahr, H., Heere-Ress, E., Mayer, B. X., Eichler, H., Pehamberger, H., Gana-Weisz, M., Ben-David, E., Kloog, Y., and Wolff, K. (1999). Novel Ras antagonist blocks human melanoma growth. *Proc. Natl. Acad. Sci. U.S.A.* **96,** 14019–14024.

Kloog, Y., and Cox, A. D. (2000). RAS inhibitors: Potential for cancer therapeutics. *Mol. Med. Today* **6,** 398–402.

Kohl, N. E., Mosser, S. D., deSolms, S. J., Giuliani, E. A., Pompliano, D. L., Graham, S. L., Smith, R. L., Scolnick, E. M., Oliff, A., and Gibbs, J. B. (1993). Selective inhibition of ras-dependent transformation by a farnesyltransferase inhibitor. *Science* **260,** 1934–1937.

Kohl, N. E., Omer, C. A., Conner, M. W., Anthony, N. J., Davide, J. P., deSolms, S. J., Giuliani, E. A., Gomez, R. P., Graham, S. L., Hamilton, K., *et al.* (1995). Inhibition of farnesyltransferase induces regression of mammary and salivary carcinomas in ras transgenic mice. *Nat. Med.* **1,** 792–797.

Lee, P. W. K., and Gilmore, R. (1998). Reovirus cell attachment protein σ1: Structure–function relationships and biogenesis. *Curr. Top. Microbiol. Immunol.* **233,** 137–153.

Levitzki, A. (1994). Signal-transduction therapy. A novel approach to disease management. *Eur. J. Biochem.* **226,** 1–13.

Liu, M., Bryant, M. S., Chen, J., *et al.* (1998). Antitumor activity of SCH 66336, an orally bioavailable tricyclic inhibitor of farnesyl protein transferase, in human tumor xenograft models and wap-ras transgenic mice. *Cancer Res.* **58,** 4947–4956.

Lloyd, R. M., and Shatkin, A. J. (1992). Translational stimulation by reovirus polypeptide σ3: Substitution for VAI RNA and inhibition of phosphorylation of the α-subunit of eukaryotic initiation factor 2. *J. Virol.* **66,** 6878–6884.

Lowe, P. N., and Skinner, R. H. (1994). Regulation of Ras signal transduction in normal and transformed cells. *Cell Signalling* **6,** 109–123.

Marom, M., Haklai, R., Ben-Baruch, G., Marciano, D., Egozi, Y., and Kloog, Y. (1995). Selective inhibition of Ras-dependent cell growth by farnesylthiosalisylic acid. *J. Biol. Chem.* **270,** 22263–22270.

Martuza, R. L. (2000). Conditionally replicating herpes vectors for cancer therapy. *J. Clin. Invest.* **105,** 841–846.

Mundschau, L. J., and Faller, D. V. (1992). Oncogenic ras induces an inhibitor of double-stranded RNA-dependent eukaryotic initiation factor 2 α-kinase activation. *J. Biol. Chem.* **267,** 23092–23098.

Nister, M., Claesson-Welsh, L., Eriksson, A., Heldin, C. H., and Westermark, B. (1991). Differential expression of platelet-derived growth factor receptors in human malignant glioma cell lines. *J. Biol. Chem.* **266,** 16755–16763.

Olson, M. F., and Marais, R. (2000). Ras protein signalling. *Semin. Immunol.* **12,** 63–73.

Paul, R. W., Choi, A. H., and Lee, P. W. K. (1989). The α-anomeric form of sialic acid is the minimal receptor determinant recognized by reovirus. *Virology* **172,** 382–385.

Rosen, L., Evans, H. E., and Spickard, A. (1963). Reovirus infections in human volunteers. *Am. J. Hyg.* **77,** 29–37.

Reuther, G. W., and Der, C. J. (2000). the Ras branch of small GTPases: Ras family members don't fall far from the tree. *Curr. Opin. Cell Biol.* **12,** 157–165.

Sabin, A. B. (1959). Reoviruses: A new group of respiratory and enteric viruses formerly classified as ECHO type 10 is described. *Science* **130,** 1387–1389.

Sepp-Lorenzino, L., Ma, Z., Rands, E., Kohl, N. E., Gibbs, J. B., Oliff, A., and Rozen, N. (1995). A peptidomimetic inhibitor of farnesyl: Protein transferase blocks the anchorage-dependent and -independent growth of human tumor cell lines. *Cancer Res.* **55,** 5302–5309.

Sherry, B., and Fields, B. N. (1989). The M1 gene is associated with the novel myocarditic phenotype of a reovirus mutant. *J. Virol.* **63,** 4850–4856.

Slamon, D. J., Clark, G. M., Wong, S. G., Levin, W. J., Ullrich, A., and McGuire, W. L. (1987). Human breast cancer: Correlation of relapse and survival with amplification of the HER-2/neu oncogene. *Science* **235,** 177–182.

Slamon, D. J., Godolphin, W., Jones, L. A, Holt, J. A., Wong, S. G., Keith, D. E., Levin, W. J., Stuart, S. G., Udove, J., Ullrich, A., *et al.* (1989). Studies of the HER-2/neu proto-oncogene in human breast and ovarian cancer. *Science* **244,** 707–712.

Strong, J. E., and Lee, P. W. K. (1996). The *v-erbB* oncogene confers enhanced cellular susceptibility to reovirus infection. *J. Virol.* **70,** 612–616.

Strong, J. E., Tang, D., and Lee, P. W. K. (1993). Evidence that the epidermal growth factor receptor on host cells confers reovirus infection efficiency. *Virology* **197,** 405–411.

Strong, J. E., Coffey, M. C., Tang, D., Sabinin, P., and Lee, P. W. K. (1998). The molecular basis of viral oncolysis: Usurpation of the Ras signaling pathway by reovirus. *EMBO J.* **17,** 3351–3362.

Tang, D., Strong, J. E., and Lee, P. W. K. (1993). Recognition of the epidermal growth factor receptor by reovirus. *Virology* **179,** 412–414.

Tyler, K. L., and Fields, B. N. (1996). Reoviruses. *In* "Fields Virology" (B. N. Fields, D. M. Knipe, and P. M. Howley, eds.), 3rd Ed., pp. 1597–1623. Lippincott-Raven, Philadelphia.

Weber, W., Gill, G. N., and Spiess, J. (1984). Production of an epidermal growth factor receptor-related protein. *Science* **224,** 294–297.

Weiner, H. L., Drayna, D., Averill, D. R. Jr., and Fields, B. N. (1977). Molecular basis of reovirus virulence: Role of the S1 gene. *Proc. Natl. Acad. Sci. U.S.A.* **74,** 5744–5748.

Weiner, H. L., Powers, M. L., and Fields, B. N. (1980). Absolute linkage of virulence and central nervous system tropism of reoviruses to viral hemagglutinin. *J. Infect. Dis.* **141,** 609–616.

Whyte, D. B., *et al.* (1997). K- and N-Ras are geranylgeranylated in cells treated with farnesyl protein transferase inhibitors. *J. Biol. Chem.* **272,** 14459–14464.

Williams, B. R. (1999). PKR; a sentinel kinase for cellular stress. *Oncogene* **18,** 6112–6120.

3

Oncolytic Herpes Simplex Virus (G207) Therapy: From Basic to Clinical

Tomoki Todo,[1],* Michael I. Ebright,[†] Yuman Fong,[†] and Samuel D. Rabkin*

*Molecular Neurosurgery Laboratory
Massachusetts General Hospital
Harvard Medical School
Charlestown, Massachusetts 02129

[†]Department of Surgery
Memorial Sloan-Kettering Cancer Center
New York, New York 10021

I. INTRODUCTION

Standard methods of treating tumors include surgical resection, chemotherapy, and radiation. Because these modalities are often unsuccessful for aggressive cancers, novel therapeutics are needed. Oncolytic viral therapy is a promising new strategy due to the creation of vectors that can selectively replicate, spread *in situ*, and lyse malignant cells while sparing normal tissue. Herpes simplex

[1]Correspondence should be addressed to: Tomoki Todo, Molecular Neurosurgery Laboratory, MGH-East, Box 17, 149 13th Street, Charlestown, MA 02129.

Tumor-Suppressing Viruses, Genes, and Drugs
Innovative Cancer Therapy Approaches
Copyright © 2002 by Academic Press.

virus type 1 (HSV-1) is especially suited for cancer therapy due to features of its basic biology: HSV-1 infects most tumor cell types, in most mammals; the life cycle of HSV-1 is well studied; the HSV-1 genomic sequence is available; the function of the majority of genes has been identified; and the genome can be manipulated (Hunter *et al.*, 1995; Jacobs *et al.*, 1999; Martuza, 2000; Yeung and Tufaro, 2000). The HSV-1 virion contains ~152 kb of DNA, containing over 80 open reading frames, of which only 45 are necessary for viral infection and replication (Roizman, 1996). This leaves ample room for genetic manipulation. In addition, HSV-1 has advantages for clinical application that are not available with other currently used viral vectors: Antiviral drugs are available that can terminate viral replication; only a relatively low multiplicity of infection (MOI) is needed to achieve total cell killing; and there are HSV-1-sensitive mouse strains and nonhuman primates that permit preclinical *in vivo* evaluation of safety and efficacy. HSV-1 is naturally neurotropic, and neuropathogenic when it replicates in the nervous system, yet is capable of establishing latency in certain neuronal cells (Roizman and Sears, 1995). Due to its neurotropism, much of the original interest in HSV-1 vectors involved the treatment of tumors of the central nervous system (CNS); however, animal studies have demonstrated a broad applicability of these vectors to non-neurological malignancies.

The most important requirement for replication-competent viral vectors is to attenuate the virus, so that it elicits minimal toxicity in normal tissue while retaining its capability to replicate within the tumor. First-generation, replication-competent HSV-1 vectors were genetically engineered with an alteration in one nonessential gene, with mutations in genes encoding proteins associated with viral DNA synthesis, or those associated with virulence (Jacobs *et al.*, 1999). The earliest mutants chosen for study involved deletions in the *UL23* gene encoding thymidine kinase (TK) (Boviatsis *et al.*, 1994a; Jia *et al.*, 1994; Kaplitt *et al.*, 1994; Martuza *et al.*, 1991) and inactivation of the *UL39* gene encoding infected-cell protein (ICP) 6, the large subunit of ribonucleotide reductase (RR) (Boviatsis *et al.*, 1994b; Kaplitt *et al.*, 1994; Kogishi *et al.*, 1999; Kramm *et al.*, 1996; Mineta *et al.*, 1994; Yoon *et al.*, 1998) (Table 1). These are both key enzymes for nucleotide metabolism and viral DNA synthesis, and as such, mutants can only replicate in dividing cells where the upregulated, corresponding host enzyme can compensate.

A third viral gene that has been investigated with some enthusiasm is γ34.5 or RL1, coding for ICP34.5. The γ34.5 gene is the major determinant of HSV-1 neurovirulence, and its product blocks host cell induced shutoff of protein synthesis in response to viral infection (Chou and Roizman, 1992). γ34.5-deficient HSV-1 vectors retain their ability to replicate within neoplastic cells, yet are non- neuropathogenic (Chambers *et al.*, 1995; Kesari *et al.*, 1995; Markert *et al.*,

Table 1 HSV-1 Mutants Tested for Oncolytic Activity

Virus	TK[a]	RR[b]	γ34.5	Note and reference
*dl*sptk	−	+	+/+	Martuza *et al.* (1991)
d18.36tk	−	+	+/+	Kaplitt *et al.* (1994)
RH105	−	+	+/+	Boviatsis *et al.* (1994a)
KOS-SB	−	+	+/+	Jia *et al.* (1994)
ICP6Δ	+	−	+/+	Kaplitt *et al.* (1994)
hrR3	+	−	+/+	Boviatsis *et al.* (1994b); Mineta *et al.* (1994)
R7020 (NV1020)	+	+	+/−	Advani *et al.* (1999)
1716	+	+	−/−	Randazzo *et al.* (1995)
R3616	+	+	−/−	Chambers *et al.* (1995); Markert *et al.* (1993)
R4009	+	+	−/−	Andreansky *et al.* (1996)
G207	+	−	−/−	Mineta *et al.* (1995)
MGH-1	+	−	−/−	Kramm *et al.* (1997)
3616UB	+	+	−/−	Uracil N-glycosylase (−) Pyles *et al.* (1997)
G47Δ	+	−	−/−	α47(−) Todo *et al.* (2001a)

[a] TK, thymidine kinase.
[b] RR, ribonucleotide reductase.

1993). The HSV-1 genome contains two copies of this gene. HSV-1 mutants R3616, R4009, and 1716 have deletions (R3616 and 1716) or stop codon insertions (R4009) in both copies of the γ34.5 gene of wild-type HSV-1 strains F (R3616 and R4009) or strain 17 (1716) (Andreansky *et al.*, 1996, 1997; Chou *et al.*, 1990; MacLean *et al.*, 1991) (Table 1). R7020 (NV1020) is a different type of attenuated HSV-1, initially designed for HSV vaccination, with one set of the inverted repeat regions (containing γ34.5) replaced with the *TK* gene under a stronger promoter and HSV-2 glycoproteins G, J, D, and I gene sequences (Meignier *et al.*, 1988).

The initial studies of Martuza *et al.* (1991) with HSV-1 mutants for oncolytic therapy used a TK deletion mutant *dl*sptk. They showed that human glioblastoma cells were exquisitely sensitive to infection. As little as one plaque forming unit (pfu) for every 10,000 cells produced nearly total destruction of the cancer cells *in vitro*. When brain tumors were induced in immunodeficient mice, direct stereotactic injection of a relatively small dose of *dl*sptk allowed increased survival (Martuza *et al.*, 1991). Other TK-deleted mutants (*dl*8.36tk, KOS-SB, RH105) proved similarly efficacious (Boviatsis *et al.*, 1994a; Jia *et al.*, 1994; Kaplitt *et al.*, 1994). A concern about these mutants was that the lack of TK would render a toxic herpetic infection insensitive to treatment with acyclovir or ganciclovir. Attention then turned to HSV-1 mutants lacking RR, which are hypersensitive to acyclovir and ganciclovir (Coen *et al.*, 1989; Mineta *et al.*, 1994). HrR3, containing an inactivating *lacZ* insertion in ICP6 (Goldstein and Weller, 1988), was highly efficacious against a variety of human tumor xenografts (Kogishi *et al.*, 1999; Mineta *et al.*, 1994; Yoon *et al.*, 1998). The first studies to test HSV oncolytic therapy outside of the central nervous system used 1716, a $\gamma34.5$ deletion mutant, in mouse models of malignant mesothelioma (Kucharczuk *et al.*, 1997) and melanoma (Randazzo *et al.*, 1997). The specificity of 1716 for malignant tissue was demonstrated after intraperitoneal injection, by the lack of evidence of virus in nonmalignant tissues including liver, kidney, spleen, and brain, by either immunohistochemistry or PCR (Kucharczuk *et al.*, 1997).

These first-generation vectors often were insufficiently attenuated in normal tissue, leaving concerns about their safety for clinical use. Also, vectors having mutations in only one gene could, although at a very low probability, revert to a pathogenic phenotype. In an attempt to overcome these safety weaknesses, second-generation HSV-1 vectors were developed with multiple mutations; deletions in the $\gamma34.5$ gene together with a second mutation in the ICP6 gene (G207 and MGH-1) (Kramm *et al.*, 1997; Mineta *et al.*, 1995) or in uracyl N-glycosylase (UNG, UL2) (3616UB) (Pyles *et al.*, 1997). This review will focus on the studies with G207.

II. PRECLINICAL STUDIES OF G207

A. Structure of G207

For the clinical use of replication-competent HSV-1 vectors, it is essential that ample safeguards be employed. G207 is the first of the second-generation HSV-1 vectors and was developed with such an emphasis (Mineta *et al.*, 1995). G207 was created by inserting the *Escherichia coli lacZ* gene in the ICP6 coding region (UL39) of R3616. Therefore, it has deletions in both copies of the $\gamma34.5$ gene and lacks RR activity (Figure 1). These widely separated mutations minimize the risk of G207 reverting to a pathogenic phenotype. Because G207 is derived from wild-type HSV-1 strain F, it also contains a temperature-sensitive mutation in the immediate-early $\alpha4$ gene that is required for early and late gene expression (Ejercito *et al.*, 1968; Leopardi and Roizman, 1996). As a result, the virus can actively replicate at 37°C, but not at 39.5°C. The multiple mutations further confer favorable properties on the virus for treating human cancers: G207 replicates preferentially in tumor cells and is harmless in normal tissue due to attenuated virulence; G207 is about 10-fold more sensitive to ganciclovir/acyclovir than its parent virus R3616 (Mineta *et al.*, 1995); and the reporter gene *lacZ* allows easy histochemical detection of G207-infected cells.

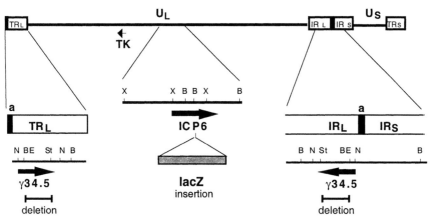

Figure 1 Structure of G207. (Top) schematic of the HSV genome with unique long (U_L) and short (U_S) sequences bracketed by inverted repeats (TR_L, IR_L, IR_S, TR_S). The location of the thymidine kinase (TK) gene is illustrated. (Bottom) The γ34.5 gene, the major determinant of HSV-1 neurovirulence, is located in the long inverted repeats and therefore is diploid. Arrows indicate the orientation and location of genes. G207 has 1-kb deletions in both copies of the γ34.5 gene and an insertion of the *E. coli lacZ* gene in the ICP6 coding region (UL39) inactivating ribonucleotide reductase. Restriction endonuclease sites; B, BamHI; BE, BstEII; N, NcoI; St, StuI; X, XhoI.

B. Application of G207 for cancer

1. *In vitro* cytotoxicity

In theory, G207 should replicate more actively in cells with higher RR activity, that is, more rapidly dividing cells. Testing of G207 in various cell lines revealed that the vast majority, although not all, of human tumor cell lines are susceptible to G207 infection (Table 2). Those tumors include breast cancer (Toda *et al.*, 1998a), prostate cancer (Walker *et al.*, 1999), colorectal cancer (Kooby *et al.*, 1999), gastric cancer (Bennett *et al.*, 2000), hepatic cancer, gallbladder cancer (Nakano *et al.*, 2001), urinary bladder cancer (Oyama *et al.*, 2000), head and neck cancer (Chahlavi *et al.*, 1999a), ovarian cancer (Coukos *et al.*, 2000), malignant melanoma (Todo *et al.*, 2001a), and neuroblastoma (Mashour *et al.*, 2000). Different cancer cell lines originating from the same organ type can vary significantly in viral sensitivity. For instance, when exposed to identical doses of G207 *in vitro*, one breast cancer line (MDA-MB-435) exhibited over 99% cell kill, while another (MDA-MB-231) exhibited less than 5% (Toda *et al.*, 1998a). Kooby *et al.* (1999) exposed five different human colorectal cell lines to G207, also discovering vastly different viral sensitivities. Even at a low MOI, three of the cell lines

Table 2 Cell Susceptibility to G207 Cytotoxicity[a]

Human cell line	Cell type	Susceptibility[b]
U87MG	Glioma	+++
T98MG	Glioma	+++
U373MG	Glioma	+++
U138MG	Glioma	+++
A172	Glioma	++
F5	Malignant meningioma	++++
GPSM4	Malignant meningioma	++++
GPSM5	Malignant meningioma	++++
Hela	Cervical epitheloid carcinoma	++++
Hep3B	Hepatocellular carcinoma	+++
HuH7	Hepatocellular carcinoma	+++
MDA-MB-231	Breast adenocarcinoma	−
MDA-MB-435	Breast adenocarcinoma	++++
MCF7	Breast adenocarcinoma	+++
T47D	Breast adenocarcinoma	+++
UM-SCC 22A	Head and neck squamous cell carcinoma	++++
UM-SCC 38	Head and neck squamous cell carcinoma	++++
PCI 51	Head and neck squamous cell carcinoma	+++
SCC-25/CP	Head and neck squamous cell carcinoma	++++
SQ20B	Head and neck squamous cell carcinoma	++++
SCC1483	Head and neck squamous cell carcinoma	+[c]
LNCap	Prostate adenocarcinoma	++++
DU-145	Prostate adenocarcinoma	++++
PC3	Prostate adenocarcinoma	+++
TSUPR-1	Prostate adenocarcinoma	++
RT-4	Bladder transitional cell carcinoma (grade I)	+++
MGH-U3	Bladder transitional cell carcinoma (grade I)	+++
5637	Bladder transitional cell carcinoma (grade II)	+++
EJ	Bladder transitional cell carcinoma (grade III)	+
KU19-19	Bladder transitional cell carcinoma	+++[d]
T24	Bladder transitional cell carcinoma	+++[d]
UKF-NB-3(VCR[r])	Neuroblastoma	++++
UKF-NB-3 (Dox[r])	Neuroblastoma	++
IMR-32	Neuroblastoma	++
IMR-32 (cisPt[r])	Neuroblastoma	+++
UKF-NB-4	Neuroblastoma	+++
UKF-NB-4 (dox[r])	Neuroblastoma	+++
K562	Leukemia	−
SK-MEL3	Melanoma	+
RPMI 7951	Melanoma	+++
Mel888	Melanoma	++++
Mel624	Melanoma	+++
MKN-74	Gastric cancer	+++[e]
OCUM-2MD3	Gastric cancer	+++[e]
AGS	Gastric cancer	+++[e]
MKN-1	Gastric cancer	++[e]
MKN-45-P	Gastric cancer	++[e]
C86	Colorectal cancer	+++[f]
C85	Colorectal cancer	+++[f]
HCT 8	Colorectal cancer	+++[f]

(continues)

Table 2 (*Continued*)

Human cell line	Cell type	Susceptibility[b]
C29	Colorectal cancer	$++^f$
C18	Colorectal cancer	$+^f$
A2780	Epithelial ovarian carcinoma	$++^g$
SKOV3	Epithelial ovarian carcinoma	$++^g$
NIH:OVCAR3	Epithelial ovarian carcinoma	$++^g$
CaOV3	Epithelial ovarian carcinoma	$++^g$
Primary cultures	Epithelial ovarian carcinoma	$+++^g$
GB-d15	Gallbladder carcinoma	$++++^h$
HAG-1	Gallbladder carcinoma	$++^h$
TGBC1TKB	Gallbladder carcinoma	$++^h$
TGBC2TKB	Gallbladder carcinoma	$++++^h$
Astrocytes	Astrocytes	$+++$
Bone marrow	Bone marrow	$-$
TIL888	T lymphocytes	$-$
TIL1413	T lymphocytes	$-$

Mouse

168FARN	Mammary tumor	$-$
4T1	Mammary tumor (metastatic)	$-$
66c14	Mammary tumor (metastatic)	$-$
Cloudman S91 clone M3	Melanoma	$+$
Harding-Passey	Melanoma	$-$
MCA207	Fibrosarcoma	$+$
CT26	Colon adenocarcinoma	$+$
203GL	Glioma	$-$
SR-B10.A	Glioma	$+$
N18	Neuroblastoma	$++$
Neuro2a	Neuroblastoma	$+$
TRAMP	Prostate adenocarcinoma	$++$
$PR14_2$	Prostate adenocarcinoma	$+$

Rat

Astrocytes	Astrocytes	$+$

Hamster

KIGB-5	Gallbladder carcinoma	$++^h$

[a]Cell susceptibility to G207 is based on *in vitro* cytotoxicity studies at a MOI of 0.1. All data were produced in the laboratories of T. Todo and S. D. Rabkin unless noted otherwise.
[b] $++++$ Total cytopathic effect achieved within 1–2 days; $+++$ Total cytopathic effect achieved within 3–6 days; $++$ About 40–70% cell killing at 3–4 days postinfection; $+$ About 10–40% cell killing at 3–4 days postinfection; $-$ Minimal or no cell killing at 3–4 days postinfection.
[c]Data based on Carew *et al*. (1999).
[d]Data based on Oyama *et al*. (2000).
[e]Data based on Bennett *et al*. (2000).
[f]Data based on Kooby *et al*. (1999).
[g]Data based on Coukos *et al*. (2000).
[h]Data based on Nakano *et al*. (2001).

were over 76% eradicated after 3 days, and completely killed within 1 week. The cytotoxicity data correlated with both infection rates as detected by the *lacZ* reporter gene, as well as cellular S-phase fraction. Human epithelial ovarian cancer (EOC) cells were effectively killed by G207 *in vitro*, with primary EOC cells more sensitive than established cell lines (Coukos *et al.*, 2000). Normal human peritoneal mesothelial cells were spared: Though infected by G207, mesothelial cells poorly supported replication of the virus (Coukos *et al.*, 2000). The only human cell types that consistently showed resistance to G207 infection were hematopoietic-derived cell lines such as K562 (leukemia) and tumor infiltrating lymphocyte (TIL) lines. Human bone marrow cells, though proliferating, are also resistant to G207 replication and cytotoxicity (Wu *et al.*, 2001). Because occult cancer cells can be killed by G207 at a low MOI, G207 can be used as a purging agent for autologous bone marrow transplantation in advanced cancer patients. Murine cell lines in general are relatively resistant to G207 infection, with a few exceptions such as N18 (Todo *et al.*, 1999a) (Table 2).

2. Brain tumors

G207 was initially designed for use in malignant brain tumors. Screening of human glioma cell lines (U87MG, T98G, U373MG, U138MG) for G207 susceptibility revealed that G207 at a multiplicity of infection (MOI) of 0.1 causes total cell killing within 3–6 days (Mineta *et al.*, 1995). In contrast, G207 at the same dose caused no cytopathic effect in primary cultures of rat cortical astrocytes and cerebellar neurons. Treatment of established U87MG tumors (subcutaneous or intracerebral) in athymic mice with a single intraneoplastic inoculation of G207 [5×10^7 pfu (subcutaneous) or 2×10^6 pfu (intracerebral)] successfully inhibited tumor growth and prolonged survival (Mineta *et al.*, 1995). A similar cytolytic activity of G207 was observed in human malignant meningioma cell lines, F5, GPSM4, and GPSM5 (Yazaki *et al.*, 1995). A single intraneoplastic G207 inoculation of subcutaneous F5 tumors in athymic mice resulted in significant inhibition of tumor growth, that was dose dependent, 5×10^7 pfu being significantly better than 5×10^6 pfu (Yazaki *et al.*, 1995). Prominent *lacZ* expression from G207 replication within tumors could still be observed 24 days postinoculation. G207 inoculation (5×10^6 pfu) significantly prolonged the survival of athymic mice with intracranial F5 tumors.

3. Animal tumor models

G207 has been tested in a variety of animal tumor models (Table 3). Besides the subcutaneous region and the brain, tumors have been generated in various organs including the liver, peritoneum, urinary bladder, and cheek pouch. In immunocompromised animals, the *in vivo* therapeutic efficacy of G207 correlates well with the cytopathic activity exhibited in cell culture. This was first demonstrated using two human breast cancer cell lines, G207-sensitive MDA-MB-435 and G207-resistant MDA-MB-231: Intraneoplastic G207 inoculation was efficacious only for MDA-MB-435 subcutaneous or intracerebral tumors (Toda *et al.*, 1998a). A similar *in vitro* and *in vivo* correlation was observed with subcutaneous tumors of human head and neck cancer (Carew *et al.*, 1999). In these tumor models, it was further observed that the efficacy of G207 for resistant tumors could be increased by giving multiple inoculations, thereby increasing the total administration dose. In subcutaneous tumor models of human colorectal cancer, the results indicated that tumor S-phase fraction, as well as *in vitro* cytotoxicity assay, could be used to predict response to treatment *in vivo* (Kooby *et al.*, 1999). In our experience, results from G207 cytotoxicity assays *in vitro* at a MOI of 0.1 best predict the *in vivo* outcome.

G207 has been evaluated for use in the treatment of genitourinary malignancies. *In vivo*, human prostate subcutaneous tumors are highly sensitive to a single intraneoplastic viral injection (Walker *et al.*, 1999). When previously irradiated tumors were tested after regrowth, two intraneoplastic viral injections significantly attenuated growth. Furthermore, two intravenous doses of virus significantly attenuated growth, and resulted in a 25% cure rate (Walker *et al.*, 1999). This was the first study to show viral efficacy through intravenous administration. Oyama *et al.* (2000) demonstrated *in vitro* efficacy of G207 against two human transitional cell carcinoma lines. When treated as subcutaneous tumors in athymic mice, intraneoplastic injection reduced tumor volume. Intravenous treatment was also successful in treating these tumors; however, three administrations of very high doses were required. The success of intravenous therapy is of great importance, as the intravenous route may be the best approach for diffuse metastatic disease that is otherwise inaccessible. Oyama *et al.* (2000) also treated an orthotopic model of bladder cancer by intravesical administration of G207, significantly reducing tumor burden. Similar results in an orthotopic bladder cancer model have been observed in our laboratory (Y. Fong) using both G207 and NV1020. Several complete responses were seen with weekly intravesical administration in groups treated with either virus (Cozzi *et al.*, 2001). These studies demonstrate the great promise of oncolytic therapy in the treatment of genitourinary malignancies.

Human squamous cell carcinoma is among the fastest growing tumors, and subcutaneous xenografts are very sensitive to G207 treatment (Carew *et al.*, 1999; Chahlavi *et al.*, 1999a). In a more clinically relevant model, cheek pouch tumors were induced in hamsters by direct application of carcinogen. When treated with regional viral administration directly into the external carotid artery, tumor growth was inhibited over 50% compared with controls (Carew *et al.*, 1999). As in other tumor models, viral replication spared adjacent normal tissue.

Table 3 Animal Tumor Models Studied for G207 Efficacy

Cell line	Cell type	Cell species	Host strain	Host species	Model type	Tumor site	Reference
U87MG	Glioma	Human	Athymic	Mouse	Xenogeneic	Subcutaneous intracerebral	Mineta et al. (1995); Todo et al. (2001a)
F5	Malignant meningioma	Human	Athymic	Mouse	Xenogeneic	Subcutaneous intracranial	Yazaki et al. (1995)
MDA-MB-435 MDA-MB-231	Breast adenocarcinoma	Human	Athymic	Mouse	Xenogeneic	Subcutaneous intracerebral	Toda et al. (1998a)
CT26	Colon epithelial carcinoma	Mouse	Balb/c	Mouse	Syngeneic	Subcutaneous	Chahlavi et al. (1999b); Toda et al. (1999)
Cloudman S91 clone M3	Melanoma	Mouse	DBA/2	Mouse	Syngeneic	Subcutaneous	Toda et al. (1999)
N18	Neuroblastoma	Mouse	A/J	Mouse	Syngeneic	Intracerebral subcutaneous	Chahlavi et al. (1999b); Todo et al. (1999a,b, 2000a)
Neuro2a	Neuroblastoma	Mouse	A/J	Mouse	Syngeneic	Intracerebral subcutaneous	Todo et al. (2001a,b)
LNCap	Prostate cancer (androgen responsive)	Human	Athymic	Mouse	Xenogeneic	Subcutaneous	Walker et al. (1999)
DU-145	Prostate cancer (androgen unresponsive)						
TRAMP	Prostate adenocarcinoma	Mouse	C57BL/6	Mouse	Syngeneic	Subcutaneous	Jorgensen et al. (2001)

Cell line	Tumor type	Species	Host	Graft type	Site	Reference
PR-14$_2$	Prostate adenocarcinoma	Mouse	C3(1)/Tag-transgenic FVB/N	Mouse	Syngeneic	Subcutaneous	S. Varghese, et al. unpublished data
UM-SCC 22A UM-SCC 38 SQ20B	Head and neck squamous cell carcinoma	Human	Athymic	Mouse	Xenogeneic	Subcutaneous	Chahlavi et al. (1999a)
MSK QLL2 SCC 15 SCC 1483	Head and neck squamous cell carcinoma	Human	Athymic	Rat	Xenogeneic	Subcutaneous	Carew et al. (1999)
Carcinogen induction	Squamous cell carcinoma	Hamster	Syrian golden	Hamster	Autologous	Cheek pouch	Carew et al. (1999)
HCT8, C86, C85	Colorectal carcinoma	Human	Athymic	Rat	Xenogeneic	Subcutaneous	Kooby et al. (1999)
McA-RH7777	Morris hepatoma	Rat	Buffalo	Rat	Syngeneic	Hepatic metastases	Kooby et al. (1999)
SKOV3	Epithelial ovarian carcinoma	Human	CB17 SCID	Mouse	Xenogeneic	Intraperitoneal	Coukos et al. (2000)
OCUM-2MD3	Gastric carcinoma	Human	Athymic	Mouse	Xenogeneic	Intraperitoneal	Bennett et al. (2000)
KU19-9	Bladder transitional cell carcinoma	Human	Athymic	Mouse	Xenogeneic	Subcutaneous intravesical	Oyama et al. (2000)
T24	Bladder transitional cell carcinoma	Human	Athymic	Mouse	Xenogeneic	Subcutaneous	Oyama et al. (2000)
IMR32	Neuroblastoma	Human	Athymic	Mouse	Xenogeneic	Sciatic nerve	Mashour et al. (2001)
KIGB-5	Gallbladder carcinoma	Hamster	Syrian golden	Hamster	Syngeneic	Subcutaneous	Nakano et al. (2001)

G207 may prove to be a useful therapeutic adjunct in the challenging management of head and neck cancer.

The efficacy of G207 has been evaluated in a clinically relevant model of diffuse hepatic metastases. When virus was administered through an intraportal route, treated livers contained significantly fewer tumor nodules than controls (Kooby *et al.*, 1999). When stained for the presence of the *lacZ* reporter gene, tumor nodules in the liver stained strongly, while there was only scant positive expression in surrounding normal parenchyma. These experiments demonstrate the remarkable specificity of viral infection in a diffuse tumor model. In practice, after hepatic metastases are surgically resected, nearly 65% of patients experience local tumor recurrence (Fong *et al.*, 1999; Kemeny *et al.*, 1999; Panis *et al.*, 1992). Increases in hepatocyte mitotic activity associated with liver regeneration are thought to stimulate growth of microscopic residual disease. It would be ideal to deliver virus through a regional arterial route shortly after surgical resection. It has been observed by Yoon *et al.* (2000) that nonmalignant liver parenchyma supports viral replication when virus is administered shortly after 75% hepatectomy. Further experiments from our laboratory (Y. Fong) sought to determine the optimal time of viral delivery after hepatic resection. In mice, we found that viral replication is supported by non-neoplastic hepatocytes during the peak of regeneration when the mitotic index is high. Additionally, we observed that animals given virus during the peak of hepatic regeneration suffered higher mortality, while animals receiving virus beyond this peak were unharmed (Y. Fong, unpublished results, 2000). When timed appropriately, oncolytic virus may prove to be very effective in the treatment of hepatic metastases.

Ovarian cancer is an ideal candidate for regional oncolytic viral therapy, as it is usually diagnosed at a later stage, and usually remains localized within the peritoneal cavity. Intraperitoneal delivery of G207 significantly reduced tumor burden and spread of intraperitoneal human epithelial ovarian cancer (EOC) tumors (Coukos *et al.*, 2000). Another appropriate candidate for intraperitoneal oncolytic therapy is peritoneally disseminated gastric cancer. In athymic mice inoculated with peritoneal gastric cancer cells, regional oncolytic viral therapy (intraperitoneal administration) both reduced tumor burden and conferred significant survival advantage (Bennett *et al.*, 2000). A temporary but significant increase in G207 DNA was detected in tumor specimens, using real-time quantitative PCR, showing evidence for *in vivo* intraneoplastic viral proliferation (Bennett *et al.*, 2000a). By 3 days after intraperitoneal injection of G207 (5×10^7 pfu), viral DNA increased to 7.5×10^4 genomic equivalents which then diminished considerably by day 7. This was, however, about 50-fold lower than the virus increase observed after intraperitoneal injection of wild-type HSV-1 KOS, which reached 3.5×10^6 genomes by day 3. Additional unpublished results from our laboratory (Y. Fong, 2000) reveal superior cell killing *in vitro* by NV1020 over G207 in gastric cancer. Furthermore, when given in low doses, NV1020 affected a significantly greater reduction in tumor burden *in vivo* compared with G207. This difference was also seen when treating a more established peritoneal disease. Neither virus displayed any evidence of replication in normal tissues such as brain, kidney, or liver, even when administered at high doses.

G207 kills cancer cells, but spares the surrounding normal tissues, including peripheral nerves. This is a great advantage in treating cancers of the prostate or peripheral nerve sheaths, in which damage to involved nerves during surgery can be a major complication. To evaluate the nerve-sparing efficacy of G207, a unique peripheral nerve sheath tumor model was utilized, in which IMR32 human neuroblastoma cells were injected into the sciatic nerve of athymic mice to form tumors (Mashour *et al.*, 2001). A single intraneoplastic inoculation of G207 (10^6 pfu) not only caused significant inhibition of tumor growth, but also reduction of nerve function impairment, leading to prolonged survival. Direct inoculation of the healthy sciatic nerve with wild-type HSV-1 strain F (5×10^6 pfu) caused paralysis of the ipsilateral or both hind limbs by day 7, whereas G207 (5×10^6 pfu) caused no neurological abnormalities. Although both strain F and G207 caused some Wallerian degeneration of the inoculated nerve, strain F further caused inflammatory changes in the spinal cord and the brain, which was absent with G207 (Mashour *et al.*, 2001).

Overall, the route of delivery appears to have significant impact on antitumor efficacy of HSV-based oncolytic therapy. Most tumor models that have demonstrated efficacy have delivered the virus regionally to the tumor site. Direct injection of brain or subcutaneous tumors have been the most commonly used models (Hunter *et al.*, 1995). Intraarterial delivery into the hepatic artery has been used to successfully treat liver tumors (Kooby *et al.*, 1999). Intraperitoneal delivery has been the most successful in treating abdominal disease (Coukos *et al.*, 1999, 2000; Kucharczuk *et al.*, 1997). Bladder irrigation has been used in attempts to treat local bladder cancer (Oyama *et al.*, 2000). The superior efficacy of locally delivered virus is not surprising. However, there are some emerging data that intravenous delivery may also be effective in treating certain experimental models of cancer, but it requires larger doses of virus (Oyama *et al.*, 2000; Walker *et al.*, 1999). This route of delivery would greatly simplify treatment. For the time being, local delivery will likely be the method by which virus will be delivered in the first clinical trials. This is because directing the viral agent immediately to the tumor targets will maximize tumor exposure, minimize exposure of noncancerous organs, and minimize exposure of virus to possible neutralizing effects of circulating antibodies. The net effect will be to maximize the likelihood of detecting antitumor efficacy with the smallest doses of virus possible. Of greatest importance in the future will be studies to examine combined use of systemic viral delivery with other therapies that may potentiate the effects of oncolytic viral therapy, so that the systemic delivery may become clinically applicable.

C. Induction of antitumor immunity by G207

An important question needing answering prior to human trials was the effect of the host immune system on the antitumor activity of G207. Efficacy of replication-competent HSV-1 vectors had been examined in athymic or SCID mice (Chambers *et al.*, 1995; Kesari *et al.*, 1995; Lasner *et al.*, 1996; Mineta *et al.*, 1995; Pyles *et al.*, 1997; Yazaki *et al.*, 1995), or rodent models which are relatively HSV-resistant (Boviatsis *et al.*, 1994b; Kramm *et al.*, 1996, 1997; Randazzo *et al.*, 1995). In an immunocompetent condition, there are several issues to be considered: (1) Replication of the virus within a tumor may be inhibited by an antiviral immune response, resulting in reduction of the cytopathic effect. With adenovirus vectors, it has been shown that a strong immune response against the vector itself acts adversely to the therapeutic efficacy (Yang *et al.*, 1994, 1995). (2) On the other hand, an immune response may be associated with elicitation of an antitumor immune response that provides an enhancement of the antitumor activity, as observed in "suicide" gene therapy with HSV-TK gene transfer followed by a ganciclovir administration (Barba *et al.*, 1994). For brain tumors, these issues are further complicated by the immunologically privileged environment of the brain. One difficulty in investigating the immune effect has been the lack of suitable animal tumor models that are susceptible to HSV infection. Many mouse strains and a majority of murine cell lines are relatively resistant to HSV-1 (Lopez, 1975) (Table 2).

To address the role of the host immune response in oncolytic HSV tumor therapy, a number of syngeneic mouse tumors have been examined. N18 murine neuroblastoma cells were used as an immunocompetent mouse brain tumor model (Todo *et al.*, 1999a). N18 is one of the more susceptible murine cell lines tested for G207 susceptibility, although less sensitive than most human cell lines (Table 2). N18 cells are capable of forming both intracerebral and subcutaneous tumors in syngeneic A/J mice, the most susceptible inbred mouse strain to HSV infection (Lopez, 1975; Zawatzky *et al.*, 1981). In A/J mice harboring established N18 tumors subcutaneously or in the brain, intraneoplastic inoculation with G207 caused a significant reduction of tumor growth or prolongation of survival (Todo *et al.*, 1999a). In A/J mice bearing bilateral subcutaneous N18 tumors, intraneoplastic G207 inoculation into one tumor alone caused growth reduction and/or regression of both the inoculated and the noninoculated contralateral tumor. Animals that were cured of their subcutaneous tumors by G207 were protected against tumor rechallenge, either in the periphery or the brain. The antitumor immunity was associated with cytotoxic T lymphocyte (CTL) activity *in vitro* that was specific to N18 tumor cells and persisted for at least 13 months. In A/J mice bearing subcutaneous and intracerebral N18 tumors simultaneously, G207 inoculation into the subcutaneous tumor alone caused growth suppression of both the subcutaneous and the remote brain tumor, leading to a significant prolongation of survival (Todo *et al.*, 1999a). G207-induced, systemic antitumor immunity was also observed in BALB/c mice bearing subcutaneous CT26 (colon carcinoma) tumors and DBA/2 mice bearing subcutaneous M3 (melanoma) tumors (Toda *et al.*, 1999). In the CT26 model, intraneoplastic inoculation of G207 induced CTL activity that recognized a dominant, tumor-specific, major histocompatibility complex (MHC) class I-restricted epitope (AH1) from CT26 cells. Similar systemic antitumor immunity induction by G207 was observed in Syrian hamsters bearing subcutaneous KIGB-5 (gallbladder carcinoma) tumors (Nakano *et al.*, 2001). Thus, in an immunocompetent condition, the oncolytic activity of G207 can be augmented by induction of specific and systemic antitumor immunity effective both in the periphery and in the brain.

The role of host immune responses in the antitumor activity of G207 led to the investigation of the consequences of corticosteroid-induced immunousuppression (Todo *et al.*, 1999b). This has particular clinical significance for patients with malignant brain tumors, because they are often treated with corticosteroids to reduce vasogenic brain edema surrounding the tumor. In A/J mice bearing subcutaneous N18 tumors, intraneoplastic inoculation of G207 (10^7 pfu) induced significant suppression of tumor growth whether dexamethasone (5 mg/kg) was given or not (Todo *et al.*, 1999b). However, when dexamethasone was given for an extensive time (16 days starting at day -2) as would be the case in patients, all immunosuppressed mice treated with G207 showed tumor regrowth despite initial shrinkage, whereas 50% of the G207-treated mice not immunosuppressed were cured. Dexamethasone administration significantly reduced serum neutralizing antibodies against G207 after intraneoplastic G207 inoculation, but this did not affect the amount of infectious G207 isolated from tumors. Infectious G207 was isolated from these tumors even 10 days after inoculation, longer than reported previously with 1716 inoculated tumors in immunocompetent mice (Randazzo *et al.*, 1995). The most striking effect of dexamethasone administration was that it completely abolished G207-induced CTL activity against N18 cells (Todo *et al.*, 1999b). These results indicated that corticosteroid-induced immunosuppression does not affect the oncolytic activity of G207, but may diminish the long-term efficacy due to suppression of tumor-specific CTL induction. In a rat model of focal HSV-1 encephalitis, dexamethasone administration (5 mg/kg) did not increase, but rather decreased, HSV-1 replication and dissemination (Thompson *et al.*, 2000).

In considering clinical application, the effect of prior immunity to HSV-1 on antitumor efficacy needs to be addressed because of the approximately 60% of the adult population that is seropositive for HSV-1 (Langenberg *et al.*, 1999; Oliver *et al.*, 1995; Siegel *et al.*, 1992). The presence of humoral immunity does not prevent recurrence of HSV-1 infection and reactivation, which has created problems in developing vaccines (Whitley, 1996).

HSV-1-infected cells are relatively resistant to antibody-dependent cell-mediated cytotoxicity (Dubin *et al.*, 1991; Rager-Zisman and Allison, 1976). A/J and BALB/c mice were immunized by repeated intraperitoneal inoculations of wild-type HSV-1 KOS, and then the antitumor efficacy of G207 on established subcutaneous N18 and CT26 tumors was determined (Chahlavi *et al.*, 1999b). In both tumor models, the antitumor efficacy of G207 was the same whether the mice were immunized or not for HSV-1. Similar to G207, preexisting HSV-1 immunity had no effect on the antitumor efficacy of strain 1716 for established intraperitoneal tumors in BALB/c mice when injected intraperioneally (Lambright *et al.*, 2000). In studies with hrR3, preexisting HSV-1 immunity caused a reduction in reporter gene expression and an increase in inflammatory cell infiltrate within a day of intraneoplastic inoculation in Fischer rat brain, which was no longer observed at 2–5 days postinoculation (Herrlinger *et al.*, 1998). Because HSV-1 predominantly spreads cell-to-cell, circulating antibodies known to neutralize free virus may have little effect on HSV-1 directly inoculated into tumors.

D. Immunotherapy using transgene-expressing HSV vectors

Cancer immunotherapy strategies involve the upregulation of host defenses to target and destroy tumor cells. Different methods have been developed to engineer tumor cell vaccines *ex vivo* to secrete immunostimulatory cytokines (Pardoll, 1995). One promising immunotherapy strategy is to engineer HSV-1 oncolytic vectors to express transgenes encoding immunostimulatory cytokines. There are two vector strategies that have been employed to this end; recombinant HSV containing cytokine transgenes or defective HSV vectors containing cytokine transgenes in combination with G207 as helper virus. Our laboratory (Y. Fong) has explored the use of replication-competent HSV-1 vectors with transgenes encoding either murine granulocyte–macrophage colony stimulating factor (GM-CSF) or interleukin (IL)-12. These vectors produce high quantities of cytokines in infected cells, and have been tested *in vivo* in a murine squamous cell carcinoma flank tumor model (Wong *et al.*, 2001). Tumors treated with HSV-IL-12 vectors showed significantly better tumor volume reduction than either HSV-GM-CSF or parent vector. HSV-IL-12 infected tumors showed substantially more CD8$^+$ T lymphocyte infiltration than tumors treated with the other vectors. In addition, when the mice were rechallenged with cancer cells, significantly fewer mice in the HSV-IL-12 group developed tumors. In an orthotopic hepatoma model, rats with tumor initially treated with HSV-IL-12 also demonstrated significant immunity to tumor rechallenge when compared to rats initially treated with the parent virus. When HSV-1 oncolytic vectors are supplemented with cytokine transgenes, it may be possible to decrease potential toxicity by administering lower viral doses to obtain a similar response. This effect was observed in a murine colorectal tumor flank model, where HSV-IL-12 at low doses demonstrated significant tumor volume reduction while the parent virus at the same dose was ineffective (Bennett *et al.*, 2001).

We (T. Todo and S. D. Rabkin) have used defective HSV vectors expressing a variety of immunomodulatory transgenes (IL-12, GM-CSF, B7-Ig, TK) in combination with G207 (Miyatake *et al.*, 1997; Toda *et al.*, 1998b; Todo *et al.*, 2001b). The defective vector genome contains only the transgene-expressing plasmid and no viral genes. Because a viral genome length of DNA (~152 kb) is packaged, each defective particle contains ~10–20 copies of the transgene-expressing plasmid, depending on its length (Spaete and Frenkel, 1982). This combination strategy is also attractive, because it is likely that different tumor cells will be infected by either G207 or the defective vector, and therefore transgene expression will not be terminated by HSV replication and cytotoxicity as would occur with the recombinant HSV vectors. *In situ* expression of IL-12 or soluble B7-1, a potent, secretory-type costimulatory factor, using defective HSV in combination with G207 was significantly better in inhibiting tumor growth than G207 alone (Toda *et al.*, 1998b; Todo *et al.*, 2001b). This augmentation of G207 activity occurs at doses well below the efficacious dose of G207 alone and significantly enhances the induction of tumor cell-specific CTL activity.

E. Enhancement of G207 antitumor activity

The therapeutic efficacy of G207 depends on the dose, the extent of intratumoral viral replication, and the host immune response. The dose is limited by the concentration of virus stocks and the injectable volume depending on the route of administration. Some tumor cells do not support G207 replication. There are various ways, however, to enhance the antitumor activity of G207 without further manipulating the viral genome. Multiple inoculations of G207 that increase the total administration dosage have been shown to be effective for tumors that are relatively resistant to G207 (Carew *et al.*, 1999). In BALB/c mice bearing subcutaneous CT26 tumors, biweekly intraneoplastic inoculations of low-titer G207 for 3 weeks was significantly better than a single inoculation with a higher total dose (Chahlavi *et al.*, 1999b). This may reflect better distribution of the virus, better induction of antitumor immune responses, and/or better cell killing due to multiple tumor cell infections over a period of time.

Most cancer therapy consists of multiple treatment modalities, and it is likely that oncolytic viral therapy will be similar. Therefore, many groups have focused efforts on combining the use of oncolytic viruses with conventional nonsurgical treatments such as chemotherapy and radiation. Because viral therapy works via independent mechanisms, the opportunity exists to significantly enhance established cancer treatment regimens. Several studies have begun to explore the use of HSV-1 vectors in combination with chemotherapeutic agents. The efficacy of G207 in combination with chemotherapy was evaluated using human head and neck cancer cells (Chahlavi *et al.*, 1999a). Cisplatin did not inhibit the cytopathic effect of G207 *in vitro*. In athymic mice bearing subcutaneous UMSCC-38 cisplatin-sensitive tumors, combination therapy resulted in 100% cures in contrast to 42% with G207 (10^7 pfu) or 14% with cisplatin (4 mg/kg × 5 days) alone. Chase *et al.* (1998) developed a RR-defective virus with a transgene expressing rat cytochrome P450 2B1 (CYP2B1). CYP2B1 is responsible for activation of the prodrug cyclophosphamide. After confirming that CYP2B1 was expressed in infected cells, and that cyclophosphamide metabolites did not interfere with viral replication, a regimen of virus and cyclophosphamide was tested against flank tumor models of three different glioma lines. In all three cell lines, virus plus cyclophosphamide exhibited a significantly better response than virus alone (Chase *et al.*, 1998). Toyoizumi *et al.* (1999) explored the use of strain 1716 with four different chemotherapeutic agents. Although the cytotoxic effect was additive for three of the drugs tested, mitomycin C was synergistic in two of five non-small cell lung cancer (NSCLC) cell lines. When tested in a flank tumor model, strain 1716 and mitomycin C in combination was significantly more effective (in an additive manner) than either treatment alone.

A few studies have also explored the combination of radiation therapy and oncolytic HSV. Advani *et al.* (1998) determined that the combination of R3616 and radiation was significantly more effective than either treatment alone for subcutaneous human U87MG glioblastoma flank tumors in athymic mice. Moreover, they revealed that viral replication was enhanced two- to five-fold in tumors that were irradiated. A similar enhancing effect was observed when R3616 was combined with fractionated radiation in subcutaneous or intracranial xenograft models of U87MG, suggesting a synergistic effect (Bradley *et al.*, 1999). Additional studies by Advani *et al.* (1999) demonstrated the efficacy of R7020 in radio- resistant SQ-20B human epidermoid tumors. However, when combination therapy of G207 and fractionated ionizing radiation was tested in prostate cancer models (human LNCaP tumors in athymic mice and murine TRAMP tumors in either athymic mice or syngeneic C57BL/6 mice), no enhancing effect of the combination was observed (Jorgensen *et al.*, 2001). Subcutaneous LNCaP tumors that recurred after radiation therapy remained susceptible to G207 therapy, indicating that G207 may be useful as salvage therapy for recurrent tumors after radiation therapy (Walker *et al.*, 1999). The effect of radiation on HSV-1 replication and antitumor activity seems dependent on the dose and timing of radiation, the mutations in the virus vector and the animal models used, and further evaluation is needed to fully understand these interactions.

It has been reported that ganciclovir administration in the 9L rat gliosarcoma brain tumor model significantly increased the antitumor action of hrR3 (Boviatsis *et al.*, 1994b; Kramm *et al.*, 1996). In cells infected with an

HSV-1 vectors retaining the *TK* gene (Table 1), ganciclovir is phosphorylated by HSV-TK, and further anabolized by cellular kinases to ganciclovir triphosphate which inhibits host cell DNA replication by chain termination (Cheng et al., 1983). Presumably, enhanced antitumor activity was due to the so-called bystander effect (Culver et al., 1992; Freeman et al., 1993; Ram et al., 1993) which is associated with gap junctional intercellular communication of phophorylated ganciclovir (Bi et al., 1993; Samejima and Meruelo, 1995; Mesnil et al., 1996; Shinoura et al., 1996; Vrionis et al., 1997; Estin et al., 1999; Rubsam et al., 1999) and induction of immune responses *in vivo* (Barba et al., 1994; Colombo et al., 1995; Gagandeep et al., 1996; Ramesh et al., 1996; Vile et al., 1994; Yamamoto et al., 1997). In culture, addition of ganciclovir either had no effect or inhibited the cytocidal action of G207 at replication-permissive temperatures (34.5°C), while it significantly increased the cell killing in three of four cell lines studied when the virus replication was inhibited at nonpermissive temperatures (39.5°C) (Todo et al., 2000a). Ganciclovir treatment had no effect on G207-mediated N18 tumor growth inhibition in A/J mice at a variety of viral doses (10^5, 10^7, $10^7 \times 2$ pfu). It did not enhance the antitumor activity of R3616 in GL261 glioma tumors in C57BL/6 mice (Miyatake et al., 1997), MGH-1 in 9L tumors in Fisher 344 CD rats (Kramm et al., 1997), or HT29 colon carcinoma tumors in BALB/c athymic mice (Yoon et al., 1998). Because the potential HSV-TK/ganciclovir-mediated enhancement of G207 oncolytic activity may be balanced out by the inhibitory action of ganciclovir on viral replication, ganciclovir administration may not be beneficial to the efficacy of G207 tumor therapy. Rather, ganciclovir is useful for controlling G207 replication when termination of the therapy is necessary.

F. Safety evaluation of G207

HSV-1 is the most common cause of central nervous system viral infections, and HSV-1 encephalitis is known for its high mortality and long-term morbidity rate (Corey and Spear, 1986). G207 was initially developed for brain tumor therapy, and therefore it has been extensively evaluated for its toxicity in the brain. In BALB/c mice, the highest dose of G207 (10^7 pfu) caused no symptoms for over 20 weeks when inoculated intracerebrally or intraventricularly (Sundaresan et al., 2000). In A/J mice, the most susceptible mouse strain to HSV-1 infection (Lopez, 1975; Zawatzky et al., 1981), intracerebral inoculation of clinical-grade G207 at 2×10^6 pfu caused only a temporary and slight hunching in two of eight mice (Todo et al., 2001a). Furthermore, in BALB/c mice that survived an intracerebral inoculation of wild-type HSV-1 KOS at $5 \times 10^2 - 2 \times 10^3$ pfu, a subsequent challenge with an intracerebral inoculation of G207 (10^7 pfu) at the same stereotactic coordinates did not result in reactivation of detectable infectious virus or symptoms of disease (Sundaresan et al., 2000).

Aotus nancymae (New World owl monkeys) are among the more sensitive animals to HSV infection (Katzin *et al.*, 1967). A total of 22 *Aotus* primates have been used for safety evaluation of G207 (intracerebral and/or intraprostatic) (Hunter *et al.*, 1999; Mineta *et al.*, 1995; Todo *et al.*, 2000b; Varghese *et al.*, 2001). In *Aotus*, a single intracerebral inoculation of G207, up to 10^9 pfu, or repeat inoculations of 10^7 pfu caused neither virus-related disease nor detectable changes in the brain as assessed by magnetic resonance imaging (MRI) and pathological studies (Hunter *et al.*, 1999). In contrast, an intracerebral inoculation of 10^3 pfu of wild-type HSV-1 strain F caused viral encephalitis with the animal becoming moribund within 5 days of inoculation. Intracerebral G207 inoculation, however, caused measurable increases in serum anti-HSV titers in all *Aotus* studied. Four *Aotus* were used to evaluate the shedding and biodistribution of G207 after intracerebral inoculation of clinical-grade, column-purified G207 (3×10^7 pfu) (Todo *et al.*, 2000b). Using PCR analyses and viral culture, neither infectious virus nor viral DNA was detected from tear, saliva, or vaginal secretion samples at any time point up to 1 month postinoculation. Analyses of tissues obtained at necropsy at 1 month from two of the four animals, plus one monkey inoculated with laboratory-grade G207 (10^9 pfu) 2 years earlier, showed G207 DNA distribution restricted to the brain, with no infectious virus being isolated. Histopathology revealed normal brain tissues including the sites of inoculation. A detailed time course study of serum anti-HSV antibody titers showed an increase as early as 21 days postinoculation.

This high attenuation of neurovirulence in G207, while retaining oncolytic activity, is caused by mutations in both the $\gamma 34.5$ and ICP6 genes: Neither mutation alone fully diminishes the virulence. Strain 1716 ($\gamma 34.5\Delta$) has been shown to cause severe inflammation in the brain that led to death in some animals when inoculated intracerebrally into AO rats or BALB/c mice at a dose of 10^6 pfu, intracerebrally into athymic mice at a dose of 10^5 pfu, or intraventricularly into athymic mice at a low dose of 10^3 pfu (Lasner *et al.*, 1998; McMenamin *et al.*, 1998). In these animals, immunohistochemistry for HSV-1 antigens demonstrated active viral replication in neurons of the basal ganglia and cerebral cortex. R3616 ($\gamma 34.5\Delta$) was reported to have caused fatal encephalitis in one out of six Swiss Webster mice when inoculated intracerebrally at a dose of 10^6 pfu, although no symptoms were observed in six SCID mice (Pyles *et al.*, 1997). Intrathecal injection of hrR3 (ICP6$^-$) at a dose of 10^8 pfu into Fisher rats has been shown to cause prolonged morbidity for up to 7 days and deaths in 2 of 10 animals (Kramm *et al.*, 1996). In these animals, leptomeningial blood vessels were highly infected by hrR3 within the first $2-5$ days of intrathecal virus inoculation. In the same animal model, MGH-1 ($\gamma 34.5\Delta$/ICP6$^-$) caused no morbidity or mortality (Kramm *et al.*, 1997).

As a step toward the clinical application of G207 for prostate cancer therapy, G207 was further evaluated for safety after inoculation into the prostate

(Varghese *et al.*, 2001). None of the BALB/c mice receiving intraprostatic G207 (10^7 pfu) inoculation manifested abnormal signs or symptoms. In contrast, four of eight mice receiving wild-type HSV-1 strain F (10^6 pfu) died within 13 days of intraprostatic inoculation. Histopathology of G207-inoculated prostate was normal, whereas strain F-inoculated prostate showed epithelial flattening, sloughing, and stromal edema. Four *Aotus*, including two animals that had received prior intracerebral G207 inoculations, were inoculated intraprostatically with clinical-grade G207 (10^7 pfu). None of these animals manifested abnormal signs or symptoms. Shedding of the virus was not detected from urine, serum, tear, saliva, or swab samples from urethra and rectum at any time point up to 56 days postinoculation (Varghese *et al.*, 2001). Animals were sacrificed at 14, 21, or 56 days postinoculation, and no histopathological abnormalities due to virus infection were found in the prostate or other organs. Thus, G207 was shown to be safe for prostate inoculation.

III. G207 CLINICAL TRIAL

A phase I clinical trial of G207 in recurrent malignant glioma patients was performed at Georgetown University Medical Center (Washington, D.C.) and the University of Alabama at Birmingham (Markert *et al.*, 2000). Between February 1998 and May 1999, 21 patients were enrolled (15 male, 6 female; mean age 54.1, range 38–72), according to the inclusion and exclusion criteria listed in Table 4. Sixteen patients had a diagnosis of glioblastoma and five of anaplastic astrocytoma at the time of enrollment. All patients had received radiation therapy at least 4 weeks prior to G207 inoculation. The trial was designed to assess the safety of G207 inoculated in the brain in a dose escalation study. There were three patients at each dose, starting from 10^6 pfu and increasing to 3×10^9 pfu. G207 was inoculated stereotactically at a single site, except for the highest dose which was inoculated at five sites, into an enhancing region of the tumor visualized by a computerized tomography (CT) scan with contrast enhancement (Markert *et al.*, 2000). While this clinical trial was ongoing, another phase I clinical trial in patients with recurrent glioma was being carried out in Glasgow, United Kingdom, using strain 1716 ($\gamma34.5\Delta$) (Rampling *et al.*, 2000). In the Glasgow trial, patients received from 10^3 to 10^5 pfu 1716 at multiple sites by stereotactic injection. There was no evidence of virally induced adverse clinical events, and four of nine patients were alive 1 year after treatment (Rampling *et al.*, 2000).

In the G207 trial, no acute, moderate to severe toxicity attributable to G207 was observed (Markert *et al.*, 2000). Minor adverse events included seizure (two cases) and brain edema (one case). One patient manifested mental status deterioration and dysphasia within 24 h of inoculation (3×10^8 pfu).

Table 4 Patient Criteria for the G207 Phase I Clinical Trial

Inclusion criteria
1. Pathologically proven residual glioblastoma multiforme, anaplastic astrocytoma, or gliosarcoma.
2. Tumor growth on MRI after the last treatment undertaken.
3. Failed radiation therapy ($\geq$5000 cGy) at least 4 weeks prior to inoculation.
4. Enhancing lesion must be 1 cm or greater in diameter as determined by MRI.
5. Normal hematological, renal, and liver function.
6. Karnofsky score greater than or equal to 70.
7. 18 years of age or older and able to give informed consent.
8. Willing to practice birth control.

Exclusion criteria
1. Surgical resection within 4 weeks of inoculation.
2. Acute infection, granulocytopenia, or medical condition precluding surgery.
3. Pregnant or lactating females.
4. History of encephalitis, multiple sclerosis, or other CNS infection.
5. Tumor that requires ventricular, brainstem, basal ganglia, or posterior fossa inoculation, or requires access through a ventricle for inoculation.
6. Prior participation in viral therapy protocol.
7. Prior participation in chemotherapy, cytotoxic therapy, immunotherapy, or gene therapy protocol within 6 weeks of inoculation.
8. Required increase in steroid dose within 2 weeks of inoculation.
9. HIV seropositive.
10. Concurrent treatment with any medication active against HSV (e.g., acyclovir).
11. Active oral herpes lesion.

A biopsy performed in this patient 14 days postinoculation showed histology of viable tumor with increased cellularity, without inflammation suggestive of encephalitis, and was negative for HSV immunostaining. PCR analysis of the biopsy specimen for G207 was equivocal (positive for glycoprotein B and DNA polymerase sequences of HSV-1 DNA, but negative for *lacZ*). Two other cases manifested neurological deterioration more than 3 months after G207 treatment that could not be explained by tumor progression as detected by MRI. Biopsies were performed and tumor specimens were negative for HSV immunostaining in both cases. Among seven biopsied or resected tumor specimens analyzed, including the three described above, specimens from two patients were positive for G207 DNA by PCR analysis (56 and 157 days postinoculation).

The physical and neurological status of the patients was followed by Karnofsky performance score and mini-mental status exam (Markert *et al.*, 2000). The mean ($\pm$SD) Karnofsky score dropped from 84.3 $\pm$12.5 preinoculation to 81.7 $\pm$18.5 3 months postinoculation (12 patients remaining). An improvement in Karnofsky score was observed in 6 of 21 patients (29%) at some time after G207 inoculation. Five of 19 patients were negative for serum anti-HSV-1 antibody prior to G207 treatment. Despite corticosteroid treatment of these patients, one patient seroconverted after G207 inoculation. Serial MRI evaluations were performed in 20 cases. Eight of 20 patients had a decrease in tumor volume (enhancing area) between 4 days and 1 month postinoculation. All patients, except one that died from cerebral infarction 10 months after G207 treatment, eventually showed tumor progression. Autopsy was performed in five cases, and histology of the brains showed no evidence of encephalitis, white matter degeneration, or inflammatory changes, and all were negative for HSV-1 immunoreactivity. In three cases, the tumor was localized to one region of the brain without significant tumor cell invasion into the surrounding brain tissue as usually observed with typical glioblastoma cases. One glioblastoma patient that died from cerebral infarction had no evidence of residual tumor at autopsy. Overall, the phase I clinical trial confirmed the safety of G207 inoculated into the brain at doses up to 3×10^9 pfu. Radiographic and neuropathological findings were suggestive of antitumor activity in some patients. The results warrant proceeding to further clinical trials to evaluate G207 efficacy in brain tumors and other tumor types.

IV. CONCLUSIONS

The use of replication-competent HSV vectors for cancer therapy has gone from an interesting concept to clinical trial in less than 10 years. The potential of this therapeutic approach is born out by the development of conditionally replicating viruses from other viral families such as adenovirus (Heise and Kirn, 2000; Khuri et al., 2000), reovirus (Norman and Lee, 2000), Newcastle disease virus (Lorence et al., 1994), vaccinia virus (Mastrangelo et al., 2000), autonomous parvoviruses (Dupont et al., 2000; Haag et al., 2000), vesicular stomatitis virus (Stojdl et al., 2000), and poliovirus (Gromeier et al., 2000). This approach to cancer therapy has many features that make it very attractive: (1) amplification of the therapeutic agent (HSV) in situ, rather than metabolism and excretion as occurs with most drugs; (2) the potential to target tumor cells; (3) induction of an antitumor immune response by the oncolytic viral infection; (4) the ability of the vector to deliver single or multiple therapeutic genes; (5) targeting of the tumor cell pathways that are likely independent of those targeted by traditional cancer therapies, and therefore combinations of viral vectors with chemotherapy, radiation, or surgery should provide added benefit; (6) the ability to terminate an adverse HSV infection with antiviral drugs; and (7) the nonpathogenic phenotype of such a cytotoxic agent. Whether the promise demonstrated by oncolytic viruses in animal models will be translated to patients awaits further clinical study. We are just in the infancy of this approach to cancer therapy, and there are still multiple genetic manipulations that can be performed on HSV to enhance its efficacy.

Acknowledgments

We thank all the members of the Molecular Neurosurgery Laboratory who participated in these studies, in particular Robert L. Martuza who first developed the strategy of replication-competent HSV vectors for cancer therapy and has been instrumental in all facets of this research and its translation to the clinic. Samuel D. Rabkin is a consultant for MediGene, Inc., which has a license from Georgetown University for G207. The studies of Drs. Todo and Rabkin have been supported in part by grants from NINDS, the Department of Defense, CapCure Foundation, and NeuroVir, Inc.

References

Andreansky, S. S., Soroceanu, L., Flotte, E. R., Chou, J., Markert, J. M., Gillespie, G. Y., Roizman, B., and Whitley, R. J. (1997). Evaluation of genetically engineered herpes simplex viruses as oncolytic agents for human malignant brain tumors. *Cancer Res.* **57**, 1502–1509.

Andreansky, S. S., He, B., Gillespie, G. Y., Soroceanu, L., Markert, J., Chou, J., Roizman, B., and Whitley, R. J. (1996). The application of genetically engineered herpes simplex virus to the treatment of experimental brain tumors. *Proc. Natl. Acad. Sci. U.S.A.* **93**, 11313–11318.

Advani, S. J., Sibley, G. S., Song, P. Y., Hallahan, D. E., Kataoka, Y., Roizman, B., and Weichselbaum, R. R. (1998). Enhancement of replication of genetically engineered herpes simplex viruses by ionizing radiation: A new paradigm for destruction of therapeutically intractable tumors. *Gene Ther.* **5**, 160–165.

Advani, S. J., Chung, S. M., Yan, S. Y., Gillespie, G. Y., Markert, J. M., Whitley, R. J., Roizman, B., and Weichselbaum, R. R. (1999). Replication-competent, nonneuroinvasive genetically engineered herpes virus is highly effective in the treatment of therapy-resistant experimental human tumors. *Cancer Res.* **59**, 2055–2058.

Barba, D., Joseph, H., Sadelain, M., and Gage, F. H. (1994). Development of anti-tumor immunity following thymidine kinase-mediated killing of experimental brain tumors. *Proc. Natl. Acad. Sci. U.S.A.* **91**, 4348–4352.

Bennett, J. J., Kooby, D. A., Delman, K., McAuliffe, P., Halterman, M. W., Federoff, H., and Fong, Y. (2000). Antitumor efficacy of regional oncolytic viral therapy for peritoneally disseminated cancer. *J. Mol. Med.* **78**, 166–174.

Bennett, J. J., Malhotra, S. Wong, R. J., Delman, K., Zager, J., St.-Louis, M., Johnson, P. A., and Fong, Y. (2001). IL-12 secretion enhances antitumor efficacy of oncolytic herpes simplex viral therapy for colorectal cancer. *Ann. Surg.* **233**, 819–826.

Bi, W. L., Parysek, L. M., Warnick, R., and Stambrook, P. J. (1993). *In vitro* evidence that metabolic cooperation is responsible for the bystander effect observed with HSV tk retroviral gene therapy. *Hum. Gene Ther.* **4**, 725–731.

Boviatsis, E. J., Scharf, J. M., Chase, M., Harrington, K., Kowall, N. W., Breakefield, X. O., and Chiocca, E. A. (1994a). Antitumor activity and reporter gene transfer into rat brain neoplasms inoculated with herpes simplex virus vectors defective in thymidine kinase or ribonucleotide reductase. *Gene Ther.* **1**, 323–331.

Boviatsis, E. J., Park, J. S., Sena-Esteves, M., Kramm, C. M., Chase, M., Efird, J. T., Wei, M. X., Breakefield, X. O., and Chiocca, E. A. (1994b). Long-term survival of rats harboring brain neoplasms treated with ganciclovir and a herpes simplex virus vector that retains an intact thymidine kinase gene. *Cancer Res.* **54**, 5745–5751.

Bradley, J. D., Kataoka, Y., Advani, S., Chung, S. M., Arani, R. B., Gillespie, G. Y., Whitley, R. J., Markert, J. M., Roizman, B., and Weichselbaum, R. R. (1999). Ionizing radiation improves survival in mice bearing intracranial high-grade gliomas injected with genetically modified herpes simplex virus. *Clin. Cancer Res.* **5**, 1517–1522.

Carew, J. F., Kooby, D. A., Halterman, M. W., Federoff, H. J., and Fong, Y. (1999). Selective infection and cytolysis of human head and neck squamous cell carcinoma with sparing of normal mucosa by a cytotoxic herpes simplex virus type 1 (G207). *Hum. Gene Ther.* **10,** 1599–1606.

Chahlavi, A., Todo, T., Martuza, R. L., and Rabkin, S. D. (1999a). Replication-competent herpes simplex virus vector G207 and cisplatin combination therapy for head and neck squamous cell carcinoma. *Neoplasia* **1,** 162–169.

Chahlavi, A., Rabkin, S., Todo, T., Sundaresan, P., and Martuza, R. (1999b). Effect of prior exposure to herpes simplex virus 1 on viral vector-mediated tumor therapy in immunocompetent mice. *Gene Ther.* **6,** 1751–1758.

Chambers, R., Gillespie, G. Y., Soroceanu, L., Andreansky, S., Chatterjee, S., Chou, J., Roizman, B., and Whitley, R. J. (1995). Comparison of genetically engineered herpes simplex viruses for the treatment of brain tumors in a SCID mouse model of human malignant glioma. *Proc. Natl. Acad. Sci. U.S.A.* **92,** 1411–1415.

Chase, M., Chung, R. Y., and Chiocca, E. A. (1998). An oncolytic viral mutant that delivers the CYP2B1 transgene and augments cyclophosphamide chemotherapy. *Nat. Biotechnol.* **16,** 444–448.

Cheng, Y. C., Grill, S. P., Dutschman, G. E., Nakayama, K., and Bastow, K. F. (1983). Metabolism of 9-(1,3-dihydroxy-2-propoxymethyl)guanine, a new anti-herpes virus compound, in herpes simplex virus-infected cells. *J. Biol. Chem.* **258,** 12460–12464.

Chou, J., and Roizman, B. (1992). The γ1(34.5) gene of herpes simplex virus 1 precludes neuroblastoma cells from triggering total shutoff of protein synthesis characteristic of programed cell death in neuronal cells. *Proc. Natl. Acad. Sci. U.S.A.* **89,** 3266–3270.

Chou, J., Kern, E. R., Whitley, R. J., and Roizman, B. (1990). Mapping of herpes simplex virus-1 neurovirulence to γ34.5, a gene nonessential for growth in culture. *Science* **250,** 1262–1266.

Coen, D. M., Goldstein, D. J., and Weller, S. K. (1989). Herpes simplex virus ribonucleotide reductase mutants are hypersensitive to acyclovir. *Antimicrob. Agents Chemother.* **33,** 1395–1399.

Colombo, B. M., Benedetti, S., Ottolenghi, S., Mora, M., Pollo, B., Poli, G., and Finocchiaro, G. (1995). The "bystander effect": Association of U-87 cell death with ganciclovir-mediated apoptosis of nearby cells and lack of effect in athymic mice. *Hum. Gene Ther.* **6,** 763–772.

Corey, L., and Spear, P. G. (1986). Infections with herpes simplex viruses. *N. Engl. J. Med.* **314,** 749–757.

Coukos, G., Makrigiannakis, A., Kang, E. H., Caparelli, D., Benjamin, I., Kaiser, L. R., Rubin, S. C., Albelda, S. M., and Molnar-Kimber, K. L. (1999). Use of carrier cells to deliver a replication-selective herpes simplex virus-1 mutant for the intraperitoneal therapy of epithelial ovarian cancer. *Clin. Cancer Res.* **5,** 1523–1537.

Coukos, G., Makrigiannakis, A., Montas, S., Kaiser, L. R., Toyozumi, T., Benjamin, I., Albelda, S. M., Rubin, S. C., and Molnar-Kimber, K. L. (2000). Multi-attenuated herpes simplex virus-1 mutant G207 exerts cytotoxicity against epithelial ovarian cancer but not normal mesothelium and is suitable for intraperitoneal oncolytic therapy. *Cancer Gene Ther.* **7,** 275–283.

Cozzi, P. J., Malhotra, S., McAuliffe, P., Kooby, D. A., Federoff, H. J., Huryk, B., Johnson, P., Scardino, P. T., Heston, W. D. W., and Fong, Y. (2001). Intravesical oncolytic viral therapy using attenuated replication-competent herpes simplex viruses G207 and Nv 1020 is effective in the treatment of bladder cancer in an orthotopic syngeneic model. *FASEB J.* **15,** 1306–1308.

Culver, K. W., Ram, Z., Wallbridge, S., Ishii, H., Oldfield, E. H., and Blaese, R. M. (1992). In vivo gene transfer with retroviral vector-producer cells for treatment of experimental brain tumors. *Science* **256,** 1550–1552.

Dubin, G., Socolof, E., Frank, I. and Friedman, H. M. (1991). Herpes simplex virus type 1 Fc receptor protects infected cells from antibody-dependent cellular cytotoxicity. *J. Virol.* **65,** 7046–7050.

Dupont, F., Avalosse, B., Karim, A., Mine, N., Bosseler, M., Maron, A., Van den Broeke, A. V., Ghanem, G. E., Burny, A., and Zeicher, M. (2000). Tumor-selective gene transduction and cell killing with an oncotropic autonomous parvovirus-based vector. *Gene Ther.* **7,** 790–796.

Ejercito, P. M., Kieff, E. D., and Roizman, B. (1968). Characterization of herpes simplex virus strains differing in their effects on social behaviour of infected cells. *J. Gen. Virol.* **2,** 357–364.

Estin, D., Li, M., Spray, D., and Wu, J. K. (1999). Connexins are expressed in primary brain tumors and enhance the bystander effect in gene therapy. *Neurosurgery* **44,** 361–369.

Fong, Y., Fortner, J., Sun, R. L., Brennan, M. F., and Blumgart, L. H. (1999). Clinical score for predicting recurrence after hepatic resection for metastatic colorectal cancer: Analysis of 1001 consecutive cases. *Ann. Surg.* **230,** 309–318.

Freeman, S. M., Abboud, C. N., Whartenby, K. A., Packman, C. H., Koeplin, D. S., Moolten, F. L., and Abraham, G. N. (1993). The "bystander effect": Tumor regression when a fraction of the tumor mass is genetically modified. *Cancer Res.* **53,** 5274–5283.

Gagandeep, S., Brew, R., Green, B., Christmas, S. E., Klatzmann, D., Poston, G. J., and Kinsella, A. R. (1996). Prodrug-activated gene therapy: Involvement of an immunological component in the "bystander effect". *Cancer Gene Ther.* **3,** 83–88.

Goldstein, D. J., and Weller, S. K. (1988). Herpes simplex virus type 1-induced ribonucleotide reductase activity is dispensable for virus growth and DNA synthesis: Isolation and characterization of an ICP6 *lacZ* insertion mutant. *J. Virol.* **62,** 196–205.

Gromeier, M., Lachmann, S., Rosenfeld, M. R., Gutin, P. H., and Wimmer, E. (2000). Intergeneric poliovirus recombinants for the treatment of malignant glioma [see comments]. *Proc. Natl. Acad. Sci. U.S.A.* **97,** 6803–6808.

Haag, A., Menten, P., Van Damme, J., Dinsart, C., Rommelaere, J., and Cornelis, J. J. (2000). Highly efficient transduction and expression of cytokine genes in human tumor cells by means of autonomous parvovirus vectors; generation of antitumor responses in recipient mice. *Hum. Gene Ther.* **11,** 597–609.

Heise, C., and Kirn, D. H. (2000). Replication-selective adenoviruses as oncolytic agents. *J. Clin. Invest.* **105,** 847–851.

Herrlinger, U., Kramm, C. M., Aboody-Guterman, K. S., Silver, J. S., Ikeda, K., Johnston, K. M., Pechan, P. A., Barth, R. F., Finkelstein, D., Chiocca, E. A., Louis, D. N., and Breakefield, X. O. (1998). Pre-existing herpes simplex virus 1 (HSV-1) immunity decreases, but does not abolish, gene transfer to experimental brain tumors by a HSV-1 vector. *Gene Ther.* **5,** 809–819.

Hunter, W., Rabkin, S., and Martuza, R. (1995). Brain tumor therapy using genetically engineered replication-competent virus. *In* "Viral Vectors" (M. G. Kaplitt and A. D. Loewy, eds.), pp. 259–274. Academic Press, San Diego.

Hunter, W. D., Martuza, R. L., Feigenbaum, F., Todo, T., Mineta, T., Yazaki, T., Toda, M., Newsome, J. T., Platenberg, R. C., Manz, H. J., and Rabkin, S. D. (1999). Attenuated, replication-competent herpes simplex virus type 1 mutant G207: Safety evaluation of intracerebral injection in nonhuman primates. *J. Virol.* **73,** 6319–6326.

Jacobs, A., Breakefield, X. O., and Fraefel, C. (1999). HSV-1- based vectors for gene therapy of neurological diseases and brain tumors: Part II. Vector systems and applications. *Neoplasia* **1,** 402–416.

Jia, W. W., McDermott, M., Goldie, J., Cynader, M., Tan, J., and Tufaro, F. (1994). Selective destruction of gliomas in immunocompetent rats by thymidine kinase-defective herpes simplex virus type 1. *J. Natl. Cancer Inst.* **86,** 1209–1215.

Jorgensen, T. J., Katz, S., Wittmack, E. K., Varghese, S., Todo, T., Rabkin, S. D., and Martuza, R. L. (2001). Ionizing radiation does not alter the antitumor activity of herpes simplex virus vector G207 in subcutaneous human and murine prostate tumor models. *Neoplasia.* In press.

Kaplitt, M. G., Tjuvajev, J. G., Leib, D. A., Berk, J., Pettigrew, K. D., Posner, J. B., Pfaff, D. W., Rabkin, S. D., and Blasberg, R. G. (1994). Mutant herpes simplex virus induced regression of tumors growing in immunocompetent rats. *J. Neurooncol.* **19,** 137–147.

Katzin, D. S., Connor, J. D., Wilson, L. A., and Sexton, R. S. (1967). Experimental herpes simplex infection in the owl monkey. *Proc. Soc. Exp. Biol. Med.* **125,** 391–398.

Kemeny, N., Huang, Y., Cohen, A. M., Shi, W., Conti, J. A., Brennan, M. F., Bertino, J. R., Turnbull, A. D., Sullivan, D., Stockman, J., Blumgart, L. H., and Fong, Y. (1999). Hepatic arterial infusion of chemotherapy after resection of hepatic metastases from colorectal cancer. *N. Engl. J. Med.* **341,** 2039–2048.

Kesari, S., Randazzo, B. P., Valyi-Nagy, T., Huang, Q. S., Brown, S. M., MacLean, A. R., Lee, V. M., Trojanowski, J. Q., and Fraser, N. W. (1995). Therapy of experimental human brain tumors using a neuroattenuated herpes simplex virus mutant. *Lab. Invest.* **73,** 636–648.

Khuri, F. R., Nemunaitis, J., Ganly, I., Arseneau, J., Tannock, I. F., Romel, L., Gore, M., Ironside, J., MacDougall, R. H., Heise, C., Randlev, B., Gillenwater, A. M., Bruso, P., Kaye, S. B., Hong, W. K., and Kirn, D. H. (2000). A controlled trial of intratumoral ONYX-015, a selectively-replicating adenovirus, in combination with cisplatin and 5-fluorouracil in patients with recurrent head and neck cancer. *Nat. Med.* **6,** 879–885.

Kogishi, J., Miyatake, S., Hangai, M., Akimoto, M., Okazaki, K., and Honda, Y. (1999). Mutant herpes simplex virus-mediated suppression of retinoblastoma. *Curr. Eye Res.* **18,** 321–326.

Kooby, D. A., Carew, J. F., Halterman, M. W., Mack, J. E., Bertino, J. R., Blumgart, L. H., Federoff, H. J., and Fong, Y. (1999). Oncolytic viral therapy for human colorectal cancer and liver metastases using a multi-mutated herpes simplex virus type-1 (G207). *FASEB J.* **13,** 1325–1334.

Kramm, C. M., Rainov, N. G., Sena-Esteves, M., Barnett, F. H., Chase, M., Herrlinger, U., Pechan, P. A., Chiocca, E. A., and Breakefield, X. O. (1996). Long-term survival in a rodent model of disseminated brain tumors by combined intrathecal delivery of herpes vectors and ganciclovir treatment. *Hum. Gene Ther.* **7,** 1989–1994.

Kramm, C. M., Chase, M., Herrlinger, U., Jacobs, A., Pechan, P. A., Rainov, N. G., Sena-Esteves, M., Aghi, M., Chiocca, E. A., and Breakefield, X. O. (1997). Therapeutic efficiency and safety of a second-generation replication-competent HSV1 vector for brain tumor gene therapy. *Hum. Gene Ther.* **8,** 2057–2068.

Kucharczuk, J. C., Randazzo, B., Chang, M. Y., Amin, K. M., Elshami, A. A., Sterman, D. H., Rizk, N. P., Molnar-Kimber, K. L., Brown, S. M., MacLean, A. R., Litzky, L. A., Fraser, N. W., Albelda, S. M., and Kaiser, L. R. (1997). Use of a "replication-restricted" herpes virus to treat experimental human malignant mesothelioma. *Cancer Res.* **57,** 466–471.

Lambright, E. S., Kang, E. H., Force, S., Lanuti, M., Caparrelli, D., Kaiser, L. R., Albelda, S. M., and Molnar-Kimber, K. L. (2000). Effect of preexisting anti-herpes immunity on the efficacy of herpes simplex viral therapy in a murine intraperitoneal tumor model. *Mol. Ther.* **2,** 387–393.

Langenberg, A. G., Corey, L., Ashley, R. L., Leong, W. P., and Straus, S. E. (1999). A prospective study of new infections with herpes simplex virus type 1 and type 2. Chiron HSV Vaccine Study Group. *N. Engl. J. Med.* **341,** 1432–1438.

Lasner, T. M., Kesari, S., Brown, S. M., Lee, V. M., Fraser, N. W., and Trojanowski, J. Q. (1996). Therapy of a murine model of pediatric brain tumors using a herpes simplex virus type-1 ICP34.5 mutant and demonstration of viral replication within the CNS. *J. Neuropathol. Exp. Neurol.* **55,** 1259–1269.

Lasner, T. M., Tal-Singer, R., Kesari, S., Lee, V. M., Trojanowski, J. Q., and Fraser, N. W. (1998). Toxicity and neuronal infection of a HSV-1 ICP34.5 mutant in nude mice. *J. Neurovirol.* **4,** 100–105.

Leopardi, R., and Roizman, B. (1996). The herpes simplex virus major regulatory protein ICP4 blocks apoptosis induced by the virus or by hyperthermia. *Proc. Natl. Acad. Sci. U.S.A.* **93,** 9583–9587.

Lopez, C. (1975). Genetics of natural resistance to herpesvirus infections in mice. *Nature* **258,** 152–155.

Lorence, R. M., Reichard, K. W., Katubig, B. B., Reyes, H. M., Phuangsab, A., Mitchell, B. R., Cascino, C. J., Walter, R. J., and Peeples, M. E. (1994). Complete regression of human neuroblastoma xenografts in athymic mice after local Newcastle disease virus therapy. *J. Natl. Cancer Inst.* **86,** 1228–1233.

MacLean, A. R., ul-Fareed, M., Robertson, L., Harland, J., and Brown, S. M. (1991). Herpes simplex virus type 1 deletion variants 1714 and 1716 pinpoint neurovirulence-related sequences in Glasgow strain 17⁺ between immediate early gene 1 and the 'a' sequence. *J. Gen. Virol.* **72**, 631–639.

McMenamin, M. M., Byrnes, A. P., Charlton, H. M., Coffin, R. S., Latchman, D. S., and Wood, M. J. (1998). A gamma 34.5 mutant of herpes simplex 1 causes severe inflammation in the brain. *Neuroscience* **83**, 1225–1237.

Markert, J. M., Malick, A., Coen, D. M., and Martuza, R. L. (1993). Reduction and elimination of encephalitis in an experimental glioma therapy model with attenuated herpes simplex mutants that retain susceptibility to acyclovir. *Neurosurgery* **32**, 597–603.

Markert, J. M., Medlock, M. D., Rabkin, S. D., Gillespie, G. Y., Todo, T., Hunter, W. D., Palmer, C. A., Feigenbaum, F., Tornatore, C., Tufaro, F., and Martuza, R. L. (2000). Conditionally replicating herpes simplex virus mutant, G207 for the treatment of malignant glioma: Results of a phase I trial. *Gene Ther.* **7**, 867–874.

Martuza, R. L. (2000). Conditionally replicating herpes vectors for cancer therapy. *J. Clin. Invest.* **105**, 841–846.

Martuza, R. L., Malick, A., Markert, J. M., Ruffner, K. L., and Coen, D. M. (1991). Experimental therapy of human glioma by means of a genetically engineered virus mutant. *Science* **252**, 854–856.

Mashour, G. A., Moulding, H. D., Chahlavi, A., Khan, G. A., Rabkin, S. D., Martuza, R. L., Driever, P. H., and Kurtz, A. (2001). Therapeutic efficacy of G207 in a novel peripheral nerve sheath tumor model. *Exp. Neurol.* **169**, 64–71.

Mastrangelo, M. J., Eisenlohr, L. C., Gomella, L., and Lattime, E. C. (2000). Poxvirus vectors: Orphaned and underappreciated. *J. Clin. Invest.* **105**, 1031–1034.

Meignier, B., Longnecker, R., and Roizman, B. (1988). *In vivo* behavior of genetically engineered herpes simplex viruses R7017 and R7020: Construction and evaluation in rodents. *J. Infect. Dis.* **158**, 602–614.

Mesnil, M., Piccoli, C., Tiraby, G., Willecke, K., and Yamasaki, H. (1996). Bystander killing of cancer cells by herpes simplex virus thymidine kinase gene is mediated by connexins. *Proc. Natl. Acad. Sci. U.S.A.* **93**, 1831–1835.

Mineta, T., Rabkin, S. D., and Martuza, R. L. (1994). Treatment of malignant gliomas using ganciclovir-hypersensitive, ribonucleotide reductase-deficient herpes simplex viral mutant. *Cancer Res.* **54**, 3963–3966.

Mineta, T., Rabkin, S. D., Yazaki, T., Hunter, W. D., and Martuza, R. L. (1995). Attenuated multi-mutated herpes simplex virus-1 for the treatment of malignant gliomas. *Nat. Med.* **1**, 938–943.

Miyatake, S., Martuza, R. L., and Rabkin, S. D. (1997). Defective herpes simplex virus vectors expressing thymidine kinase for the treatment of malignant glioma. *Cancer Gene Ther.* **4**, 222–228.

Nakano, K., Todo, T., Tanaka, M., and Chijiiwa, K. (2001). Therapeutic efficacy of G207, a conditionally-replicating herpes simplex virus type 1 mutant, for gallbladder carcinoma in immunocompetent hamsters. *Mol. Ther.* **3**, 431–437.

Norman, K. L., and Lee, P. W. (2000). Reovirus as a novel oncolytic agent. *J. Clin. Invest.* **105**, 1035–1038.

Oliver, L., Wald, A., Kim, M., Zeh, J., Selke, S., Ashley, R., and Corey, L. (1995). Seroprevalence of herpes simplex virus infections in a family medicine clinic [see comments]. *Arch. Fam. Med.* **4**, 228–232.

Oyama, M., Ohigashi, T., Hoshi, M., Nakashima, J., Tachibana, M., Murai, M., Uyemura, K., and Yazaki, T. (2000). Intravesical and intravenous therapy of human bladder cancer by the herpes vector G207. *Hum. Gene Ther.* **11**, 1683–1693.

Panis, Y., Ribeiro, J., Chretien, Y., and Nordlinger, B. (1992). Dormant liver metastases: An experimental study. *Br. J. Surg.* **79**, 221–223.

Pardoll, D. M. (1995). Paracrine cytokine adjuvants in cancer immunotherapy. *Annu. Rev. Immunol.* **13,** 399–415.

Pyles, R. B., Warnick, R. E., Chalk, C. L., Szanti, B. E., and Parysek, L. M. (1997). A novel multiply-mutated HSV-1 strain for the treatment of human brain tumors. *Hum. Gene Ther.* **8,** 533–544.

Rager-Zisman, B., and Allison, A. C. (1976). Mechanism of immunologic resistance to herpes simplex virus 1 (HSV-1) infection. *J. Immunol.* **116,** 35–40.

Ram, Z., Culver, K. W., Walbridge, S., Blaese, R. M., and Oldfield, E. H. (1993). In situ retroviral-mediated gene transfer for the treatment of brain tumors. *Cancer Res.* **53,** 83–88.

Ramesh, R., Marrogi, A. J., Munshi, A., Abboud, C. N., and Freeman, S. M. (1996). In vivo analysis of the 'bystander effect': A cytokine cascade. *Exp. Hematol.* **24,** 829–838.

Rampling, R., Cruickshank, G., Papanastassiou, V., Nicoll, J., Hadley, D., Brennan, D., Petty, R., MacLean, A., Harland, J., McKie, E., Mabbs, R., and Brown, M. (2000). Toxicity evaluation of replication-competent herpes simplex virus (ICP 34.5 null mutant 1716) in patients with recurrent malignant glioma [see comments]. *Gene Ther.* **7,** 859–866.

Randazzo, B. P., Kesari, S., Gesser, R. M., Alsop, D., Ford, J. C., Brown, S. M., Maclean, A., and Fraser, N. W. (1995). Treatment of experimental intracranial murine melanoma with a neuroattenuated herpes simplex virus 1 mutant. *Virology* **211,** 94–101.

Randazzo, B. P., Bhat, M. G., Kesari, S., Fraser, N. W., and Brown, S. M. (1997). Treatment of experimental subcutaneous human melanoma with a replication-restricted herpes simplex virus mutant. *J. Invest. Dermatol.* **108,** 933–937.

Roizman, B. (1996). The function of herpes simplex virus genes: A primer for genetic engineering of novel vectors. *Proc. Natl. Acad. Sci. U.S.A.* **93,** 11307–11312.

Roizman, B., and Sears, A. E. (1995). Herpes simplex viruses and their replication. In "Fields Virology" (B. N. Fields, D. M. Knipe, and P. M. Howley, eds.), pp. 2231–2296. Lippincott-Raven, Philadelphia.

Rubsam, L. Z., Boucher, P. D., Murphy, P. J., KuKuruga, M., and Shewach, S. (1999). Cytotoxicity and accumulation of ganciclovir triphosphate in bystander cells cocultured with herpes simplex virus type 1 thymidine kinase-expressing human glioblastoma cells. *Cancer Res.* **59,** 669–675.

Samejima, Y., and Meruelo, D. (1995). Bystander killing induces apoptosis and is inhibited by forskolin. *Gene Ther.* **2,** 50–58.

Shinoura, N., Chen, L., Wani, M. A., Kim, Y. G., Larson, J. J., Warnick, R. E., Simon, M., Menon, A. G., Bi, W. L., and Stambrook, P. J. (1996). Protein and messenger RNA expression of connexin43 in astrocytomas: Implications in brain tumor gene therapy. *J. Neurosurg.* **84,** 839–845.

Siegel, D., Golden, E., Washington, A. E., Morse, S. A., Fullilove, M. T., Catania, J. A., Marin, B., and Hulley, S. B. (1992). Prevalence and correlates of herpes simplex infections. The population-based AIDS in Multiethnic Neighborhoods Study. *JAMA* **268,** 1702–1708.

Spaete, R. R., and Frenkel, N. (1982). The herpes simplex virus amplicon: A new eucaryotic defective-virus cloning-amplifying vector. *Cell* **30,** 295–304.

Stojdl, D. F., Lichty, B., Knowles, S., Marius, R., Atkins, H., Sonenberg, N., and Bell, J. C. (2000). Exploiting tumor-specific defects in the interferon pathway with a previously unknown oncolytic virus. *Nat. Med.* **6,** 821–825.

Sundaresan, P., Hunter, W. D., Martuza, R. L., and Rabkin, S. D. (2000). Attenuated, replication-competent herpes simplex virus type 1 mutant G207: Safety evaluation in mice. *J. Virol.* **74,** 3832–3841.

Thompson, K. A., Blessing, W. W., and Wesselingh, S. L. (2000). Herpes simplex replication and dissemination is not increased by corticosteroid treatment in a rat model of focal Herpes encephalitis. *J. Neurovirol.* **6,** 25–32.

Toda, M., Rabkin, S. D., and Martuza, R. L. (1998a). Treatment of human breast cancer in a brain metastatic model by G207, a replication-competent multimutated herpes simplex virus 1. *Hum. Gene Ther.* **9,** 2177–2185.

Toda, M., Martuza, R. L., Kojima, H., and Rabkin, S. D. (1998b). *In situ* cancer vaccination: An IL-12 defective vector/replication-competent herpes simplex virus combination induces local and systemic antitumor activity. *J. Immunol.* 160, 4457–4464.

Toda, M., Rabkin, S. D., Kojima, H., and Martuza, R. L. (1999). Herpes simplex virus as an 'in situ cancer vaccine' for the induction of specific anti-tumor immunity. *Hum. Gene Ther.* 10, 385–393.

Todo, T., Rabkin, S. D., Sundaresan, P., Wu, A., Meehan, K. R., Herscowitz, H. B., and Martuza, R. L. (1999a). Systemic antitumor immunity in experimental brain tumor therapy using a multi-mutated, replication-competent herpes simplex virus. *Hum. Gene Ther.* 10, 2741–2755.

Todo, T., Rabkin, S. D., Chahlavi, A., and Martuza, R. L. (1999b). Corticosteroid administration does not affect viral oncolytic activity, but inhibits antitumor immunity in replication-competent herpes simplex virus tumor therapy. *Hum. Gene Ther.* 10, 2869–2878.

Todo, T., Rabkin, S. D., and Martuza, R. L. (2000a). Evaluation of ganciclovir-mediated enhancement of the antitumoral effect in oncolytic, multimutated herpes simplex virus type 1 (G207) therapy of brain tumors. *Cancer Gene Ther.* 7, 939–946.

Todo, T., Feigenbaum, F., Rabkin, S. D., Lakeman, F., Newsome, J. T., Johnson, P. A., Mitchell, E., Belliveau, D., Ostrove, J. M., and Martuza, R. L. (2000b). Viral shedding and biodistribution of G207, a multimutated, conditionally-replicating herpes simplex virus type 1, after intracerebral inoculation in *Aotus*. *Mol. Ther.* 2, 588–595.

Todo, T., Martuza, R. L., Rabkin, S. D., and Johnson, P. A. (2001a). Oncolytic herpes simplex virus vector with enhanced MHC class 1 presentation and tumor cell killing. *Proc. Natl. Acad. Sci. U.S.A.* 98, 6396–6401.

Todo, T., Martuza, R. L., Dallman, M. J., and Rabkin, S. D. (2001b). *In situ* expression of soluble B7-1 in the context of oncolytic herpes simplex virus induces a potent antitumor immunity. *Cancer Res.* 61, 153–161.

Toyoizumi, T., Mick, R., Abbas, A. E., Kang, E. H., Kaiser, L. R., and Molnar-Kimber, K. L. (1999). Combined therapy with chemotherapeutic agents and herpes simplex virus type 1 ICP34.5 mutant (HSV-1716) in human non-small cell lung cancer. *Hum. Gene Ther.* 10, 3013–3029.

Varghese, S., Newsome, J. T., Rabkin, S. D., McGeagh, K., Mahoney, D., Nielsen, P., Todo, T., and Martuza, R. L. (2001). Safety evaluation of G207, a conditional replication virus, injected intraprostatically in mice and owl monkeys. *Hum. Gene Ther.* 12, 999–1010.

Vile, R. G., Nelson, J. A., Castleden, S., Chong, H., and Hart, I. R. (1994). Systemic gene therapy of murine melanoma using tissue specific expression of the *HSVtk* gene involves an immune component. *Cancer Res.* 54, 6228–6234.

Vrionis, F. D., Wu, J. K., Qi, P., Waltzman, M., Cherington, V., and Spray, D. C. (1997). The bystander effect exerted by tumor cells expressing the herpes simplex virus thymidine kinase (*HSVtk*) gene is dependent on connexin expression and cell communication via gap junctions. *Gene Ther.* 4, 577–585.

Walker, J. R., McGeagh, K. G., Sundaresan, P., Jorgensen, T. J., Rabkin, S. D., and Martuza, R. L. (1999). Local and systemic therapy of human prostate adenocarcinoma with the conditionally replicating herpes simplex virus vector G207. *Hum. Gene Ther.* 10, 2237–2243.

Whitley, R. J. (1996). Herpes simplex viruses. *In* "Fields Virology" (B. N. Fields, D. M. Knipe, and P. M. Howley, eds.), Vol. 2, pp. 2297–2342. Lippincott-Raven, Philadelphia.

Wong, R. J., Patel, S. G., Kim, S. H., DeMatteo, R. P., Malhotra, S. M., Bennett, J. J., St.-Louis, M., Shah, J. P., Johnson, P. A., and Fong, Y. (2001). Cytokine gene transfer enhances herpes oncolytic therapy in murine squamous cell carcinoma. *Hum. Gene Ther.* 12, 253–265.

Wu, A., Mazumder, A., Martuza, R. L., Liu, X., Thein, M., Meehan, K. R., and Rabkin, S. D. (2001). Biological purging of breast cancer cells using an attenuated replication-competent herpes simplex virus in human hematopoietic stem cell transplantion. *Cancer Res.* 61, 3009–3015.

Yamamoto, S., Suzuki, S., Hoshino, A., Akimoto, M., and Shimada, T. (1997). Herpes simplex virus thymidine kinase/ganciclovir-mediated killing of tumor cell induces tumor-specific cytotoxic T cells in mice. *Cancer Gene Ther.* **4**, 91–96.

Yang, Y., Nunes, F. A., Berencsi, K., Furth, E. E., Gönczöl, E., and Wilson, J. M. (1994). Cellular immunity to viral antigens limits E1-deleted adenoviruses for gene transfer. *Proc. Natl. Acad. Sci. U.S.A.* **91**, 4407–4411.

Yang, Y., Li, Q., Ertl, H. C. J., and Wilson, J. M. (1995). Cellular and humoral immune responses to viral antigens create barriers to lung-directed gene therapy with recombinant adenoviruses. *J. Virol.* **69**, 2004–2015.

Yazaki, T., Manz, H. J., Rabkin, S. D., and Martuza, R. L. (1995). Treatment of human malignant meningiomas by G207, a replication-competent multimutated herpes simplex virus 1. *Cancer Res.* **55**, 4752–4756.

Yeung, S. N., and Tufaro, F. (2000). Replicating herpes simplex virus vectors for cancer therapy. *Exp. Opin. Pharmacother.* **1**, 623–631.

Yoon, S. S., Carroll, N. M., Chiocca, E. A., and Tanabe, K. K. (1998). Cancer gene therapy using a replication-competent herpes simplex virus type 1 vector. *Ann. Surg.* **228**, 366–374.

Yoon, S. S., Nakamura, H., Carroll, N. M., Bode, B. P., Chiocca, E. A., and Tanabe, K. K. (2000). An oncolytic herpes simplex virus type 1 selectively destroys diffuse liver metastases from colon carcinoma. *FASEB J.* **14**, 301–311.

Zawatzky, R., Hilfenhaus, J., Marcucci, F., and Kirchner, H. (1981). Experimental infection of inbred mice with herpes simplex virus type 1: I. Investigation of humoral and cellular immunity and of interferon induction. *J. Gen. Virol.* **53**, 31–38.

4

p53 and Its Targets

Rishu Takimoto and Wafik S. El-Deiry
Laboratory of Molecular Oncology and Cell Cycle Regulation
Howard Hughes Medical Institute
Department of Medicine, Genetics, and Pharmacology
University of Pennsylvania School of Medicine
Philadelphia, Pennsylvania 19104

I. INTRODUCTION

The p53 tumor suppressor gene is the most frequently mutated gene in human cancer. Although knowledge about p53 function has accumulated, its entire biochemical function is still unclear. p53 protein can be stabilized in response to certain stimuli such as DNA damage, hypoxia, viral infection, or oncogene activation resulting in diverse biological effects like cell cycle arrest, apoptosis, senescence, differentiation, and antiangiogenesis. Activation of p53 has been shown to be augmented by phosphorylation, dephosphorylation, and acetylation, yielding a potent sequence-specific DNA-binding transcription factor. The wide range of p53s biological effect can be understood by its activation of a number of target genes including $p21^{WAF1}$, GADD45, 14-3-3σ, bax, Fas/Apo1, KILLER/DR5, PIG's, Tsp1, IGF-BP3, and others. In this review, the transcriptional targets of p53, their regulation by p53, and their importance in carrying out the biological effects of p53 are discussed.

II. ACTIVATION OF p53

p53 is a short-lived protein that is generally undetectable in normal cells. p53 may also exist in an inactive form, which could be activated by low doses of ultraviolet (UV) irradiation (Hupp *et al.*, 1995). It has been shown that ubiquitin-mediated proteolysis plays a role in the rapid turnover of p53 protein (Chowdary *et al.*, 1994). *In vitro* studies suggest that p53 may be negatively autoregulated by specifically inhibiting translation of its own mRNA (Mosner *et al.*, 1995).

Several stressful conditions can signal p53 stabilization and activation (Figure 1). Several types of DNA damage, for example, double-strand DNA breaks following ionizing irradiation (IR), thymine dimers produced by UV irradiation or chemical damage to DNA bases can lead to p53 activation. Hypoxia, heat-shock, radioactive chemicals, DNA transfection, and expression of viral and cellular oncogenes have also been shown to activate p53 (Kastan *et al.*, 1991; Donehower and Bradley, 1993; Zhan *et al.*, 1993; Graeber *et al.*, 1994; Yeargin and Haas, 1995; Lowe and Ruley, 1993; Demers *et al.*, 1994;

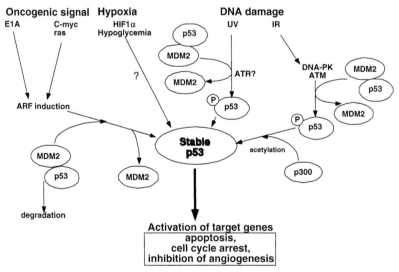

Figure 1 Mechanism of activation of stabilization of p53. Specific phosphorylations and acetylation of p53 appear to be involved in the activation of p53.

Maheswaran *et al.*, 1995). The precise mechanism by which p53 is activated by various distinct cellular stresses is not yet entirely elucidated. It is believed that p53 becomes activated due to both an increase in p53 protein levels as well as posttranslational modifications (Ko and Prives, 1996). An increase in the p53 protein level may be achieved by inhibiting directly or indirectly the degradation of p53, as well as enhancing the rate of translational initiation of p53 mRNA. Posttranslational modifications of the C-terminus of p53 including phosphorylation, dephosphorylation, acetylation, antibody binding, deletion of the C-terminus, or addition of a C-terminus peptide can induce sequence-specific DNA-binding of p53. One important question, and an area of active investigation, is how the cells sense the DNA damage and transmit the signal to stabilize p53 protein. One model is that cellular proteins recognize the damaged DNA and communicate with p53 protein. This model is based in part on studies showing cells from patients with ataxia telangiectasia (AT) have a delayed p53 response to DNA damage. It has been reported that p53 may recognize the damaged DNA through binding to DNA ends, excision-repair damage sites, or internal deletion loops (Lee *et al.*, 1995), as well as to short single-stranded DNA of 16–40 nucleotides long (Jayaraman and Prives, 1995). Specific phosphorylation, dephosphorylation, and acetylation events have been reported to activate p53 (reviewed in El-Deiry, 1998). MDM2 protein, which was originally found to interact with and inhibit p53-dependent transcriptional activity (Oliner *et al.*, 1993), has been found to promote rapid degradation of p53. It has become clear that this MDM2-dependent degradative pathway contributes to the maintenance of low levels of p53 in normal cells (Kubbutat *et al.*, 1997; Haupt *et al.*, 1997). In cancer cells with p53 mutation, abnormally low levels of MDM2 contribute to a rise in p53 protein levels. Because MDM2 is a target of p53-dependent transcriptional activation, those degradative pathways become disrupted in tumor cells. Phosphorylation of the amino-terminus of p53, which occurs after DNA damage, appears to reduce its affinity for binding to MDM2, which leads to stabilization of p53 (Shieh *et al.*, 1997). Site specific mutational analysis of some phosphorylation sites has suggested that certain phosphorylations may not be essential for DNA damage-induced stabilization of p53 (Ashcroft *et al.*, 1999). DNA damage causes phosphorylation of serine residues in the amino terminus of p53. In particular, serine-15 has been found to be phosphorylated in response to DNA damage by IR or UV irradiation. Ataxia telangiectasia cells show delayed phosphorylation of serine-15 in response to IR but show normal phosphorylation after UV irradiation suggesting that ATM kinase is involved in the serine-15 phosphorylation after IR, although it is not absolutely required (Canman *et al.*, 1998). However, more recent data suggests that serine-20 may be the critical residue regulating p53 stability, whereas serine-15 phosphorylation may be activating transcription by p53 without directly affecting protein stability (Hirao *et al.*, 2000).

On DNA damage, ATM kinase phosphorylates the product of c-Abl, which results in the activation of its tyrosine kinase activity (Kharbanda et al., 1995; Shafman et al., 1997; Baskaran et al., 1997). Activated c-Abl may bind to p53 and enhance its transcriptional activity (Goga et al., 1995). Experiments have revealed that ATM may associate with p53 and phosphorylates its amino-terminus directly (Khanna et al., 1998). Other kinases which can phosphorylate p53 include cyclin-dependent kinase (CDKs), casein kinase I (CK I), casein kinase II (CK II), protein kinase C (PKC), mitogen activated protein kinase (Milne et al., 1994), Jun amino-terminal kinase (JNK) (Milne et al., 1995), and Raf kinase (Jamal and Ziff, 1995). DNA-dependent protein kinase (DNA-PK), which is activated by DNA strand breaks, may be required to activate sequence-specific DNA-binding by p53 following DNA damage (Woo et al., 1998). In the other cases, it has been shown that serine/threonine protein phosphatase type 5 (PP5) can modulate the phosphorylation and DNA binding activity of p53 (Zuo et al., 1998). ATM also appears to be required for IR-induced dephosphorylation of p53 serine-376 which allows specific binding of 14-3-3σ proteins to p53 and leads to an increase in the sequence-specific DNA-binding activity of p53 (Waterman et al., 1998). Acetylation of the C-terminus of p53 by CREB-binding protein (p300/CBP) was shown to enhance sequence-specific DNA-binding by p53 (Gu and Roeder, 1997). p300/CBP are closely related histone acetyl transferases (HATs) (Bannister and Kouzarides, 1996; Ogryzko et al., 1996) that interact with p53 and function as coactivators for p53-mediated transcription (Avantaggiati et al., 1997; Gu et al., 1997; Lill et al., 1997). The activation of sequence-specific DNA-binding by p53 following DNA damage may involve sequential amino-terminal phosphorylation followed by carboxyl-terminal acetylation by the coactivator p300 following DNA damage (Sakaguchi et al., 1998). With the use of phosphorylated or acetylated peptide specific antibodies, p300 has been shown to acetylate Lys-382, whereas the p300/CBP-associated factor (PCAF) acetylates Lys-320 of p53. Either acetylation leads to enhanced sequence-specific DNA-binding in vitro. p53 was found to be acetylated at Lys-382 and phosphorylated at Ser-33 and Ser-37 in vivo after exposure of cells to UV light or ionizing radiation. Interestingly, acetylation of p53 by p300 and PCAF was strongly inhibited by phosphopeptides corresponding to the amino terminus of p53 phosphorylated at Ser-37 and/or Ser-33, suggesting that phosphorylation in response to DNA damage may enhance the interaction of p300 and PCAF, thereby driving p53 acetylation. Studies also suggest that HDAC1-dependent deacetylation of p53 can reduce its transcriptional activity as well as its growth suppressive and death promoting actions (Luo et al., 2000).

Several viral and cellular oncogenes have been shown to stabilize p53. Viral oncogenes including SV40 T antigen, adenovirus E1A, and human papillomavirus 16 E7 stabilize p53 (Lowe and Ruley, 1993; Demers et al., 1994). This stabilization does not translate into the activation of p53, but rather a physiologically inactive p53 which is functionally inhibited. On the other hand, adenovirus has other cooperating oncogenes like E1B and E4, which bind to p53 and inhibit apoptosis and/or growth arrest thereby leading to successful viral DNA replication and cellular transformation. E1A can also inhibit transcriptional activation by p53 through interaction with CBP/p300, which may inhibit p53-mediated growth arrest and/or apoptosis (Steegenga et al., 1996; Somasundaram and El-Deiry, 1997). The E6 gene product encoded by HPV16 binds to p53, which results in degradation of p53 and suppression of negative growth signals from p53.

The mechanism of p53 stabilization in response to viral oncogene expression has not been clearly understood until more recently when p19ARF (p14ARF in humans), a product of INK4a/ARF locus translated in an alternate reading frame (Kamijo et al., 1997; Stone et al., 1995) was identified. It was found that the ability of E1A to stabilize p53 is severely compromised in p19ARF-null cells (de Stanchina et al., 1998). p19ARF is a tumor suppressor, which can induce cell-cycle arrest in a p53-dependent as well as a p53-independent manner (Weber et al., 2000).

p19ARF can physically associate with p53 itself and/or Mdm2 to alter p53 levels and activity (Kamijo et al., 1998; Pomerantz et al., 1998; Y. Zhang et al., 1998). E2F1, which was found to directly activate p14ARF transcription, linked E1A to p14ARF and then to p53 (Bates et al., 1998). Certain cellular oncogenes like c-myc and ras have also been shown to stabilize p53 in a p19ARF-dependent manner (Zindy et al., 1998; Palmero et al., 1998). Furthermore, the tumor suppressors WT1 and BRCA1 also have been shown to stabilize p53 and modulate p53-mediated transcription (H. Zhang et al., 1998; Somasundaram et al., 1999).

III. DOWNSTREAM MEDIATORS OF p53

A. Downstream molecules of p53-dependent cell cycle arrest

p53 functions in the cellular response to DNA damage thereby preventing accumulation of potentially oncogenic mutations and genomic instability (Lane, 1992). Activation of p53 leads to suppression of cell growth or apoptosis to prevent the propagation of mutated genes (Levine, 1997). p53 has also been implicated in differentiation (Aloni-Grinstein et al., 1995), senescence (Atadja et al., 1995), and inhibition of angiogenesis (Dameron et al., 1994). Although we do not yet fully understand how p53 elicits its effects on cells, it is clear that the transcriptional activation function of p53 is a major component of its biological effects (Crook et al., 1994; Pietenpol et al., 1994). Activated p53 binds to a specific DNA sequence and activates transcription. The importance of the

DNA-binding is underscored by the fact that the vast majority of p53 mutations derived from tumors map within the domain required for sequence-specific DNA-binding. p53 normally recognizes a 20 base-pair response element that has an internal symmetry. The consensus DNA sequence for p53 binding is 5′-PuPuPuC(A/T-A/T)GPyPyPy-N(0-13)-PuPuPuC(A/T-A/T)GPyPyPy-3′ with the third C and seventh G being highly conserved in the 10 base-pair half-sites (El-Deiry *et al.*, 1992). Identification of transcriptional targets of p53 has been critical in dissecting pathways by which p53 functions (reviewed in Vogelstein, *et al.*, 2000). A growing number of genes have been found to contain p53-binding sites and/or response elements and thus to have the potential to mediate the effects of p53 on cells, through upregulation of their expression and function (Table 1).

Table 1 Transcriptional Targets of p53

Target gene	Interacting molecule	Effect
Cell cycle related		
p21$^{WAF1/CIP1}$	Cyclin-cdks and PCNA	G1 and G2/M arrest
GADD45	cdc2-CyclinB and PCNA	G2/M arrest and DNA repair
14-3-3s	Phosphorylated cdc2	Inhibition of dephosphorylation of cdc 25
B99	Unknown	G2 arrest
p53R2	Unknown	G2 arrest
Apoptosis related		
Bax	Release of cytochrome *c*	Apoptosis
FAS/APO1	Activation of caspases via FADD	Apoptosis
KILLER/DR5	Activation of caspases via FADD	Apoptosis
Noxa	Unknown	Apoptosis
Ei24/PIG8	Unknown	Apoptosis
PERP	Unknown	Apoptosis and cell cycle arrest
PAG608/wig-1	Unknown	Apoptosis and cell cycle arrest?
PIG's	Oxidative stress	Apoptosis
p85	Unknown	Apoptosis
IGF-BP3	Inhibition of growth signaling	Apoptosis
MCG10	Poly(C), RNA	Apoptosis and G2/M arrest
Angiogenesis related		
TSP1	Unknown	Inhibit angiogenesis
BAI1	Unknown	Inhibit angiogenesis
GD-AiF	Unknown	Inhibit angiogenesis
Unknown function		
Cyclin G	Unknown	Apoptosis?
Wip 1	Unknown	Phosphatase for unknown substrate
HIC-1	Unknown	Cell growth inhibition?
TP53TG5	Unknown	Cell growth inhibition?
Cathepsin D	Unknown	Increase chemosensitivity, apoptosis
wig-1	Unknown	Apoptosis, cell cycle arrest?
PA26	Unknown	Growth inhibition

1. p21^{WAFI/CIP1}

Stabilization of p53 in many cell types results in arrest of progression through the cell cycle, with evidence for both a G1 and G2/M phase accumulation (Bunz *et al.*, 1998; Levine, 1997). The ability of p53 to induce a growth arrest is correlated with its ability to function as a sequence-specific transcriptional coactivator (Figure 2). In particular, G1 arrest induced by p53 has been well studied. *p21*^{WAFI/CIP1}, which encodes a cyclin-dependent kinase inhibitor (CKI) is a critical target of p53 in facilitating G1 arrest (El-Deiry *et al.*, 1993; Harper *et al.*, 1993; Waldman *et al.*, 1995). p21 binds to a number of cyclin and Cdk complexes: cyclin D1-Cdk4, cyclin E-Cdk2, cyclin A-Cdk2, cyclin A-cdc2, and cyclin B-cdc2, which results in the inhibition of their kinase activity (Harper *et al.*, 1995). This allows accumulation of hypophosphorylated Rb, which remains associated with transcription factors such as E2F, and the resultant failure to activate E2F-responsive gene results in G1 arrest. p21^{WAFI/CIP1} also binds to PCNA and blocks its role in DNA replication, at least *in vitro*. Thus, p21^{WAFI/CIP1} inhibits DNA replication by binding to cyclin–Cdk complexes and PCNA (Li *et al.*, 1994). Mice with homozygous deletion of the *p21* gene develop normally, but the mouse embryo fibroblast derived from these mice are only partially deficient in their ability to arrest cells in G1 in response to DNA damage, suggesting the existence of a p21-independent pathway that contributes to the p53-mediated G1 arrest (Deng *et al.*, 1995). However, the removal of both alleles of *p21* from a colon cancer cell line, which harbors wild-type p53, results in complete elimination of the DNA damage-induced G1 arrest in these cells (Waldman *et al.*, 1995). p53 has also been implicated in the control of a G2/M

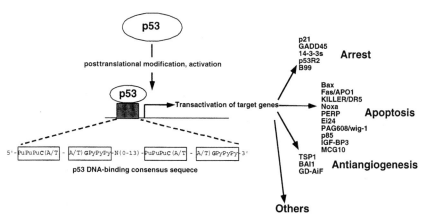

Figure 2 Transactivation of various p53 target genes through p53 DNA-binding consensus sequence.

checkpoint. Transcriptional downregulation of cyclin B1 by p53 (Agarwal *et al.*, 1995; Stewart *et al.*, 1995) may be involved in p53-mediated G2/M arrest as cyclin B1/Cdc2 complex is the major regulatory factor required for entry into mitosis (Elledge, 1996; O'Connor, 1997). More recently, the cyclin B1 promoter has shown to be downregulated by p53 (Innocente *et al.*, 1999).

2. GADD45

GADD45 (growth arrest and DNA-damage inducible gene number 45), another p53-induced gene, may be involved in G2/M arrest because overexpression of GADD45 results in a G2/M arrest, which can be attenuated by overexpression of cyclin B1 and Cdc2, and GADD45 inhibits the activity of the cyclin B1/Cdc2 complex *in vitro* possibly through disruption of the complex or by direct binding and inhibition of cyclin B/Cdc2 kinase activity (Zhan *et al.*, 1999). However, there is evidence that the induction of GADD45 can occur in a p53-independent manner (Zhan *et al.*, 1998). Studies have demonstrated that GADD45-null cells have genomic instability, but GADD45-null mice only developed tumors if exposed to γ-radiation. These studies have established a role for GADD45 in maintaining genomic integrity in the p53 response (Hollander *et al.*, 1999).

3. 14-3-3σ

14-3-3σ, another IR-inducible p53-target gene, may also be involved in G2/M arrest (Hermeking *et al.*, 1997). It appears to act by binding and sequestering phosphorylated Cdc25c in the cytoplasm, thereby preventing its phosphorylation and activation of cyclin B/Cdc2 (Peng *et al.*, 1997). However, in 14-3-3σ-null cells, it is clear that Cdc25c can be sequestered in the cytoplasm by other 14-3-3 proteins following DNA damage, whereas Cdc2 localization may be critically affected (Chan *et al.*, 1999).

4. B99

By using a differential subtractive hybridization approach in a murine cell line stably transfected with a temperature-sensitive p53 mutant (Val135), this gene was isolated as a wild-type p53-inducible gene which may play a role in the microtubule network in the G2 phase of cell cycle. Ectopic overexpression of B99 in p53-null fibroblasts leads to G2 arrest (Utrera *et al.*, 1998).

B. Downstream molecules of p53-dependent apoptosis

p53 plays a major role in triggering apoptosis under many different physiological conditions. Unlike the cell cycle arrest function of p53, a dependence of apoptosis on the transcriptional activity of p53 is not entirely clear. Evidence suggests that apoptosis could be induced by p53 using both transcription-dependent and -independent mechanisms (Figure 1). For example, apoptosis has been shown to be induced by p53 in the presence of actinomycin D or cycloheximide, which block RNA or protein synthesis, respectively (Caelles *et al.*, 1994; Wagner *et al.*, 1994). Similarly, p53 mutants which are deficient in transcriptional activation have been shown to activate apoptosis in Hela cells (Haupt *et al.*, 1995). However, in another report, one of the mutants used was found to be defective for inducing apoptosis in baby rat kidney cells (Sabbatini *et al.*, 1995). Hence, p53 may induce apoptosis by both transcription-dependent and -independent mechanisms, and cell type might be an important factor in determining the mechanism of apoptosis induction.

1. Bax

Bax is a p53-target gene whose expression is upregulated through a p53 DNA-binding response element located within the *Bax* promoter (Miyashita and Reed, 1995). The *Bax* gene encodes a 21 kDa protein proapoptotic member of the Bcl-2 family (Oltvai *et al.*, 1993). Bax signaling involves steps that involve translocation of Bax to the mitochondria and release of cytochrome *c* from the mitochondria, which triggers activation of caspase 9 in concert with Apaf1 and the downstream caspase cascade (Goping *et al.*, 1998; Eskes *et al.*, 1998). *Bax* knockout thymocytes are not defective in undergoing apoptosis following DNA damage (Knudson *et al.*, 1995). However, neuronal cell apoptosis is reduced in Bax-null cells (Chong *et al.*, 2000) and in mutant T antigen expressing choroid plexus (Yin *et al.*, 1997).

2. Fas/APO1

Fas/APO1 is a potent inducer of apoptosis in hematopoietic or liver cells exposed to Fas ligand (Nagata, 1997). On Fas ligand binding to the Fas receptor, the Fas receptor trimerizes. The death domain of Fas/APO1 recruits the FADD (Fas-associated death domain) adaptor which recruits initiator caspase 8 to the DISC (death-inducing signaling complex), resulting in the activation of the caspase cascade. Fas/APO1 is not absolutely required for p53 to induce apoptosis because cells which are deficient for Fas/APO1 are proficient in inducing apoptosis on p53 activation (Fuchs *et al.*, 1997).

3. KILLER/DR5

KILLER/DR5 is a death-domain containing proapoptotic member of a more recently discovered family of TRAIL [tumor necrosis factor (TNF)-related apoptosis inducing ligand] receptors (Ashkenazi and Dixit, 1998). Expression of KILLER/DR5 appears to be increased following exposure of wild-type p53-expressing cells to cytotoxic DNA damaging agents such as γ-radiation, doxorubicin, or etoposide (Wu et al., 1997). Indeed, a p53 responsive element was found within intron 1 in the KILLER/DR5 genomic locus (Takimoto and El-Diery, 2000). Like the Fas/APO1 receptor, signaling through proapoptotic TRAIL receptors involves downstream caspase activation (Pan et al., 1997; Sheridan et al., 1997).

4. Noxa

Noxa was identified by using mRNA differential display to determine how expression profiles differed between the X-ray-irradiated wild-type and IRF-1/p53 doubly deficient mouse embryonic fibroblasts (MEFs) (E. Oda et al., 2000). Induction of Noxa in irradiated primary mouse cells was dependent on p53. Noxa encodes a Bcl-2 homology 3 (BH3)-only member of the Bcl-2 family of proteins; this member contains the BH3 region but not other BH domains. When ectopically expressed, Noxa underwent BH3 motif-dependent localization to the mitochondria and interacted with antiapoptotic Bcl-2 family members, resulting in the activation of caspase-9. Inhibition of the endogenous Noxa induction results in the suppression of apoptosis. Noxa may thus represent a mediator of p53-dependent apoptosis (E. Oda et al., 2000).

5. Ei24 (PIG8)

Differential display led to the isolation of genes whose expression was upregulated in NIH3T3 cells exposed to the cytotoxic chemotherapeutic agent etoposide (Lehar et al., 1996). Ei24/PIG8 was found to be induced by ionizing irradiation or following overexpression of wild-type p53 as well as exposure to etoposide. Ectopic Ei24/PIG8 expression was found to markedly inhibit cell colony formation, to induce the morphological features of apoptosis, and to reduce the number of β-galactosidase-marked cells, which is efficiently blocked by coexpression of Bcl-X(L). The Ei24/PIG8 gene is localized on human chromosome 11q23, a region frequently altered in human cancers (Gu et al., 2000). Ei24 may play an important role in negative cell growth control by functioning as an apoptotic effector of p53 tumor suppressor activities.

6. PERP

Using a subtractive cloning strategy, PERP (p53 apoptosis effector related to PMP-22) was found to be expressed in a p53-dependent manner and at high levels in apoptotic cells compared with G1-arrested cells (Attardi et al., 2000). PERP induction has been linked to p53-dependent apoptosis, including in the response to E2F-1-driven hyperproliferation. PERP shows sequence similarity to the PMP-22/gas3 tetraspan membrane protein implicated in hereditary human neuropathies such as Charcot-Marie-Tooth. The mechanism of how PERP induces apoptosis in a p53-dependent manner is still unclear.

7. p53AIP1

Recently *p53AIP* (p53-regulated apoptosis-inducing protein 1), a novel apoptosis inducing gene, was isolated through direct cloning of human genomic DNA. Overexpression of this gene leads to apoptotic cell death accompanied with its mictochondrial translocalization. Interestingly, induction of this gene was only observed when phosphorylation of p53 occurred at Ser46 on severe DNA damage (K. Oda *et al.*, 2000).

8. Other p53 target genes for apoptosis

Other p53 target genes, which are believed to be involved in inducing apoptosis are PAG608/Wig-1 (Israeli *et al.*, 1997; Varmeh-Ziaie *et al.*, 1997), PIG-s (p53-induced genes) (Polyak *et al.*, 1997), p85 (Yin *et al.*, 1998), IGF-BP3 (Buckbinder *et al.*, 1995; Rajah *et al.*, 1997), and more recently identified MCG10 (Zhu and Chen, 2000). Induction of apoptosis by PIGs involves formation of reactive oxygen species (Polyak *et al.*, 1997). IGF-BP3 (insulin-like growth factor-binding protein 3) may induce apoptosis by inhibiting growth factor-associated signaling or through other mechanisms (Buckbinder *et al.*, 1995; Grimberg, 2000). Repression of apoptotic protectors such as Bcl-2 may also contribute to apoptosis induction by p53 (Miyashita *et al.*, 1994).

C. p53-Dependent inhibition of angiogenesis

Overexpression of p53 has been found to inhibit angiogenesis possibly through the upregulation of Tsp1 (Dameron *et al.*, 1994), BAI1 (Nishimori *et al.*, 1997), and/or GD-AiF (Van Meir *et al.*, 1994).

1. *Tsp1*

Early passage Li-Fraumeni cells which carry one wild-type p53 allele were found to secrete large amounts of a potent angiogenic inhibitor, TSP1 (Dameron *et al.*, 1994). However, the late passage cells underwent an angiogenic switch associated with their immortalization, loss or mutation of the wild-type p53 allele, and reduced expression of TSP1 (Dameron *et al.*, 1994). It was also shown that the promoter of TSP1 is regulated positively by p53 (Dameron *et al.*, 1994).

2. BAI1

A novel p53-inducible gene that encodes a 1584-amino acid product containing five thrombospondin type 1 (TSP-type 1) repeats was isolated, and was found to be specifically expressed in the brain (Nishimori *et al.*, 1997). Overexpression of this gene, designated *BAI1* (brain-specific angiogenesis inhibitor 1), inhibited *in vivo* neovascularization induced by bFGF in the rat cornea.

3. GD-AiF

Introducing a tetracycline-regulated wild-type p53 into p53-null glioblastoma cells, GD-AiF was purified as a factor that can neutralize the angiogenicity by the parental cells as well as of basic fibroblast growth factor. It remains unclear how GD-AiF functions to inhibit angiogenesis (Van Meir *et al.*, 1994).

D. Other p53 target genes of unknown function

1. Cyclin G

Cyclin G was previously identified as a target gene of the p53 tumor suppresser protein, and levels of cyclin G are increased after induction of p53 by DNA damage (Okamoto and Beach, 1994). Although the function of cyclin G has not been well established, it is clear that cyclin G plays a facilitating role in modulating apoptosis induced by different stimuli (Okamoto and Prives, 1999).

2. Wip 1

Wip 1 was isolated as a wild-type p53-inducible gene using differential display (Fiscella *et al.*, 1997). Wip 1 has high homology to the type 2C protein phosphatases and appears to accumulate following DNA damage in a wild-type p53-dependent manner. Similar to waf1, ectopic expression of Wip 1 in human cells suppressed colony formation. However, the mechanism of Wip 1 induction, growth inhibition following Wip 1 overexpression, and the cellular substrates for phosphatase activity remain unclear.

3. HIC-1

A novel zinc-finger transcription factor gene, *HIC-1* (Hypermethylated in cancer 1), which is ubiquitously expressed in normal tissues but underexpressed in different tumor cells where it is hypermethylated, was isolated (Wales *et al.*, 1995). Multiple characteristics of this gene, including the presence of a p53 binding site in the 5' flanking region, activation of the gene by expression of a wild-type *p53* gene, and suppression of G418 selectability of cultured brain, breast, and colon cancer cells following insertion of the gene, make HIC-1 gene a candidate for a tumor suppressor gene in the region 17p13.3 (Wales *et al.*, 1995). However, its precise role is still unclear.

4. *TP53TG5*

Through a strategy of direct cloning of TP53-binding DNA sequences from human genomic DNA, a novel TP53-target gene, termed *TP53TG5* (TP53-target gene 5), was isolated (Isaka *et al.*, 2000). This gene, which was localized to chromosome band 20q13.1, encodes a 290-amino acid peptide with no significant homology with any other proteins. A colony-formation assay using the human glioblastoma cell line T98G, which lacks wild-type TP53 and expresses no endogenous TP53TG5, revealed a growth-suppressive effect of the TP53TG5 gene product. Furthermore, immunohistochemical studies, following transfection of T98G with plasmid designed to express green fluorescent protein-fused TP53TG5, revealed cell-cycle-dependent intracellular localization of this protein (Isaka *et al.*, 2000). However, the function of TP53TG5 downstream of p53 remains unclear.

5. *Cathepsin D*

Cathepsin D (*CD*) was isolated as a p53-inducible gene by subtractive hybridization technique (Wu *et al.*, 1998). CD, the major intracellular aspartyl protease, has been previously implicated as a mediator of interferon (IFN)-γ- and TNF-α-induced apoptosis. Overexpression of CD inhibited growth of colon, liver, and ovarian cancer cells (Wu *et al.*, 1998). CD protein expression was increased by exposure of ML1 cells to various DNA damaging agents such as etoposide, adriamycin, or γ-radiation. Inhibition of CD protease activity with Pepstatin A suppressed p53-dependent apoptosis in lymphoid cells, suggesting a possible role for CD in p53-dependent cell death. $CD^{-/-}$ fibroblasts were found to be more resistant to killing by adriamycin and etoposide, as compared to $CD^{+/+}$ cells. CD protease may provide a link to p53-dependent tumor suppression and metastasis suppression.

6. GML

GML (GPI-anchored molecule-like protein) was isolated as a new target of p53, and its expression correlates with p53 status in esophageal cancer cells (Furuhata *et al.*, 1996). As the predicted amino acid sequence showed a high degree of homology to the family of glycosylphosphatidylinositol (GPI)-anchored membrane proteins, this gene was designated as *GML* (GPI-anchored molecule-like protein). Introduction of GML cDNA suppressed the growth of esophageal cancer cells in culture. A correlation between the presence of GML expression and the sensitivity of esophageal cancer cells to anticancer drugs implied that the gene product may play a significant role in the apoptotic pathway or cell-cycle regulation induced by p53 after DNA damage.

7. *p53R2*

A recently identified DNA damage-inducible p53-target gene is *p53R2*, which was isolated by using differential display. P53R2 has striking similarity to the ribonucleotide reductase small subunit which is important in DNA synthesis during cell division. Induction of p53R2 in p53-deficient cells caused G2/M arrest and prevented cells from death in response to adriamycin. Inhibition of endogenous p53R2 expression in cells that have an intact p53-dependent DNA damage checkpoint reduced ribonucleotide reductase activity, DNA repair, and cell survival after exposure to various genotoxins (Tanaka *et al.*, 2000).

8. *PA26*

PA26 has been found to be induced by genotoxic stress such as UV, γ-irradiation, and cytotoxic drugs in a p53-dependent manner. This gene is supposed to be a member of the GADD family, but the precise function of PA26 remains unclear (Velasco-Miguel *et al.*, 1999)

Acknowledgment

W. S. E.-D is an Assistant Investigator of the Howard Hughes Medical Institute.

References

Agarwal, M. L., Agarwal, A., Taylor, W. R., and Stark, G. R. (1995). p53 controls both the G2/M and the G1 cell cycle check-point and mediates reversible growth arrest in human fibroblasts. *Proc. Natl. Acad. Sci. U.S.A.* **92,** 8493–8497.

Aloni-Grinstein, R., Schwartz, D., and Rotter, V. (1995). Accumulation of wild-type p53 protein upon γ-irradiation induces a G2 arrest-dependent immunoglobulin κ light chain expression. *EMBO J.* **14,** 1392–1401.

Ashcroft, M., Kubbutat, M. H., and Vousden, K. H. (1999). Regulation of p53 function and stability by phosphorylation. *Mol. Cell. Biol.* **19,** 1751–1758.

Ashkenazi, A., and Dixit, V. M. (1998). Death receptors: Signaling and modulation. *Science* **281,** 1305–1308.

Atadja, P., Wong, H., Garkavtsev, I., Veillette, C., and Riabowol, K. (1995). Increased activity of p53 in senescing fibroblasts. *Proc. Natl. Acad. Sci. U.S.A.* **92,** 8348–8352.

Attardi, L. D., Reczek, E. E., Cosmas, C., Demicco, E. G., McCurrach, M. E., Lowe, S. W., and Jacks, T. (2000). PERP, an apoptosis-associated target of p53, is a novel member of the PMP-22/gas3 family. *Genes Dev.* **14,** 704–718.

Avantaggiati, M. L., Ogryzko, V., Gardner, K., Giordano, A. A., Levine, A. S., and Kelly, K. (1997). Recruitment of p300/CBP in p53-dependent signal pathways. *Cell* **89,** 1175–1184.

Bannister, A. J., and Kouzarides, T. (1996). The CBP co-activator is a histone acetyl transferase. *Nature* **384,** 641–643.

Baskaran R., Wood, L. D., Whitaker, L. L., Canman, C. E., Morgan, S. E., Xu, Y., Barlow, C., Baltimore, D., Wynshaw-Boris, A., Katan, M. B., and Wang, J. Y. (1997). Ataxia telangiectasia mutant protein activates c-Abl tyrosine kinase in response to ionizing radiation. *Nature* **387,** 516–519.

Bates, S., Phillips, A. C., Clark, P. A., Stott, F., Peters, G., Ludwig, R. L., and Vousden, K. H. (1998). p14ARF links the tumor suppressors RB and p53. *Nature* **395,** 124–125.

Buckbinder, L., Talbott, R., Velascomiguel, S., Takenaka, I., Faha, B., Seizinger, B. R., and Kley, N. (1995). Induction of the growth inhibitor IGF-binding protein-3 by p53. *Nature* **377,** 646–649.

Bunz, F., Dutriaux, A., Lengauer, C., Waldman, T., Zhou, S., Brown, J. P., Sedivy, J. M., Kinzler, K. W., and Vogelstein, B. (1998). Requirement of p53 and p21 to sustain G2 arrest after DNA damage. *Science* **282,** 1497–1501.

Caelles, C., Heimberg, A., and Karin, M. (1994). p53-dependent apoptosis in the absence of p53-target genes. *Nature* **370,** 220–223.

Canman, C. E., Lim, D. S., Cimprich, K. A., Taya, Y., Tamai, K., Sakaguchi, K., Appella, E., Kastan, M. B., and Siliciano, D. D. (1998). Activation of the ATM kinase by ionizing radiation and phosphorylation of p53. *Science* **281,** 1677–1679.

Chan, T. A., Hermeking, H., Lengauer, C., Kinzler, K. W., and Vogelstein, B. (1999). 14-3-3σ is required to prevent mitotic catastrophe after DNA damage. *Nature* **401,** 616–620.

Chong, M. J., Murray, M. R., Gosink, E. C., Russell, H. R., Srinivasan, A., Kapsetaki, M., Korsmeyer, S. J., and McKinnon, P. J. (2000). Atm and Bax cooperate in ionizing radiation-induced apoptosis in the central nervous system. *Proc. Natl. Acad. Sci. U.S.A.* **97,** 889–894.

Chowdary, D. R., Dermody, J. J., Jha, K. K., and Ozer, H. L. (1994). Accumulation of p53 in a mutant cell line defective in the ubiquitin pathway. *Mol. Cell. Biol.* **14,** 1997–2003.

Crook, T., Marston, N. J., Sara, E. A., and Vousden, K. H. (1994). Transcriptional activation by p53 correlates with suppression of growth but not transformation. *Cell* **79,** 817–827.

Dameron, K. M., Volpert, O. V., Tainsky, M. A., and Bouck, N. (1994). Control of angiogenesis in fibroblasts by p53 regulation of angiogenesis of thrombospondin-1. *Science* **265,** 1582–1584.

Demers, G. W., Halbert, C. L., and Galloway, D. A. (1994). Elevated wild-type p53 protein levels in human epithelial cell lines immortalized by human papillomavirus type 16 E7 gene. *Virology* **198,** 169–174.

Deng, C., Zhang, P., Harper, J. W., Elledge, S. J., and Leder, P. (1995). Mice lacking p21CIP1/WAF1 undergo normal development, but are defective in G1 checkpoint control. *Cell* **82,** 675–784.

de Stanchina, E., McCurrach, M. E., Zindy, F., Shieh, S.-Y., Ferbeyre, G., Samuelson, A. V., Prives, C., Roussel, M. F., Sherr, C. H., and Lowe, S. W. (1998). E1A signaling to p53 involves the p19ARF tumor suppressor. *Genes Dev.* **12,** 2434–2442.

Donehower, L. A., and Bradley, A. (1993). The tumor suppressor p53. *Biochem. Biophys. Acta* **1155,** 181–205.

El-Deiry, W. S. (1998). Regulation of p53 downstream genes. *Semin. Cancer Biol.* **8,** 345–357.

El-Deiry, W. S., Kern, S. E., Pientenpol, J. A., Kinzler, K. W., and Vogelstein, B. (1992). Definition of a consensus binding site for p53. *Nat. Genet.* **1,** 45–49.

El-Deiry, W. S., Tokino, T., Velculescu, V. E., Levy, D. B., Parsons, R., Trent, J. M., Lin, D., Mercer, E. W., Kinzler, K. W., and Vogelstein, B. (1993). WAF1, a potential mediator of p53 tumor suppression. *Cell* **75,** 817–825.

Elledge, S. J. (1996). Cell cycle checkpoints: Preventing an identity crisis. *Science* **274,** 1664–1672.

Eskes, R., Antonsson, B., Osen-sand, A., Montessait, S., Richter, C., Sadoul, R., Mazzei, G., Nichols, A., and Martinou, J. C. (1998). Bax-induced cytochrome C release from mitochondria is independent of the permeability transition pore but highly dependent on Mg^{2+} ions. *J. Cell Biol.* **143**, 217–224.

Fiscella, M., Zhang, H., Fan, S., Sakaguchi, K., Shen, S., Mercer, W. E., Vande Woude, G. F., O'Connor, P. M., and Appella, E. (1997). Wip1, a novel human protein phosphatase that is induced in response to ionizing radiation in a p53-dependent manner. *Proc. Natl. Acad. Sci. U.S.A.* **94**, 6048–6053.

Fuchs, E. J., McKenna, K. A., and Bedi, A. (1997). p53-dependent DNA-damage induced apoptosis requires Fas/APO-1-independent activation of CPP32beta. *Cancer Res.* **57**, 2550–2554.

Furuhata, T., Tokino, T., Urano, T., and Nakamura, Y. (1996). Isolation of a novel GPI-anchored gene specifically regulated by p53; correlation between its expression and anti-cancer drug sensitivity. *Oncogene* **13**, 1965–1970.

Goga, A., Liu, X., Hambuch, T. M., Senechal, K., Major, E., Berk, A. J., Witte, O. N., and Sawyers, C. L. (1995). p53 dependent growth suppression by the c-Abl nuclear kinase. *Oncogene* **11**, 791–799.

Goping, I. S., Gross, A., Lavoie, J. N., Nguyen, M., Jemmerson, R., Roth, K., Korsmeyer, S. J., and Shore, G. C. (1998). Regulated targeting of BAX to mitochondria. *J. Cell Biol.* **143**, 207–215.

Graeber, T. G., Peterson, J. F., Tsai, M., Monica, K., Fornace, A. J., Jr., and Giaccia, A. J. (1994). Hypoxia induces accumulation of p53 protein, but activation of a G1-phase checkpoint by low-oxygen conditions is independent of p53 status. *Mol. Cell. Biol.* **14**, 6264–6277.

Grimberg, A. (2000). P53 and IGFBP-3: Apoptosis and cancer protection. *Mol. Genet. Metab.* **70**, 85–98.

Gu, W., and Roeder, R. G. (1997). Activation of p53 sequence-specific DNA binding by acetylation of the p53 C-terminal domain. *Cell* **90**, 595–606.

Gu, W., Shi, X. L., and Roeder, R. G. (1997). Synergistic activation of transcription by CBP and p53. *Nature* **387**, 819–823.

Gu, Z., Flemington, C., Chittenden, T., and Zambetti, G. P. (2000). ei24, a p53 response gene involved in growth suppression and apoptosis. *Mol. Cell. Biol.* **20**, 233–241.

Harper, J. W., Adami, G. R., Wei, N., Keymarsi, K. M., and Elledge, S. (1993). The p21 Cdk interacting protein Cip1 is a potent inhibitor of G1 cyclin-dependent kinases. *Cell* **75**, 805–816.

Harper, J. W., Elledge, S. J., Keyomarsi, K., Dynlacht, B., Tsai, L.-H., Zhang, P., Dobrowolski, S., Bai, C., Connell-Croley, L., Swindell, E., Fox, M. P., and Wei, N. (1995). Inhibition of cyclin-dependent kinases by p21. *Mol. Biol. Cell* **6**, 387–400.

Haupt, Y., Rowen, S., Shaulian, E., Vousden, K., and Oren, M. (1995). Induction of apoptosis in HeLa cells by transactivation-deficient p53. *Genes Dev.* **9**, 2170–2183.

Haupt, Y., Maya, R., Kaza, A., and Oren, M. (1997). Mdm2 promotes the rapid degradation of p53. *Nature* **387**, 296–299.

Hermeking, H., Lengauer, C., Polyak, K., He, T. C., Zhang, L., Thiagalingam, S., Kinzler, K. W., and Vogelstein, B. (1997). 14-3-3 sigma is a p53-regulated inhibitor of G2/M progression. *Mol. Cell* **1**, 3–11.

Hirao, A., Kong, Y. Y., Matsuoka, S., Wakeham, A., Ruland, J., Yoshida, H., Liu, D., Elledge, S. J., and Mak, T. W. (2000). DNA damage-induced activation of p53 by the checkpoint kinase Chk2. *Science* **287**, 1824–1827.

Hollander, M. C., Sheikh, M. S., Bulavin, D. V., Lundgren, K., Augeri-Henmueller, L., Shehee, R., Molinaro, T. A., Kim, K. E., Tolosa, E., Ashwell, J. D., Rosenberg, M. P., Zhan, Q., Fernandez-Salguero, P. M., Morgan, W. F., Deng, C. X., and Fornace, A. J., Jr. (1999). Genomic instability in Gadd45a-deficient mice. *Nat. Genet.* **23**, 176–184.

Hupp, T. R., Sparks, A., and Lane, D. P. (1995). Small peptides activate the latent sequence-specific DNA binding function of p53. *Cell* **83**, 237–245.

Innocente, S. A., Abrahamson, J. L. A., Cogswell, J. P., and Lee, J. M. (1999). p53 regulates a G2 checkpoint through cyclin B1. *Proc. Natl. Acad. Sci. U.S.A.* **96**, 2147–2152.

Isaka, S., Takei, Y., Tokino, T., Koyama, K., Miyoshi, Y., Suzuki, M., Takahashi, E., Azuma, C., Murata, Y., and Nakamura, Y. (2000). Isolation and characterization of a novel TP53-inducible gene, TP53TG5, which suppresses growth and shows cell cycle-dependent transition of expression. *Genes Chromosomes Cancer* **27**, 345–352.

Israeli, D., Tessler, E., Haupt, Y., Elkeles, A., Wilder, S., Amson, R., Telerman, A., and Oren, M. (1997). A novel p53-inducible gene, PAG608, encodes a nuclear zinc finger protein whose overexpression promotes apoptosis. *EMBO J.* **16**, 4384–4392.

Jamal, S., and Ziff, E. B. (1995). Raf phosphorylates p53 *in vitro* and potentiates p53-dependent transcriptional activation *in vivo*. *Oncogene* **10**, 2095–2101.

Jayaraman, L., and Prives, C. (1995). Activation of p53 sequence-specific DNA-binding by short single strands of DNA requires the p53 C-terminus. *Cell* **81**, 1021–1029.

Kamijo, T., Zindy, F., Roussel, M. F., Quelle, D. E., Downing, J. R., Ashmun, R. A., Grosveld, G., and Sherr, C. J. (1997). Tumor suppressor at the mouse INK4a locus mediated by the alternative reading frame product p19ARF. *Cell* **91**, 649–659.

Kamijo, T., Weber, J. D., Zambetti, G., Zindy, F., Roussel, M. F., and Sherr, C. J. (1998). Functional and physical interactions of the ARF tumor suppressor with p53 and Mdm2. *Proc. Natl. Acad. Sci. U.S.A.* **95**, 8292–8297.

Kastan, M. B., Onyekwere, O., Sidransky, D., Vogelstein, B., and Craig, R. W. (1991). Participation of p53 in the cellular response to DNA damage. *Cancer Res.* **51**, 6304–6311.

Khanna, K. K., Keating, K. E., Kozlov, S., Scott, S., Gatei, M., Hobson, K., Taya., Y., Gabrielli, B., Chan, D., Lees-Miller S. P., and Lavin, M. F. (1998). ATM associates with and phosphorylates p53: Mapping the region of interaction. *Nat. Genet.* **20**, 398–400.

Kharbanda, S., Ren, R., Pandey, P., Shafman, T. D., Feller, S. M., Weichselbaum, R. R., and Kufe, D. W. (1995). Activation of the c-Abl tyrosine kinase in the stress response to DNA-damaging agents. *Nature* **376**, 785–788.

Ko, L. J., and Prives, C. (1996). p53: puzzle and paradigm. *Genes Dev.* **10**, 1054–1072.

Knudson, C. M., Tung, K. S., Tourtellotte, W. G., Brown, G. A., and Korsmeyer, S. J. (1995). Bax-deficient mice with lymphoid hyperplasia and male germ cell death. *Science* **270**, 96–99.

Kubbutat, M. H. G., Jones, S. N., and Vousden, K. (1997). Regulation of p53 stability by MDM2. *Nature* **387**, 299–303.

Lane, D. P. (1992). p53, guardian of the genome. *Nature* **358**, 15–16.

Lee, S., Elenbaas, B., Levine, A., and Griffith, J (1995). p53 and its 14 kDa C-terminal domain recognize primary DNA damage in the form of insertion/deletion mismatches. *Cell* **81**, 1013–1020.

Lehar, S. M., Nacht, M., Jacks, T., Vater, C. A., Chittenden, T., and Guild, B. C. (1996). Identification and cloning of EI24, a gene induced by p53 in etoposide-treated cells. *Oncogene* **12**, 1181–1187.

Levine, A. J. (1997). p53, the cellular gatekeeper for growth and division. *Cell* **88**, 323–331.

Li, R., Waga, S., Hannon, G. J., Beach, D., and Stillman, B. (1994). Differential effects by the p21 CDK inhibitor on PCNA-dependent DNA replication and repair. *Nature* **371**, 534–537.

Lill, N. L., Grossman, S. R., Ginsberg, S. R., DeCaprio, J., and Livingston, D. M. (1997). Binding and modulation of p53 by p300/CBP coactivators. *Nature* **387**, 823–827.

Lowe, S., and Ruley, H. E. (1993). Stabilization of the p53 tumor suppressor is induced by adenovirus 5 E1A and accompanies apoptosis. *Genes Dev.* **7**, 535–545.

Luo, J., Su, F., Chen, D., Shiloh, A., and Gu, W. (2000). Deacetylation of p53 modulates its effect on cell growth and apoptosis. *Nature* **408**, 377–381.

Milne, D. M., Campbell, D. G., Caudwell, F. B., and Meek, D. W. (1994). Phosphorylation of the tumor suppressor protein p53 by mitogen-activated protein kinases. *J. Biol. Chem.* **269**, 9253–9260.

Milne, D. M., Campbell, D. G., Caudwell, F. B., and Meek, D. W. (1995). p53 is phosphorylated *in vitro* and *in-vivo* by an ultra-violet radiation-induced protein kinase characteristic of the c-jun kinase. *J. Biol. Chem.* **270,** 5511–5518.

Miyashita, T., and Reed, J. C. (1995). Tumor suppressor p53 is a direct transcriptional activator of the human bax gene. *Cell* **80,** 293–299.

Miyashita, T., Harigai, M., Hanada, M., and Reed, J. C. (1994). Identification of a p53-dependent negative response element in the Bcl-2 gene. *Cancer Res.* **54,** 3131–3135.

Mosner, J. T., Mummenbrauer, T., Bauer, C., Sczakiel, G., Grossee, F., and Deppert, W. (1995). Negative feedback regulation of wild-type p53 biosynthesis. *EMBO J.* **15,** 4442–4449.

Nagata, S. (1997). Apoptosis by death factor. *Cell* **88,** 355–365.

Nishimori, H., Shiratsuchi, T., Urano, T., Kimura, K., Kiyona, K., Tatsumi, K., Yoshida, S., Ono, M., Kuwano, M., Nakamura, Y., and Tokino, T. (1997). A novel brain-specific p53-target gene, BAI1, containing thrombospondin type I repeats inhibits experimental angiogenesis. *Oncogene* **15,** 2145–2150.

Oda, E., Ohki, R., Murasawa, H., Nemoto, J., Shibue, T., Yamashita, T., Tokino, T., Taniguchi, T., and Tanaka, N. (2000). Noxa, a BH3-only member of the Bcl-2 family and candidate mediator of p53-induced apoptosis. *Science* **288,** 1053–1058.

Oda, K., Arakawa, H., Tanaka, T., Matsuda, K., Tanikawa, C., Mori, T., Nishimori, H., Tamai, K., Tokino, T., Nakamura, Y., and Taya, Y. (2000). p53AIP1, a potential mediator of p53-dependent apoptosis, and its regulation by Ser-46-phosphorylated p53. *Cell* **102,** 849–862.

O'Connor, P. M. (1997). Mammalian G1 and G2 phase checkpoints. *Cancer Surv.* **29,** 151–182.

Ogryzko, V. V., Schiltz, R. L., Russanova, V., Howard, B. H., Evans, R., and Nakatani, Y. (1996). The transcriptional coactivators p300 and CBP are histone acetyl transferases. *Cell* **87,** 953–959.

Okamoto, K., and Beach, D. (1994). Cyclin G is a transcriptional target of the p53 tumor suppressor protein. *EMBO J.* **13,** 4816–4822.

Okamoto, K., and Prives, C. (1999). A role of cyclin G in the process of apoptosis. *Oncogene* **18,** 4606–4615.

Oliner, J. D., Pientenpol, J. A., Thiagalingam, S., Gyuris, J., Kinzler, K. W., and Vogelstein, B. (1993). Oncoprotien MDM2 conceals the activation domain of tumor suppressor p53. *Nature* **362,** 857–860.

Oltvai, Z. N., Milliman, C. L., and Korsmeyer, S. J. (1993). Bcl-2 heterodimerizes *in vivo* with a conserved homolog, Bax, that accelerates programmed cell death. *Cell* **74,** 609–619.

Palmero, I., Pantoja, C., and Serrano, M. (1998). p19ARF links the tumor suppressor p53 and Ras. *Nature* **395,** 125–126.

Pan, G., Ni, J., Wei, Y.-F., Yu, G.-I., Gentz, R., and Dixit, V. M. (1997). An antagonist decoy receptor and a death domain-containing receptor for TRAIL. *Science* **277,** 815–818.

Peng, C. Y., Graves, P. R., Thoma, R. S., Wu, Z., Shaw, A. S., and Piwnica-Worms, H. (1997). Mitotic and G2 checkpoint control: Regulation of 14-3-3σ protein binding by phosphorylation of cdc25C on serine-216. *Science* **277,** 1501–1505.

Pietenpol, J. A., Tokino, T., Thiagalingam, S., El-Deiry, W. S., Kinzler, K. W., and Vogelstein, B. (1994). Sequence-specific transcriptional activation is essential for growth suppression by p53. *Proc. Natl. Acad. Sci. U.S.A.* **91,** 1998–2002.

Polyak, K., Xia, Y., Zwier, J. K., Kinzler, K. W., and Vogelstein, B. (1997). A model for p53-induced apoptosis. *Nature* **389,** 300–305.

Pomerantz, J., Schreiber-Agus, N., Liegeois, N. J., Silverman, A., Allan, L., Chin, L., Potes, J., Chen, K. Orlow, I., Lee, H.-W., Cordon-Cardo, C., and DePinho, R. A. (1998). The Ink4a tumor suppressor gene product, p19ARF, interacts with MDM2 and neutralizes MDM2's inhibition of p53. *Cell* **92,** 713–723.

Rajah, R., Valentinis, B., and Cohen, P. (1997). Insulin like growth factor (IGF)-binding protein-3 induces apoptosis and mediates the effect of transforming growth factor-beta 1 on programmed cell death through a p53- and IGF-independent mechanism. *J. Biol. Chem.* **272**, 12181–12188.

Sabbatini, P., Lin, J., Levine, A. J., and White, E. (1995). Essential role for p53-mediated transcription in E1A-induced apoptosis. *Genes Dev.* **9**, 2184–2192.

Sakaguchi, K., Herrera, J. E., Saito, S., Miki, T., Bustin, M., Vassilev, A., Anderson, C. W., and Appella, E. (1998). DNA damage activates p53 through a phosphorylation–acetylation cascade. *Genes Dev.* **12**, 2831–2841.

Shafman, T., Khanna, K. K., Kedar, P., Spring, K., Kozlov, S., Yen, T., Hobson, K., Gatei, K., Zhang, N., Watters, D., Egerton, M., Shiloh, Y., Kharbanda, S., Kufe D., and Lavin, M. F. (1997). Interaction between ATM protein and c-Abl in response to DNA damage. *Nature* **387**, 520–523.

Sheridan, J. P., Marsters, S. A., Pitti, R. M., Gurney, A., Skubatch, M., Baldwin, D., Ramakrishnan, L., Gray, C. L., Baker, K., Wood, W. I., Goddard, A. D., Godowski, P., and Ashkenazi, A. (1997). Control of TRAIL-induced apoptosis by a family of signaling and decoy receptors. *Science* **277**, 818–821.

Shieh, S. Y., Ikeda, M., Taya, Y., and Prives, C. (1997). DNA damage-induced phosphorylation of p53 alleviates inhibition by MDM2. *Cell* **91**, 325–334.

Somasundaram, K., and El-Deiry W. S. (1997). Inhibition of p53-mediated transactivation and cell cycle arrest by E1A through its p300/CBP-interacting region. *Oncogene* **14**, 1047–1057.

Somasundaram, K., MacLachlan, T. K., Burns, T. F., Sgagias, M., Cowan, K. H., Weber, B. L., and El-Deiry, W. S. (1999). BRCA1 signals ARF-dependent stabilization and coactivation of p53. *Oncogene* **18**, 6605–6614.

Steegenga, W. T., van Laar, T., Riteco, N., Mandarino, A., Shvarts, A., van der Eb, A., and Jochensen, A. G. (1996). Adenovirus E1A proteins inhibit activation of transcription by p53. *Mol. Cell Biol.* **16**, 2101–2109.

Stewart, N., Hicks, G. G., Parakevas, F., and Mowat, M. (1995). Evidence for a second cell cycle block at G2/M by p53. *Oncogene* **10**, 109–115.

Stone, S., Jiang, P., Dayanath, P., Tavtgian, S. V., Katcher, H., Parry, D., Peters, G., and Kamb, A. (1995). Complex structure and regulation of the p16 (MTS1) locus. *Cancer Res.* **55**, 2988–2994.

Takimoto, R., and El-Deiry, W. (2000). Wild-type p53 transactivates the KILLER/DR5 gene through an intronic sequence-specific DNA-binding site. *Oncogene* **19**, 1735–1743.

Tanaka, H., Arakawa, H., Yamaguchi, T., Shiraishi, K., Fukuda, S., Matsui, K., Takei, Y., and Nakamura, Y. (2000). A ribonucleotide reductase gene involved in a p53-dependent cell-cycle checkpoint for DNA damage. *Nature* **404**, 42–49.

Utrera, R., Collavin, L., Lazarevic, D., Delia, D., and Schneider, C. A. (1998). Novel p53-inducible gene coding for a microtubule-localized protein with G2-phase-specific expression. *EMBO J.* **17**, 5015–5025.

Van Meir, E. G., Polverini, P. J., Chazin, V. R., Su Huang, H. J., de Tribolet, N., Cavenee, W. K. (1994). Release of an inhibitor of angiogenesis upon induction of wild type p53 expression in glioblastoma cells. *Nat. Genet.* **8**, 171–176.

Varmeh-Ziaie, S., Okan, I., Wang, Y., Magnusson, K. P., Warthoe, P., Strauss, M., and Wiman, K. G. (1997). Wig-1, a new p53-induced gene encoding a zinc finger protein. *Oncogene* **15**, 2699–2704.

Velasco-Miguel, S., Buckbinder, L., Jean, P., Gelbert, L., Talbott, R., Laidlaw, J., Seizinger, B., and Kley, N. (1999). PA26, a novel target of the p53 tumor suppressor and member of the GADD family of DNA damage and growth arrest inducible genes. *Oncogene* **18**, 127–137.

Vogelstein, B., Lane, D., and Levine, A. J. (2000). Surfing the p53 network. *Nature* **408**, 307–310.

Wagner, A. J., Kokontis, J. M., and Hay, N. (1994). MYC-mediated apoptosis requires wild-type p53 in a manner independent of cell cycle arrest and the ability of p53 to induce p21waf1/cip1. *Genes Dev.* **8**, 2817–2830.

Waldman, T., Kinzler, K., and Vogelstein, B. (1995). p21WAF1 is necessary for the p53-mediated G1 arrest in human cancer cells. *Cancer Res.* **55,** 5187–5190.

Wales, M. M., Biel, M. A., El-Deiry, W., Nelkin, B. D., Issa, J. P., Cavence, W. K., Kuerbitz, S. J., and Baylin, S. B. (1995). p53 activates expression of HIC-1, a new candidate tumour suppressor gene on 17p13.3. *Nat. Med.* **1,** 570–577.

Waterman, M. J. F., Stavridi, E. S., Waterman, J. L. F., and Halazonetis, T. D. (1998). ATM-dependent activation of p53 involves dephosphorylation and association with 14-3-3 proteins. *Nat. Genet.* **19,** 175–178.

Weber, J. D., Jeffers, J. R., Rehg, J. E., Randle, D. H., Lozano, G., Roussel, M. F., Sherr, C. J., and Zambetti, G. P. (2000). p53-independent functions of the p19(ARF) tumor suppressor. *Genes Dev.* **14,** 2358–2365.

Woo, R. A., McLure, K. G., Lees-Miller, S. P., Rancourt, D. E., and Lee, P. W. K. (1998). DNA-dependent protein kinase acts upstream of p53 in response to DNA damage. *Nature* **394,** 700–704.

Wu, G. S., Burns, T. F., McDonald III, E. R., Jiang, W., Meng, W., Krantz, I. D., Kao, G., Gan, D. D., Zhou, J.-Y., Muschel, R., Hamilton, S. R., Spinner, N. B., Markowitz, S., Wu, G., and El-Deiry, W. S. (1997). KILLER/DR5 is a DNA-damage-inducible p53-regulated death receptor gene. *Nat. Genet.* **17,** 141–143.

Wu, G. S., Saftig, P., Peters, C., and El-Deiry, W. S. (1998). Potential role for cathepsin D in p53-dependent tumor suppression and chemosensitivity. *Oncogene* **16,** 2177–2183.

Yeargin, J., and Haas, M. (1995). Elevated levels of wild-type p53 induced by radiolabeling of cells leads to apoptosis or sustained growth arrest. *Curr. Biol.* **5,** 423–431.

Yin, C., Knudson, C. M., Korsmeyer, S. J., and Van Dyke, T. (1997). Bax suppresses tumorigenesis and stimulates apoptosis *in vivo. Nature* **385,** 637–640.

Yin, Y., Terauchi, Y., Solomon, G. G., Aizawa, S., Rangarajan, P. N., Yazaki, Y., Kadowaki, T., and Barrett, J. C. (1998). Involvement of p85 in p53-dependent apoptotic response to oxidative stress. *Nature* **391,** 707–710.

Zhan, Q., Carrier, F., and Fornace, A. J., Jr. (1993). Induction of cellular p53 activity by DNA-damaging agents and growth arrest. *Mol. Cell. Biol.* **13,** 4242–4250.

Zhan, Q. M., Chen, I. T., Antimore, M. J., and Fornace, A. J., Jr. (1998). Tumor suppressor p53 can participate in transcriptional induction of the GADD45 promoter in the absence of direct DNA binding. *Mol. Cell. Biol.* **18,** 2768–2778.

Zhan, Q., Antinore, M. J., Wang, X. W., Carrier, F., Smith, M. L., Harris, C. C., and Fornace, A. J., Jr. (1999). Association with Cdc2 and inhibition of Cdc2/cyclin B1 kinase activity by the p53-regulated protein Gadd45. *Oncogene* **18,** 2892–2900.

Zhang, H., Somasundaram, K., Peng, Y., Tian, H., Zhang, H., Bi, D., Weber, B. L., and El-Deiry, W. S. (1998). BRCA1 physically associates with p53 and stimulates its transcriptional activity. *Oncogene* **16,** 1713–1721.

Zhang, Y., Xiong, Y., and Yarbrough, W. G. (1998). ARF promotes MDM2 degradation and stabilizes p53: ARF-INK4a locus deletion impairs both the Rb and p53 tumor suppression pathways. *Cell* **92,** 725–734.

Zhu, J., and Chen, X. (2000). MCG10, a novel p53 target gene that encodes a KH domain RNA-binding protein, is capable of inducing apoptosis and cell cycle arrest in G(2)-M. *Mol. Cell. Biol.* **20,** 5602–5618.

Zindy, F., Eischen, D. H., Randle, D. H., Kamijo, T., Cleveland, J. L., Sherr, C. J., and Roussel, M. F. (1998). MYC-induced immortalization and apoptosis targets the ARF–p53 pathway. *Genes Dev.* **12,** 2424–2433.

Zuo, Z., Dean, N. M., and Honkanen, R. E. (1998). Serine/threonine protein phosphatase type 5 acts upstream of p53 to regulate the induction of p21(WAF1/CIP1) and mediate growth arrest. *J. Biol. Chem.* **273,** 12250–12258.

5

Prospects for Tumor Suppressor Gene Therapy: RB as an Example

Daniel J. Riley and Wen-Hwa Lee[1]

Departments of Medicine and Molecular Medicine
The University of Texas Health Science Center at San Antonio
San Antonio, Texas 78245

I. INTRODUCTION

In order for a normal cell or a population of normal cells to become malignant, growth must become deregulated from or hyperactive to appropriate stimuli. Unregulated tumor growth is characterized by a number of interrelated features, which include: (1) progression of the cell cycle with few or no restrictions; (2) failure of cells to differentiate fully and to withdraw from the mitotic pool when they should; (3) failure of genetically damaged cells to undergo programmed cell death; and (4) accumulation of genetic mutations that provide certain clones with selective growth advantages and result in aberrant distribution of chromosomes among daughter cells. The latter feature may be the most crucial ultimately after tumor initiation, for it results in the accumulation of

---

[1] Correspondence should be addressed to: Wen-Hwa Lee, Dept. of Molecular Medicine, Institute of Biotechnology, 15355 Lambda Dr., San Antonio, TX 78245.

Tumor-Suppressing Viruses, Genes, and Drugs
Innovative Cancer Therapy Approaches
Copyright © 2002 by Academic Press.

more mutations and in the progression of cellular atypia. All these abnormalities, of course, are interrelated in carcinogenesis and tumor progression: when one process is abnormal, the chances increase that other abnormalities develop.

Genetic instability is ultimately the engine of both tumor progression and tumor heterogeneity (Lengauer et al., 1998). It guarantees that no two tumors are exactly alike and that no single tumor is composed of genetically identical cells. This heterogeneity makes therapeutic strategies aimed at a single genetic mutation very difficult in treatment of different tumors, and often prevents one form of specific antitumor therapy from being generalizable. Most antitumor pharmaceutical agents like DNA intercalators, microtubular assembly inhibitors, and antimetabolites, block steps in cell proliferation. They work in treating several tumors because the cells in the tumors are some of the most rapidly proliferating cells in the host. By their very nature, however, they hinder the propagation of *all* rapidly proliferating cells, not just specific tumor cells, and therefore are limited in their effectiveness largely by toxicity. In addition, multiple drug resistance can develop when certain cells within the tumor mass clonally expand after being selected by several courses of therapy to eliminate less hardy competitors. Without correcting the fundamental problem, that is, the loss of function or gain of function mutations caused by gene mutations, tumor growth is often masked temporarily, but not cured.

Rationally designed and ideal antitumor therapy should target precisely the roots of the carcinogenesis and tumor progression. Therefore the goals of effective antitumor treatment should include specificity for cancer cells with sparing of toxic side effects for normal cells, and, ideally, correction of the fundamental problem rather than masking of it. In the case of tumor suppressor genes, true correction of the mutational inactivation of the gene itself or the functional equivalent of restoring tumor suppressor activity occurs. To date, tumor suppressor gene therapy for human tumors remains in nascent stages, in large part because gene delivery vectors and techniques are not good enough yet to completely restore normal tumor suppressor gene activity in all tumor cells for long periods of time. Nonetheless, knowledge about the precise functions of tumor suppressor gene products is growing rapidly. Furthermore, the concept of cancer suppression by tumor suppressor gene therapy has been proved beyond a doubt in animal studies that accurately model some spontaneous human tumors. The more we know about specific tumor suppressor genes, the functions of their gene products, and their patterns of mutation in human tumors, the better position we will be in when gene delivery technology eventually makes human tumor suppressor gene therapy useful and practical.

Because the retinoblastoma gene product is the prototype, RB will be used here as an example to elucidate the potential of tumor suppressors as fundamental antitumor agents. First, RB has been shown to be an effective tumor suppressor in cell culture and *in vivo*. Second, mutations of RB itself or of other proteins involved directly in the pathways of RB function are purported to occur in *all* human tumors, such that inactivation of RB or one of its crucial regulatory proteins (e.g., cyclin D/Cdk4) or targets (e.g., E2F-1) is considered a requirement for carcinogenesis (Sellers *et al.*, 1998; Weinberg, 1995). Moreover, RB has multiple distinct but intersecting roles in regulating cell cycle progression (Goodrich *et al.*, 1991; Herwig and Strauss, 1997; Weinberg, 1995) and apoptosis (Favrot *et al.*, 1998; Herwig and Strauss, 1997), in causing or maintaining terminal cellular differentiation (Chen *et al.*, 1995; Herwig and Strauss, 1997; Lipinski and Jacks, 1999), and in guarding the fidelity of DNA replication or chromosome segregation (Zheng *et al.*, 2000). RB therefore illustrates well how therapy aimed at restoring tumor suppressor gene function can be rationally used to treat individual tumors with distinct patterns of tumor suppressor gene inactivation. Therapies to restore functions of tumor suppressors other than RB should be designed using similar rationale to treat cancers with different cell-type specific mutation patterns, in which distinct molecular mechanisms lead to neoplastic cell growth.

The retinoblastoma gene (*RB*) was the first tumor suppressor gene cloned (Friend *et al.*, 1987; Fung *et al.*, 1987; Lee *et al.*, 1987). The search for the recessive gene associated with a relatively rare childhood eye tumor retinoblastoma was important because it provided the first proof of the concept of tumor suppression by recessive alleles. The concept dated back to studies of the kinetics of benzpyrene-induced papillomas in mice (Charles and Luce-Clausen, 1942) and to studies in which normal cells could prevent neoplastic properties when fused with tumor cells (Harris *et al.*, 1969). Later the different incidences and ages at diagnosis of unilateral and bilateral retinoblastomas were used in mathematical thought experiments to predict that two genetic "hits" are required to generate inherited and sporadic forms of retinoblastoma (Knudson, 1971). One of the hits is a mutation in the germline for inherited forms (Comings, 1973). Retinoblastoma was then linked to a genetic locus on chromosome 13q14 (Francke, 1976; Hagstrom and Dryja, 1999; Strong *et al.*, 1981), and loss of heterozygosity (LOH) for that locus was proven in retinoblastomas compared to normal cells from the same patients (Strong *et al.*, 1981). Moreover, LOH was shown to occur in human retinoblastomas predominantly by chromosomal mechanisms such as nondysjunction, chromosomal loss, and reduplication (Cavenee *et al.*, 1983). Positional cloning of the human *RB-1* gene, demonstration of its specific mutation in all forms of human retinoblastoma (Lee *et al.*, 1987), and reversal of the neoplastic phenotype in cultured retinoblastoma cells by retroviral-mediated restoration of wild-type *RB* (Huang *et al.*, 1988), proved once and for all that *RB* was indeed a tumor suppressor in the truest sense of the term (see more below).

II. FUNCTIONS OF RB

A. G1 gatekeeping and repression of transcription

Cellular and molecular functions of the *RB* gene product (RB protein) have since been characterized extensively. In the normal cell cycle, "active" RB serves as a gatekeeper to restrict access past the restriction ("R") point in the first cellular growth phase (G1) that commits the cell to DNA synthesis and to subsequent phases of the cell division cycle (Ford and Pardee, 1999; Knudson, 1993). Transforming DNA tumor virus oncoproteins such as SV40 large T antigen, adenovirus E1A, and human papillomavirus E7 can bind tightly to RB and thereby inactivate its major function in G1 growth suppression (Chellappan *et al.*, 1992; DeCaprio *et al.*, 1988; Dyson *et al.*, 1989; Whyte *et al.*, 1988). Indeed, the inactivation of RB, RB-related proteins, and p53 is the mechanism by which these viruses are oncogenic: they keep RB in its "inactive" state in all phases of the cell cycle, and thereby mimic constitutive, unregulated RB phosphorylation (Nevins, 1991).

In normal cells, RB is regulated primarily by phosphorylation. Hypophosphorylated forms predominate in terminally differentiated cells, in G0, and in G1 phases of the cell cycle (Buchkovich *et al.*, 1989; Chen *et al.*, 1989; DeCaprio *et al.*, 1989). Beginning in early or mid-G1 phase, RB is phosphorylated on multiple but specific serine and threonine residues, probably by cyclin-dependent kinases Cdk4 and Cdk6 in complexes with D cyclins (Dowdy *et al.*, 1993; Ewen *et al.*, 1993; Hatakeyama *et al.*, 1994; Hinds *et al.*, 1992; Kato *et al.*, 1993). Later in mid-G1, additional specific phosphorylations occur sequentially, probably by Cdk2/cyclin E, and fully inactivate RB with respect to its G1 functions (Lundberg and Weinberg, 1998). The first phosphorylations may be necessary to produce conformational changes in RB and to allow later G1 cyclin–cdk complexes access to different phosphorylation sites (Connell-Crowley *et al.*, 1997; Harbour *et al.*, 1999; Kitagawa *et al.*, 1996; Knudsen and Wang, 1996; Lundberg and Weinberg, 1998). Further distinct phosphorylations may then occur in S, G2, and M phases (Chen *et al.*, 1989). The kinases responsible for these phosphorylations are not yet known, but cdc2 complexed with cyclin A may be one of them (Lees *et al.*, 1991; Ludlow *et al.*, 1993; Zarakowska and Mittnacht, 1997). The catalytic α subunit of protein phosphatase 1 associates with RB, and may be responsible in part for the dephosphorylation of RB during G2/M (Durfee *et al.*, 1993), before hypophosphorylated RB again predominates during the next G1 phase.

A major mechanism by which RB controls progression of the cell cycle past G1 to S phase is by binding and sequestering a family of transcription factors know collectively as E2F (Flemington *et al.*, 1993; Harbour and Dean, 2000; Helin *et al.*, 1993; Lees *et al.*, 1993; Nevins, 1991; Shan *et al.*, 1992;

Strong *et al.*, 1981; Trouche *et al.*, 1997). Since the initial discovery of E2F (now E2F-1), at least five members of a family have been cloned (Sardet *et al.*, 1995). They bind to promoter DNA as heterodimers with either DP1 or DP2, and activate the transcription of several genes important for S-phase progression, including DNA polymerase α, dihydrofolate reductase, c-jun, cyclin E, and cdk2 (Charles and Luce-Clausen, 1942; Shan *et al.*, 1992). Almost all naturally occurring, tumor-derived mutants tested to date fail to bind E2F (Weinberg, 1995). Furthermore, deregulated expression of E2F-1, disruption of the Rb–E2F-1 interaction, and direct transcriptional activation of E2F-responsive promoters can bypass the block to G1/S-phase transition induced by RB (Sellers *et al.*, 1998; Shan and Lee, 1994). These data, taken together, have clearly shown that RB controls G1/S-phase cell cycle progression by sequestering E2F and repressing transcription of key genes required for S phase.

More recent studies have more clearly defined how RB represses transcription by modulating chromatin architecture at the site of E2F-bound promoters (Brehm *et al.*, 1998; Dahiya *et al.*, 2000; Hirsch *et al.*, 2000; Lai *et al.*, 1999; Luo *et al.*, 1998; Magnaghi-Jaulin *et al.*, 1998; Nicolas *et al.*, 2000; Sutcliffe *et al.*, 2000; Zhang *et al.*, 2000). RB associates through a standard LxCxE motif to histone deacetylases and brings E2F-1, which binds to a different region on RB, into close association with histone deacetylases (HDAC). Histone deacetylases then facilitate removal of highly charged acetyl groups from core histones, tightening the association between DNA and nucleosomes, and preventing other transcription factors from gaining access to unwound DNA. In effect RB acts as a tether or chaperone for bringing HDAC to E2F-bound DNA; without RB, specific DNA sequences transcribed by E2F would not otherwise bind HDAC and allow HDAC to repress transcription. Mutations in the RB A and B domains, which abolish binding to E2F, also reduce binding to HDAC1 and abrogate RB-mediated transcriptional repression. Viral oncoproteins such as HPV-16 E7, which bind to the same A/B domain site in RB, can displace HDAC and thereby allow transcription to occur from E2F-bound promoters, as can full phosphorylation of RB during G1 (Magnaghi-Jaulin *et al.*, 1998).

Other studies have suggested that RB, through its T-antigen binding A and B domains, not only inhibits the transactivating function of E2F, but also directly represses transcription of certain promoters when bound to them via E2F (Knudsen *et al.*, 1998; Lukas *et al.*, 1997; Ross *et al.*, 1999; Sellers *et al.*, 1995; Weintraub *et al.*, 1992). One of the most elegant of these studies showed that a chimeric protein, in which the E2F1 transactivation domain was replaced with the RB T-antigen binding domains, could induce G1 cell cycle arrest and repress transcription of a reporter construct driven by pertinent E2F-responsive promoters (Lukas *et al.*, 1997). E2F-responsive promoters included those in the 5′-flanking sequences of genes encoding DHFR, DNA polymerase α, and B-myb. Fusion of the same RB T-antigen binding domains to a DNA-binding domain unrelated to E2F likewise generated a transcriptional repressor, proving the involvement of RB, exclusive of E2F, in transcriptional repression. RB has another important function in relation to E2F: it protects E2F-1 from degradation by the ubiquitin–proteasome pathway during a precise period when RB/E2F complex formation is essential for regulating cell cycle progression (Hateboer et al., 1996; Hofmann et al., 1996). Finally, RB also represses transcription by regulating RNA polymerase III expression (Hirsch et al., 2000) or by preventing interactions required for RNA polymerase III access to transcribed DNA (Sutcliffe et al., 2000).

E2F-independent mechanisms of RB-mediated transcriptional repression have more recently been demonstrated. Certain RB mutants (RB arginine 661 to tryptophan and RB Δ480) specifically engineered to disrupt E2F-1 binding while leaving the rest of the molecule largely intact (Sellers et al., 1998) or to mimic constitutive phosphorylation by changing several serine and threonine Cdk phosphorylation sites to glutamic acid (Barrientes et al., 2000) were analyzed for their ability when introduced into $Rb^{-/-}$ cells to form "flat" (G1 arrested) cells, to suppress anchorage-independent colony formation, and to coactivate specific programs of differentiation. Several of these mutants failed to bind E2F-1 or repress E2F-1-mediated transcription of a reporter construct with E2F-1- responsive promoter elements, but were nonetheless able to block G1/S phase progression and anchorage-independent growth. These studies showed that E2F binding is not necessarily required for RB to cause growth or tumor suppression, and that growth suppression and differentiation are not always affected equivalently (see more below).

B. Direct regulation of differentiation by specific transcriptional activation

Transcriptional repression of proliferation-promoting transcription factors like E2F and its S-phase target genes is not the only mechanism by which RB functions in the normal cell; it also positively regulates transcription of genes that result in specific patterns of cellular differentiation. The first evidence that RB was involved functionally in this process was indirect. In human and developing mouse tissues stained with antibodies against RB, the most prominent and persistent staining was in the most highly differentiated cells, such as neurons (Cordon-Cardo and Riele, 1994; Szekely et al., 1992). Furthermore, RB remains primarily in its most hypophosphorylated state during defined states of cellular differentiation (Chen et al., 1995a, 1996a,b); it is modified by hyperphosphorylation only when cells are stimulated to reenter G1 and subsequent committed phases of the division cycle. The generation of mice null for Rb advanced the idea that RB was crucial for cellular differentiation. Such mice do not survive embryonic stages; they die during mid-gestation (embryonic days 13.5–15.5), precisely at the time when crucial cell types in the developing brain, spinal ganglia, liver, and hematopoietic system begin to differentiate terminally (Clarke et al., 1992; Jacks et al., 1992; Lee et al., 1992, 1994). The neuronal defects in mice lacking functional Rb are consistent with the importance of RB predicted by the abundant expression of RB in highly differentiated, postmitotic, mouse and human neurons.

The proof of the key role of RB in terminal differentiation came when specific cells from $Rb^{-/-}$ and $Rb^{+/-}$ mice were examined thoroughly. The abnormal development of the hindbrain and dorsal root ganglia in RB null mouse embryos manifests as aberrant mitosis and apoptosis in regions that should differentiate into postmitotic neurons at mid-gestational stages (Callaghan et al., 1999; Lee et al., 1994). Likewise, erythrocytes in the fetal liver fail to extrude their nuclei and never achieve the fully differentiated state (Jacks et al., 1992; Lee et al., 1992). A more recent report has suggested that RB promotes the differentiation of neurons and erythrocytes in mice by specific interaction with Id2, a basic helix-loop-helix (bHLH) transcription factor. Interaction between Rb and Id2 is purported to sequester Id2, thereby preventing it from interacting with other bHLH factors (Ivarone et al., 1994; Lassorella et al., 2000). Activation of Id2 transcription by Myc can lead to expression of excessive amounts of Id2 and can bypass the inhibitory effects of RB on G1 cell cycle progression, presumably by titrating and occupying functional, hypophosphorylated RB.

In $Rb^{+/-}$ mice, which develop normally during gestation but suffer from pituitary neuroendocrine tumors with nearly complete penetrance later in life, loss of heterozygosity for the remaining wild-type Rb allele occurs early in the genesis of tumors (Nikitin and Lee, 1996). The loss is probably an initiating event in carcinogenesis. It results in failure of the pituitary intermediate lobe neuroendocrine cells to differentiate, owing to loss of proper dopaminergic innervation. As detailed below, RB gene therapy restores dopaminergic innervation of pituitary melanotroph tumor cells, and allows them to differentiate into a more normal phenotype (Riley et al., 1996). Thus, restoration of normal RB functions in tumor cells occurs not only by cell cycle gatekeeping and antiproliferative effects, but by restoration of a more normal pattern of differentiation.

Analysis of cells derived from $Rb^{-/-}$ embryos has shown key roles for Rb protein in the differentiation of several cell types that begin to differentiate terminally at the time during gestation when the embryos die or shortly thereafter. Skeletal muscle cells remain in the proliferative pool of myoblasts, and cannot terminally differentiate into mature myotubes without functional Rb (Gu et al., 1993; Schneider et al., 1994). In myocytes, RB is crucial for setting in motion a series transcriptional programs that begins through direct interaction with myoD (Novitch et al., 1996; Zacksenhaus et al., 1996). The abnormal muscle development in $Rb^{-/-}$ embryos would normally be masked, because it occurs mostly after gestational day 15.5, that is, after a time point when the embryos die of other defects. The more recent report of the nearly complete rescue of abnormal neuronal and erythropoietic phenotypes in $Rb^{-/-}$ by crossing $Rb^{+/-}$ with mice deficient for Id2 has shown indeed that skeletal muscle development without functional Rb is aberrant in late gestational stages: Rb-Id2 double knockout embryos die at birth with severe abnormalities in skeletal muscle development (Lassorella et al., 2000).

In a similar manner, $Rb^{-/-}$ fibroblasts are unable to differentiate into adipocytes like their wild-type counterparts (Chen et al., 1996b; Classon et al., 2000). Restoration of functional Rb by transfection allows these fibroblasts, when properly stimulated, to differentiate normally by direct interaction with and activation of the C/EBP family of transcription factors that coordinately controls early stages of adipocyte differentiation. It appears that RB protein functions in this process like a chaperone (Chen et al., 1995a, 1996a,b), and that it has effects on C/EBP activation different than its family member p107 (Classon et al., 2000). Interestingly, direct activation of PPARγ, a transcription factor distal in the coordinate sequential pathway required for adipocyte differentiation, can bypass the requirement for Rb (Hansen et al., 1999), perhaps showing that Rb is not involved in the latest steps of terminal differentiation, but rather in some earlier or intermediate steps. RB can also activate transcription by c-Jun (Nead et al., 1998).

In addition to roles in the differentiation of neurons, skeletal muscle, and adipocytes, analogous roles for RB in the differentiation of erythrocytes (Jacks et al., 1992; Lassorella et al., 2000; Lee et al., 1992), monocytes (Chen et al., 1996a), and perhaps osteoblasts (Sellers et al., 1998; Sidle et al., 1996) by activation of transcription factors have been demonstrated. Thus RB has a positive role in activating transcription factors important for several distinct patterns of terminal differentiation. The function of RB in positive activation of transcription and differentiation may be an even more important consequence of RB therapy in treating tumors than the simple cystostatic effects of RB, but both functions are important in making tumor cells more like normal cells.

G1 cell cycle arrest and terminal differentiation can be dissociated in certain RB mutants (Barrientes et al., 2000; Riley et al., 1997; Sellers et al., 1998; Sterner et al., 1996; Whitaker et al., 1998). Some naturally occurring and specifically manufactured RB mutations result in low penetrance retinoblastomas and relatively mild phenotypes (Dryja et al., 1993; Otterson et al., 1997). For example, small in frame deletions in the portion of the RB gene encoding the N-terminus of RB can maintain the ability to control progression through the restriction point of G1, but cannot substitute for all the functions of RB during embryonic development (Riley et al., 1997). Other synthetic RB mutants nonetheless retain abilities to suppress passage past the G1 restriction point and to coactivate limited but precisely defined measures of terminal differentiation such as bone mineralization in reconstituted $Rb^{-/-}$ osteosarcoma cells (Sellers et al., 1998). Mutation of certain Cdk4/6 and Cdk2 phosphorylation sites in RB also results in cells with differential ability to bind E2F-1, cause G1 cell cycle arrest, and promote terminal differentiation in cell culture assays. No specific, individual phosphorylation events could be correlated with release of E2F or with activation of a specific program of differentiation (Barrientes et al., 2000). It is clear, however, that phosphorylation during G1 phase of the cell cycle occurs sequentially and on different phospho-acceptor sites (Knudsen and Wang, 1996), first by Cdk4/6–cyclin D complexes, then later by Cdk2–cyclin E complexes (Harbour et al., 1999; Kitagawa et al., 1996; Lundberg and Weinberg, 1998). The early G1 phosphorylations are not sufficient by themselves to release RB from binding to E2F or to commit the cell to passage into S phase. Additional phosphorylations of RB in mid-G1 by Cdk2 are required for these latter events. Specific deletion or phosphorylation mutants of RB might eventually be designed to differentially affect cell cycle progression and terminal differentiation, or to restore one function without restoring the other.

C. RB reduces apoptosis

Some tumor suppressor proteins, most notably p53, function directly in endogenous pathways of apopotosis. When cells are treated with DNA damaging agents, p53 expression is upregulated and the cells with damaged DNA are targeted for apoptosis in order to prevent passage of inaccurately replicated DNA or chromosomes to daughter cells. Thus p53 has been called the "guardian of the genome" (Lane, 1992). Tumor cells with null or activating, oncogenic mutations in such a scenerio receive conflicting internal signals to proliferate without appropriate external and internal stimuli. Gene therapy with p53 has received attention not only because p53 mutations are very common in a variety of human tumors (Roth et al., 1996; Roth and Cristiano, 1997), but because restoration of p53 in rapidly proliferating tumor cells missing p53 causes many of the cells to die. p53 is thus proapoptotic.

RB, in contrast, is antiapoptotic (Berry, 1996; Herwig and Strauss, 1997; Slack et al., 1995; Wang et al., 1997). $Rb^{-/-}$ embryos have aberrant mitoses in key areas of the central nervous system, hematopoietic cells in the liver, and developing skeletal muscles (Lee et al., 1994; Riley et al., 1997). Reconstitution of RB function, either during embryonic development or in carcinogenesis, is associated not with increased cell death, but with more normal withdrawal from the cell cycle and appropriate differentiation. When RB is reintroduced into cells lacking it, either during embryonic development (Chang et al., 1993; Riley et al., 1997) or during different stages of carcinogenesis (Nikitin et al., 1998, 1999; Riley et al., 1996), it allows them to respond more appropriately to regulatory signals and to assume a more normal, differentiated phenotype rather than die. Precisely how RB causes regression of large tumors late in their development is not yet clear. It may indirectly increase rates of tumor cell death in the long run if certain tumor cells are so dysregulated that they can no longer execute programs of terminal differentiation. Although it works by mechanisms quite different and in some ways opposite to those used by p53 (Pendergast, 1999; Stanbolic et al., 1999), RB gene replacement therapy nonetheless influences apoptosis.

D. M phase and faithful chromosome segregation

Perhaps the most important hallmark of cancerous cells, in contrast to rapidly proliferating but nonmalignant cells, is genomic instability associated with aberrant or improperly checked chromosome segregation. Genetic instability leads directly to the accumulation of further mutations in subsequent tumor cell generations and to an increasingly anaplastic tumor. Such instability can occur at two different levels: (1) defective DNA repair resulting in missense, nonsense, and other small but functionally important mutations that can be passed on to daughter cells; and (2) missegregation of whole chromosomes or pieces of chromosomes during mitosis, resulting in aneuploidy and translocations (Lengauer et al., 1998; Solomon et al., 1991). Both processes are guarded by checkpoint control systems (Hartwell, 1992). Chromosome segregation is controlled by a group of proteins that together coordinate the components of M phase (Hartwell, 1992; He et al., 1995). Deranged function or loss of key proteins in the structure of the mitotic chromosome would be catastrophic and lethal, since it would not allow a cell to divide and pass its genetic material on to its daughters. An example of such a lethal chromosomal abnormality during rapid cell division is early and spontaneous abortion associated with gross karyotypic abnormalities in pleuripotent stem cells of a fetus. Loss of function of proteins that play more subtle regulatory roles during mitosis, including components of the checkpoint apparatus or modulators of centromere structure, on the other hand, would be more relevant to cancer. Mutation of such proteins might not be lethal to daughter cells but might instead lead to high frequencies of chromosomal abnormalities and to neoplasia, by imparting selective growth advantage compared with other cells. Loss of function of DNA repair proteins, regulators of mitotic chromosome separation, or checkpoint controls after DNA synthesis and mitosis therefore could lead to the so-called "mutator phenotype" (Loeb and Loeb, 2000). Indeed, some familial cancer syndromes have been linked to loss of function of proteins like Msh2 in nonpolypoid colon cancer or BRCA1 and BRCA2 in breast cancer, whose primary functions are in DNA repair pathways (Dasika et al., 1999).

Although the functions of RB have been most thoroughly characterized in G1 and G0 phases, that is, in regulating cell cycle passage, tethering transcription factors to DNA, and in directly controlling terminal differentiation, they are not its only functions. Increasing evidence suggests that RB also plays important roles in guarding the genome in higher mammals. The role of another tumor suppressor gene p53 in guarding the integrity of the genome is already well accepted (Cross *et al.*, 1995; Fukasawa *et al.*, 1996; Gualberto *et al.*, 1998; Lane, 1992; Lanni and Jacks, 1998). When treated with genotoxic agents, cells lacking functional RB or p53 fail to finish mitosis correctly before entering a new cell cycle, and thus become aneuploid (DiLeonardo *et al.*, 1997). p53 has been found to be crucial for apoptosis after errors in DNA repair, for centrosome duplication activity, and for mitotic or postmitotic checkpoint control (DiLeonardo *et al.*, 1997; Harvey *et al.*, 1993; Khan and Wahl, 1998). Similar evidence is also emerging to link RB to M-phase functions through its interaction with mitotic proteins.

The hypophosphorylated form of RB exists not only during G1/G0, but also during M phase (Chen *et al.*, 1996; Chen *et al.*, 1989). Hypophosphorylated RB causes G2 as well as G1 cell cycle arrest (Karantza *et al.*, 1993), and is a critical determinant in preventing DNA endoreduplication during S phase (Niculescu *et al.*, 1998). RB also redistributes within the nucleus during mitosis (Thomas *et al.*, 1996) and associates with several proteins that have crucial functions in M-phase progression. The human Nuc2 protein (a.k.a. hCDC27) is a subunit of the anaphase-promoting complex that controls the onset of sister chromatid separation and metaphase-to-anaphase transition by degrading specific substrates on the metaphase chromosome (King *et al.*, 1995; Tugendreich *et al.*, 1995). It associates specifically with hypophosphorylated RB during M phase (Chen *et al.*, 1995b). Mitosin (a.k.a. CENP-F), a component of the kinetochore, also interacts specifically with RB during M phase (Zhu *et al.*, 1995). The significance of this association has not yet been fully explored, but mitosin expression has been correlated with the aggressiveness of certain human tumors (Clark *et al.*, 1997). A third protein, the catalytic subunit of protein phosphatase 1α (pp1α), was also identified in a screen for RB-associated proteins (Durfee *et al.*, 1993). Although human PP1α has not been definitively shown to dephosphorylate RB in higher eukaryotic cells, it is a potential phosphatase for RB. Moreover, PP1α is crucial for kinetochore function, chromosome segregation, and M-phase progression, at least in yeast (Durfee *et al.*, 1993; Ludlow *et al.*, 1993). The association between mitotic proteins and RB has provided circumstantial evidence that RB has a role in mitosis and chromosome segregation, but more functional data to show how the RB association influences chromosome segregation or checkpoint control are also beginning to emerge.

The mitotic RB-associated protein that has most directly linked RB to the process of faithful chromosome segregation is the more recently cloned

Hec1 (highly expressed in cancer). It was discovered as an RB-associated protein through yeast two-hybrid screening of a human lymphocyte cDNA library
using p56 RB as bait (Chen et al., 1997a). The tumor virus oncoprotein-bind-
ing A/B region of RB interacts with Homo sapiens Hec1 (hsHec1) through an
IxCxE motif (Chen et al., 1997b). Hec1 is a 76-kDa coiled-coil protein crucial
for faithful chromosome segregation. Mammalian cells microinjected with neutralizing anti-Hec antibodies undergo aberrant mitosis, with disordered metaphase plate alignment and grossly inequitable, apparently random distribution
of chromosomes to daughter cells (Chen et al., 1997a). The yeast homologue
of hsHec1, Ndc80 (a.k.a. Tid3 and scHec1) (Zheng et al., 1999), has a similarly
crucial function in Saccharomyces cerevesiae mitosis, since temperature sensitive
mutants null for ndc80/hec1 die after one subsequent cell division cycle. As
in mammalian cells with hsHec1 functionally inactivated, yeast with mutated
ndc80/hec1 suffer grossly disordered chromosome segregation, as well as
mitotic delay and decreased viability in subsequent generations (Chen et al.,
2001).

        Similarities and differences between the yeast and human homologue of
Hec1 have allowed the demonstration of a role for RB in enhancing the fidelity of
chromosome segregation. Although hsHec1 and Ndc80/Hec1 are only ~18%
identical in overall amino acid sequence and share only ~37% of the same amino
acids in the critical N-terminus, they are definite functional homologues. Wild-
type human Hec1 can fully complement the lethal yeast ndc80/hec1 mutation
(Chen et al., 2001). Yeast cells, which contain the same basic structural components of the mitotic apparatus as higher eukaryotes, are much more amenable
than mammalian cells to monitoring the dynamics of chromosome segregation.
Because of this property, yeast have provided the experimental conditions necessary to explore the function of RB in chromosome segregation without interference from either endogenous Hec1 or RB. Because yeast Ndc80/Hec1 does not
contain the IxCxE motif required for RB binding (Chen et al., 2001; Dahiya
et al., 2000) and because S. cerevisiae has no RB counterpart, the cells were also
effectively null for RB.

        Random mutagenesis was used to generate a yeast strain in which
a temperature-sensitive (ts) hsHec1 mutant could rescue the ndc80/hec1 null
mutation. This hshec1-113 ts mutant has a single nonsense mutation resulting in a premature stop at codon 395 and expresses a C-terminally truncated
45-kDa protein. The mutation is sufficiently subtle to maintain binding to RB
and most other Hec1-associated proteins, as well as to allow rescue of the
ndc80/hec1 null yeast from grossly disordered mitosis and lethality at the permissive temperature. Nonetheless it results in mitotic delay and in an abnormally
high rate of chromosome loss and nondysjunction as determined by colony sec-
toring assays (Zheng et al., 2000). When wild-type human RB was expressed in
this same segregation error-prone yeast strain at the permissive temperature, the

incidence of chromosomal segregation errors decreased dramatically, by fivefold (Zheng *et al.*, 2000). Expression of a human tumor-derived H209 RB mutant, in contrast, had no effect on chromosomal segregation errors. Likewise, further mutation of the ts hshec1-113 mutant (hshec1-113EK) to alter its RB binding domain prevented the enhancing effect of even wild-type RB on the fidelity of chromosome segregation. These results proved that fully functional RB was required to enhance the fidelity of chromosome segregation mediated by Hec1.

One significant protein–protein interaction disrupted in the C-terminally truncated 45-kDa hshec1-113 protein was with Smc1 (structural maintenance of chromosomes 1), a protein essential for chromosome cohesion (Zheng *et al.*, 1999). RB acted at least in part to enhance the fidelity of chromosome segregation by increasing the DNA binding activity of Smc1, which specifically associates with hsHec1 using a binding motif distinct from the one used by RB. In its interaction with Hec1 and Smc1 during mitosis, then, RB once again seems to function like a chaperone, as it does in bringing HDAC and certain transcription factors to appropriate cis-acting DNA elements during G1 cell cycle arrest and terminal differentiation.

The presence of RB in a complex with Hec1 and Smc1 in reconstituted yeast cells suggests similar functions for RB might operate in higher order eukaryotes. While not present normally in yeast, RB is present in other eukaryotes. It is interesting to speculate that RB evolved in part to improve the fidelity of the more complex chromosome segregation in higher organisms compared to segregation in yeast. By the same token, loss of RB function in mammalian cells might similarly result in subtly disordered chromosomal segregation or mitotic checkpoint control, precisely the type of disorder that could lead to nonlethal aneuploidy or translocations in tumor cells. Whether loss of RB function leads to impairment of the fidelity of chromosome segregation in mammalian tumor cells remains to be explored more directly once the tools for following subtle chromosomal derangements in mammalian systems can be improved.

G2/M phase abnormalities and mitotic errors in $Rb^{-/-}$ fibroblasts and tumor cells have also suggested that RB has an important role in assuring the fidelity of chromosome segregation, and that Rb-deficient cells are more prone to mitotic errors. $Rb^{-/-}$ cells exposed to ionized radiation, DNA damaging chemicals, and mitotic inhibitors have higher rates of p53-dependent apoptosis, G1 and G2/M phase delays, and chromosomal abnormalities than similar $Rb^{+/+}$ cells or null cells transfected with a wild-type *RB* construct (DiLeonardo *et al.*, 1997; Khan and Wahl, 1998).

Because RB does appear to have a role in improving the fidelity of chromosome segregation errors in higher eukaryotes, restoration of RB functions in mitosis by *RB* gene therapy in tumor cells might conceivably prevent further oncogenic gene mutations in tumors already missing functional RB protein. RB reconstitution could also make tumor cells preferentially more susceptible to conventional chemotherapeutic agents or radiation therapy. Thus, in addition to roles in regulating proliferation, promoting terminal differentiation, and affecting cell survival, *RB* gene therapy for cancer might fundamentally improve the chances that a given tumor cell clone will not develop more mutations by chromosomal mechanisms resulting in aneuploidy, even late in the course of carcinogenesis. The actual effects of RB replacement on genomic stablity in cancer cells, however, remains to be explored more thoroughly.

III. SUCCESSES WITH *RB* GENE THERAPY

The concept of cancer suppression by replacement of a normal retinoblastoma gene into cells lacking it was first proven in cultured human retinoblastoma cells and in xenograft models. When RB-deficient WERl-27 cells were reconstituted with wild-type *RB* gene delivered via a retroviral vector, their neoplastic properties were suppressed in both soft agar assays and when the cells were injected into nude mice (Huang *et al.*, 1988). Similar studies in several other RB-deficient cancer cell lines derived from osteosarcomas, prostate, transitional cell bladder, breast, and lung tumors showed similar results and made the concept of *RB* gene therapy for cancer more generalized (Bookstein and Lee, 1991).

In order to take the concept of tumor suppression by RB to the next step, immunocompetent animal models of spontaneous carcinogenesis owing to *Rb* mutation were generated (Clarke *et al.*, 1992; Jacks *et al.*, 1992; Lee *et al.*, 1992). These models effectively mimic patients with heterozygous germline *RB* mutations, at least conceptually, for the tumors that develop predictably in them lose heterozygosity for the remaining wild-type allele as the seminal event in the initiation of carcinogenesis (Nikitin and Lee, 1996). As mentioned earlier, $Rb^{-/-}$ embryos die during mid-to late gestation with defects related to failure to differentiate highly specialized cells like neurons in the midbrain and dorsal root ganglia (Lee *et al.*, 1994). $Rb^{+/-}$ mice, in contrast, survive gestation and mature normally, but develop multiple neuroendocrine tumors, including those arising from pituitary intermediate lobe melanotrophs, medullary thyroid C-cells, anterior pituitary, and adrenal medulla (Nikitin *et al.*, 1999). Despite some variations related to backcross strain, the development of the characteristic neuroendocrine tumors, especially pituitary intermediate lobe melanotroph tumors, has been shown to be very predictable, nearly 100% penetrant, and relatively synchronous (Chang *et al.*, 1993; Hu *et al.*, 1994). $Rb^{+/-}$ mice therefore proved useful as models for treatment of spontaneous tumors. The pituitary intermediate lobe tumors served as the mechanistic equivalents in mice of retinoblastomas in humans with heterozygous germline *RB* mutations.

To prove that RB replacement could indeed be used to treat spontaneous Rb-deficient tumors *in vivo*, a human *RB* transgene was first shown to

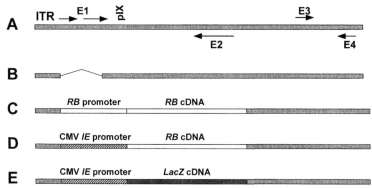

Figure 1 Recombinant adenoviruses (B–D) (Demers *et al.*, 1998; Riley *et al.*, 1996) and comparison with the wild-type adenovirus 5 single-stranded DNA genome (A); internal terminal repeat (ITR), genes encoding E1a, E1b, protein IX, and E2-E4. (B) rAd5.0, a replication deficient control virus with *e1a* and most of the *e1b* genes deleted. (C). rAd5.R.RB, carrying the 2.8-kb human *RB* cDNA driven by its own 0.9-kb promoter sequence in place of the *e1a* region. (D) rAd.C.RB, carrying the same *RB* cDNA driven by CMV IE1 promoter/enhancer. (E) rAd5.C.Z, carrying *Escherichia coli LacZ* driven by the CMV IE1 promoter/enhancer.

rescue the phenotypes arising from Rb deficiency in mice (Bignon *et al.*, 1993; Chang *et al.*, 1993; Lee *et al.*, 1992). A transgene consisting of the 2.8-kb *RB* cDNA driven by a short 1.6-kb human *RB* promoter (Hong *et al.*, 1989) sequence was sufficient when introduced by interbreeding to rescue genetically both the embryonic lethal phenotype in $Rb^{-/-}$ mice and the tumor phenotype in $Rb^{+/-}$ mice (Chang *et al.*, 1993). Using the same transgene, gene delivery vectors for *in vivo* *RB* gene therapy were then generated. Recombinant adenoviruses (Demers *et al.*, 1998; Riley *et al.*, 1996) (Figure 1) resulted in abundant RB protein expression in a wide range of cultured tumor cells, including those derived from mouse pituitary melanotroph tumors. When the adenovirus expressing RB via its own promoter was injected directly into $Rb^{+/-}$ mouse pituitary intermediate lobes or into established pituitary tumors, it resulted in delayed progression or actual regression of established tumors (Figure 2) (Riley *et al.*, 1996). Local, adenoviral-mediated *RB* gene therapy also reduced rates of tumor cell proliferation and restored in the melanotroph cells a more normal, differentiated, neuroendocrine phenotype characterized by appropriate connections to controlling neurons and by decreased rates of apoptosis. Finally, the therapy significantly delayed tumor progression and prolonged the life spans of mice compared to littermates treated with adenoviral vector alone or with nothing at all. Treatment of primary, spontaneous tumors *in vivo* by delivery of *RB* gene therapy was thus accomplished in a preclinical animal model.

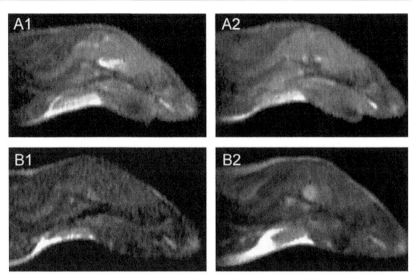

Figure 2 Representative T1-weighted magnetic resonance images through the pituitary melanotroph tumors of mice before (A1, B1) and 28 days after (A2, B2) injection with
adenoviral vectors. The gross tumor in the mouse depicted in A decreased in size after
injection with a recombinant adenovirus expressing RB under control of its own promoter (Ad5.R.RB). The tumor in a sex-matched littermate (B) injected with an inert
control virus (Ad5.0) continued to grow (Riley et al., 1996).

Although pituitary intermediate lobe melanotroph tumors in $Rb^{+/-}$
mice model human retinoblastomas mechanistically, precisely equivalent
tumors in humans are extremely rare. The more generalized applicablity of RB
gene therapy was taken a step further when $Rb^{+/-}$ mice were found to have a
form of multiple endocrine neoplasia. Locally invasive tumors of the intermediate pituitary lobe were universal in these mice, and most often accounted by
brainstem compression for the premature death of the animals. Tumors of calci-
tonin producing cells of the thyroid, anterior pituitary, and adrenal medulla,
however, were also found with variable degrees of penetrance (Nikitin et al.,
1998, 1999). Like the pituitary melanotroph tumors, they lost heterozygosity
early during the course of carcinogenesis, and thus could potentially benefit
from RB gene replacement. Lung metastases from the C cell thyroid tumors
were common. Therefore $Rb^{+/-}$ mice proved to be good models to show the
potential effectiveness of RB gene replacement therapy in metastatic as well as
primary tumors. To treat lung metastases, liposome-mediated RB gene therapy
was used. Like the RB adenoviruses, RB mini-transgenes encapsulated in
cationic liposomes were also effective for in vivo gene therapy. The advantage of
liposomes was that they could be delivered in high doses intravenously, where

they homed to the lungs and expressed functional RB protein abundantly. Using this method and only a brief treatment course, the incidence of lung metastases from Rb-deficient medullary thyroid tumors was reduced from 84 to 12% in autopsy studies. These studies showed that restoration of normal RB function was therapeutic even in very late stages of tumor progression, after the tumor had undoubtedly acquired a sufficient number of mutations in genes other than *RB* to become widely metastatic.

IV. PERSPECTIVES

Adenoviruses and liposomes expressing full-length RB were remarkably effective in treating both primary and metastatic Rb-deficient tumors in mice, but could not be completely effective because they could not be delivered to every cell, nor could they result in permanent *RB* gene expression. Neither adenoviruses nor liposomes, despite their ability to deliver high doses of *RB* gene to target cells, allow efficient incorporation of the delivered gene into the host cell genome, where it could be passed on to daughter cells. The problem with *RB* gene therapy at present lies not with the concept, but with practical deficiencies in available gene delivery vectors. When the *RB* transgene is stably integrated into the host genome by selective breeding of transgenic mice with $Rb^{+/-}$ mice, even small amounts of RB protein expression are sufficient to suppress the development of tumors (Chang *et al.*, 1993). Technology for gene regulation in experimental models has advanced to the point that RB expression can be regulated in time and space (Utomo *et al.*, 1999). This method again requires manipulation of the embryonic stem cell or blastocyst and several generations of interbreeding, however, and is applicable only for studying mechanisms in animal models, not yet for human gene therapy. Current gene delivery vectors just cannot recapitulate the efficient and long-term regulation of gene dosage that accompanies transgene incorporation into the genome from the outset of gestation.

If more effective delivery vectors could be developed, *RB* gene therapy could be used in humans to treat late stage tumors, including retinoblastomas and lung tumors. Retinoblastomas, if discovered early, are isolated in a single eye or to discrete foci in both eyes. These primary tumors could potentially be treated with local gene delivery. Such treatment could spare a young child unilateral or complete blindness that might ensue from enucleation, the current standard of care. Other tumors even more common than childhood retinoblastomas, for example, small cell lung carcinomas (SCLCs), might also be targets for treatment by *RB* gene therapy. SCLCs arise from a type of neuroendocrine cell; they are notoriously aggressive and difficult to cure. Nearly all SCLCs examined to date carry inactivating *RB* mutations (Harbour *et al.*, 1988; Hensel *et al.*, 1990; Shimizu *et al.*, 1994).

Liposome- or adenovirus-mediated *RB* gene delivery to the lungs could conceivably be accomplished by inhalation or by intravenous injection, as has been attempted with limited success in the treatment of cystic fibrosis by *CFTR* gene therapy (Zabner *et al.*, 1996). Furthermore, since RB is mutated or otherwise functionally inactivated in a number of other tumor types (Bookstein and Lee, 1991), more locally accessible tumors such as isolated retinoblastomas themselves, hepatocellular carcinomas, and transitional cell carcinomas of the bladder might be appropriate human tumors for initial trials of *RB* gene therapy in humans. Protocols using herpes virus thymidine kinase "suicide gene" therapy to avoid eye enucleation and radiation to the head in children with primary retinoblastoma have already been discussed and encouraged in national meetings (Department of Health and Human Services, 2000; Hurwitz *et al.*, 1999). Local retinoblastoma gene replacement therapy for such tumors might be just as practical in the current era, and might offer more fundamental therapy by restoring normal RB functions in a tumor that arose precisely because of homozygous *RB* mutation in a population of retinoblasts.

Limitations of current vector efficiency, transient rather than permanent protein expression with most vectors, problems involved with diminishing effectiveness after repeated dosing, and a current moratorium on adenovirus-mediated human gene trials because of vector-related toxicities have dampened enthusiasm recently for human gene therapy protocols. These problems will eventually be overcome, however. When they are, tumor suppressor gene therapy will be used for treating cancer (Roth and Cristiano, 1997). *RB* gene therapy has worked in preclincal animal trails and to a limited extent in a few human trials (Department of Health and Human Services, 2000). Because *RB* mutations or inactivation of RB functions are probably involved in the majority of human tumors, *RB* gene therapy can serve as the prototype for rationally designed tumor suppressor gene protocols in the future. One can envision a future in which genetic profiles of specific tumor suppressor mutations will be known for each tumor, based on a biopsy and on directed genotyping for *RB*, *p53*, *BRCA1*, *BRCA2*, *APC*, *msh2*, *pten*, and other known and characterized tumor suppressor genes. According to this profile, tumor suppressor gene therapy or modification of tumor suppressor gene product function could be tailored for the individual tumor and patient. One tumor suppressor therapy would not work for all tumors, of course, but similar methods and strategies might be employed for delivering different tumor suppressors to specific tumors for which they would be appropriate. When as much is known about the molecular and cellular biology of the other tumor suppressors as is known about RB in cell cycle progression, transcriptional regulation, differentiation, apoptosis, and mitotic events, tumor suppressor gene therapy might one day be a rationally designed reality.

References

Barrientes, S., Cooke, C., and Goodrich, D. W. (2000). Glutamic acid mutagenesis of retinoblastoma protein phosphorylation sites has diverse effects on function. *Oncogene* **19**, 562–570.

Berry, D. E., Lu., Y., Schmidt, B., Fallon, P. G., O'Connell, C., Hu, S. X., Xu, H. J., and Blanck, G. (1996). Retinoblastoma protein inhibits IFN-γ-induced apoptosis. *Oncogene* **12**, 1809–1819.

Bignon, Y.-J., Chen, Y., Chang, C.-Y., Riley, D. J., Windle, J. J., Mellon, P. L., and Lee, W.-H. (1993). Expression of a retinoblastoma transgene results in dwarf mice. *Genes Dev.* **7**, 1654–1662.

Bookstein, R., and Lee, W.-H. (1991). Molecular genetics of the retinoblastoma tumor suppressor gene. *Crit. Rev. Oncog.* **2**, 211–227.

Brehm, A., Miska, E. A., McCance, D. J., Reid, J. L., Bannister, A. J., and Kouzarides, T. (1998). Retinoblastoma protein recruits histone deacetylase to repress transcription. *Nature* **391**, 597–601.

Buchkovich, K., Duffy, L. A., and Harlow, E. (1989). The retinoblastoma protein is phosphorylated during specific phases of the cell cycle. *Cell* **58**, 1097–1105.

Callaghan, D. A., Dong, L., Callaghan, S. M., Hou, Y. X., Dagnino, L., and Slack, R. S. (1999). Neuronal precursor cells differentiating in the absence of Rb exhibit delayed terminal mitosis and deregulated E2F 1 and 3 activity. *Dev. Biol.* **207**, 257–270.

Cavenee, W. K., Dryja, T. P., Phillips, R. A., Benedict, W. F., Godbout, R., Gallie, B. L., Murphree, A. L., Strong, L. C., and White, R. L. (1983). Expression of recessive alleles by chromosomal mechanisms in retinoblastoma. *Nature* **305**, 779–784.

Chang, C.-Y., Riley, D. J., Lee, E. Y.-H. P., and Lee, W.-H. (1993). Quantitative effects of the retinoblastoma gene in mouse development and tissue-specific tumorigenesis. *Cell Growth Differ.* **4**, 1057–1064.

Charles, D. R., and Luce-Clausen, E. M. (1942). The kinetics of papilloma formation in benzpyrine-treated mice. *Cancer Res.* **2**, 261–263.

Chellappan, S., Kraus, V. B., Kroger, B., Munger, K., Howley, P. M., Phelps, W. C., and Nevins, J. R. (1992). Adenovirus E1A, simian virus 40 tumor antigen, and human papilloma virus E7 share the capacity to disrupt the interaction between E2F and the retinoblastoma gene product. *Proc. Natl. Acad. Sci. U.S.A.* **89**, 4549–4553.

Chen, C.-F., Chen, Y., Chen, P.-L., Dai, K., Riley, D. J., and Lee, W.-H. (1996). A new member of the hsp90 family of molecular chaperones interacts with the retinoblastoma protein during mitosis and after heat shock. *Mol. Cell. Biol.* **16**, 4691–4699.

Chen, P.-L., Scully, P., Shew, J. Y., Wang, J. Y., and Lee, W.-H. (1989). Phosphorylation of the retinoblastoma gene product is modulated during cell cycle and cellular differentiation. *Cell* **58**, 1193–1198.

Chen, P.-L., Riley, D. J., and Lee, W.-H. (1995a). The retinoblastoma protein as a fundamental mediator of growth and differentiation signals. *Crit. Rev. Eukaryotic Gene Express.* **2**, 79–95.

Chen, P.-L., Ueng, Y. C., Durfee, T., Chen, C.-F., Yang-Feng, T., and Lee, W.-H. (1995b). Identification of a human homologue of yeast nuc2 which interacts with the retinoblastoma protein in a specific manner. *Cell Growth Differ.* **6**, 199–210.

Chen, P.-L., Riley, D. J., Chen-Kiang, S., and Lee, W.-H. (1996a). Retinoblastoma protein directly interacts with and activates the transcription factor NF-IL6. *Proc. Natl. Acad. Sci. U.S.A.* **93**, 465–469.

Chen, P.-L., Riley, D. J., Chen, Y., and Lee, W.-H. (1996b). Retinoblastoma protein positively regulates terminal adipocyte differentiation through direct interaction with C/EBPs. *Genes Dev.* **10**, 2794–2804.

Chen, Y., Riley, D. J., Chen, P.-L., and Lee, W.-H. (1997a). HEC, a novel nuclear protein rich in leucine heptad repeats specifically involved in mitosis. *Mol. Cell. Biol.* **17**, 6049–6056.

Chen, Y., Sharp, Z. D., and Lee, W.-H. (1997b). Hec binds to the seventh regulatory subunit of the 26S proteasome and modulates the proteolysis of mitotic cyclins. *J. Biol. Chem.* **272**, 24081–24087.

Chen, Y., Riley, D. J., Chen, P.-L., and Lee, W.-H. (2001). Faithful chromosome segregation requires phosphorylation of the coiled-coil protein Hec1 on serine 165 by the centrosomal kinase Nek2. Submitted.

Clark, G. M., Allred, D. C., Hilsenbeck, S. G., Chamness, G. C., Osborne, C. K., Jones, D., and Lee, W.-H. (1997). Mitosin (a new proliferation marker) correlates with clinical outcome in node-negative breast cancer. *Cancer Res.* **57**, 5505–5508.

Clarke, A. R., Maandag, E. R., van Roon, M., van der Lugt, N. M., van der Valk, M., Hooper, M. L., Berns, A., and te Riele, H. (1992). Requirement for a functional Rb-1 gene in murine development. *Nature* **359**, 328–330.

Classon, M., Kennedy, B. K., Mulloy, R., and Harlow, E. (2000). Opposing roles of pRb and p107 in adipocyte differentiation. *Proc. Natl. Acad. Sci. U.S.A.* **97**, 10826–10831.

Comings, D. E. (1973). A general theory of carcinogenesis. *Proc. Natl. Acad. Sci. U.S.A.* **70**, 3324–3328.

Connell-Crowley, L., Harper, J. W., and Goodrich, D. W. (1997). Cyclin D1/Cdk4 regulates retinoblastoma protein-mediated cell cycle arrest by site-specific phosphorylation. *Mol. Biol. Cell* **8**, 287–301.

Cordon-Cardo, C., and Riele, V. M. (1994). Expression of the retinoblastoma protein is regulated in normal human tissues. *Am. J. Pathol.* **144**, 500–510.

Cross, S. M., Sanchez, C. A., Morgan, C. A., Schimke, M. K., Ramel, S., Idzerda, R. L., Raskind, W. H., and Reid, B. J. (1995). A p53-dependent mouse spindle checkpoint. *Science* **267**, 1353–1356.

Dahiya, A., Gavin, M. R., Luo, R. X., and Dean, D. C. (2000). Role of the LXCXE binding site in RB function. *Mol. Cell. Biol.* **20**, 6799–6805.

Dasika, G. K., Lin, S. C., Zhao, S., Sung, P., Tompkinson, A., and Lee, E. Y. (1999). DNA damage-induced cell cycle checkpoints and DNA strand break repair in development and tumorigenesis. *Oncogene* **18**, 7883–7899.

DeCaprio, J. A., Ludlow, J. W., Figge, J., Shew, J.-Y., Huang, C.-M., Lee, W.-H., Marsilio, E., Paucha, P., and Livingston, D. M. (1988). SV40 large tumor antigen forms a specific complex with the product of the retinoblastoma susceptibility gene. *Cell* **54**, 275–283.

DeCaprio, J. A., Ludlow, J. W., Lynch, D., Furukawa, Y., Griffin, J., Pinwica-Worms, H., Huang, C. M., and Livingston, D. M. (1989). The product of the reinoblastoma susceptibility gene has properties of a cell cycle regulatory element. *Cell* **58**, 1085–1095.

Demers, G. W., Harris, M. P., Wen, S. F., Engler, H., Nielsen, L. L., and Maneval, D. C. (1998). A recombinant adenoviral vector expressing full-length human retinoblastoma susceptibility gene inhibits human tumor cell growth. *Cancer Gene Ther.* **5**, 207–214.

Department of Health and Human Services (2000). Recombinant DNA Advisory Committee. National Institutes of Health. Minutes of meeting. June 14, 1999. *Hum. Gene Ther.* **11**, 1369–1382.

DiLeonardo, A. S., Khan, S. H., Linke, S. P., Greco, V., Seidita, G., and Wahl, G. M. (1997). DNA replication in the presence of mitotic spindle inhibitors in human and mouse fibroblasts lacking either p53 or pRb function. *Cancer Res.* **57**, 1013–1019.

Dowdy, S. F., Hinds, P. W., Louie, K., Reed, S. I., Arnold, A., and Weinberg, R. A. (1993). Physical interaction of the retinoblastoma protein with human cyclins. *Cell* **73**, 499–511.

Dryja, T. P., Rapaport, J., McGee, T. L., Nork, T. M., and Schwatrz, T. L. (1993). Molecular etiology of low penetrance retinoblastoma in two pedigrees. *Am. J. Hum. Genet.* **52**, 1122–1128.

Durfee, T., Becherer, K., Chen, P.-L., Yeh, S.-H., Yang, Y., Kilburn, A. E., Lee, W.-H., and Elledge, S. J. (1993). The retinoblastoma protein associates with the protein phosphatase type 1 catalytic subunit. *Genes Dev.* **7**, 555–569.

Dyson, N., Howley, P. M., Munger, K., and Harlow, E. (1989). The human papillomavirus-16 E7 oncoprotein is able to bind to the retinoblastoma gene product. *Science* **243**, 934–937.

Ewen, M. E., Sluss, H. K., Sherr, C. J., Matsushime, H., Kato, J. Y., and Livingston, D. M. (1993). Functional interactions of the retinoblastoma protein with mammalian D-type cyclins. *Cell* **73**, 487–497.

Favrot, M., Coll, J.-L., Louis, N., and Negolsecu, A. (1998). Cell death and cancer: Replacement of apoptotic genes and inactivation of death suppressor genes in therapy. *Gene Ther.* **5**, 728–739.

Flemington, E. K., Speck, S. H., and Kaelin, W. G., Jr. (1993). E2F-1-mediated transactivation is inhibited by complex formation with the retinoblastoma susceptibility gene product. *Proc. Natl. Acad. Sci. U.S.A.* **90**, 6914–6918.

Ford, H. L., and Pardee, A. B. (1999). Cancer and the cell cycle. *J. Cell. Biochem. (Suppl.)* **32/33**, 166–172.

Francke, U. (1976). Retinoblastoma and chromosome 13. *Birth Defects* **12**, 131–134.

Friend, S. H., Bernards, R., Rogeli, S., Weinberg, R. A., Rappaport, J. M., Albert, D. M., and Dryja, T. P. (1986). A human DNA segment with properties of the gene that predisposes to retinoblastoma and osteosarcoma. *Nature* **323**, 643–646.

Fukasawa, K., Choi, T., Kuriyama, R., Rulong, S., and Vande Woude, G. F. (1996). Abnormal centrosome amplification in the absence of p53. *Science* **271**, 1744–1747.

Fung, Y.-K. T., Murphee, A. L., T' Ang, A., Qian, J., Hinrichs, S. H., and Benedict, W. F. (1987). Structural evidence for the authenticity of the human retinoblastoma gene. *Science* **236**, 1657–1661.

Goodrich, D. W., Wang, N. P., Qian, Y.-W., Lee, E. Y.-H. P., and Lee, W.-H. (1991). The retinoblastoma gene product regulates progression through the G1 phase of the cell cycle. *Cell* **67**, 293–302.

Gu, W., Schneider, J. W., Condorelli, G., Kaushal, S., Mahdavi, V., and Nadal-Ginard, B. (1993). Interaction of myogenic factors and the retinoblastoma protein mediates muscle cell commitment and differentiation. *Cell* **72**, 309–324.

Gualberto, A., Aldape, K., Kozakiewicz, K., and Tisty, T. D. (1998). An oncogenic form of p53 confers a dominant, gain-of-function phenotype that disrupts spindle checkpoint control. *Proc. Natl. Acad. Sci. U.S.A.* **95**, 5166–5171.

Hagstrom, S. A., and Dryja, T. P. (1999). Mitotic recombination map of 13cen-13q14 derived from an investigation of loss of heterozygosity in retinoblastomas. *Proc. Natl. Acad. Sci. U.S.A.* **96**, 2952–2957.

Hansen, J. B., Petersen, R. K., Larsen, B. M., Bartkova, J., Alsner, J., and Kristiansen, K. (1999). Activation of peroxisome proliferator-activated receptor γ bypasses the function of the retinoblastoma protein in adipocyte differentiation. *J. Biol. Chem.* **274**, 2386–2393.

Harbour, J. W., and Dean, D. C. (2000). The Rb/E2F pathway: Expanding roles and emerging paradigms. *Genes Dev.* **14**, 2393–2409.

Harbour, J. W., Lai, S. L., Whang-Peng, J., Gazdar, A. F., Minna, J. D., and Kaye, F. J. (1988). Abnormalities in structure and expression of the human retinoblastoma gene in SCLC. *Science* **241**, 353–357.

Harbour, J. W., Luo, R. X., Dei Santi, A., Postigo, A. A., and Dean, D. C. (1999). Cdk phosphorylation triggers sequential intramolecular interactions that progressively block Rb functions as cells move through G1. *Cell* **98**, 859–869.

Harris, H., Miller, O. J., Klein, G., Worst, P., and Tachibana, T. (1969). Suppression of malignancy by cell fusion. *Nature* **223**, 363–368.

Hartwell, L. (1992). Defects in a cell cycle checkpoint may be responsible for the genomic instability of cancer cells. *Cell* **71**, 543–546.

Harvey, M., Sands, A. T., Weiss, R. S., Hegi, M. E., Wiseman, R. W., Pantazis, P., Giovanella, B. C., Tainsky, M. A., Bradley, A., and Donehower, L. A. (1993). *In vitro* growth characteristics of embryo fibroblasts isolated from p53-deficient mice. *Oncogene* **8**, 2457–2467.

Hatakeyama, M., Brill, J. A., Fink, G. R., and Weinberg, R. A. (1994). Collaboration of G1 cyclins in the functional activation of the retinoblastoma protein. *Genes Dev.* **8**, 1759–1771.

Hateboer, G., Kerkhoven, R. M., Shvarts, A., Bernards, R., and Beijersbergen, R. L. (1996). Degradation of E2F by the ubiquitin–proteasome pathway: Regulation by retinoblastoma family proteins and adenovirus transforming proteins. *Genes Dev.* **10**, 2960–2970.

He, D., Zeng, C., and Brinkley, B. R. (1995). Nuclear matrix proteins and the structural and functional components of the mitotic apparatus. *Int. Rev. Cytol.* **162B**, 1–74.

Helin, K., Harlow, E., and Fattaey, A. (1993). Inhibition of E2F-1 transactivation by direct binding of the retinoblastoma protein. *Mol. Cell. Biol.* **13**, 6501–6508.

Hensel, C. H., Hsieh, C. L., Gazdar, A. F., Johnson, B. E., Sakaguchi, A. Y., Naylor, S. L., Lee, W. H., and Lee, E. Y. (1990). Altered structure and expression of the human retinoblastoma susceptibility gene in small cell lung cancer. *Cancer Res.* **50**, 3067–3072.

Herwig, S., and Strauss, M. (1997). The retinoblastoma protein: A master regulator of cell cycle, differentiation and apoptosis. *Eur. J. Biochem.* **246**, 581–601.

Hinds, P. W., Mittnacht, S., Dulic, V., Arnold, A., Reed, S. I., and Weinberg, R. A. (1992). Regulation of retinoblastoma protein functions by ectopic expression of human cyclins. *Cell* **70**, 993–1006.

Hirsch, H. A., Gu, L., and Henry, R. W. (2000). The retinoblastoma tumor suppressor protein targets distinct general transcription factors to regulate RNA polymerase III gene expression. *Mol. Cell. Biol.* **20**, 9182–9191.

Hofmann, F., Martelli, F., Livingston, D. M., and Wang, Z. (1996). The retinoblastoma gene product protects E2F-1 from degradation by the ubiquitin–proteasome pathway. *Genes Dev.* **10**, 2949–2959.

Hong, F. D., Huang, H.-J. S., To, H., Young, L.-J. S., Oro, A., Bookstein, R., Lee, E. Y., and Lee, W.-H. (1989). Structure of the human retinoblastoma gene. *Proc. Natl. Acad. Sci. U.S.A.* **86**, 5502–5506.

Hu, N., Gutsmann, A., Herbert, D. C., Bradley, A., Lee, W.-H., and Lee, E. Y.-H. P. (1994). Heterozygous Rb-1Δ20/+ mice are predisposed to tumors of the pituitary gland with nearly complete penetrance. *Oncogene* **9**, 1021–1027.

Huang, H.-J. S., Yee, J.-K., Shew, J.-Y., Chen, P.-L., Bookstein, R., Friedmann, T., Lee, E. Y.-H. P., and Lee, W.-H. (1988). Suppression of the neoplastic phenotype by replacement of the human RB gene in human cancer cells. *Science* **242**, 1563–1566.

Hurwitz, R. L., Brenner, M. K., Poplack, D. G., and Horowitz, M. C. (1999). Retinoblastoma treatment (letter). *Science* **285**, 663–664.

Iavarone, A., Garg, P., Lasorella, A., Hsu, J., and Israel, M. A. (1994). The helix-loop-helix protein Id-2 enhances cell proliferation and binds to the retinoblastoma protein. *Genes Dev.* **8**, 1270–1284.

Jacks, T., Fazeli, A., Schmitt, E. M., Bronson, R. T., Goodell, M. A., and Weinberg, R. A. (1992). Effects of an Rb mutation in the mouse. *Nature* **359**, 295–300.

Karantza, V., Maroo, A., Fay, D., and Sedivy, J. M. (1993). Overproduction of Rb protein after the G1/S boundary causes G2 arrest. *Mol. Cell. Biol.* **13**, 6640–6652.

Kato, J., Matsushime, H., Heibert, S. W., Ewen, M. E., and Sherr, C. J. (1993). Direct binding of cyclin D to the retinoblastoma gene product (pRb) and pRb phosphorylation by the cyclin D-dependent kinase, CDK4. *Genes Dev.* **7**, 331–342.

Khan, S. H., and Wahl, G. M. (1998). p53 and pRb prevent rereplication in response to microtubule inhibitors by mediating reversible G1 arrest. *Cancer Res.* **58**, 396–401.

King, R. W., Peters, J. M., Tugendreich, S., Rolfe, M., Heiter, P., and Kirschner, M. W. (1995). A 20S complex containing CDC27 and CDC16 catalyzes the mitosis-specific conjugation of ubiquitin to cyclin B. *Cell* **81**, 279–288.

Kitagawa, M., Higashi, H., Jung, H. K., Suzuki-Takahashi, I., Ikeda, M., Tamai, K., Kato, J., Segawa, K., Yoshida, E., Nishimura, S., and Taya, Y. (1996). The consensus motif for phosphorylation by cyclin D1-Cdk4 is different from that for phosphorylation by cyclinA/E-Cdk2. *EMBO J.* **15**, 7060–7069.

Knudsen, E. S., and Wang, J. Y. J. (1996). Differential regulation of retinoblastoma protein function by specific Cdk phosphorylation sites. *J. Biol. Chem.* **271**, 8313–8320.

Knudsen, E. S., Buckmaster, C., Chen, T.-T., Feramisco, J. R., and Wang, J. Y. J. (1998). Inhibition of DNA synthesis by RB: Effects on G1/S transition and S-phase progression. *Genes Dev.* **12**, 2278–2292.

Knudson, A. G. (1971). Mutation and cancer: Statistical study of retinoblastoma. *Proc. Natl. Acad. Sci. U.S.A.* **68**, 820–823.

Knudson, A. G. (1993). Anti-oncogenes and human cancer. *Proc. Natl. Acad. Sci. U.S.A.* **90,** 10914–10921.

Lai, A., Lee, J. M., Yang, W. M., DeCaprio, J. A., Kaelin, W. G. Jr., Seto, E., and Branton, P. E. (1999). RBP1 recruits both histone deacetylase-dependent and -independent repression activities to retinoblastoma family proteins. *Mol. Cell. Biol.* **19,** 6632–6641.

Lane, D. P. (1992). p53, guardian of the genome. *Nature* **358,** 15–16.

Lanni, J. S., and Jacks, T. (1998). Characterization of the p53-dependent postmitotic checkpoint following spindle disruption. *Mol. Cell. Biol.* **18,** 1055–1064.

Lassorella, A., Noseda, M., Beyna, M., and Iavarone, A. (2000). Id2 is a retinoblastoma protein target and mediates signaling by Myc oncoproteins. *Nature* **407,** 592–598.

Lee, E. Y.-H. P., Chang, C.-Y., Hu, N., Wang, Y. C., Lai, C. C., Herrup, K., Lee, W.-H., and Bradley, A. (1992). Mice deficient for Rb are nonviable and show defects in neurogenesis and hematopoiesis. *Nature* **359,** 288–294.

Lee, E. Y.-H. P., Hu, N., Yuan, S.-S., Cox, L. A., Bradley, A., Lee, W.-H., and Herrup, K. (1994). Dual roles of the retinoblastoma protein in cell cycle regulation and neuron differentiation. *Genes Dev.* **8,** 2008–2021.

Lee, W.-H., Bookstein, R., Hong, F., Young, L. J., Shew, J. Y., and Lee, E. Y. (1987). Human retinoblastoma susceptibility gene: Cloning, identification, and sequence. *Science* **235,** 1394–1399.

Lees, J. A., Buchkovich, K. J., Marshak, D. R., Anderson, C. W., and Harlow, E. (1991). The retinoblastoma protein is phosphorylated on multiple sites by cdc2. *EMBO J.* **10,** 4279–4290.

Lees, J. A., Saito, M., Vidal, M., Valentine, M., Look, T., Harlow, E., Dyson, N., and Helin, K. (1993). The retinoblastoma protein binds to a family of E2F transcription factors. *Mol. Cell. Biol.* **13,** 7813–7825.

Lengauer, C., Kinzler, K. W., and Vogelstein, B. (1998). Genetic instablities in human cancers. *Nature* **396,** 643–649.

Loeb, K. R., and Loeb, L. A. (2000). Significance of multiple mutations in cancer. *Carcinogenesis* **21,** 379–385.

Lipinski, M. M., and Jacks, T. (1999). The retinoblastoma gene family in differentiation and development. *Oncogene* **18,** 7873–7882.

Ludlow, J. W., Glendening, C. L., Livingston, D. M., and DeCaprio, J. A. (1993). Specific enzymatic dephosphorylation of the retinoblastoma protein. *Mol. Cell. Biol.* **13,** 367–372.

Lukas, J., Herzinger, T., Hansen, K., Moroni, M. C., Resnitzky, D., Helin, K., Reed, S. I., and Bartek, J. (1997). Cyclin E-induced S phase without activation of the pRb/E2F pathway. *Genes Dev.* **11,** 1479–1492.

Lundberg, A. S., and Weinberg, R. A. (1998). Functional inactivation of the retinoblastoma protein requires sequential modification by at least two distinct cyclin–cdk complexes. *Mol. Cell. Biol.* **18,** 753–761.

Luo, R. X., Postigo, A. A., and Dean, D. C. (1998). Rb interacts with histone deacetylase to repress transcription. *Cell* **92,** 463–473.

Magnaghi-Jaulin, L., Groisman, R., Naguibneva, I., Robin, P., Lorain, S., Le Villain, J. P., Troalen, F., Trouche, D., and Harel-Bellan, A. (1998). Retinoblastoma protein represses transcription by recruiting a histone deacetylase. *Nature* **391,** 601–605.

Nead, M. A., Baglia, L. A., Antinore, M. J., Ludlow, J. W., and McCance, D. J. (1998). Rb binds c-Jun and activates transcription. *EMBO J.* **17,** 2342–2352.

Nevins, J. R. (1992). E2F: A link between the Rb tumor suppressor and the viral oncoproteins. *Science* **258,** 512–515.

Nicolas, E., Morales, V., Magnaghi-Jaulin, L., Harel-Bellan, A., Richard-Foy, H., and Trouche, D. (2000). RbAp48 belongs to the histone deacetylase complex that associates with the retinoblastoma protein. *J. Biol. Chem.* **275,** 9797–9804.

Niculescu III, A. B., Chen, X., Smeets, M., Hengst, L., Prives, C., and Reed, S. I. (1998). Effects of p21(Cip1/Waf1) at both the G1/S and G2/M cell cycle transitions: pRb is a critical determinant in blocking DNA replication and in preventing endoreplication. *Mol. Cell. Biol.* **18,** 629–643.

Nikitin, A. Yu., and Lee, W.-H. (1996). Early loss of the retinoblastoma gene is associated with impaired growth-inhibitory innervation during melanotroph carcinogenesis in Rb$^{+/-}$ mice. *Genes Dev.* **10,** 1870–1879.

Nikitin, A. Yu., Riley, D. J., and Lee, W.-H. (1999). A paradigm for cancer treatment using the retinoblastoma gene in a mouse model. *Ann. N.Y. Acad. Sci.* **886,** 12–22.

Nikitin, A. Yu., Juarez-Perez, M. I., Li, S., Huang, L., and Lee, W.- H. (1999). RB-mediated suppression of spontaneous multiple endocrine neoplasia and lung metastases in Rb$^{+/-}$ mice. *Proc. Natl. Acad. Sci. U.S.A.* **96,** 3916–3921.

Novitch, B. G., Mulligan, G. J., Jacks, T., and Lassar, A. B. (1996). Skeletal muscle cells lacking the retinoblastoma protein display defects in muscle gene expression and accumulate in S and G2 phases of the cell cycle. *J. Cell Biol.* **135,** 441–456.

Otterson, G. A., Chen, W., Coxon, A. B., Khleif, S. N., and Kaye, F. J. (1997). Incomplete penetrance of familial retinoblastoma linked to germ-line mutations that result in partial loss of RB function. *Proc. Natl. Acad. Sci. U.S.A.* **94,** 12036–12040.

Prendergast, G. C. (1999). Mechanisms of apoptosis by c-myc. *Oncogene* **18,** 2967–2987.

Riley, D. J., Nikitin, A. Yu., and Lee, W.-H. (1996). Adenovirus-mediated retinoblastoma gene therapy suppreses spontaneous pituitary melanotroph tumors in Rb$^{+/-}$ mice. *Nat. Med.* **2,** 1316–1321.

Riley, D. J., Liu, C.-Y., and Lee, W.-H. (1997). Mutations of N-terminal regions render the retinoblastoma protein insufficient for functions in development and tumor suppression. *Mol. Cell. Biol.* **17,** 7342–7352.

Ross, J. F., Liu, X., and Dynlacht, B. D. (1999). Mechanism of transcriptional repression of E2F by the retinoblastoma tumor suppressor protein. *Mol. Cell* **3,** 195–205.

Roth, J. A., and Cristiano, R. J. (1997). Gene therapy for cancer: What have we done and where are we going? *J. Natl. Cancer Inst.* **89,** 21–39.

Roth, J. A., Nguyen, D., Lawrence, D. D., Kemp, B. L., Carrasco, C. H., Ferson, D. Z., Hong W. K., Komaki, R., Lee, J. J., Nesbitt, J. C., Pisters, K. M., Putnam, J. P., Schea, R., Shin, D. M., Walsh, G. L., Dolormente, M. M., Han, C. I., Martin, F. D., Yen, N., Xu, K., Stephens, L. C., McDonell, T. J., Mukhopadhyay, T., and Cai, D., (1996). Retrovirus-mediated wild-type p53 gene transfer to tumors of patients with lung cancer. *Nat. Med.* **2,** 985–991.

Sardet, C., Vidal, M., Cobrinik, D., Geng, Y., Onufryk, C., Chen, A., and Weinberg, R. A. (1995). E2F-4 and E2F-5, two novel members of the E2F family, are expressed in early phases of the cell cycle. *Proc. Natl. Acad. Sci. U.S.A.* **92,** 2403–2407.

Schneider, J. W., Gu, W., Zhu, L., Mahdavi, V., and Nadal-Ginard, B. (1994). Reversal of terminal differentiation mediated by p107 in Rb$^{-/-}$ muscle cells. *Science* **264,** 1467–1471.

Sellers, W. R., Rodgers, J. W., and Kaelin, W. G., Jr. (1995). A potent transrepression domain in the retinoblastoma protein induces a cell cycle arrest when bound to E2F sites. *Proc. Natl. Acad. Sci. U.S.A.* **92,** 11544–11548.

Sellers, W. R., Novitch, B. G., Miyake, S., Heith, A., Otterson, G. A., Kaye, F. J., Lassar, A. B., and Kealin, W. G., Jr. (1998). Stable binding to E2F is not required for the retinoblastoma protein to activate transcription, promote differentiation, and suppress tumor cell growth. *Genes Dev.* **12,** 95–106.

Shan, B., and Lee, W.-H. (1994). Deregulated expression of E2F-1 induces S-phase entry and leads to apoptosis. *Mol. Cell. Biol.* **14,** 8166–8173.

Shan, B., Zhu, X. Chen, P.-L., Durfee, T., Yang, Y., Sharp, D., and Lee, W.-H. (1992). Molecular cloning of cellular genes encoding retinoblastoma-associated proteins: Identification of a gene with properties of the transcription factor E2F. *Mol. Cell. Biol.* **12,** 5620–5631.

Shimizu, E., Coxon, A., Otterson, G. A., Steinberg, S. M., Kratzke, R. A., Kim, Y. W., Fedorko, J., Oie, H., Johnson, B. E., and Mulshine, J. L. (1994). RB protein status and clinical correlation from 171 cell lines representing lung cancer, extrapulmonary small cell carcinoma, and mesothelioma. *Oncogene* **9**, 2441–2448.

Sidle, A., Palaty, C., Dirks, P., Wiggan, O., Kiess, M., Gill, R. M., Wong, A. K., and Hamel, P. A. (1996). Activity of the retinoblastoma family proteins, pRB, p107, and p130, during cellular proliferation and differentiation. *Crit. Rev. Biochem. Mol. Biol.* **31**, 237–271.

Slack, R. S., Skerjanc, I. S., Lach, B., Craig, J., Jardine, K., and McBurney, M. W. (1995). Cells differentiating into neuroectoderm undergo apoptosis in the absence of functional retinoblastoma proteins. *J. Cell Biol.* **129**, 779–788.

Solomon, E., Borrow, J., and Goddard, A. D. (1991). Chromosome aberrations and cancer. *Science* **254**, 1153–1160.

Stambolic, V., Mak, T. W. and Woodgett, J. R. (1999). Modulation of cellular apoptotic potential: Contributions to oncogenesis. *Oncogene* **18**, 6094–6103.

Sterner, J. M., Tao, Y., Kennett, S. B., Kim, H. G., and Horowitz, J. M. (1996). The amino terminus of the retinoblastoma (Rb) protein associates with a cyclin-dependent kinase-like kinase via Rb amino acids required for growth suppression. *Cell Growth Differ.* **7**, 53–64.

Strong, L. C., Riccardi, V. M., Ferrell, R. E., and Sparkes, R. S. (1981). Familial retinoblastoma and chromosome 13 deletion transmitted via insertional translocation. *Science* **213**, 1501–1505.

Sutcliffe, J. E., Brown, T. R., Allison, S. J., Scott, P. H., and White, R. J. (2000). Retinoblastoma protein disrupts interactions required for RNA polymerase III transcription. *Mol. Cell. Biol.* **20**, 9192–9202.

Szekely, L., Jiang, W. Q., Bulic-Jakus, F., Rosen, A., Ringertz, N., Klein, G., and Wiman, K. G. (1992). Cell type and differentiation dependent heterogeneity in retinoblastoma protein expression in SCID mouse fetuses. *Cell Growth Differ.* **3**, 149–156.

Thomas, R. C., Edwards, M. J., and Marks, R. (1996). Translocation of the retinoblastoma gene product during mitosis. *Exp. Cell Res.* **223**, 227–232.

Trouche, D., LeChalony, C., Muchardt, C., Yaniv, M., and Kouzarides, T. (1997). RB and hbrm cooperate to repress the activation functions of E2F1. *Proc. Natl. Acad. Sci. U.S.A.* **94**, 11268–11273.

Tugendreich, S., Tomkiel, J., Ernshaw, W., and Heiter, P. (1995). CDC27Hs colocalizes with CDC16Hs to the centrosome and mitotic spindle and is essential for the metaphase to anaphase transition. *Cell* **81**, 261–268.

Utomo, A. R., Nikitin, A. Yu., and Lee, W.-H. (1999). Temporal, spatial, and cell-type specific control of Cre-mediated DNA recombination in transgenic mice. *Nat. Biotechnol.* **17**, 1091–1096.

Wang, J., Guo, K., Wills, K. N., and Walsh, K. (1997). Rb functions to inhibit apoptosis during myocyte differentiation. *Cancer Res.* **57**, 351–354.

Weinberg, R. A. (1995). The retinoblastoma protein and cell cycle control. *Cell* **81**, 323–330.

Weintraub, S. J., Prater, C. A., and Dean, D. C. (1992). Retinoblastoma protein switches the E2F site from positive to negative element. *Nature* **358**, 259–261.

Whitaker, L. L., Su, H., Baskaran, R., Knudsen, E. S., and Wang, J. Y. (1998). Growth suppression by an E2F-binding defective retinoblastoma (RB): Contribution from the RB C pocket. *Mol. Cell. Biol.* **18**, 4032–4042.

Whyte, P., Buchkovich, K. J., Horowitz, J. M., Friend, S. H., Raybuck, M., Weinberg, R. A., and Harlow, E. (1988). Association between an oncogene an anti-oncogene: The adenovirus E1A protein binds to the retinoblastoma gene product. *Nature* **334**, 124–129.

Zabner, J., Ramsey, B. W., Meeker, D. P., Aitken, M. L., Balfour, R. P., Gibson, R. L., Launspach, J., Moscicki, R. A., Richards, S. M., and Standaert, T. A. (1996). Repeat administration of an adenovirus vector encoding cystic fibrosis transmembrane conductance regulator to the nasal epithelium of patients with cystic fibrosis. *J. Clin. Invest.* **97**, 1504–1511.

Zacksenhaus, E., Jiang, Z., Chung, D., Marth, J. D., Phillips, R. A., and Gallie, B. L. (1996). pRb controls proliferation, differentiation, and death of skeletal muscle cells and other lineages during embryogenesis. *Genes Dev.* **10**, 3051–3064.

Zarakowska, T., and Mittnacht, S. (1997). Differential phosphorylation of the retinoblastoma protein by G1/S cyclin-dependent kinases. *J. Biol. Chem.* **272**, 12738–12746.

Zhang, H. S., Gavin, M., Dahiya, A., Postigo, A. A., Ma, D., Luo, R. X., Harbour, J. W., and Dean, D. C. (2000). Exit from G1 and S phase of the cell cycle is regulated by repressor complexes containing HDAC-Rb-hSW1/SNF and Rb-hSW1/SNF. *Cell* **101**, 79–89.

Zheng, L., Chen, Y., and Lee, W.-H. (1999). Hec1p, an evolutionarily conserved coiled-coil protein, modulates chromosome segregation through interaction with SMC proteins. *Mol. Cell. Biol.* **19**, 5417–5428.

Zheng, L., Chen, Y., Riley, D. J., Chen, P.-L., and Lee, W.-H. (2000). Retinoblastoma protein enhances the fidelity of chromosome segregation mediated by hsHec1p. *Mol. Cell. Biol.* **20**, 3529–3537.

Zhu, X., Mancini, M. A., Chang, K.-H., Liu, C.-Y., Chen, C.-F., Shan, B., Jones, D., Yang-Feng, T. L., and Lee, W.-H. (1995). Characterization of a novel 350-kilodalton nuclear phosphoprotein that is specifically involved in mitotic-phase progression. *Mol. Cell. Biol.* **15**, 5017–5029.

6

CDK Inhibitors: Genes and Drugs

Jun-ya Kato, Kiichiro Tomoda, Yukinobu Arata, Toshiaki Tanaka, and Noriko Yoneda-Kato
Graduate School of Biological Sciences
Nara Institute of Science and Technology
Ikoma, Nara 630-0101, Japan

I. INTRODUCTION

Cell proliferation (cell cycle progression) is promoted by the sequential activation of a series of highly conserved protein kinase complexes composed of a regulatory subunit, cyclin, and a catalytic subunit, the cyclin-dependent kinase (Cdk). Because mammalian cell proliferation depends on the extracellular factors, and the G1 phase is the only period of time in which cells are sensitized to transmit signals extracellularly, regulators of cyclin–Cdk complexes that function during G1 play a central role in determination of cell proliferation. Antimitogenic signals block G1 cyclin–Cdk kinase activities and cell proliferation either by preventing the expression of G1 cyclins or by inducing small polypeptides that tightly associate with and inhibit the activity of G1 cyclin–Cdk kinases. These Cdk inhibitory polypeptides are generally called Cdk inhibitors, however, in this chapter, they are referred to as Cdk inhibitory proteins in order to discriminate these polypeptides from the chemicals that selectively inhibit the Cdk activity. The concept of a Cdk inhibitory protein is well conserved from yeast to human, but the number of the gene family and the actual amino acid sequences diverge from species to species. Mammalian cells contain seven Cdk inhibitory proteins, which can be classified into two groups, the Cip/Kip family (p21^{Cip1}, p27^{Kip1}, and p57^{Kip2}) and the INK4 family (p16^{INK4a}, p15^{INK4b}, p18^{INK4c}, and p19^{INK4d}). Because ectopic expression of any Cdk inhibitory proteins inhibits normal cell growth, they can all be considered as a candidate for tumor suppressor proteins. However, only a few are actually found to be altered in human cancers. In this chapter, we will focus on those candidates that are closely related to human cancers and describe the possible mechanism of tumor development and future prospect of diagnosis and therapy.

II. G1 REGULATION

A. Positive regulators of the cell cycle — G1 cyclin – Cdk complexes

Mammalian cell proliferation depends on the presence and absence of certain extracellular factors (Pardee, 1989). Mitogenic signals such as growth factor stimulation are absolutely required for cells to enter the cell cycle and proliferate. On the contrary, certain antimitogenic factors [e.g., transforming growth factor beta (TGF-β)] inhibit cell growth and induce other cell responses such as terminal cell differentiation, cell senescence, and apoptosis. An important point is that these extracellular factors are effective to cells only when they are in the G1 phase of the cell cycle. For example, cells continuously require growth factors while progressing through almost the entire G1 phase. However, once they initiate DNA replication, they no longer require mitogenic stimuli. Cells continue replicating chromosomal DNA and complete cell division in the absence of growth factors. Because cyclin–Cdk complex is a key factor that drives the cell cycle, and the G1 phase is the only period of time in which cells are sensitive to extracellular signals, cyclin–Cdk complexes that function during the G1 phase play a central role in determining whether cells are to proliferate or not (Sherr, 1993).

Mammalian cells contain two types of G1 cyclins, cyclins D and E. Cyclin D consists of three subtypes (D1, D2, and D3), expression of which is induced on stimulation with growth factors in a tissue-specific manner (for reviews, see Sherr, 1993, 1994). Cyclin D can form complexes with Cdk2, 4, 5, and 6 (Xiong *et al.*, 1992), among which Cdk4 and Cdk6 are the major catalytic partners in normally proliferating cells (Matsushime *et al.*, 1992; Meyerson and Harlow, 1994). All three D-type cyclins are equally capable of activating Cdk4 and Cdk6, and these two Cdks can be activated only by D-type cyclins (Kato *et al.*, 1993), thereby they are referred to as the cyclin D-dependent kinases. Although cyclins D2 and D3, but not D1, can activate Cdk2 *in vitro* and on a certain occasion (Ewen *et al.*, 1993), its physiological relevance remains to be clarified. It is not possible for Cdk5 to be activated by D-type cyclins. The expression patterns of three D-type cyclins and Cdk4/6 are tissue specific. The cyclin D1–Cdk4 kinase is the major cyclin D–Cdk complex in fibroblasts and macrophages (Matsushime *et al.*, 1994), whereas cyclin D2/D3–Cdk6 plays a central role in nonadherent hematopietic cells such as interleukin-3 (IL-3)-dependent myeloid cells, or T- and B-lympho-cytes (Meyerson and Harlow, 1994). Although the induction mechanism and physiological significance of the different cyclin D–Cdk complexes are not clear yet, cyclin D expression and the cyclin D–Cdk activity generally become maximum in the mid G1 (Matsushime *et al.*, 1994), and cyclin D–Cdk function is required in mid to late G1 in a variety of proliferating cells (Baldin *et al.*, 1993; Quelle *et al.*, 1993). However, it is interesting to note that only cyclin D1 is highly involved in human cancers (for reviews, see Hall and Peters, 1996; Sherr, 1996).

The expression of another G1 cyclin, cyclin E, occurs after that of cyclin D and leads to the activation of its catalytic partner, Cdk2 (Koff *et al.*, 1991, 1992). In contrast to cyclin D being the growth factor-inducible gene, cyclin E is a cell cycle-dependent factor, being rapidly induced to express before the onset of DNA replication in a transcription-dependent manner (Ohtani *et al.*, 1995), and swiftly declining after entry into the S phase via the ubiquitin-dependent proteolytic pathway (Singer *et al.*, 1999; Nakayama *et al.*, 2000). Both G1 cyclin–Cdk complexes share the common substrate, the Rb protein, but the sites of phosphorylation and their subsequent effects are different (Wein-berg, 1995; Sherr, 1996; Taya, 1997). Phosphorylation by cyclin D–Cdk4 kinase changes the conformation of the Rb protein to allow intramolecular interaction between the C-terminal region and the central pocket, which, in turn, displaces histone deacetylase from the pocket and blocks active transcrip-tional repression by Rb. This also facilitates phosphorylation of Rb protein by cyclin E–Cdk2 kinase, resulting in dissociation of Rb from E2F and transcrip-tional activation of E2F-inducible genes (Harbour *et al.*, 1999). Rb is the major substrate for cyclin D–Cdk complexes because they do not phosphorylate other proteins such as the canonical CDK substrate, histone H1, *in vitro* (Matsushime *et al.*, 1992; Ewen *et al.*, 1993; Kato *et al.*, 1993), and because inhibition of cyclin D–Cdk activity leads to the cell cycle arrest in G1 phase in an Rb-depen-dent manner (Baldin *et al.*, 1993; Quelle *et al.*, 1993; Guan *et al.*, 1994; Koh *et al.*, 1995; Lukas *et al.*, 1994, 1995a,b; Medema *et al.*, 1995), while cyclin E–Cdk2 kinase has its own substrates that seem to be more closely connected with the initiation of DNA replication (Fang and Newport, 1991; Krude *et al.*,

1997; Lukas *et al.*, 1997; Zhao *et al.*, 1998). The Rb protein binds to and negatively regulates the activities of a series of different proteins (Weinberg, 1995),
among which the transcription factor E2F is the most important target for transition into the S phase, because ectopic overexpression of this factor itself is sufficient to initiate DNA replication in quiescent cells (Johnson *et al.*, 1993; Qin
et al., 1994; Shan and Lee, 1994). The target genes for E2F contain a variety of
genes involved in cell cycle regulation, DNA replication, oncogenic transformation, and apoptosis (Dyson, 1998; Nevins, 1998), and interestingly the cyclin E
gene is one of them (Ohtani *et al.*, 1995). Since cyclin E–Cdk2 enhances the
E2F activity through phosphorylation of the Rb protein, a positive feedback
loop is created between cyclin E and E2F, which allows both activities to
increase rapidly in late G1. For the onset of S phase, E2F-specific target gene
products and cyclin E–Cdk2-specific target substrates cooperate with each other
(Zhang *et al.*, 1999; Arata *et al.*, 2000) to activate the complex formed around
the origin of DNA replication on chromosomal DNA (Stillman, 1996). After
entry into S phase, the G1 regulatory factors are destroyed or inactivated,
preparing for the next G1 phase.

B. Negative regulators of the cell cycle — Mammalian Cdk inhibitory proteins

When cells undergo different cellular responses such as proliferative senescence or terminal differentiation, the activity of the proliferation machinery
(mostly the G1 cyclin–Cdk activity) needs to be shut down. The activities of
G1 cyclin–Cdk kinase can be negatively regulated in multiple ways: reduction
of the G1 cyclin expression (Matsushime *et al.*, 1991), inhibitory phosphorylation on the Cdk catalytic subunit (Gu *et al.*, 1992; Terada *et al.*, 1995;
Iavarone and Massague, 1997), alteration of the subcellular localization from
the nucleus to the cytoplasm, and by a group of small polypeptides collectively
called Cdk inhibitory proteins (for reviews, see Elledge and Harper, 1994;
Sherr and Roberts, 1995). Cdk inhibitory proteins function in a stoichiometric
way. They block the activity of the G1 cyclin–Cdk kinases by tightly binding
to them. Cdk inhibitory proteins are found in mammals, frogs, worms, flies,
plants, and yeasts, but their gene family and their amino acid sequences are
quite different from species to species. Therefore, there may not be the common regulatory mechanism for different Cdk inhibitory proteins, which is conserved among species.

In mammals, seven Cdk inhibitory proteins have been identified and
characterized. They can be divided into two different categories; one is the
Cip/Kip family, and the other is the INK4 family. The Cip/Kip family includes
$p21^{Cip1}$, $p27^{Kip1}$, and $p57^{Kip2}$, and inhibits a wide variety of cyclin–Cdk kinases
(cyclin D–Cdk4/Cdk6, cyclin E–Cdk2, and cyclin A–Cdk2) by binding to the

cyclin–Cdk complex without breaking it up. The Cip/Kip proteins contain a highly conserved domain in the N-terminal half of the molecule, through which they bind to and sufficiently inactivate cyclin–Cdk complexes. The C-terminal half of the molecule is diverged among the family and is proposed to contribute to the function specific to the individual molecule (Chen *et al.*, 1995, 1996; Nakanishi *et al.*, 1995; Russo *et al.*, 1996). More recent work has shown that the Cip/Kip proteins not only possess the inhibitory function but also act as positive regulators of cyclin D-dependent kinases. This concept is reviewed in detail by Sherr and Roberts (1999).

1. p21^{Cip1}

p21 was originally identified as an interactor of the cyclin–Cdk complex (Gu *et al.*, 1993; Harper *et al.*, 1993; Xiong *et al.*, 1993), one of the p53 tumor suppressor-inducible genes (El-Deiry *et al.*, 1993), and a senescence-inducible gene (Noda *et al.*, 1994). Furthermore, it is suggested to play an important role in p53-mediated G1 checkpoint control initiated by DNA damage (Dulic *et al.*, 1994). However, p21-knockout mice are born and grown normally, and do not exhibit higher frequency of tumor development than normal littermates (Brugarolas *et al.*, 1995; Deng *et al.*, 1995). Cells isolated from p21-knockout mice exhibit only a partial deficiency in G1 checkpoint control (Brugarolas *et al.*, 1995; Deng *et al.*, 1995), indicating that p21 functions in collaboration with other p53-target gene products to execute a G1 checkpoint control program. Such other target gene products include 14-3-3 and Gadd45. More recently it has been shown that p21 plays a role in G2 checkpoint control as well and prevents polyploidy of the chromosomal DNA (Bunz *et al.*, 1998). Detailed analysis of the p21-knockout mice reveals that although most Cdk inhibitory functions of p21 can be complemented by other Cip/Kip proteins, p21-knockout cells are slightly impaired in their ability of terminal differentiation and the maintenance of resting hematopoietic cells (Cheng *et al.*, 2000).

p21 proteins are ubiquitinated and degraded by 26S proteasome. However, a more recent report shows that ubiquitination is not absolutely required for p21 degradation (Sheaff *et al.*, 2000). Whatever the mechanism, degradation of p21 is constant during the cell cycle and differentiation, suggesting that p21 regulation is transcriptional. In fact, p21 mRNA is induced by a variety of transcription factors including STATs, RAR, vitamin D-receptor, MyoD, and BRCA1 in addition to p53 in response to DNA damage (Dulic *et al.*, 1994), terminal differentiation, and replicative senescence (Noda *et al.*, 1994), and it disappears on inactivation of the transcription factors.

p21 interacts with PCNA (proliferating cell nuclear antigen) in the unique C-terminal domain (Waga *et al.*, 1994; Chen *et al.*, 1995) and modulates the activities of DNA polymerase δ (Li *et al.*, 1994) and DNA-(cytosine-5) methyltransferase (Chuang *et al.*, 1997). p57 is also capable of interacting with PCNA, but p27 lacks this ability. Besides p21 blocking the cyclin–Cdk kinase activity through the domain located in the N-terminus, p21 binds to and inhibits stress-activated JNK (Shim *et al.*, 1996) and apoptosis-inducing ASK kinases (Asada *et al.*, 1999), suggesting a role outside of the cell cycle regulation.

2. p27[Kip1]

p27 was first identified as an inhibitory activity toward cyclin E–Cdk2 kinase in cells arrested by the treatment of TGF-β (Koff *et al.*, 1993). Gel filtration analysis coupled with various biochemical techniques revealed that this inhibitory activity is originated from a protein with an approximate molecular mass of 27 kDa. With affinity chromatography, the protein was purified and sequenced (Polyak *et al.*, 1994a,b). Almost at the same time, the *p27* gene was cloned by the yeast two hybrid screen with a bait of cyclin D (Toyoshima and Hunter, 1994). p27 is shown to be induced in response to a wide variety of antimitogenic signals such as growth factor-deprivation (Nourse *et al.*, 1994), contact inhibition (Polyak *et al.*, 1994a), and treatment with cAMP (Kato *et al.*, 1994) or immunosuppressant rapamycin (Nourse *et al.*, 1994). Although p21 is largely induced by the intracellular signals such as DNA damage, p27 plays an important role in growth control in response to extracellular signals. p27-knockout mice are larger than normal littermates because of the increase in cell number of each organ, and at a later period, they develop pituitary tumors (Kiyokawa *et al.*, 1996; Nakayama *et al.*, 1996; Fero *et al.*, 1996). These findings indicate that p27 plays a pivotal role in organ formation during early development and tumor suppression as well as in cell cycle control. In addition, several reports imply that p27 is involved in apoptotic regulation in some instances (e.g., see Hiromura *et al.*, 1999).

Although p27 can be transcriptionally activated by the cAMP-response-element-binding protein (CREB)-binding protein (CBP) coactivator in response to retinoic acid treatment (Kawasaki *et al.*, 1998) or by the Ah receptor (Kolluri *et al.*, 1999), the main regulatory mechanism for p27 is posttranslational. p27 protein level fluctuates during the cell cycle (high in G0/G1 and low in S, G2, and M phases), while p27 mRNA is constant throughout the cell cycle. This is mostly due to the activation and inactivation of substrate-specific and cell cycle-dependent proteolysis (Pagano *et al.*, 1995). Although there are some data which imply the translational control of p27 expression (Hengst and Reed, 1996; Millard *et al.*, 1997, 2000), little is known about this regulation. Among several p27-specific proteolytic activities reported so far, degradation through the ubiquitination and 26S proteasome-mediated mechanism currently attracts most attention of the researchers. The SCF complex composed of Skp2, Skp1, Cul1, and CDC34 (SCF^{Skp2}) functions as a ubiquitin ligase to p27 (Carrano et al., 1999; Sutterluty et al., 1999; Tsvetkov et al., 1999). The Skp2 component contains an F-box domain, directly interacts with the targeting molecule, and can be replaced with other F-box proteins for different substrates. For ubiquitination, p27 needs to be phosphorylated because the F-box protein interacts with a target molecule once it is phosphorylated. Although several residues in p27 are phosphorylated in vivo, threonine at 187 seems to be the key residue whose phosphorylation triggers ubiquitination and subsequent degradation (Sheaff et al., 1997; Vlach et al., 1997).

The cyclin E–Cdk2 complex has been proposed to be the kinase that phosphorylate threonine 187 (Sheaff et al., 1997; Vlach et al., 1997). Therefore, p27 normally inhibits the activity of this kinase complex but, on a certain occasion, becomes a substrate of this kinase. Therefore, the simplest scenario is as follows. During early and mid G1, p27 suppresses low levels of cyclin E–Cdk2 kinase. Once the cyclin E–Cdk2 complex becomes dominant in late G1, it phosphorylates p27 to initiate association with and ubiquitination by SCF^{Skp2}, and degradation by 26S proteasome. The results from the analysis of Skp2-knockout mice support the idea that SCF^{Skp2} functions as a negative regulator of p27 in vivo (Nakayama et al., 2000). However, the complexity stems from the findings that the target molecules of SCF^{Skp2} include cyclin D, p21, and E2F in addition to p27. Furthermore, detailed time course experiments reveal that p27 is degraded far before the maximum activation of the cyclin E–Cdk2 kinase. Thus, it seems plausible that p27 downregulation is controlled by the multistep mechanisms, which may result in the complication of the regulation. Other proteolytic activities for p27 reported so far include a ubiquitin-independent processing enzyme and caspases (including caspase-3-like activity), but further analysis will be required to evaluate the physiological relevance of these activities.

Another negative regulator of p27 is Jab1 (Tomoda et al., 1999), which specifically interacts with p27, but not with cyclins B1 and E, Cdk2, or closely related p21. Overexpression of Jab1 in mammalian cells translocates p27 from the nucleus to the cytoplasm, decreasing the amount of p27 by accelerating its degradation. Ectopic Jab1 expression in mouse fibroblasts partially overcomes p27-mediated G1 arrest and markedly reduces their dependence on serum. It seems likely that Jab1 functions as a nuclear–cytoplasmic transporter and in fact, blockade of nuclear export by leptomycin B inhibits Jab1-mediated downregulation of p27. However, the p27 molecule lacking a nuclear localization signal sequence and the mutant fused to artificial nuclear export signal sequence stably remain in the cytoplasm, suggesting that unknown functions of Jab1 other than p27-nuclear export are required for p27 downregulation. Jab1 was originally identified as a coactivator of c-Jun and Jun D, but an increase in AP-1 activity is neither necessary nor sufficient for Jab1-mediated p27 degradation. Furthermore, addition of recombinant Jab1 proteins into the in vitro ubiquitination or degradation assays does not affect the kinetics of the reaction, but it remains to be clarified whether the ubiquitin system, especially the SCF^{Skp2} ubiquitin ligase complex, is involved in Jab1-mediated downregulation of p27.

Jab1 is also identified as the fifth component (CSN5) of the COP9 signalosome complex (Deng *et al.*, 2000). The COP9 signalosome was first identified as a repressor of photomorphogenesis in *Arabidopsis* (Wei and Deng, 1999). Identification of the mammalian COP9 signalosome complex with the same composition of the components as in plants suggests that this complex plays a pivotal role in a more fundamental biological function. Although two forms of Jab1-containing complexes are found, some COP9 signalosome subunits other than Jab1 are capable of reducing p27 expression, suggesting that p27 downregulation by Jab1 is through the COP9 signalosome. In *Arabidopsis*, COP1 and HY5 are the downstream components of the COP9 signalosome. HY5 is a bZIP transcription factor that controls the expression of light-inducible genes, and COP1 is a RING-finger protein with WD-40 repeats whose nuclear abundance is negatively regulated by light. Because COP1 regulates stability of HY5 through the proteasome- dependent degradation and a COP1 homolog exists in mammals, it is possible that COP1 mediates p27 degradation through Jab1. The hypothesis needs to be further investigated by adequate experiments.

In some cases, the activity of p27 protein can be regulated without degradation. c-Myc oncoprotein induces the expression of cyclin D2, which, presumably in a complex with Cdk4, sequesters and inactivates p27 (Perez-Roger *et al.*, 1999; Bouchard *et al.*, 1999), facilitating progression through the G1 phase. Similarly, the adenovirus-encoded E1A and human papilloma virus-encoded E7 oncoproteins bind to the C-terminal half of p27 and neutralize the Cdk inhibitory activity without inducing degradation (Mal *et al.*, 1996). In proliferating cells, low level of the remaining p27 is sequestered to the cyclin D–Cdk4 complexes. In tumor cells with the amplified or transcriptionally activated cyclin D gene, or cancer cells infected with these viruses, nullification of the p27 function may lead to uncontrolled growth of transformed cells.

3. p57^{Kip2}

p57 was identified as the third member of the Cip/Kip family of Cdk inhibitory proteins (Lee *et al.*, 1995; Matsuoka *et al.*, 1995). In contrast to p21 and p27, which are expressed in a wide variety of tissues, p57 expression is tissue-specific. The *p57* gene is located on the human chromosome 11p15.5, which is often deleted in Bechwith-Weidemann syndrome and Wilms tumors and is under the control of genomic imprinting. In fact, p57 is regulated by imprinting and expresses only from the maternal genome. p57-knockout mice exhibit some phenotypes characteristic to the above-mentioned syndromes (Zhang *et al.*,

1997), but the *p57* gene mutation is rarely found in Wilms' tumor and in the Bechwith-Weidemann syndrome, implying that these diseases are due to the loss of more than one (p57) factor mapped to this region of the chromosome. From the analysis of p57-single-knockout and p27/p57-double-knockout mice, p57 is shown to play an important role in organ development and cellular differentiation (P. Zhang *et al.*, 1998).

4. INK4

The INK4 family contains four members, p16[INK4a] (Serrano *et al.*, 1993), p15[INK4b] (Hannon and Beach, 1994), p18[INK4c] (Guan *et al.*, 1994; Hirai *et al.*, 1995), and p19[INK4d] (Chan *et al.*, 1995; Hirai *et al.*, 1995), and they all consist of four ankyrin repeats. The ankyrin repeats were originally identified in a transcription inhibitor IκB, and proposed to function in a protein–protein interaction. The target of INK4 proteins is restricted only to the cyclin D-dependent kinases, Cdk4 and Cdk6. INK4 tightly binds to Cdk4 and Cdk6, eventually preventing their complex formation with cyclin D (Parry *et al.*, 1995). The precise signals and mechanisms that induce each of the INK4 proteins remain poorly understood. p16[INK4a] is not normally expressed in living animals, but the gene is transcriptionally activated once the cells are transferred to the *in vitro* culture system. p16[INK4a] is induced to express when cells undergo senescence, and it is highly expressed in some types of tumor cells, implying involvement in tumor suppression (see below). It is clearly evident that p15[INK4b] is induced by TGF-β and contributes to its ability to induce G1 arrest (Hannon and Beach, 1994). p18[INK4c] and p19[INK4d] are moderately expressed in some cultured cell lines, but *in vivo* they are focally expressed during fetal development, suggesting a role in terminal differentiation.

III. p16[INK4a] AND THE Rb PATHWAY

Among the seven mammalian Cdk inhibitory proteins, p16[INK4a] is the only locus in which genetic alterations are highly connected with human tumors (for review, see Hall and Peters, 1996). Deletions and mutations of the *INK4a* gene locus was originally identified as the tumor suppressor gene *MTS1* in melanomas, but later, alterations of this gene locus were discovered in a wide variety of human cancers (Elledge and Harper, 1994; Sherr and Roberts, 1995). Inactivation of *INK4a* includes (1) minimum or global deletion including spanning, covering, and encompassing this region, (2) missense and nonsense point mutations within the coding sequences, (3) transcriptional blockade by methylation in the vicinity of the promoter region, and (4) neutralization of the Cdk inhibitory activity of the p16 protein by interaction with Tax, the viral oncoprotein encoded by HTLV-1 (Suzuki *et al.*, 1996). An interesting point is that there is a reversed correlation between mutations in the *INK4a* gene and other G1 phase-regulatory genes. For example, the *Rb* gene mutation is rare in cancer cells that contain deletions in the *INK4a* gene. Conversely, the *INK4a* gene is usually intact in tumor cells harboring inactivating mutations in the *Rb* gene (Hall and Peters, 1996). Taking it into consideration that the cyclin D1 or Cdk4 gene is amplified or overexpressed, which is frequently observed in tumor cells retaining wild-type *INK4a* and *Rb* genes, almost all types of human cancer cells contain alterations in one of the *INK4a*, *cyclin D1*, *Cdk4*, and *Rb* genes. Moreover, the most paradoxical observation that expression of tumor suppressor p16 is elevated in some tumor cells can be explained by the genetic inactivation of Rb or the enhancement of cyclin D–Cdk4 kinase in those particular tumor cells. Another example of such genetic interaction is seen in ARF, MDM2, and p53 (see below).

Biochemically, p16 directly binds to Cdk4 and Cdk6 and inhibits their associated protein kinase activity (Serrano *et al.*, 1993). Because the major target substrate of cyclin D–Cdk4 kinase is the Rb protein, overexpression of p16 only arrests proliferation of cells containing functional Rb proteins (Koh *et al.*, 1995; Lukas *et al.*, 1995b; Medema *et al.*, 1995). Thus, the signaling pathway consisting of p16^{INK4a}/cyclin D–Cdk4/Rb is proven to be both genetically and biochemically functioning in proliferation control and tumor suppression (Hall and Peters, 1996). Because p16 (the INK4 protein) is exclusively found to be associated with Cdk4 and Cdk6 monomers *in vivo*, one may imagine that p16 (the INK4 protein) competes with cyclins for binding to Cdks. However, in a test tube or under a certain *in vitro* cell culture condition, a triple complex of INK4 protein–cyclinD–Cdk4 can be formed (Hirai *et al.*, 1995; Reynisdottir and Massague, 1997), indicating that the contact sites for p16 and cyclin are different in Cdk4. Although the precise mechanism of p16 action in normal cells remains to be clarified, it is suggested that p16 bound-Cdk4 is more resistant to interaction with cyclin D than monomeric Cdk4 and that free cyclin D is more unstable than the Cdk4-bound form. Therefore, the current most reasonable model is that at the beginning of p16 induction, p16 binds to and inhibits the cyclin D–Cdk4 dimeric complex, and later p16 additionally impairs complex formation between cyclin D and Cdk4.

Although p16 directly inhibits the cyclin D–Cdk4 activity, the Cip/Kip family of Cdk inhibitory protein is required to shut down the cyclin E–Cdk2 kinase and to arrest the cell cycle in G1 (Sherr and Roberts, 1999). In proliferating cells, the Cip/Kip family of Cdk inhibitory proteins (p21 and p27) are bound to abundantly existing cyclin D–Cdk4 complexes, and their inhibitory activities are suppressed. In other words, cyclin D–Cdk4 complexes have a second function other than phosphorylating the Rb protein, which is to sequester and neutralize p21 and p27. Since p16 eventually breaks down cyclin D–Cdk4 complexes, in cells with high p16^{INK4a} expression, p21 and p27 are released from cyclin D–Cdk4 complexes and inhibit the cyclin E–Cdk2 kinase. This is the mechanism in which Cdk2 activities are suppressed in cells arrested by overexpression of Cdk4-specific inhibitory proteins.

IV. p19^{ARF} AND p53 PATHWAY

One of the paradoxical observations concerning the INK4 family is that only INK4a [but not the other INK4s (INK4b, c, and d)] is exclusively involved in human cancers (Hall and Peters, 1996; Elledge and Harper, 1994; Sherr and Roberts, 1995). The simplest interpretation is that INK4a is a part of the checkpoint mechanism and is transcriptionally activated to avoid unscheduled proliferation. In fact, cells undergoing senescence stimulate p16^{INK4a} expression. An alternative explanation is that the *INK4a* locus, but not other *INK4* loci, encodes for another protein, p19^{ARF} (Quelle *et al.*, 1995). p16^{INK4a} and p19^{ARF} transcripts contain the common exons 2 and 3, but utilize the first exons specific to each of them (exon 1a and 1b for p19^{ARF} and p16^{INK4a}, respectively). Because their open reading frames are not the same, their amino acid sequences are completely different, although they share common second and third exons. Ectopic overexpression of p19^{ARF} arrests normally proliferating fibroblasts at G1 and G2 phases, but fails to do so in the cells harboring mutations or deletions within the *p53* gene locus or containing viral oncoproteins that neutralize the function of p53, indicating that the effect of p19^{ARF} is dependent on the p53 status. Thus, the *INK4a* gene locus encodes two different proteins with no homology, each of which is capable of negatively regulating the cell cycle progression by independently functioning upstream of two independent major tumor suppressor gene products, Rb and p53.

p53 is the molecule playing a central role in checkpoint control mechanism (for review, see Levine, 1997; Sherr, 1998; Giaccia and Kastan, 1998). p53 is regulated in multiple ways and functions as a transcription factor to enhance expression of genes involved in the control of cell cycle and apoptosis. One of the p53-inducible genes is *MDM2*, which, in turn, negatively regulates p53 by promoting its degradation, thereby creating a negative feedback loop to allow the effect of p53 to be temporal. p19^{ARF} binds to MDM2 and interferes with its function as a negative regulator of p53 (Pomerantz *et al.*, 1998; Y. Zhang *et al.*, 1998). Currently, two models, although they are not mutually exclusive, are proposed. One is to inhibit the enzymatic activity associated with MDM2 (Honda and Yasuda, 1999), the ubiquitin ligase E3 specific to p53, which is required for downregulation of p53. The other is to sequester MDM2 by transporting MDM2 from the nucleoplasm to the nucleolus (Weber *et al.*, 1999). p19^{ARF} contains its own nucleolar localization signal sequences, but it seems that the transportation of the MDM2–p19^{ARF} complex to the nucleolus is mediated by the signal sequence present in the C-terminus of MDM2.

Expression of p19^{ARF} is induced when cells undergo senescence (Kamijo *et al.*, 1997). The detailed activation mechanism of the ARF promoter remains to be clarified, but several factors are reported to induce ARF expression. One such factor is the transcription factor E2F-1 (Bates *et al.*, 1998). E2F is one of the most important targets of the Rb protein. The Rb–E2F complex binds to the E2F-site and strongly represses transcription of the genes containing E2F-sites in the promoter. Phosphorylation of the Rb protein by the G1 cyclin–Cdk kinase releases E2F from Rb-mediated restraint, and the resultant "free E2F" activates transcription of a subset of E2F-inducible genes, some of which encode proteins functioning to initiate and promote chromosomal DNA replication. Among the six members of the E2F gene family, only the first member (E2F-1) is functionally involved in induction of apoptosis as well as in cell cycle progression. E2F-1-mediated apoptosis includes both p53-dependent and -independent mechanisms, and the former seems to be mediated by p19^{ARF}. The promoter of ARF contains the E2F-binding site, and E2F does bind to this site *in vitro*. However, it is suggested that E2F-1-mediated ARF induction includes the indirect mechanism as well. Whatever the major mechanism, the observation that E2F-1 induces p19^{ARF} links the Rb pathway with the p53 pathway. One can easily imagine that mitogenic signals activate the Rb pathway (cyclin D–Cdk4/Rb/E2F) to induce S-phase entry and at the same time activate the p53 pathway through E2F-1/p19^{ARF} to monitor and judge whether the timing of S-phase initiation is adequate for the cells. Besides E2F-1, several transcription factors (Jacobs *et al.*, 1999a,b) and other factors involved in the signal transduction pathway have been reported in the literature.

Most nucleotide alterations found within the *INK4/ARF* locus locate to the second exon and cause missense or nonsense mutations in the p16^{INK4a} coding frame but do not affect the amino acid sequences of p19^{ARF} (Quelle *et al.*, 1997), suggesting that it is p16^{INK4a} that mainly functions as a tumor suppressor in human cancers. However, p19^{ARF}-specific exon 1b knockout mice were highly susceptible to spontaneous tumor development, and the challenge to carcinogens further increased the frequency of spontaneous tumor development (Kamijo *et al.*, 1997). Very interestingly, exon 2 knockout, which is supposed to impair both p16^{INK4a} and p19^{ARF}, provides mice with the same phenotype as that of ARF-specific knockout (Serrano *et al.*, 1996). Although p16^{INK4a}-specific knockout mice have not been fully explored, these results suggest that p19^{ARF} can function as a tumor suppressor protein. In fact, some types of tumors contain mutations that exclusively affect p19^{ARF} expression (or functions). Thus, mutations and deletions in both *INK4a* and *ARF* genes need to be examined to find out whether this locus is inactivated.

V. p27 AND HUMAN CANCER

In contrast to p16^{INK4a}, none of the Cip/Kip family of Cdk inhibitory proteins are found to be genetically involved in human cancers. Although p21^{Cip1} is one of the p53 inducible gene products and plays an important role in the p53 signaling pathway, mutations or deletions of the *p21* gene locus have not been found. Similarly, mutations of the *p27* gene are rare except for that when the p27 locus is occasionally found to be hemizygously deleted in acute lymphoblastic leukemia, non-Hodgkin's lymphoma, adult T-cell leukemia/lymphoma, myelodysplastic syndrome, and acute myeloid leukemia. In these cases, the remaining locus is intact and produces detectable mRNA and proteins, and, therefore, it was believed that it was not *p27* but some other genes mapped to the vicinity of *p27* (human chromosome 12p12-13) that was responsible for tumor suppression. On the contrary, analysis of p27-deficient animals showed that the *p27* gene can function as a tumor suppressor gene, at least in the mouse (Kiyokawa *et al.*, 1996; Nakayama *et al.*, 1996; Fero *et al.*, 1996). p27-knockout mice (p27$^{-/-}$) develop pituitary tumors and are more susceptible to malignant tumor development in multiple tissues when exposed to radiation and chemical carcinogens. Interestingly, p27 heterozygous mice (p27$^{+/-}$) also show an increased predisposition to malignant tumor induction by carcinogenic insults (Fero *et al.*, 1998). Therefore, the *p27* gene is not the classic tumor-suppressor gene but is haploinsufficient for tumor suppression. However, total rates of *p27* gene mutations and deletions are still not very high in human cancers, precluding the possibility that p27 itself is the major genetic target for malignant transformation.

In 1997, three independent groups reported new insights into the relationship between p27 and cancers. They found that cancers with low p27 protein expression are well correlated with a poor prognosis (Porter *et al.*, 1997; Catzavelos *et al.*, 1997; Loda *et al.*, 1997). They first found this correlation in breast and colorectal carcinomas, but now this is the case for a wide variety of human tumors. Because in most of these cases the *p27* gene is not altered, the genetic target seems to be functioning upstream of p27. In the case of p27$^{+/-}$-mice mentioned above, it seems plausible that inactivation of one allele, leading to a moderate decrease in p27 protein expression, is sufficient to predispose to tumorigenesis. The next obvious question would be what is the genetic target that is responsible for low p27 protein expression in malignant tumors. The satisfying answer to this question is not ready yet, but the candidates include (1) Skp2 and Cul1, components of the SCF ubiquitin ligase, (2) cyclin E or its upstream regulator such as Rho, Tsc1, and ERM, and (3) Jab1 and other COP9 signalosome components. The *Skp2* gene is amplified in certain tumors and often overexpresses in transformed cells (Zhang *et al.*, 1995), which makes the gene a good candidate as the genetic target of malignant transformation. The cyclin E gene is overexpressed in some tumors although the gene amplification is rare, and Tsc1 (the tuberous sclerosis complex gene-2, also called hamartin, see Soucek *et al.*, 1998)/ERM/Rho functions to regulate cyclin E–Cdk2 activity. In another example, cyclin K, Kaposi's sarcoma virus (KSHV or human herpes virus-8)-encoded cyclin, activates Cdk6 and, in contrast to cyclin D-activated Cdk6, phosphorylates p27 *in vivo* to induce downregulation of p27 (Ellis *et al.*, 1999; Mann *et al.*, 1999). Leukemia-associated chimeric protein NPM-MLF1 binds to SGN3 and reduces p27 expression through the COP9 signalosome.

Because G1 cyclin–Cdk complexes mainly locate in the nucleus to phosphorylate specific substrates that mostly happen to be the nuclear protein, p27 needs to be in the nucleus to inhibit the cyclin–Cdk kinase activity and subsequently arrest the cell cycle in G1. As a matter of fact, p27 is a nuclear protein in normal cells and is found in the nucleus of most cancer cells. However, in some tumor cells, p27-specific signals are detected in the cytoplasm as well. The biological relevance has not been determined yet, but cytoplasmic location of p27 is associated with a more malignant phenotype of the tumors compared with those with nuclear p27 and is often seen during the early stages of carcinogenesis. Therefore, it appears that p27 is partially inactivated at the early stages by the cytoplasmic translocation, and as the tumor progression proceeds to the later stages, p27 is more perfectly neutralized by accelerated degradation. So far, two factors, Jab1 (Tomoda *et al.*, 1999) and mNPAP60/Nup50 (Smitherman *et al.*, 2000; Muller *et al.*, 2000), are known to modulate the subcellular localization. Jab1 accelerates nuclear export and mNPAP60/Nup50 is involved in nuclear import. It would be interesting to know whether either of the two factors participate in cytoplasmic location of p27 in malignant tumors.

Spontaneous pituitary tumors are extremely rare in mice but are characteristic of p27-knockout mice (Kiyokawa *et al.*, 1996; Nakayama *et al.*, 1996; Fero *et al.*, 1996). It is of interest that they also arise in p18^{INK4c}-deficient mice and in Rb-knockout mice and that pituitary tumorigenesis is greatly accelerated in mice lacking both p18 and p27 (Franklin *et al.*, 1998). These results suggest a functional collaboration between these two Cdk inhibitory proteins, and it is proposed that p18 and p27 mediate two separate pathways of pituitary tumorigenesis, likely by controlling the functions of Rb. The phenotype of p18-deficient mice suggests a role for p18 in tumor suppression in humans, but inactivation of the p18 gene locus by mutations, deletions, or hypermethylation is not frequently found in human cancers.

VI. CONCLUSIONS AND FUTURE PERSPECTIVES

Because Cdk inhibitory proteins possess the ability to inhibit cell growth, they can all be thought of as the component of tumor suppressive mechanisms. In fact, some are genetically perturbed, or their expression is altered in human cancers, which renders them good prognostic markers for malignant transformation. Because of rapid biotechnological development such as specific suppression of target gene expression by antisense oligos or ribozymes, and high-efficiency introduction of the protein into cells by the TAT system, it is suggested that they themselves can be a beneficial antitumor drug. Considering that the study of Cdk inhibitory proteins just started in 1993, it would not be too difficult to imagine that all the mystery on Cdk inhibitory proteins will be solved during the next decade, which surely will lead to the development of ultimate diagnostic methods and antitumor remedies with the help of human genomic information.

References

Arata, Y., Fujita, M., Ohtani, K., Kijima, S., and Kato, J.-Y. (2000). Cdk2-dependent and-independent pathways in E2F-mediated S phase induction. *J. Biol. Chem.* **275,** 6337–6345.

Asada, M., Yamada, T., Ichijo, H., Delia, D., Miyazono, K., Fukumuro, K., and Mizutani, S. (1999). Apoptosis inhibitory activity of cytoplasmic p21$^{Cip1/WAF1}$ in monocytic differentiation. *EMBO J.* **18,** 1223–1234.

Baldin, V., Lukas, J., Marcote, M. J., Pagano, M., and Draetta, G. (1993). Cyclin D1 is a nuclear protein required for cell cycle progression in G1. *Genes Dev.* **7,** 812–821.

Bates, S., Phillips, A. C., Clark, P. A., Stott, F., Peters, G., Ludwig, R. L., and Vousden, K. H. (1998). p14ARF links the tumour suppressors RB and p53. *Nature* **395,** 124–125.

Bouchard, C., Thieke, K., Maier, A., Saffrich, R., Hanley- Hyde, J., Ansorge, W., Reed, S., Sicinski, P., Bartek, J., and Eilers, M. (1999). Direct induction of cyclin D2 by Myc contributes to cell cycle progression and sequestration of p27. *EMBO J.* **18,** 5321–5333.

Brugarolas, J., Chandrasekaran, C., Gordon, J. I., Beach, D., Jacks, T., and Hannon, G. J. (1995). Radiation-induced cell cycle arrest compromised by p21 deficiency. *Nature* **377,** 552–557.

Bunz, F., Dutriaux, A., Lengauer, C., Waldman, T., Zhou, S., Brown, J. P., Sedivy, J. M., Kinzler, K. W., and Vogelstein, B. (1998). Requirement for p53 and p21 to sustain G2 arrest after DNA damage. *Science* **282,** 1497–1501.

Carrano, A. C., Eytan, E., Hershko, A., and Pagano, M. (1999). SKP2 is required for ubiquitin-mediated degradation of the CDK inhibitor p27. *Nat. Cell Biol.* **1,** 193–199.

Catzavelos, C., Bhattacharya, N., Ung, Y. C., Wilson, J. A., Roncari, L., Sandhu, C., Shaw, P., Yeger, H., Morava-Protzner, I., Kapusta, L., Franssen, E., Pritchard, K. I., and Slingerland, J. M. (1997). Decreased levels of the cell-cycle inhibitor p27^{Kip1} protein: Prognostic implications in primary breast cancer. *Nat. Med.* **3,** 227–230.

Chan, F. K. M., Zhang, J., Chen, L., Shapiro, D. N., and Winoto, A. (1995). Identification of human/mouse p19, a novel CDK4/CDK6 inhibitor with homology to p16^{INK4}. *Mol. Cell. Biol.* **15,** 2682–2688.

Chen, J., Jackson, P. K., Kirschner, M. W., and Dutta, A. (1995). Separate domains of p21 involved in the inhibition of cdk kinase and PCNA. *Nature* **374,** 386–388.

Chen, J., Saha, P., Kornbluth, S., Dynlacht, B. D., and Dutta, A. (1996). Cyclin-binding motifs are essential for the function of p21 Cip1. *Mol. Cell. Biol.* **16,** 4673–4682.

Cheng, T., Rodrigues, N., Shen, H., Yang, Y., Dombkowski, D., Sykes, M., and Scadden, D. T. (2000). Hematopoietic stem cell quiescence maintained by p21$^{Cip1/Waf1}$. *Science* **287,** 1804–1808.

Chuang, L. S., Ian, H. I., Koh, T. W., Ng, H. H., Xu, G., and Li, B. F. (1997). Human DNA-(cytosine-5) methyltransferase–PCNA complex as a target for p21^{WAF1}. *Science* **277,** 1996–2000.

Deng, C., Zhang, P., Harper, J. W., Elledge, S. J., and Leder, P. (1995). Mice lacking p21[Cip1/Waf1] undergo normal development, but are defective in G1 checkpoint control. *Cell* **82,** 675–684.

Deng, X. W., Dubiel, W., Wei, N., Hofmann, K., Mundt, K., Colicelli, J., Kato, J.-Y., Naumann, M., Segal, D., Seeger, M., Glickman, M., Chamovitz, D. A., and Carr, A. (2000). Unified nomenclature for the COP9 signalosome and its subunits: An essential regulator of development. *Trends Genet.* **16,** 202–203.

Dulic, V., Kaufmann, W. K., Wilson, S. J., Tlsty, T. D., Lees, E., Harper, J. W., Elledge, S. J., and Reed, S. I. (1994). p53-Dependent inhibition of cyclin-dependent kinase activities in human fibroblasts during radiation-induced G1 arrest. *Cell* **76,** 1013–1023.

Dyson, N. (1998). The regulation of E2F by pRB-family proteins. *Genes Dev.* **12,** 2245–2262.

El-Deiry, W. S., Tokino, T., Velculescu, V. E., Levy, D. B., Parsons, R., Trent, J. M., Lin, D., Mercer, E., Kinzler, K. W., and Vogelstein, B. (1993). WAF1, a potential mediator of p53 tumor suppression. *Cell* **75,** 817–825.

Elledge, S. J., and Harper, J. W. (1994). CDK inhibitors: On the threshold of check points and development. *Curr. Opin. Cell Biol.* **6,** 847–852.

Ellis, M., Chew, Y. P., Fallis, L., Freddersdorf, S., Boshoff, C., Weiss, R. A., Lu, X., and Mittnacht, S. (1999). Degradation of p27[Kip1] cdk inhibitor triggered by Kaposi s sarcoma virus cyclin–cdk6 complex. *EMBO J.* **18,** 644–653.

Ewen, M. E., Sluss, H. K., Sherr, C. J., Matsushime, H., Kato, J.-Y., and Livingston, D. M. (1993). Functional interactions of the retinoblastoma protein with mammalian D-type cyclins. *Cell* **73,** 487–497.

Fang, F., and Newport, J. W. (1991). Evidence that the G1–S and G2–M transitions are controlled by different cdc2 proteins in higher eukaryotes. *Cell* **66,** 731–742.

Fero, M. L., Rivkin, M., Tasch, M., Porter, P., Carow, C. E., Firpo, E., Polyak, K., Tsai, L. H., Broudy, V., Perlmutter, R. M., Kaushansky, K., and Roberts, J. M. (1996). A syndrome of multiorgan hyperplasia with features of gigantism, tumorigenesis, and female sterility in p27[Kip1]-deficient mice. *Cell* **85,** 733–744.

Fero, M. L., Randel, E., Gurley, K. E., Roberts, J. M., and Kemp, C. J. (1998). The murine gene p27[Kip1] is haplo-insufficient for tumor suppression. *Nature* **396,** 177–180.

Franklin, D. S., Godfrey, V. L., Lee, H., Kovalev, G. I., Schoonhoven, R., Chen-Kiang, S., Su, L., and Xinog, Y. (1998). CDK inhibitors p18[INK4c] and p27[Kip1] mediate two separate pathways to collaboratively suppress pituitary tumorigenesis. *Genes Dev.* **12,** 2899–2911.

Giaccia, A. J., and Kastan, M. B. (1998). The complexity of p53 modulation: Emerging patterns from divergent signals. *Genes Dev.* **12,** 2973–2983.

Gu, Y., Rosenblatt, J., and Morgan, D. O. (1992). Cell cycle regulation of CDK2 activity by phosphorylation of Thr160 and Tyr15. *EMBO J.* **11,** 3995–4005.

Gu, Y., Turek, C. W., and Morgan, D. O. (1993). Inhibition of CDK2 activity *in vivo* by an associated 20K regulatory subunit. *Nature* **366,** 707–710.

Guan, K.-L., Jenkins, C. W., Li, Y., Nichols, M. A., Wu, X., O'Keefe, C. L., Matera, A. G., and Xiong, Y. (1994). Growth suppression by p18, a p16[INK4/MTS1]- and p15[INK4b/MTS2]-related CDK6 inhibitor, correlates with wild-type pRb function. *Genes Dev.* **8,** 2939–2952.

Hall, M., and Peters, G. (1996). Genetic alterations of cyclins, cyclin-dependent kinases, and Cdk inhibitors in human cancer. *Adv. Cancer Res.* **68,** 67–108.

Hannon, G. J., and Beach, D. (1994). p15[INK4b] is a potential effector of TGF-induced cell cycle arrest. *Nature* **371,** 257–261.

Harbour, J. W., Luo R. X., Dei Santi, A., Postigo, A. A., and Dean, D. C. (1999). Cdk phosphorylation triggers sequential intramolecular interactions that progressively block Rb functions as cells move through G1. *Cell* **98,** 859–869.

Harper, J. W., Adami, G. R., Wei, N., Keyomarsi, K., and Elledge, S. J. (1993). The p21 cdk-interacting protein Cip1 is a potent inhibitor of G1 cyclin-dependent kinases. *Cell* **75,** 805–816.

Hengst, L. and Reed, S. I. (1996). Translational control of p27^{Kip1} accumulation during the cell cycle. *Science* **271,** 1861–1864.

Hirai, H., Roussel, M. F., Kato, J., Ashmun, R. A., and Sherr, C. J. (1995). Novel INK4 proteins, p19 and p18, are specific inhibitors of the cyclin D-dependent kinases CDK4 and CDK6. *Mol. Cell. Biol.* **15,** 2672–2681.

Hiromura, K., Pippin, J. W., Fero, M. L., Roberts, J. M., and Shankland, S. J. (1999). Modulation of apoptosis by the cyclin-dependent kinase inhibitor p27(Kip1). *J. Clin. Invest.* **103,** 597–604.

Honda, R., and Yasuda, H. (1999). Association of p19(ARF) with Mdm2 inhibits ubiquitin ligase activity of Mdm2 for tumor suppressor p53. *EMBO J.* **18,** 22–27.

Iavarone, A., and Massague, J. (1997). Repression of the CDK activator Cdc25A and cell-cycle arrest by cytokine TGF-β in cells lacking the CDK inhibitor p15. *Nature* **387,** 417–422.

Jacobs, J. J., Kieboom, K., Marino, S., DePinho, R. A., and van Lohuizen, M. (1999a). The oncogene and Polycomb-group gene bmi-1 regulates cell proliferation and senescence through the ink4a locus. *Nature* **397,** 164–168.

Jacobs, J. J., Scheijen, B., Voncken, J. W., Kieboom, K., Berns, A., and van Lohuizen, M. (1999b). Bmi-1 collaborates with c-Myc in tumorigenesis by inhibiting c-Myc-induced apoptosis via INK4a/ARF. *Genes Dev.* **13,** 2678–2690.

Johnson, D. G., Schwarz, J. K., Cress, W. D., and Nevins, J. R. (1993). Expression of transcription factor E2F1 induces quiescent cells to enter S phase. *Nature* **365,** 349–352.

Kamijo, T., Zindy, F., Roussel, M. F., Quelle, D. E., Downing, J. R., Ashmun, R. A., Grosveld, G., and Sherr, C. J. (1997). Tumor suppression at the mouse INK4a locus mediated by the alternative reading frame product p19^{ARF}. *Cell* **91,** 649–659.

Kato, J.-Y., Matsushime, H., Hiebert, S. W., Ewen, M. E., and Sherr, C. J. (1993). Direct binding of cyclin D to the retinoblastoma gene product (pRb) and pRb phosphorylation by the cyclin D-dependent kinase, CDK4. *Genes Dev.* **7,** 331–342.

Kato, J.-Y., Matsuoka, M., Polyak, K., Massague, J., and Sherr, C. J. (1994). Cyclic AMP-induced G1 phase arrest mediated by an inhibitor (p27^{Kip1}) of cyclin-dependent kinase-4 activation. *Cell* **79,** 487–496.

Kawasaki, H., Eckner, R., Yao, T.-P., Taira, K., Chiu, R., Livingston, D. M., and Yokoyama, K. K. (1998). Distinct roles of the co-activators p300 and CBP in retinoic-acid-induced F9-cell differentiation. *Nature* **393,** 284–289.

Kiyokawa, H., Kineman, R. D., Manova-Todorova, K. O., Soares, V. C., Hoffman, E. S., Khanam, D., Hayday, A. C., Frohman, L. A., and Koff, A. (1996). Enhanced growth of mice lacking the cyclin-dependent kinase inhibitor function of p27^{Kip1}. *Cell* **85,** 721–732.

Koff, A., Cross, F., Fisher, A., Schumacher, J., Leguellec, K., Philippe, M., and Roberts, J. M. (1991). Human cyclin E, a new cyclin that interacts with two members of the CDC2 gene family. *Cell* **66,** 1217–1228.

Koff, A., Giordano, A., Desai, D., Yamashita, K., Harper, J. W., Elledge, S., Nishimoto, T., Morgan, D. O., Franza, B. R., and Roberts, J. M. (1992). Formation and activation of a cyclin E–CDK2 complex during the G1 phase of the human cell cycle. *Science* **257,** 1689–1694.

Koff, A., Ohtsuki, M., Polyak, K., Roberts, J. M., and Massague, J. (1993). Negative regulation of G1 in mammalian cells: Inhibition of cyclin E-dependent kinase by TGF-β. *Science* **260,** 536–539.

Koh, J., Enders, G. H., Dynlacht, B. D., and Harlow, E. (1995). Tumour-derived p16 alleles encoding proteins defective in cell cycle inhibition. *Nature* **375,** 506–510.

Kolluri, S. K., Weiss, C., Koff, A., and Gottlicher, M. (1999). p27^{Kip1} induction and inhibition of proliferation by the intracellular Ah receptor in developing thymus and hepatoma cells. *Genes Dev.* **13,** 1742–1753.

Krude, T., Jackman, M., Pines, J., and Laskey, R. A. (1997). Cyclin/cdk-dependent initiation of DNA replication in a human cell-free system. *Cell* **88,** 109–119.

Lee, M. H., Reynisdottir, I., and Massague, J. (1995). Cloning of p57^{Kip2}, a cyclin-dependent kinase inhibitor with unique domain structure and tissue distribution. *Genes Dev.* **9,** 639–649.

Levine, A. J. (1997). p53, the cellular gatekeeper for growth and division. *Cell* **88,** 323–331.

Li, R., Waga, S., Hannon, G. J., Beach, D., and Stillman, B. (1994). Differential effects by the p21 CDK inhibitor on PCNA-dependent DNA replication and repair. *Nature* **371,** 534–537.

Loda, M., Cukor, B., Tam, S. W., Lavin, P., Fiorentino, M., Draetta, G. F., Jessup, M., and Pagano, M. (1997). Increased proteasome-dependent degradation of the cyclin-dependent kinase inhibitor p27 in aggressive colorectal carcinomas. *Nat. Med.* **3,** 231–234.

Lukas, J., Muller, H., Bartkova, J., Spitkovsky, D., Kjerulff, A. A., Jansen-Durr, P., Strauss, M., and Bartek, J. (1994). DNA tumor virus oncoproteins and retinoblastoma gene mutations share the ability to relieve the cell's requirement for cyclin D1 function in G1. *J. Cell Biol.* **125,** 625–638.

Lukas, J., Bartkova, J., Rohde, M., Strauss, M., and Bartek, J. (1995a). Cyclin D1 is dispensable for G1 control in retinoblastoma gene-deficient cells, independent of CDK4 activity. *Mol. Cell. Biol.* **15,** 2600–2611.

Lukas, J., Parry, D., Aagaard, L., Mann, D. J., Bartkova, J., Strauss, M., Peters, G., and Bartek, J. (1995b). Retinoblastoma protein-dependent cell cycle inhibition by the tumor suppressor p16. *Nature* **375,** 503–506.

Lukas, J., Herzinger, T., Hansen, K., Moroni, M. C., Resnitzky, D., Helin, K., Reed, S. I., and Bartek, J. (1997). Cyclin E-induced S phase without activation of the Rb/E2F pathway. *Genes Dev.* **11,** 1479–1492.

Mal, A., Poon, R. Y., Howe, P. H., Toyoshima, H., Hunter, T., and Harter, M. L. (1996). Inactivation of p27^{Kip1} by the viral E1A oncoprotein in TGF β-treated cells. *Nature* **380,** 262–265.

Mann, D. J., Child, E. S., Swanton, C., Laman, H., and Jones, N. (1999). Modulation of p27^{Kip1} levels by the cyclin encoded by Kaposi s sarcoma-associated herpesvirus. *EMBO J.* **18,** 654–663.

Matsuoka, S., Edwards, M., Bai, C., Parker, S., Zhang, P., Baldini, A., Harper, J. W., and Elledge, S. J. (1995). p57^{KIP2}, a structurally distinct member of the p21CIP1 cdk inhibitor family, is a candidate tumor suppressor gene. *Genes Dev.* **9,** 650–662.

Matsushime, H., Roussel, M. F., Ashmun, R. A., and Sherr, C. J. (1991). Colony-stimulating factor 1 regulates novel cyclins during the G1 phase of the cell cycle. *Cell* **65,** 701–713.

Matsushime, H., Ewen, M. E., Strom, D. K., Kato, J.-Y., Hanks, S. K., Roussel, M. F., and Sherr, C. J. (1992). Identification and properties of an atypical catalytic subunit (p34PSKJ3/CDK4) for mammalian D-type G1 cyclins. *Cell* **71,** 323–334.

Matsushime, H., Quelle, D. E., Shurtleff, S. A., Shibuya, M., Sherr, C. J., and Kato, J.-Y. (1994). D-type cyclin-dependent kinase activity in mammalian cells. *Mol. Cell. Biol.* **14,** 2066–2076.

Medema, R. H., Herrera, R. E., Lam, F., and Weinberg, R. A. (1995). Growth suppression by p16ink4 requires functional retinoblastoma protein. *Proc. Natl. Acad. Sci. U.S.A.* **92,** 6289–6293.

Meyerson, M., and Harlow, E. (1994). Identification of a G1 kinase activity for cdk6, a novel cyclin D partner. *Mol. Cell. Biol.* **14,** 2077–2086.

Millard, S. S., Yan, J. S., Nguyen, H., Pagano, M., Kiyokawa, H., and Koff, A. (1997). Enhanced ribosomal association of p27(Kip1) mRNA is a mechanism contributing to accumulation during growth arrest. *J. Biol. Chem.* **272,** 7093–7098.

Millard, S. S., Vidal, A., Markus, M., and Koff, A. (2000). A U-rich element in the 5′ untranslated region is necessary for the translation of p27 mRNA. *Mol. Cell. Biol.* **20,** 5947–5959.

Muller, D., Thieke, K., Burgin, A., Dickmanns, A., and Eilers, M. (2000). Cyclin E-mediated elimination of p27 requires its interaction with the nuclear pore-associated protein mNPAP60. *EMBO J.* **19,** 2168–2180.

Nakanishi, M., Robetorge, R. S., Adami, G. R., Pereira-Smith, O. M., and Smith, J. R. (1995). Identification of the active region of the DNA synthesis inhibitory gene p21Sdi1/CIP1/WAF1. *EMBO J.* **14,** 555–563.

Nakayama, K., Ishida, N., Shirane, M., Inomata, A., Inoue, T., Shishido, N., Horii, I., Loh, D. Y., and Nakayama, K. (1996). Mice lacking p27[Kip1] display increased body size, multiple organ hyperplasia, retinal dysplasia, and pituitary tumors. *Cell* **85**, 707–720.

Nakayama, K., Nagahama, H., Minamishima, Y. A., Matsumoto, M., Nakamichi, I., Kitagawa, K., Shirane, M., Tsunematsu, R., Tsukiyama, T., Ishida, N., Kitagawa, M., Nakayama, K., and Hatakeyama, S. (2000). Targeted disruption of Skp2 results in accumulation of cyclin E and p27(Kip1), polyploidy and centrosome overduplication. *EMBO J.* **19**, 2069–2081.

Nevins, J. R. (1998). Toward an understanding of the functional complexity of the E2F and retinoblastoma families. *Cell Growth Differ.* **9**, 585–593.

Noda, A., Ning, Y., Venable, S. F., Pereira-Smith, O. M., and Smith, J. R. (1994). Cloning of senescent cell-derived inhibitors of DNA synthesis using an expression screen. *Exp. Cell Res.* **211**, 90–98.

Nourse, J., Firpo, E., Flanagan, W. M., Coats, S., Polyak, K., Lee, M.-H., Massague, J., Crabtree, G. R., and Roberts, J. M. (1994). Interleukin-2-mediated elimination of the p27[Kip1] cyclin-dependent kinase inhibitor prevented by rapamycin. *Nature* **372**, 570–573.

Ohtani, K., DeGregori, J., and Nevins, J. R. (1995). Regulation of the cyclin E gene by transcription factor E2F1. *Proc. Natl. Acad. Sci. U.S.A.* **92**, 12146–12150.

Pagano, M., Tam, S. W., Theodoras, A. M., Beer-Romero, P., Del Sal, G., Chau, V., Yew, P. R., Draetta, G. F., and Rolfe, M. (1995). Role of the ubiquitin–proteasome pathway in regulating abundance of the cyclin- dependent kinase inhibitor p27. *Science* **269**, 682–685.

Pardee, A. B. (1989). G1 events and regulation of cell proliferation. *Science* **246**, 603–608.

Parry, D., Bates, S., Mann, D. J., and Peters, G. (1995). Lack of cyclin D–cdk complexes in Rb-negative cells correlates with high levels of p16INK4/MTS1 tumor suppressor gene product. *EMBO J.* **14**, 503–511.

Perez-Roger, I., Kim, S. H., Griffiths, B., Sewing, A., and Land, H. (1999). Cyclins D1 and D2 mediate myc-induced proliferation via sequestration of p27(Kip1) and p21(Cip1). *EMBO J.* **18**, 5310–5320.

Polyak, K., Kato, J.-Y., Solomon, M. J., Sherr, C. J., Massague, J., Roberts, J. M., and Koff, A. (1994a). p27Kip1, a cyclin–cdk inhibitor, links transforming growth factor- and contact inhibition to cell cycle arrest. *Genes Dev.* **8**, 9–22.

Polyak, K., Lee, M.-H., Erdjument-Bromage, H., Koff, A., Roberts, J. M., Tempst, P., and Massague, J. (1994b). Cloning of p27Kip1, a cyclin-dependent kinase inhibitor and a potential mediator of extracellular antimitogenic signals. *Cell* **78**, 59–66.

Pomerantz, J., Schreiber-Agus, N., Liegeois, N. J., Silverman, L. Alland, L., Chin, L., Potes, J., Chen, K., Orlow, I., Lee, H. W., Cordon-Cardo, C., and DePinho, R. A. (1998). The Ink4a tumor suppressor gene product, p19[Arf], interacts with MDM2 and neutralizes MDM2 s inhibition of p53. *Cell* **92**, 713–723.

Porter P. L., Malone K. E., Heagerty P. J., Alexander, G. M., Gatti, L. A., Firpo, E. J., Daling, J. R., and Roberts, J. M. (1997). Expression of cell-cycle regulators p27[Kip1] and cyclin E, alone and in combination, correlate with survival in young breast cancer patients. *Nat. Med.* **3**, 222–225.

Qin, X. Q., Livingston, D. M., Kaelin, W. G., Jr., and Adams, P. D. (1994). Deregulated transcription factor E2F-1 expression leads to S-phase entry and p53-mediated apoptosis. *Proc. Natl. Acad. Sci. U.S.A.* **91**, 10918–19022.

Quelle, D. E., Ashmun, R. A., Shurtleff, S. E., Kato, J.-Y., Bar-Sagi, D., Roussel, M., and Sherr, C. J. (1993). Overexpression of mouse D-type cyclins accelerates G1 phase in rodent fibroblasts. *Genes Dev.* **7**, 1559–1571.

Quelle, D. E., Zindy, F., Ashmun, R. A., and Sherr, C. J. (1995). Alternative reading frames of INK4a tumor suppressor gene encode two unrelated proteins capable of inducing cell cycle arrest. *Cell* **83**, 993–1000.

Quelle, D. E., Cheng, M., Ashmun, R. A., and Sherr, C. J. (1997). Cancer-associated mutations at the INK4a locus cancel cell cycle arrest by p16INK4a but not by the alternative reading frame protein p19ARF. *Proc. Natl. Acad. Sci. U.S.A.* **94,** 669–673.

Reynisdottir, I., and Massague, J. (1997). The subcellular locations of p15^{Ink4b} and p27^{Kip1} coordinate their inhibitory interactions with cdk4 and cdk2. *Genes Dev.* **11,** 492–503.

Russo, A. A., Jeffrey, P. D., Patten, A. K., Massagué, J., and Pavletich, N. P. (1996). Crystal structure of the p27^{Kip1} cyclin-dependent kinase inhibitor bound to the cyclin A–cdk2 complex. *Nature* **382,** 325–331.

Serrano, M., Hannon, G. J., and Beach, D. (1993). A new regulatory motif in cell cycle control causing specific inhibition of cyclin D/CDK4. *Nature* **366,** 704–707.

Serrano, M., Lee, H., Chin, L., Cordon-Cardo, C., Beach, D., and DePinho, R. A. (1996). Role of the INK4a locus in tumor suppression and cell mortality. *Cell* **85,** 27–37.

Shan, B., and Lee, W. H. (1994). Deregulated expression of E2F-1 induces S-phase entry and leads to apoptosis. *Mol. Cell. Biol.* **14,** 8166–8173.

Sheaff, R. J., Groudine, M., Gordon, M., Roberts, J. M., and Clurman, B. E. (1997). Cyclin E–CDK2 is a regulator of p27^{Kip1}. *Genes Dev.* **11,** 1464–1478.

Sheaff, R. J., Singer, J. D., Swanger, J., Smitherman, M., Roberts, J. M., and Clurman, B. E. (2000). Proteasomal turnover of p21^{Cip1} does not require p21^{Cip1} ubiquitination. *Mol. Cell* **5,** 403–410.

Sherr, C. J. (1993). Mammalian G1 cyclins. *Cell* **73,** 1059–1065.

Sherr, C. J. (1994). G1 phase progression: Cycling on cue. *Cell* **79,** 551–555.

Sherr, C. J. (1996). Cancer cell cycles. *Science* **274,** 1672–1677.

Sherr, C. J. (1998). Tumor surveillance via the ARF–p53 pathway. *Genes Dev.* **12,** 2984–2991.

Sherr, C. J., and Roberts, J. M. (1995). Inhibitors of mammalian G1 cyclin-dependent kinases. *Genes Dev.* **9,** 1149–1163.

Sherr, C. J., and Roberts, J. M. (1999). CDK inhibitors: Positive and negative regulators of G1-phase progression. *Genes Dev.* **13,** 1501–1512.

Shim, J., Lee, H., Park, J., Kim, H., and Choi, E. J. (1996). A non-enzymatic p21 protein inhibitor of stress-activated protein kinases. *Nature* **381,** 804–806.

Singer, J. D., Gurian-West, M., Clurman, B., and Roberts, J. M. (1999). Cullin-3 targets cyclin E for ubiquitination and controls S phase in mammalian cells. *Genes Dev.* **13,** 2375–2387.

Smitherman, M., Lee, K., Swanger, J., Kapur, R., and Clurman, B. E. (2000). Characterization and targeted disruption of murine Nup50, a p27(Kip1)-interacting component of the nuclear pore complex. *Mol. Cell. Biol.* **20,** 5631–5642.

Soucek, T., Yeung, R. S., and Hengstschlager, M. (1998). Inactivation of the cyclin-dependent kinase inhibitor p27 upon loss of the tuberous sclerosis complex gene-2. *Proc. Natl. Acad. Sci. U.S.A.* **95,** 15653–15658.

Stillman, B. (1996). Cell cycle control of DNA replication. *Science* **274,** 1659–1664.

Sutterluty, H., Chatelain, E., Marti, A., Wirbelauer, C., Senften, M., Muller, U., and Krek, W. (1999). p45^{SKP2} promotes p27^{Kip1} degradation and induces S phase in quiescent cells. *Nat. Cell Biol.* **1,** 207–214.

Suzuki, T., Kitao, S., Matsushime, H., and Yoshida, M. (1996). HTLV-1 Tax protein interacts with cyclin-dependent kinase inhibitor p16^{INK4A} and counteracts its inhibitory activity towards CDK4. *EMBO J.* **15,** 1607–1614.

Taya, Y. (1997). RB kinases and RB-binding proteins: New points of view. *Trends Biochem. Sci.* **22,** 14–17.

Terada, Y., Tatsuka, M., Jinno, S., and Okayama, H. (1995). Requirement for tyrosine phosphorylation of Cdk4 in G1 arrest induced by ultraviolet irradiation. *Nature* **376,** 358–362.

Tomoda, K., Kubota, Y., and Kato, J. (1999). Degradation of the cyclin-dependent-kinase inhibitor p27^{Kip1} is instigated by Jab1. *Nature* **398,** 160–165.

Toyoshima, H., and Hunter, T. (1994). p27, a novel inhibitor of G1 cyclin/cdk protein kinase activity, is related to p21. *Cell* **78,** 67–74.

Tsvetkov L. M., Yeh, K. H., Lee, S. J., Sun, H., and Zang, H. (1999). p27^{Kip1} ubiquitination and degradation is regulated by the SCF^{Skp2} complex through phosphorylated Thr187 in p27. *Curr. Biol.* **9,** 661–664.

Vlach, J., Hennecke, S., and Amati, B. (1997). Phosphorylation-dependent degradation of the cyclin-dependent kinase inhibitor p27^{Kip1}. *EMBO J.* **16,** 5334–5344.

Waga, S., Hannon, G. J., Beach, D., and Stillman, B. (1994). The p21 inhibitor of cyclin-dependent kinases controls DNA replication by interaction with PCNA. *Nature* **369,** 574–578.

Weber, J. D., Taylor, L. J., Roussel, M. F., Sherr, C. J., and Bar-Sagi, D. (1999). Nucleolar Arf sequesters Mdm2 and activates p53. *Nat. Cell Biol.* **1,** 20–26.

Wei, N., and Deng, X. W. (1999). Making sense of the COP9 signalosome. A regulatory protein complex conserved from *Arabidopsis* to human. *Trends Genet.* **15,** 98–103.

Weinberg, R. A. (1995). The retinoblastoma protein and cell cycle control. *Cell* **81,** 323–330.

Xiong, Y., Zhang, H., and Beach, D. (1992). D type cyclins associate with multiple protein kinases and the DNA replication and repair factor PCNA. *Cell* **71,** 505–514.

Xiong, Y., Hannon, G. J., Zhang, H., Casso, D., Kobayashi, R., and Beach, D. (1993). p21 is a universal inhibitor of cyclin kinases. *Nature* **366,** 701–704.

Zhang, H., Kobayashi, R., Galaktionov, K., and Beach, D. (1995). p19^{Skp1} and p45^{Skp2} are essential elements of the cyclin A–CDK2 S phase kinase. *Cell* **82,** 915–925.

Zhang, P., Liegeois, N. J., Wong, C., Finegold, M., Hou, H., Thompson, J. C., Silverman, A., Harper, J. W., DePinho, R. A., and Elledge, S. J. (1997). Altered cell differentitation and proliferation in mice lacking p57^{Kip2} indicates a role in Beckwith-Wiedemann syndrome. *Nature* **387,** 151–158.

Zhang, P., Wong, C., DePinho, R. A., Harper, J. W., and Elledge, S. J. (1998). Cooperation between the Cdk inhibitors p27^{Kip1} and p57^{Kip2} in the control of tissue growth and development. *Genes Dev.* **12,** 3162–3167.

Zhang, Y., Xiong, Y., and Yarbrough, W. G. (1998). ARF promotes MDM2 degradation and stabilizes p53: ARF-INK4a locus deletion impairs both the Rb and p53 tumor suppression pathways. *Cell* **92,** 725–734.

Zhang, H. S., Postigo, A. A., and Dean, D. C. (1999). Active transcriptional repression by the Rb-E2F complex mediates G1 arrest triggered by 16INK4a, TGF, and contact inhibition. *Cell* **97,** 53–61.

Zhao, J., Dynlacht, B., Imai, T., Hori, T., and Harlow, E. (1998). Expression of NPAT, a novel substrate of cyclin E–CDK2, promotes S-phase entry. *Genes Dev.* **12,** 456–461.

CDK Inhibitors: Small Molecular Weight Compounds

Laurent Meijer and Eve Damiens

CNRS, Station Biologique
29682 Roscoff Cedex, Bretagne, France

I. INTRODUCTION

Cyclin-dependent kinases (CDKs) regulate the cell division cycle phases, transcription, apoptosis, and neuronal cell, and thymocyte functions. Intensive screening has led in the last few years to the identification of several families of chemical inhibitors of CDKs. Some of these compounds display a high selectivity and efficiency ($IC_{50} < 5$ nM). Many have been cocrystallized with CDK2, and their atomic interactions with the kinase have been analyzed in detail: all are located in the ATP-binding pocket of the enzyme. Despite high selectivity, most CDK inhibitors (except purines) are potent inhibitors of glycogen synthase kinase-3 (GSK-3). Whether this GSK-3 inhibitory property is favorable to the antimitotic properties of CDK inhibitors remains to be established.

Tumor-Suppressing Viruses, Genes, and Drugs
Innovative Cancer Therapy Approaches
Copyright © 2002 by Academic Press.
All rights of reproduction in any form reserved.

Determination of the selectivity of kinase inhibitors, by affinity chromatography purification of their targets on immobilized inhibitors, is anticipated to contribute significantly to the understanding of the cellular effects of CDK inhibitors. CDK inhibitors display antiproliferative properties; they arrest cells in G1 and in G2/M. Furthermore, they facilitate or even trigger apoptosis in proliferating cells. In contrast, they protect neuronal cells and thymocytes from apoptosis. This suggests that CDKs may be involved both in triggering and inhibiting apoptosis. The consequences of this dual and conflicting effect need to be evaluated. The potential of CDK inhibitors is being extensively evaluated for cancer chemotherapy (clinical trials, phase I and II).

II. CYCLIN-DEPENDENT KINASES, THE CELL CYCLE, AND CANCER

Cancer, with an estimated 6 million new cases diagnosed each year, is becoming the leading disease-related cause of death of the human population. The main curative therapies for cancer detected at an early localized stage are surgery and radiation. Once the tumor has progressed these medical treatments are less successful. The major tool for its treatment is then chemotherapy. Although being widely employed, chemotherapy remains often ineffective and only palliative, particularly in the case of advanced solid tumors and metastatic cancers. In addition, the antineoplastic agents currently used are often characterized by several major drawbacks such as severe toxicity and resistance phenomena. Consequently, although efforts to improve classic anticancer agents continue, the current major focus involves the exploitation of novel tumor-specific molecular targets that have emerged from the intensive research carried out on the molecular mechanisms underlying the disease. In particular, pharmaceutical attention has been focused on the proteins that drive and control cell proliferation.

Among the estimated 1000 human protein kinases, cyclin-dependent kinases (CDKs) have been extensively studied because of their essential role in the regulation of cell proliferation, apoptosis, neuronal and thymus functions, and transcription (reviews in Morgan, 1997; Pavletich, 1999). The first identified CDK, cdc2, was initially discovered as a gene essential for both G1/S and G2/M transitions in yeast (Nurse and Bissett, 1981). Following the cloning of CDK1, the human cdc2 homolog (Lee and Nurse, 1987), cdc2 homologs were found to be present in all eukaryotes from plants and unicellular organisms to humans. Following the initial discovery of cyclin B in sea urchin eggs, it was also shown that cyclin homologs were present in all eukaryotes.

CDKs are typical serine/threonine kinases (molecular mass 33–40 kDa) which display the 11 subdomains shared by all protein kinases (see the protein kinase resource site: http://www.sdsc.edu/kinases). The availability of the complete genomic sequence of *Caenorhabditis elegans* provides the first complete vision of the extent of the CDK and cyclin families in an eukaryote (see overview in Plowman *et al.*, 1999): among the 19,099 predicted genes, 493 encode protein kinases. Among these, there are 14 CDKs, and the genome encodes 34 cyclins. In mammals, nine CDKs and 11 cyclins have been identified: the known CDK–cyclin complexes are presented in Figure 1. The CDKs which associate with cyclin F, G, and I have not been identified yet. In addition, there are several CDK-related kinases with no identified cyclin partner (Figure 1). These are easily recognized by their sequence homology to bona fide CDKs and by the presence of a variation of the conserved "PSTAIRE" motif, located in the cyclin-binding domain (subdomain III) (Meyerson *et al.*, 1992). Until their associated cyclin is discovered (if any is associated), these CDK-related kinases are named following the sequence of their PSTAIRE motif: PCTAIRE 1–3, PFTAIRE, PITAIRE, KKIALRE, PISSLRE, NKIAMRE, and the PITSLREs. To be active, CDK–cyclin complexes have to be phosphorylated on the

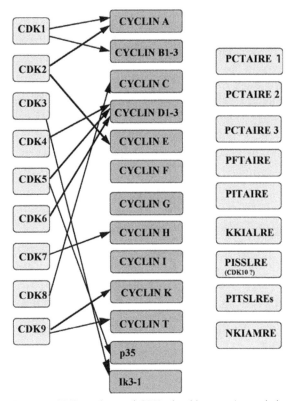

Figure 1 CDKs, cyclins, and CDK-related kinases. Arrows link cyclins to CDKs as found *in vivo*. (Left) Identified CDKs; (center) their associated cyclins; (right) CDK-like kinases (according to sequence homology).

residue corresponding to CDK2 Thr-160, located on the T-loop of the kinase. This phosphorylation is carried out by CDK7–cyclin H in association with a third protein, MAT1. The CDK subunit must also be dephosphorylated on Thr-14 and Tyr-15, two residues located at the border of the ATP-binding pocket.

The cell division cycle is divided in four tightly controlled phases (Figure 2). During the G1 phase cells become committed to enter the cell cycle and prepare the duplication of their DNA. Then cells enter the S phase to replicate their DNA. The S phase is followed by a second gap period, G2, during which cells control and repair errors that might have occurred during the S phase. Finally, cells undergo mitosis and divide in two daughter cells during the M phase. In eukaryotic cells, each phase of the cell cycle is regulated by the sequential activation of cyclin-dependent kinases (CDKs) (reviews in Vogt and Reed, 1998; Meijer et al., 1997a, 2000a; Lundberg and Weinberg, 1999; Sherr and Roberts, 1999). Different mechanisms responsible for arresting cells in the events of unfavorable conditions regulate the CDKs. These controls occur at major points termed checkpoints positioned at the G1, G2, and M phases.

Mutations in the checkpoint components may lead to aberrant cell cycle progression and consequently to genetic instability. These last few years have witnessed the discovery of the high incidence of alterations in the genes that code for CDK regulating proteins in tumors (Figure 2). Examples of a direct involvement of mutated or amplified CDKs in cancer are rare. In contrast, deregulation of their regulatory partners, cyclins CDC25 and CKIs, have been identified in numerous types of cancers. For example, rearrangements of the cyclin D1 (*PRAD1/BCL-1*)

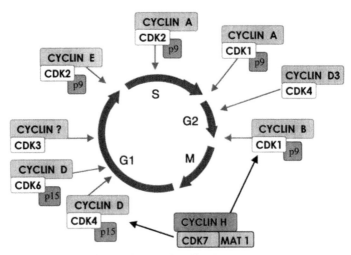

Figure 2 CDK–cyclin complexes in the cell cycle.

gene resulting in overexpression have been frequently observed in some squamous cell carcinomas, mantel cell lymphoma, and parathyroid adenomas (Lammie *et al.*, 1991; Yatabe *et al.*, 2000). Overexpression of cyclin D1 is also observed in multiple other tumors and sometimes correlated with prognosis (Barnes, 1997; Collecchi *et al.*, 1999). Other cyclins such as cyclin A and E also share oncogenic properties. Notably, overexpression of cyclin E is found in some breast carcinomas and correlates with tumor aggressiveness (Keyomarsi and Pardee, 1993, 1996; Karp and Broder, 1995). Natural CDK inhibitors also present abnormalities such as mutations and loss of protein expression in a large variety of cancers (review in Vogt and Reed, 1998; Funk, 1999). The CDK-activating CDC25A phosphatase is often overexpressed in human breast cancer, and this correlates with poor survival (Cangi *et al.*, 2000 and references therein). Finally, numerous viruses encode their own cyclins (review in Mittnacht and Boshoff, 2000). Altogether, these data suggest that CDKs constitute promising targets for the identification of new mechanism-based treatments for cancer.

III. CYCLIN-DEPENDENT KINASE INHIBITORS, A LARGE VARIETY OF STRUCTURES

The search for CDK inhibitors has been essentially based on the use of CDK1, CDK2, and CDK4 as molecular targets as described in Rialet and Meijer (1991) and Meijer and Kim (1997). Intensive screening of collections of natural and synthetic compounds, followed by structure/activity studies and molecular modeling based on the crystal structure of the inhibitor in complex with CDK2, have led to the identification and characterization of numerous inhibitors (see reviews in Meijer, 1996, 2000; Meijer and Kim, 1997; Garrett and Fattaey, 1999; Gray *et al.*, 1999; Meijer *et al.*, 1999; Fischer and Lane, 2000; Pestell *et al.*, 2000; Rosiana and Chang, 2000; Senderowicz and Sausville, 2000; Sielecki *et al.*, 2000). The structure of the best studied compounds are presented in Figure 3.

These CDK inhibitors can be divided into the following families which are representative of their chemical classes (Figure 3): (1) the purine derivatives olomoucine (Vesely *et al.*, 1994), roscovitine (De Azevedo *et al.*, 1997; Meijer *et al.*, 1997b), and purvalanols (Gray *et al.*, 1998; Chang *et al.*, 1999) and other purines (Legraverend *et al.*, 1998, 1999, 2000a,b), (2) the flavone flavopiridol and its derivative deschloroflavopiridol (Sedlacek *et al.*, 1996), (3) the indirubins, with the leading compound indirubin-3'-monoxime (Hoessel *et al.*, 1999; Leclerc *et al.*, 2000), (4) the dihydroindolo[3,2-*d*][1]benzazepin-6(5*H*)-ones paullones (Zaharevitz *et al.*, 1999; Schultz *et al.*, 1999; Leost *et al.*, 2000), (5) CGP-60474 (Zimmermann, 1995), (6) the sponge-derived hymenialdisine (Meijer *et al.*, 2000b), (7) the guanine NU 2058 and related compounds (Arris *et al.*, 2000), (8) the bisindolylmaleimide staurosporine and its derivatives UCN-01

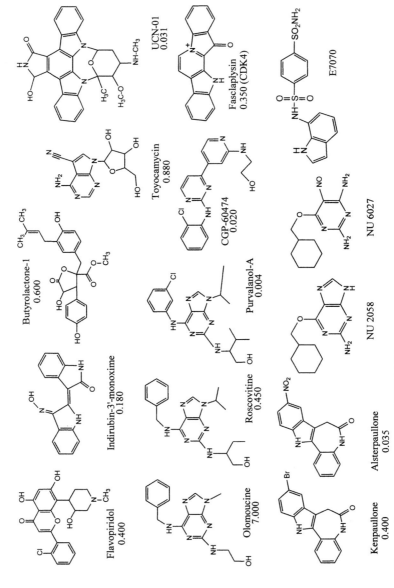

Figure 3 Structures of some reported CDK inhibitors.

and CGP 41251, (9) butyrolactone I (Kitagawa *et al.*, 1993), (10) the polysulfate suramin (Bojanowski *et al.*, 1994), (11) the nucleoside analog toyocamycin (Park *et al.*, 1996), (12) 9-hydroxyellipticine (Ohashi *et al.*, 1995), (13) fascaplysin (Soni *et al.*, 2000), and (14) E7070 (Owa *et al.*, 1999).

Despite their chemical diversity, these CDK inhibitors share several properties: a molecular weight below 600, they are flat, hydrophobic heterocycles, they all act at the same site on the CDK, the ATP-binding pocket.

IV. CYCLIN-DEPENDENT KINASE INHIBITORS, ALL COMPETING WITH ATP

Classic enzymological analysis has revealed that all identified chemical inhibitors of CDKs share a common molecular mechanism of action: they act by competing with ATP for binding in the ATP-binding pocket of the catalytic subunit of CDK–cyclin complexes. This site is located between the N-terminal, β-sheet rich, and the C-terminal, α-helices rich, lobes of the kinase (see review in Pavletich, 1999). This site of action has been confirmed by numerous cocrystallization studies (review in Gray *et al.*, 1999). In addition to ATP, nine compounds have been crystallized with CDK2 (Table 1). The high resolution crystal structures reveal that all inhibitors bind in the deep groove located between the N- and C-terminal domains of the kinase that is normally occupied by the adenine ring of ATP. The planar heterocyclic ring system in each case occupies a hydrophobic cleft surrounded by Ile-10, Val-18, Phe-82, and Leu134. Ile10, Leu83, and Leu134 are responsible for 40% of all contacts between CDK2 and the inhibitors. Two or three hydrogen bonds are consistently found between the

Table 1 CDK Inhibitors Which Have Been Cocrystallized with CDK2

Inhibitor	Reference
Isopentenyladenine	Schulze-Gahmen *et al.* (1995)
Olomoucine	Schulze-Gahmen *et al.* (1995)
Deschloroflavopiridol	de Azevedo *et al.* (1996)
Roscovitine	de Azevedo *et al.* (1997)
Staurosporine	Lawrie *et al.* (1997)
Purvalanols	Gray *et al.* (1998)
Indirubins	Hoessel *et al.* (1999)
Hymenialdisine	Meijer *et al.* (2000b)
NU2058	Arris *et al.* (2000)

inhibitors and the backbone atoms of Glu81 and Leu83 residues of CDK2. The selectivity probably comes from interactions with residues, conserved in CDKs but not in other kinases, which do not interact with ATP. The ATP-binding site of CDK2 has thus an impressive capacity to accommodate a wide variety of inhibitors.

V. CYCLIN-DEPENDENT KINASE INHIBITORS, THE SELECTIVITY PROBLEM

Following the identification of a CDK inhibitor, the question of enzymatic selectivity becomes rapidly important. There are several ways to approach this question. The first method is based on testing of the compound on an as large as possible panel of kinases. This involves the expression and purification of enzymes, the assay of their kinase activities in the presence of a concentration range of inhibitor to determine the compound's IC_{50} values. This painstaking approach is frustrating given the number of nontested kinases (a little less than 1500) and other possible targets! Nevertheless it provides an initial view on the overall selectivity of a given inhibitor. Using this method, it can readily be seen that CDK inhibitors fall into three different categories:

- Low selectivity: flavopiridol, staurosporine
- Medium selectivity: indirubins, hymenialdisine
- High selectivity: purines, paullones

This classic method has also allowed us to identify, unexpectedly, a new target for several CDK inhibitors (Table 2). Indeed it turns out that most, but not all CDK inhibitors are quite active on glycogen synthase kinase-3 (GSK-3) (Leclerc et al., 2000; Leost et al., 2000). This enzyme phosphorylates a large variety of substrates such as the tumor suppressor protein APC, axin, β-catenin, cyclin D1, and c-myc. GSK-3 is directly involved in the WNT signaling pathway (reviewed in Seidensticker and Behrens, 2000). The consequences of this dual specificity of many CDK inhibitors are described in the Discussion section of the two papers mentioned earlier.

Similarly, the selectivity of flavopiridol has been evaluated, and several unexpected targets have been identified. Flavopiridol binds to and inhibits glycogen phosphorylase; the crystal structure of glycogen phosphorylase in complex with flavopiridol has been resolved (Oikonomakos et al., 2000). Flavopiridol also inhibits GSK-3 (Leclerc et al., 2000). Furthermore, flavopiridol binds directly to duplex DNA (Bible et al., 2000). Finally, flavopiridol interacts with

Table 2 Many, but Not All, CDK Inhibitors Are Good
Inhibitors of GSK-3β[a]

Compound	GSK-3β	CDK1/cyclin B
Olomoucine	100.000	7.000
Roscovitine	130.000	0.450
Purvalanol A	13.000	0.004
Butyrolactone-I	100.000	1.100
Isopentenyladenine	60.000	55.000
Staurosporine	0.015	0.005
UCN-01	0.070	0.060
Flavopiridol	0.450	0.200
Indirubin-3[+]-monoxime	0.022	0.180
Alsterpaullone	0.004	0.035
Hymenialdisine	0.010	0.022
Toyocamycin	3.500	0.500

[a]Enzyme activities were assayed in the presence of increasing concentrations of the reported CDK inhibitors. IC$_{50}$ values were calculated from the dose–response curves and are expressed in micromoles per liter.

the multidrug resistance protein 1 (MRP1) and stimulates its ATPase activity (Hooijberg *et al.*, 1997, 1999).

These two examples illustrate how unexpected targets can be discovered for compounds initially identified and optimized as CDK inhibitors. Obviously, interaction with these targets contributes to the cellular effects of the CDK inhibitors. These examples also illustrate the limitation of running selectivity assays enzyme after enzyme. This has led us to develop another approach, the identification of targets by affinity chromatography on immobilized inhibitor (Rosania *et al.*, 1999; Knockaert *et al.*, 2000). This method is primarily based on the crystal structure of CDK–inhibitor complexes. The orientation of the inhibitor within the ATP-binding pocket tells us which side of the compound is suitable for addition of a linker which can be used to tether the inhibitor to a solid matrix. The immobilized inhibitor is then used as a chromatography matrix to purify interacting proteins from cell extracts which are then identified by protein microsequencing. Using immobilized purvalanol, we found that Erk2 is the major purvalanol-binding protein in mammalian cell lines (Knockaert *et al.*, 2000). Other targets were identified in other tissues. A similar method was more recently used in another context and led to the identification of a redox-related factor, Ref-1, as the target protein of an anti-NF-κB drug (Shimizu *et al.*, 2000). This approach can thus help to identify the range of intracellular targets of a given compound. It can also help to correlate the range of cellular actions of a family of compounds with the pattern of their protein targets.

VI. CYCLIN-DEPENDENT KINASE INHIBITORS, CELLULAR EFFECTS

The cellular effects of CDK inhibitors have been investigated in depth and will be reviewed elsewhere (Meijer et al., 2001). In brief, CDK inhibitors arrest proliferation in dividing cells, they induce or facilitate apoptosis, and, in some models, they trigger differentiation. In contrast, CDK inhibitors protect nondividing cells (neuronal, thymocytes, etc.) from apoptosis.

Cell proliferation arrest occurs in G1, in G2/M, or in both, depending on cell lines, inhibitor concentration, and other culture conditions. The G1 arrest correlates with a reduction of phosphorylation of the retinoblastoma protein (Rb), and can therefore be attributed to inhibition of CDK2–cyclin E. The G2/M arrest is associated with the suppression of vimentin and nucleolin phosphorylation at sites which are phosphorylated by CDK1–cyclin B. The G2/M arrest is thus attributed to inhibition of this enzyme. In recent work (Marko et al., 2000; Damiens et al., 2001), we investigated the mechanisms of cell cycle arrest by indirubin-3′-monoxime and found that inhibition of CDK1–cyclin B constitutes the major mechanism underlying the antitumor activity of indirubins.

The induction of apoptosis by CDK inhibitors has been described in many examples (Meijer et al., 2001). Very interestingly, this induced cell death is independent of the p53 and Rb status. Interestingly also, CDK inhibitors appear to synergize with other treatments to induce apoptosis. This is particularly well illustrated with olomoucine/roscovitine and farnesyltransferase inhibitors (Edamatsu et al., 2000), with olomoucine and mitomycin (Ongkeko et al., 1995), with flavopiridol and various antineoplastic agents (Bible and Kaufmann, 1997), and with olomoucine/roscovitine and E2F-1 (Atienza et al., 2000). For some compounds, a preferential induction of apoptosis in transformed versus normal cells has been shown. CGP60474, for example, reversibly arrests U2-OS cells in late G1 while it blocks U2-OS transformed by the SV-40 large T antigen in late G2 (Ruetz et al., 1998). Apoptosis is preferentially induced in these latter cells.

Altogether, a vast body of literature (not reviewed here) demonstrates the expected antimitotic properties of CDK inhibitors and their proapoptotic effects in dividing cells. These favorable properties have stimulated preclinical studies of CDK inhibitors as potential anticancer drugs, as reviewed below.

VII. CYCLIN-DEPENDENT KINASE INHIBITORS, ANTITUMOR ACTIVITY

A. Flavopiridol

1. Preclinical studies

Following the *in vitro* studies, antitumor effects, pharmacokinetics, and toxicology of flavopiridol have been analyzed in preclinical models. Flavopiridol inhibits the growth of human tumor xenografts originating from prostate, breast, colon, brain, stomach, leukemia, and lymphomas (Drees *et al.*, 1997; Arguello *et al.*, 1998; Christain *et al.*, 1997). *In vitro* evaluations along with animal studies first suggested that prolonged exposure was needed to observe the therapeutic effect. In fact, initial studies using colorectal (Colo205) and prostate (LnCap/DU-145) carcinoma xenograft models demonstrated that flavopiridol is cytostatic when administered frequently over a protracted period (Sedlacek *et al.*, 1996). These observations led to the first phase I trials of flavopiridol in humans as a 72-hr continuous infusion every 2 weeks (Senderowicz *et al.*, 1999) (see below).

However, subsequent experiments in lymphoma/leukemia xenografts models revealed that administration of flavopiridol, dosed at 7.5 mg/kg bolus intravenous (i.v.) or intraperitoneal on each of 5 days, is notably active and renders animals tumor free for several months. In contrast, infusional flavopiridol only delays tumor growth in similar HL-60 models. The antitumor effect appeared to be mainly due to the induction of apoptosis observed in the lymph nodes, thymus, and spleen with peak flavopiridol plasma concentration of 5–8 μM (Arguello *et al.*, 1998). Other lymphomas such as SUDHL-4, AS283, and Nalm/6 also showed significant responses to flavopiridol treatment (Arguello *et al.*, 1998). Based on these results, a phase I trial of flavopiridol administered as a bolus is currently being investigated at the National Cancer Institute (NCI) (see below).

In solid tumor models, including head and neck (HNSCC) xenografts, flavopiridol administered as a daily 5 mg/kg per day intraperitoneally for 5 consecutive days caused a 60–70% reduction in tumor size which was maintained over a period of 10 weeks. Once again this treatment led to the appearance of apoptotic cells along with cyclin D1 depletion in the tumor xenografts (Patel *et al.*, 1998). Consecutively to these observations, a phase II trial of flavopiridol in patients with refractory squamous carcinomas of the head and neck has been initiated at the NCI.

2. Preclinical pharmacokinetics and toxicology

Preclinical pharmacology and toxicology in rats and dogs reveal that flavopiridol was rapidly and largely distributed after intravenous injection. Plasma elimination was biexponential with mean α and β half-lives of 10.5 and 112 min in rats (Arnesson et al., 1995). Pharmacokinetic studies in mice showed a similar profile with α and β half-lives of about 16.4 and 201 min, respectively, and a mean total body plasma clearance of 22.6 ml/min/kg (Senderowicz, 1999). Experiments in isolated perfused rat liver indicated that glucuronidation represents the major mechanism of hepatic flavopiridol biotransformation. Metabolites are mainly excreted in the biliary tract but are also released into the systemic circulation (Jager et al., 1998). Their pharmacological and toxicological effects remain to be elucidated. Drug-related lesions in the bone marrow, spleen, and thymus were observed after intravenous administration of flavopiridol, three times a day for 3 days at a dose of 2 mg/kg/injection in rats. Pharmacokinetics and toxicity of flavopiridol given as a 72-hr continuous infusion have also been evaluated in beagle dogs. Steady-state plasma concentrations of 55 and 91 nM, which were higher than the concentration required to inhibit growth and CDK activity *in vitro*, were obtained for, respectively, doses of 0.6 mg/kg/day and 1.3 mg/kg/day. In this model the toxic dose low (TDL) of 1.3 mg/kg/day produced clinical signs of gastrointestinal toxicity (Jager et al., 1998). Gastrointestinal and bone marrow toxicity was dose-limiting. In summary, these data are in favor of the development of this useful agent to investigate for the therapy of cancer either alone or in combination with other agents.

3. Human clinical trials

Flavopiridol is the first CDK inhibitor to be tested clinically for antitumor activity. Two phase I clinical trials of flavopiridol, administered as a 72-hr continuous infusion every 2 weeks, have been completed. Significant responses were noted in patients with non-Hodgkin s lymphoma, colorectal, prostate, renal, and gastric cancer with a reduction of tumor masses that lasted for >6 months. Furthermore, a sustained complete response has been observed more than 2 years after discontinuation of flavopiridol for one patient with a refractory metastatic gastric cancer (Thomas et al., 1997). During these trials plasma flavopiridol concentration reached values of 300–500 nM, which effectively inhibit CDKs and cell proliferation *in vitro*. Dose-limiting toxicity (DLT) was secretory diarrhea due to the induction of chloride secretion by flavopiridol with a maximally tolerated dose (MTD) of 50 mg/m²/day × 3 (Shenfeld et al., 1997). Flavopiridol was well tolerated with antidiarrheal prophylaxis (combination of loperamide and cholestyramine) allowing a treatment at a MTD of 78 mg/m²/day × 3. At this higher dose level, dose-limiting toxicity consisted of symptomatic hypotension and a proinflammatory syndrome with fatigue, asthenia, hyperbilirubinemia, transient local pain, and anorexia.

A first phase I clinical trial of flavopiridol administered as a daily bolus for 5 consecutive days every 3 weeks has been more recently initiated. The maximally tolerated dose of $37.5 \, mg/m^2/day \times 5$ days established in this study was recommended for the phase II trials. Dose-limiting toxicity consisted of nausea/vomiting, neutropenia, fatigue, and diarrhea (Senderowicz et al., 1998). Phase I trials of sequential flavopiridol and standard antineoplastic agents, in particular Paclitaxel, in patients with advanced solid tumors are ongoing.

The NCI is also currently sponsoring phase II trials of infusional flavopiridol in various tumor types among others (Wright et al., 1998; Shapiro et al., 1999; Bennett et al., 1999).

In summary, flavopiridol induces growth arrest, differentiation, cytotoxic cell death, and apoptosis in a variety of tumor cell types. Interestingly, it also shows synergism with some conventional anticancer drugs in vitro. Flavopiridol is the first potent CDK inhibitor tested in human clinical trials for anticancer activity. Its antitumor activity is likely to derive from an effect on multiple targets. The objective responses obtained with this new drug will promote the development of various novel chemical inhibitors of CDKs which may provide significant therapeutic benefit for cancer patients in the future.

B. Indirubins

Danggui Longhui Wan, a mixture of 11 herbal medicines, has been traditionally utilized against certain types of leukemias. Only one of these ingredients, Qing Dai (indigo naturalis), a dark blue powder originating from various indigo-producing plants, carries the antileukemic activity. Although it is mostly constituted of indigo, a minor constituent, indirubin, was identified as the active component by the Chinese Academy of Medicine (Chen and Xie, 1984; see Hoessel et al., 1999, for references).

1. Preclinical studies

Indirubin has been tested for its antitumor activity against a number of tumor types in vivo. Administered as a daily 200 mg/kg subcutaneous (s.c.) bolus for 6–7 days in Walker carcinosarcoma 256 bearing rats, indirubin showed notable activity with an inhibition rate of 47–52%. As indicated by the inhibition rate of 50–58%, similar results were obtained when the drug was administered intraperitoneally at the same dosage, while an administration of indirubin through a gastric tube at a dosage of 500 mg/kg only reduced the tumor by 23–33%. Given orally, indirubin had a 31% inhibition rate. In Lewis cancer of lungs, the inhibition rate was 21.7% if indirubin was given intraperitoneally and 43% per os (Ma and Yao, 1983). Indirubin was also active on murine sarcoma 180 but did not prolong the survival time of mice with leukemia L7212 or P388 (Ji et al., 1981).

2. Preclinical pharmacokinetics and toxicology

Studies on indirubin metabolism in rats and mice revealed that absorption via the gastrointestinal tract was poor (Qi *et al.*, 1981). This was confirmed at the cellular level (Marko *et al.*, 2000). The half-life of absorption in mice is 5.9 hr by gastric infusion of 45.7 mg/kg. The peak plasma concentration was reached 15 hr after administration, distribution was large throughout the body, and excretion slow with an elimination half-life of 21 hr. The liver appeared to be the major organ of indirubin biotransformation, and within 96 hr, 15 and 60% of the metabolites were excreted into urine and feces, respectively.

Toxicological evaluation was undertaken in rats and dogs. Subacute toxicity experiments in rats with a dose of 100–400 mg/kg per day administered orally for 1 month did not induce any effect on hepatic and renal function nor any changes in weight and in blood picture. Anorexia and diarrhea have been observed when indirubin was given orally for 1–3 months to rats at a dose of 80 mg/kg/day. A subacute toxicity study of indirubin administered to dogs at three different doses (20, 100, or 200 mg/kg/day *per os* for 6 months) was also reported. No toxic side effect was observed at the lower dose. At the middle dosage, mild diarrhea occurs within 10–30 days, and serum glutamic–pyruvic transaminase (SGPT) was slightly elevated and accompanied by focal necrosis of the liver after 5 months in one of the three dogs treated. At the highest dosage (200 mg/kg/day) serious diarrhea and hematofecia were observed within 4–60 days, and SGPT increased after 3 months with marked toxic change on the liver section after 6 months. The bone marrow and blood indexes, renal, and cardiac functions were not affected by any of the three dosages (Sichuan Institute of Traditional Chinese Medicine, 1981). After consideration of these data, indirubin was reported to be an apparently safe anticancer agent without serious toxic side effects.

3. Human clinical trials

In a joint clinical trial, a total of 314 cases of chronic myelocytic leukemia (CML) were treated with indirubin (Cooperative Group of Clinical Therapy of Indirubin, 1980; Zhang *et al.*, 1985). Indirubin was administered orally at a dose of 150–200 mg/kg/day. About 1 week after the beginning of the treatment, most patients presented a good overall condition with increased appetite. The optimal effect of indirubin was observed after 26 to 172 days of treatment with an average of 72.2 days. Improvement of the blood picture and regression of hepatosplenomegaly was noted 2 weeks to 1 month after the beginning of the treatment. The complete remission rate reached 26% while the partial remission rate was 33%. The major toxic side effects of indirubin were abdominal pain, diarrhea, nausea, vomiting, and hematofecia, most of these being mild.

Thrombocytopenia was also observed in 20% of cases, and mild marrow suppression occurred in 4.5% of patients. Twenty patients showed an exacerbation of CML during the treatment or within 6 months after the withdrawal of indirubin (6.4%) (Ma and Yao, 1983). The long-term therapeutic effect of indirubin remains to be determined. Because of the poor intestinal absorption of indirubin associated with gastrointestinal disorders, derivatives of indirubin have been produced in order to reduce toxic side effects and improve the anticancer activity. N-Methyl-isoindigotin is a second generation drug that has been developed more recently on the basis of chemical structure and activity relationship studies. N-Methyl-isoindigotin was shown to be more active and less toxic than indirubin (Ji et al., 1991). In clinical studies, N-Methyl-isoindogotin induced 32.1% complete remission and 48.5% partial remission in CML. Side effects were minor and mainly constituted of gastrointestinal disturbance and neuralgia (Clinical Therapeutic Coordinating Group of Mesoindigo, 1988).

Although indirubin and its derivatives are clinically used for treating chronic leukemia in China, its mechanism of antiproliferative activity remained unclear for a long time. It has been discovered that indirubins selectively inhibit CDKs (Hoessel et al., 1999) and GSK-3 (Leclerc et al., 2000). To which extent these properties account for the antitumor activity of indirubins remains to be established.

C. UCN-01

UCN-01, namely, 7-hydroxy-staurosporine, exhibits antitumor activity in several in vivo preclinical models. In particular, UCN-01 is active against three human tumor xenografts (epidermoid carcinoma A431, fibrosarcoma HT1080, and acute myeloid leukemia HL-60). UCN-01 also inhibits growth of various murine models of malignancies (fibrosarcoma, K-BALB, M-MSV-BALB) which activate the v-ras and v-mos oncogenes (Akinaga et al., 1991). Xenograft model systems with renal carcinoma (A498), leukemia (MOLT-4), breast carcinoma (MCF-7), and pancreatic adenocarcinoma (PSN-1) were also sensitive to UCN-01 administered intraperitoneally or intravenously (Kurata et al., 1999; Senderowicz and Sausville, 2000). Consistent with prior in vitro studies, the best antitumor effect was obtained with continuous exposure to the drug for as long as 72 hr. In consequence, it was decided to conduct the first phase I clinical trial using a 72-hr continuous infusion every 2 weeks (see below). Furthermore, it has been demonstrated that UCN-01 might offer real advantages in cancer polytherapy by its ability to enhance the anticancer effects of several chemotherapeutic drugs such as cisplatin, mitomycin C, or 5-fluorouracil in vitro and in vivo in human tumor xenograft models as well as mouse syngeneic tumor models. For example, it was shown that a single i.v. coinjection of mitomycin C and UCN-01 exhibited greater antitumor activity than mitomycin and UCN-01

alone in xenografted A431 cells in nude mice. This synergism between mito-
mycin and UCN-01 was confirmed in the human xenografted colon carcinoma
Co-3, in the murine sarcoma 180, and in the inoculated P388 leukemia model
(Akinaga et al., 1993).

1. Preclinical pharmacokinetics and pharmacodynamics of UCN-01

After i.v. administration, UCN-01 was eliminated biphasically in mice and rats
and triphasically in dogs. In mice, the half-lives for the initial and terminal
phases of elimination in plasma after an i.v. administration at doses of 1–9 mg/kg
were 0.4–0.88 and 3.00–3.98 hr, respectively. The elimination half-lives were
similar in rats after an administration of UCN-01 by an intravenous route at
doses of 0.35–3.5 mg/kg UCN-01, and they were $(t_{1/2\alpha})$ 0.14, $(t_{1/2\beta})$ 2.99, $(t_{1/2\gamma})$
and 11.6 hr for dogs after an i.v. administration at a dose of 0.5 mg/kg. Total clearance values in mice, rats, and dogs were high and reached 1.9–2.6, 2.8–3.9, and
0.6 liter/hr/kg, respectively. The volumes of distribution at steady state were large
in the various experimental models. In beagle dogs, local (site of injection) and
gastrointestinal toxic side effects were dose-limiting at a steady-state plasma concentration of 330 nM UCN-01 after a 72-hr continuous infusion. The pharmaco-
dynamic studies performed in nude mice bearing xenografted human pancreatic
tumor cells PSN-1 revealed that UCN-01 significantly inhibits tumor growth during five consecutive i.v. administrations at a dose of 9 mg/kg/day. The inhibitory
effect of UCN-01 was maintained 3 days after the last administration. The concentration of UCN-01 in tumor tissue was much higher than this in plasma with
a tumor/plasma ratio of about 500 at 24 hr after the five consecutive administrations. This extensive distribution of UCN-01 in tumor cells may contribute to its
antineoplastic activity (Kurata et al., 1999; Sausville et al., 1998).

2. Human clinical trials

Two phase I clinical trials have been performed at the NCI and in Japan,
using 72-hr continuous infusion and 3-hr infusion every 3 weeks, respectively
(Senderowicz et al., 1999; Tamura et al., 1999). Unfortunately, the pharmacokinetics of UCN-01 displayed distinctive features as compared to the preclinical
experimental animal data, including a distribution volume and a systemic clearance extremely low in contrast to the large distribution volume and the rapid
clearance observed in mice, rats, and dogs (Kurata et al., 1999). The half-life of
UCN-01 in humans was also unusually long, mainly due to the fact that
UCN-01 binds with a high affinity to human α1-acid glycoprotein but not to
mouse, rat, and dog α1-acid glycoproteins (Fuse et al., 1998, 1999). As a consequence, careful dose escalation has been conducted, and the maximal tolerated
dose defined in the U.S. phase I clinical trial was 42.5 mg/m^2/day which consti-

tutes the recommended phase II dose of UCN-01 given on a 72-hr continuous infusion schedule (120 mg/m^2/72 hr). Dose-limiting toxicity were nausea/vomiting, hepatic transaminase elevation, symptomatic hyperglycemia in association with an insulin-resistance state, lactic acidosis, and pulmonary toxicity. In the U.S. study, it was important to note that one patient with a refractory metastatic melanoma presented a partial response that lasted 8 months. Furthermore, a few patients with leiomyosarcoma, non-Hodgkin's lymphoma, or lung cancer presented a stabilization of their tumor evolution ($\geq$6 months) (Senderowicz et al., 1999). In the Japanese phase I trial, the dose escalation was pursued to determine MTD. At a dose level up to 51.1 mg/m^2, toxicity was mild and essentially constituted by diarrhea, nausea, and arrhythmia (Tamura et al., 1999). Future trials are being envisaged in which UCN-01 would be administered in association with DNA-damaging agents and/or DNA antimetabolite drugs.

In summary, UCN-01 was originally selected as an anticancer agent on the basis of its ability to inhibit protein kinase C (PKC), but subsequent investigations revealed that PKC inhibition was not the primary mechanism of its antiproliferative action. Indeed, more recently, the interference with the CDKs has been retained to explain the influence of UCN-01 on cell growth and survival. Other targets are clearly involved in the effects of UCN-01. Preclinical and early clinical results indicated that UCN-01 was undoubtedly a useful probe in pharmacology and was able to offer real advantages in cancer therapy particularly in combination with conventional cytotoxic drugs. However, the deleterious consequences of the strong binding of UCN-01 to α1-acid glycoprotein should be overcome to develop to a large extent this new anticancer drug candidate.

VIII. CONCLUSION

These last few years have seen a spectacular development of protein kinase inhibitors (review in Adams and Lee, 1999; Toledo et al., 1999). Among these, CDK inhibitors have been of special interest in view of their potential use in cancer chemotherapy. Although many CDK inhibitors have been identified and optimized as CDK inhibitors, their effective cellular targets remain to be identified. In addition to classic pharmacokinetic properties, the issue of selectivity is the fundamental key which will allow a rational optimization of kinase inhibitors as anticancer drugs. Affinity chromatography on immobilized inhibitors appears to be the way of the future in this respect. We also believe that the combination of these kinase inhibitors, in a defined schedule, with classic anticancer drugs will increase their chance of becoming major chemotherapeutic agents. Finally, it should be stressed that all preclinical and clinical data, obtained on these compounds during the evaluation of their antitumor activity, may turn out to be useful in other therapeutic contexts, as CDKs are involved in many other physiological processes (reviewed in Meijer, 2000).

Acknowledgments

This work was supported by grants from the Association pour la Recherche sur le Cancer (ARC 5343) (L.M.) and the Conseil Régional de Bretagne (L.M.). E.D. is a recipient of a postdoctoral fellowship from the Association pour la Recherche sur le Cancer.

References

Adams, J. L., and Lee, D. (1999). Recent progress towards the identification of selective inhibitors of serine/threonine protein kinases. *Curr. Opin. Drug Discov. Dev.* **2**, 96–109.

Akinaga, S., Gomi, K., Morimoto, M., Tamaoki, T., and Okabe, M. (1991). Antitumor activity of UCN-01, a selective inhibitor of protein kinase C, in murine and human tumor models. *Cancer Res.* **51**, 4888–4892.

Akinaga, S., Nomura, K., Gomi, K. and Okabe, M. (1993). Enhancement of antitumor activity of mitomycin C *in vitro* and *in vivo* by UCN-01, a selective inhibitor of protein kinase C. *Cancer Chemother. Pharmacol.* **32**, 183–189.

Arguello, F., Alexander, M., Sterry, J. A., Tudor, G., Smith, E. M., Kalavar, N. T., Greene, J. F., Jr., Koss, W., Morgan, C. D., Stinson, S. F., Siford, T. J., Alvord, W. G., Klabansky, R. L., and Sausville, E. A. (1998). Flavopiridol induces apoptosis of normal lymphoid cells, causes immunosuppression, and has potent antitumor activity *in vivo* against human leukemia and lymphoma xenografts. *Blood* **91**, 2482–2490.

Arnesson, D., Evans, E., Kovalch, R., Moore, R., Morton, T., Tomaszewski, J., and Smith, A. C. (1995). Preclinical toxicology and pharmacology of flavopiridol (NSC-649890) in rats and dogs. *Proc. Am. Assoc. Cancer Res.* **36**, 366.

Arris, C. E., Boyle, F. T., Calvert, A. H., Curtin, N. J., Endicott, J. A., Garman, E. F., Gibson, A. E., Golding, B. T., Grant, B. T., Griffin, R. J., Jewsbury, P., Johnson, L. N., Lawrie, A. M., Newell, D. R., Noble, M. E. M., Sausville, E. A., Schultz, R., and Yu, W. (2000). Identification of novel purine and pyrimidine cycline-dependent kinase inhibitors with distinct molecular interactions and tumor cell growth inhibition profiles. *J. Med. Chem.* **43**, 2797–2804.

Atienza, C., Elliott, M. J., Dong, Y. B., Yang, H. L., Stilwell, A., Liu, T. J., and McMasters, K. M. (2000). Adenovirus-mediated E2F-1 gene transfer induces an apoptotic response in human gastric carcinoma cells that is enhanced by cyclin dependent kinase inhibitors. *Int. J. Mol. Med.* **6**, 55–63.

Barnes, D. M. (1997). Cyclin D1 in mammary carcinoma. *J. Pathol.* **181**, 267–269.

Bennett, P., Mani, S., O, Reilly, S., Wright, J., Schilsky, R., Vokes, E., and Grochow L. (1999). Phase II trial of flavopiridol in metastatic colorectal cancer: Preliminary results. *Proc. Am. Soc. Clin. Oncol.*, **18**, A1065.

Bible, K. C., and Kaufmann, S. H. (1997). Cytotoxic synergy between flavopiridol (NSC 649890, L86-8275) and various antineoplastic agents: The importance of sequence of administration. *Cancer Res.* **57**, 3375–3380.

Bible, K. C., Bible, R. H., Kottke, T. J., Svingen, P. A., Xu, K., Pang, Y. P., Hajdu, E., and Kaufmann, S. H. (2000). Flavopiridol binds to duplex DNA. *Cancer Res.* **60**, 2419–2428.

Bojanowski, K., Nishio, K., Fukuda, M., Larsen, A. G., and Saijo, N. (1994). Effect of suramin on p34[cdc2] kinase *in vitro* and in extracts from human H69 cells: Evidence for a double mechanism of action. *Biochem. Biophys. Res. Commun.* **203**, 1574–1580.

Cangi, M. G., Cukor, B., Soung, P., Signoretti, S., Moreira, G., Ranashinge, M., Cady, B., Pagano, M., and Loda, M. (2000). Role of cdc25A phosphatase in human breast cancer. *J. Clin. Invest.* **106**, 753–761.

Chang, Y. T., Gray, N. G., Rosania, G. R., Sutherlin, D. P., Kwon, S., Norman, T. C., Sarohia, R., Leost, M., Meijer, L., and Schultz, P. G. (1999). Synthesis and application of functionally diverse 2,6,9-trisubstituted purine libraries as CDK inhibitors. *Chem. Biol.* **6**, 361–375.

Chen, D. H., and Xie, J. X. (1984). Chemical constituents of traditional Chinese medicine Qing Dai. *Chin. Tradit. Herb. Drugs (Zhongcaoyao)* **15**, 6–8.

Christain, M. C., Pluda, J. M., Ho, P. T. C., Arbuck, S. G., Murgo, A. J., and Sausville, E. A. (1997). Promising new agents under development by the division of cancer treatment, diagnosis, and centers of the National Cancer Institute. *Semin. Oncol.* **2**, 219–240.

Clinical Therapeutic Coordinating Group of Mesoindigo (1988). Therapeutic effect of mesoindigo against chronic myeloid leukemia. *Acta Haematol. Sin.* **9**, 135–139.

Collecchi, P., Passoni, A., Rocchetta, M., Gnesi, E., Baldini, E., and Bevilacqua, G. (1999). Cyclin-D1 expression in node positive (N+) and node negative (N−) infiltrating human mammary carcinomas. *Int. J. Cancer* **84**, 139–144.

Cooperative Group of Clinical Therapy of Indirubin (1980). Clinical studies of 314 cases of CML treated with indirubin. *Chin. J. Intern. Med.* **1**, 132–135.

Damiens, E., Baralte, B., Marie, D., Eisenbrand, G., and Meijer, L. (2001). Anti-mitotic properties of indirubin-3′-monoxime, a CDK/GSKB inhibitor: Induction of endoreplication following prophase arrest. *Oncogene* **20**, 3786–3797.

de Azevedo, W. F., Mueller-Dieckmann, H.-J., Schultze-Gahmen, U., Worland, P. J., Sausville, E., and Kim, S.-H. (1996). Structural basis for specificity and potency of a flavonoid inhibitor of human CDK2, a cell cycle kinase. *Proc. Natl. Acad. Sci. U.S.A.* **93**, 2735–2740.

de Azevedo, W. F., Leclerc, S., Meijer, L., Havlicek, L., Strnad, M., and Kim, S.-H. (1997). Inhibition of cyclin-dependent kinases by purine analogues: Crystal structure of human cdk2 complexed with roscovitine. *Eur. J. Biochem.* **243**, 518–526.

Drees, M., Dengler, W. A., Roth, T., Labonte, H., Mayo, J., Malspeis, L., Grever, M., Sausville, E. A., and Fiebig, H. H. (1997). Flavopiridol (L86-8275): Selective antitumor activity *in vitro* and activity *in vivo* for prostate carcinoma cells. *Clin. Cancer Res.* **3**, 273–279.

Edamatsu, H., Gau, C. L., Nemoto, T., Guo, L. and Tamanoi, F. (2000). Cdk inhibitors, roscovitine and olomoucine, synergize with farnesyl transferase inhibitor (FTI) to induce efficient apoptosis of human cancer cell lines. *Oncogene* **19**, 3059–3068.

Fischer, P. M. and Lane, D. P. (2000). Inhibitors of cyclin-dependent kinases as anti-cancer therapeutics. *Curr. Med. Chem.* **7**, 1213–1245.

Funk, J. O. (1999). Cancer cell cycle control. *Anticancer Res.* **19**, 4772–4780.

Fuse, E., Tanii, H., Kurata, N., Kobayashi, H., Shimada, Y., Tamura, T., Sasaki, Y., Tanigawara, Y., Lush, R. D., Headlee, D., Figg, W. D., Arbuck, S. G., Senderowicz, A. M., Sausville, E. A., Akinaga, S., Kuwabara, T., and Kobayashi, S. (1998). Unpredicted clinical pharmacology of UCN-01 caused by specific binding to human alpha1-acid glycoprotein. *Cancer Res.* **58**, 3248–3253.

Fuse, E., Tanii, H., Takai, K., Asanome, K., Kurata, N., Kobayashi, H., Kuwabara, T., Kobayashi, S., and Sugiyama, Y. (1999). Altered pharmacokinetics of a novel anticancer drug, UCN-01, caused by specific high affinity binding to α 1-acid glycoprotein in humans. *Cancer Res.* **59**, 1054–1060.

Garrett, M. D., and Fattaey, A. (1999). Cyclin-dependent kinase inhibition and cancer therapy. *Curr. Opin. Genet. Dev.* **9**, 104–111.

Gray, N., Wodicka, L., Thunnissen, A. M., Norman, T., Kwon, S., Espinoza, F. H., Morgan, D. O., Barnes, G., Leclerc, S., Meijer, L., Kim, S. H., Lockhart, D. J., and Schultze, P. (1998). Exploiting chemical libraries, structure, and genomics in the search for new kinase inhibitors. *Science* **281**, 533–538.

Gray, N., Détivaud, L., Doerig, C., and Meijer, L. (1999). ATP-site directed inhibitors of cyclin-dependent kinases. *Curr. Med. Chem.* **6**, 859–876.

Hoessel, R., Leclerc, S., Endicott, J., Noble, M., Lawrie, A., Tunnah, P., Leost, M., Damiens, E., Marie, D., Marko, D., Niederberger, E., Tang, W., Eisenbrand, G., and Meijer, L. (1999). Indirubin, the active constituent of a Chinese antileukaemia medicine, inhibits cyclin-dependent kinases. *Nat. Cell Biol.* **1**, 60–67.

Hooijberg, J. H., Broxterman, H. J., Heijn, M., Fles, D. L., Lankelma, J., and Pinedo, H. M. (1997). Modulation by (iso)flavonoids of the ATPase activity of the multidrug resistance protein. *FEBS Lett.* **413**, 344–348.

Hooijberg, J. H., Broxterman, H. J., Scheffer, G. L., Vrasdonk, C., Heijn, M., de Jong, M. C., Schepper, R. J., Lankelma, J., and Pinedo, H. M. (1999). Potent interaction of flavopiridaol with MRP1. *Br. J. Cancer* **81**, 269–276.

Jager, W., Zembsch, B., Wolschann, P., Pittenauer, E., Senderowicz, A. M., Sausville, E. A., Sedlacek, H. H., Graf, J., and Thalhammer, T. (1998). Metabolism of the anticancer drug flavopiridol, a new inhibitor of cyclin dependent kinases, in rat liver. *Life Sci.* **62**, 1861–1873.

Ji, X. J., Zhang, F. R., Lei, L. J., and Xu, Y. T. (1981). Studies on the antineoplastic effect and toxicity of synthetic indirubin. *Acta Pharm. Sin.* **16**, 146–148.

Ji, X. J., Liu, X. M., Li, K., Chen, R. H., and Wang, L. G. (1991). Pharmacological studies of mesoindigo: Absorption and mechanism of action. *Biomed. Environ. Sci.* **4**, 332–337.

Karp, J. E., and Broder, S. (1995). Molecular foundations of cancer: New targets for intervention. *Nat. Med.* **1**, 309–320.

Keyomarsi, K., and Pardee, A. B. (1993). Redundant cyclin overexpression and gene amplification in breast cancer cells. *Proc. Natl. Acad. Sci. U.S.A.* **90**, 1112–1116.

Keyomarsi, K., and Pardee A. (1996). Cyclin E—a better prognostic marker for breast cancer than cyclin D? *Nat. Med.* **2**, 254.

Kitagawa, M., Okabe, T., Ogino, H., Matsumoto, H., Suzuki-Takahashi, I., Kokubo, T., Higashi, H., Saitoh, S., Taya, Y., Yasuda, H., Ohba, Y., Nishimura, S., Tanaka, N., and Okuyama, A. (1993). Butyrolactone I, a selective inhibitor of cdk2 and cdc2 kinase. *Oncogene* **8**, 2425–2432.

Knockaert, M., Gray, N., Damiens, E., Chang, Y. T., Grellier, P., Grant, K., Fergusson, D., Mottram, J., Soete, M., Dubremetz, J. F., LeRoch, K., Doerig, C., Schultz, P. G., and Meijer, L. (2000). Intracellular targets of cyclin-dependent kinase inhibitors: Identification by affinity chromatography using immobilised ligands. *Chem. Biol.* **7**, 411–422.

Kurata, N., Kuwabara, T., Tanii, H., Fuse, E., Akiyama, T., Akinaga, S., Kobayashi, H., Yamaguchi, K., and Kobayashi, S. (1999). Pharmacokinetics and pharmacodynamics of a novel protein kinase inhibitor, UCN-01. *Cancer Chemother. Pharmacol.* **44**, 12–18.

Lammie, G. A., Fantl, V., Smith, R., Schuring, E., Brooke, S., Michalide, R., Dickso, C., Arnold, A., and Peters, G. (1991). D11S287, a putative oncogene on chromosome 11q13, is amplified and expressed in squamous cell and mammary carcinomas and linked to BCL-1. *Oncogene* **6**, 439–444.

Lawrie, A. M., Noble, M. E., Tunnah, P., Brown, N. R., Johnson, L. N., and Endicott, J. A. (1997). Protein kinase inhibition by staurosporine revealed in details of the molecular interaction with CDK2. *Nat. Struct. Biol.* **4**, 796–800.

Leclerc, S., Garnier, M., Hoessel, R., Marko, Bibb, J. A., Snyder, G. L., Greengard, P., Biernat, J., Mandelkow, E.-M., Eisenbrand, G., and Meijer, L. (2001). Indirubins inhibit glycogen synthase kinase-3β and CDK5/p25, two kinases involved in abnormal tau phosphorylation in Alzheimer's disease—A property common to most CDK inhibitors? *J. Biol. Chem.* **276**, 251–260.

Lee, M. G., and Nurse, P. (1987). Complementation used to clone a human homologue of the fission yeast cell cycle control gene cdc2. *Nature* **237**, 31–35.

Legraverend, M., Ludwig, O., Bisagni, E., Leclerc, S., and Meijer, L. (1998). Synthesis of C2 alkynylated purines, a new family of potent inhibitors of cyclin-dependent kinases. *Bioorg. Med. Chem. Lett.* **8**, 793–798.

Legraverend, M., Ludwig, O., Bisagni, E., Leclerc, S., Meijer, L., Giocanti, N., Sadri, R. and Favaudon, V. (1999). Synthesis and *in vitro* evaluation of novel 2,6,9-trisubstituted purines acting as cyclin-dependent kinase inhibitors. *Bioorg. Med. Chem.* **7**, 1281–1293.

Legraverend, M., Noble, M., Tunnah, P., Ducrot, P., Ludwig, O., Leclerc, S., Grierson, D. S., Meijer, L., and Endicott, J. (2000). Cyclin-dependent kinase inhibition by new C-2 alkylynated purine derivatives and molecular structure of a CDK2-inhibitor complex. *J. Med. Chem.* **43**, 1282–1292.

Legraverend, M., Ludwig, O., Leclerc, S., and Meijer, L. (2001). Synthesis of a new series of purine derivatives and their anti-cyclin-dependent kinase activities. *J. Heterocyclic Chem.* **38**, 299–303.

Leost, M., Schultz, C., Link, A., Wu, Y.-Z., Biernat, J., Mandelkow, E.-M., Bibb, J. A., Snyder, G. L., Greengard, P., Zaharevitz, D. W., Gussio, R., Senderovitz, A., Sausville, E. A., Kunick, C., and Meijer, L. (2000). Paullones are potent inhibitors of glycogen synthase kinase-3β and cyclin-dependent kinase 5/p25. *Eur. J. Biochem.* **267**, 5983–5994.

Lundberg A. S., and Weinberg, R. A. (1999). Control of the cell cycle and apoptosis. *Eur. J. Cancer* **35**, 531–539.

Ma, M., and Yao, B. (1983). Progress in indirubin treatment of chronic myelocytic leukemia. *J. Tradit. Chin. Med. (Zhongcaoyao)* **3**, 245–248.

Marko, D., Schätzle, Friedel, A., Genzlinger, A., Zankl, H., Meijer, L., and Eisenbrand, G. (2000). Inhibition of cyclin-dependent kinase 1 (CDK1) by indirubin derivatives in human tumor cells. *Br. J. Cancer* **84**, 283–289.

Meijer, L. (1996). Chemical inhibitors of cyclin-dependent kinases. *Trends Cell Biol.* **6**, 393–397.

Meijer, L. (2000). Cyclin-dependent kinases inhibitors as potential anticancer, anti-neurodegenerative, anti-viral and anti-parasitic agents. *Drug Resistance Update* **3**, 83–88.

Meijer, L., and Kim, S. H. (1997). Chemical inhibitors of cyclin-dependent kinases. *Methods Enzymol.* **283**, 113–128.

Meijer, L., Guidet, S., and Philippe, M. (eds.) (1997a). "Progress in Cell Cycle Research," Vol. 3. Plenum, New York.

Meijer, L., Borgne, A., Mulner, O., Chong, J. P. J., Blow, J. J., Inagaki, N., Inagaki, M., Delcros, J. G., and Moulinoux, J. P. (1997b). Biochemical and cellular effects of roscovitine, a potent and selective inhibitor of the cyclin-dependent kinases cdc2, cdk2 and cdk5. *Eur. J. Biochem.* **243**, 527–536.

Meijer, L., Leclerc, S., and Leost, M. (1999). Properties and potential applications of chemical inhibitors of cyclin-dependent kinases. *Pharmacol. Ther.* **82**, 279–284.

Meijer, L., Jezequel, A., and Ducommun, B. (eds.) (2000a). "Progress in Cell Cycle Research," Vol. 4. Plenum, New York.

Meijer, L., Thunissen, A. M. W. H., White, A., Garnier, M., Nikolic, M., Tsai, L. H., Walter, J., Cleverley, K. E., Salinas, P. C., Wu, Y. Z., Biernat, J., Mandelkow, E. M., Kim, S.-H., and Pettit, G. R. (2000b). Inhibition of cyclin-dependent kinases, GSK-3β and casein kinase 1 by hymenialdisine, a marine sponge constituent. *Chem. Biol.* **7**, 51–63.

Meyerson, M., Enders, G. H., Wu, C. L., Su, L. K., Gorka, C., Nelson, C., Harlow, E., and Tsai, L. H. (1992). A family of human cdc2-related protein kinases. *EMBO J.* **11**, 2909–2917.

Mittnacht, S., and Boshoff, C. (2000). Viral cyclins. *Rev. Med. Virol.* **10**, 175–184.

Morgan, D. (1997). Cyclin-dependent kinases: Engines, clocks, and microprocessors. *Annu. Rev. Cell Dev. Biol.* **13**, 261–291.

Nurse, P., and Bissett, Y. (1981). Gene required in G1 for commitment to cell cycle and in G2 for control of mitosis in fission yeast. *Nature* **292**, 558–560.

Ohashi, M., Sugikawa, E., and Nakanishi, N. (1995). Inhibition of p53 phosphorylation by 9-hydroxyellipticine: A possible anticancer mechanism. *Jpn. J. Cancer Res.* **86**, 819–827.

Oikonomakos, N. G., Schnier, J. B., Zographos, S. E., Skamnaki, V. T., Tsitsanou, K. E., and Johnson, L. N. (2000). Flavopiridol inhibits glycogen phosphorylase by binding at the inhibitor site. *J. Biol. Chem.* **275**, 34566–34573.

Ongkeko, W., Ferguson, D. J. P., Harris, A. L., and Norbury, C. (1995). Inactivation of cdc2 increases the level of apoptosis induced by DNA damage. *J. Cell Sci.* **108**, 2897–2904.

Owa, T., Yoshino, H., Yoshimatsu, K., Ozawa, Y., Sugi, N. H., Nagasu, T., Koyanagi, N., and Kitoh, K. (1999). Discovery of novel antitumor sulfonamides targeting G1 phase of the cell cycle. *J. Med. Chem.* **42**, 3789–3799.

Park, S. G., Cheon, J. Y., Lee, Y. H., Park, J.-S., Lee, K. Y., Lee, C. H., and Lee, S. K. (1996). A specific inhibitor of cyclin-dependent protein kinases, CDC2 and CDK2. *Mol. Cells* **6,** 679–683.

Patel, V., Senderowicz, A. M., Pinto, D., Igishi, T., Raffeld, M., Quintanilla-Martinet, L., Ensley, J. F., Sausville, E. A., and Gutkind, J. S. (1998). Flavopiridol, a novel cyclin-dependent kinase inhibitor, suppresses the growth of head and neck squamous cell carcinomas by inducing apoptosis. *J. Clin. Invest.* **102,** 1674.

Pavletich, N. P. (1999). Mechanisms of cyclin-dependent kinase regulation: Structures of cdks, their cyclin activators, and Cip and INK4 inhibitors. *J. Mol. Biol.* **287,** 821–828.

Pestell, R., Mani, S., Wange, C., Wu, K., and Francis, R. (2000). Cyclin-dependent kinase inhibitors: Novel anticancer agents. *Expert Opin. Invest. Drugs* **9,** 1849–1870.

Plowman, G. D., Sudarsanam, S., Bingham, J., Whyte, D., and Hunter, T. (1999). The protein kinases of *Caenorhabditis elegans*: A model for signal transduction in multicellular organisms. *Proc. Natl. Acad. Sci. U.S.A.* **96,** 13603–13610.

Qi, S. B., He, G. X., and Wang Y. D. (1981). Pharmacological studies on indirubin. III. Pharmacokinetic studies on indirubin in mice. *Chin. Tradit. Herb. Drugs (Zhongcaoyao)* **12,** 23–27.

Rialet, V., and Meijer, L. (1991). A new screening test for antimitotic compounds using the universal M phase-specific protein kinase, $p34^{cdc2}$/cyclin B^{cdc13}, affinity-immobilized on $p13^{suc\ 1}$-coated microtitration plates. *Anticancer Res.* **11,** 1581–1590.

Rosiana, G. R., and Chang, Y. T. (2000). Targeting hyperproliferative disorders with cyclin dependent kinase inhibitors. *Expert Opin. Ther. Patents* **10,** 1–16.

Rosania, G. R., Merlie, J., Gray, N. S., Chang, Y. T., Schultz, P. G., and Heald, R. (1999). A cyclin-dependent kinase inhibitor inducing cancer cell differentiation: Biochemical identification using *Xenopus* egg extracts. *Proc. Natl. Acad. Sci. U.S.A.* **96,** 4797–4802.

Ruetz, S., Woods-Cook, K., Solf, R., Meyer, T., Zimmermann, J., and Fabbro, D. (1998). Effect of CGP60474 on cyclin dependent kinase (cdks), cell cycle proliferation and onset of apoptosis in normal and transformed cells. *Proc. Am. Assoc. Cancer Res.* **39,** 558.

Sausville, E. A., Lush, R. D., Headlee, D., Smith, A. C., Figg, W. D., Arbuck, S. G., Senderowicz, A. M., Fuse, E., Tanii, H., Kuwabara, T., and Kobayashi, S. (1998). Clinical pharmacology of UCN-01: Initial observations and comparison to preclinical models. *Cancer Chemother. Pharmacol.* **42,** S54–S59.

Schultz, C., Link, A., Leost, M., Zaharevitz, D. W., Gussio, R., Sausville, E. A., Meijer, L. and Kunick, C. (1999). The paullones, a series of cyclin-dependent kinase inhibitors: Synthesis, evaluation of CDK1/cyclin B inhibition, and *in vitro* antitumor activity. *J. Med. Chem.* **42,** 2909–2919.

Schulze-Gahmen, U., Brandsen, J., Jones, H. D., Morgan, D. O., Meijer, L., Vesely, J., and Kim, S. H. (1995). Multiple modes of ligand recognition: Crystal structures of cyclin-dependent protein kinase 2 in complex with ATP and two inhibitors, olomoucine and isopentenyladenine. *Proteins: Struct. Funct. Genet.* **22,** 378–391.

Sedlacek, H. H., Czech, J., Naik, R., Kaur, G., Worland, P., Losiewicz, M., Parker, B., Carlson, B., Smith, A., Senderowicz, A., and Sausville, E. (1996). Flavopiridol (L86 8275; NSC 649890), a new kinase inhibitor for tumor therapy. *Int. J. Oncol.* **9,** 1143–1168.

Seidensticker, M. J., and Behrens, J. (2000). Biochemical interactions in the wnt pathway. *Biochim. Biophys. Acta* **1495,** 168–182.

Senderowicz, A. M. (1999). Flavopiridol: The first cyclin-dependent kinase inhibitor in human clinical trials. *Invest. New Drugs* **17,** 313–320.

Senderowicz, A. M., and Sausville, E. A. (2000). Preclinical and clinical development of cyclin-dependent kinase modulators. *J. Natl. Cancer Inst.* **92,** 376–387.

Senderowicz, A. M., Headlee, D., Stinson, S. F., Lush, R. M., Kalil, N., Villalba, L., Hill, K., Steinberg, S. M., Figg, W. D., Tompkins, A., Arbuck, S. G., and Sausville, E. A. (1998). Phase I trial of continuous infusion flavopiridol, a novel cyclin-dependent kinase inhibitor, in patients with refractory neoplasms. *J. Clin. Oncol.* **16,** 2986–2999.

Senderowicz, A. M., Headlee, D., Lush, R. M., Arbuck, S., Bauer, K., Figg, W. D., Murgo, A., Inoue, K., Kobayashi, S., Kuwabara, T., and Sausville, E. A. (1999). Phase I trial of infusional UCN-01, a protein kinase inhibitor, in patients with refractory neoplasms. *Proc. Am. Soc. Clin. Oncol.* **18,** 612.

Shapiro, G., Patterson, A., Lynch, C., Lucca, J., Anderson, I., Boral, A., Elias, A., Lu, H., Salgia, R., Skarin, A., Panek-Clark, C., McKenna, R., Rabin, M., Vasconcelles, M., Eder, P., Supko Supko, J., Lynch, T., and Rollins, B. (1999). A phase II trial of flavopiridol in patients with stage IV non-small cell lung cancer. *Proc. Am. Soc. Clin. Oncol.* **18,** A2013.

Shenfeld, M., Senderowicz, A. M., Sausville, E. A., and Ke, B. (1997). A novel chemotherapeutic agent, flavopiridol, modulates intestinal epithelial chloride secretion. *Gastroenterology* **112,** 404.

Sherr, C. J., and Roberts, J. M. (1999). CDK inhibitors: Positive and negative regulators of G1-phase progression. *Genes Dev.* **13,** 1501–1512.

Shimizu, N., Sugimoto, K., Tang, J., Nishi, T., Sato, I., Hiramoto, M., Aizawa, S., Hatakeyama, M., Ohba, R., Hatori, H., Yoshikawa, T., Suzuki, F., Oomori, A., Tanaka, H., Kawaguchi, H., Watanabe, H., and Handa, H. (2000). High-performance affinity beads for identifying drug receptors. *Nat. Biotechnol.* **18,** 877–881.

Sichuan Institute of Traditional Chinese Medicine (1981). Subacute toxicity of indirubin in dogs. *Chin. Tradit. Herb. Drugs (Zhongcaoyao)* **12,** 27–29.

Sielecki, T. M., Boylan, J. F., Benfield, P. A., and Trainor, G. L. (2000). Cyclin-dependent kinase inhibitors: Useful targets in cell cycle regulation. *J. Med. Chem.* **43,** 1–18.

Soni, R., Muller, L., Furet, P., Schoepfer, J., Stephan, C., Zumstein-Mecker, S., Fretz, H., and Chaudhuri, B. (2000). Inhibition of cyclin-dependent kinase 4 (cdk4) by fascaplysin, a marine natural product. *Biochem. Biophys. Res. Commun.* **275,** 877–884.

Tamura, T., Sasaki, Y., Minami, H., Fujii, H., Ito, K., Igarashi, T., Kamiya, T., Ohtsu, T., Onosawa, Y., Yamamoto, N., Watanabe, Y., Tanigawara, Y., Fuse, E., Kuwabara, T., Kobayashi, S., and Shimada, Y. (1999). Phase I study of UCN-01 by 3-hour infusion. *Proc. Am. Soc. Clin. Oncol.* **18,** 611.

Thomas, J., Cleary, J., Tutsch, K., Arzoomanian, R., Alberti, D., Simon, K., Feirabend, D., Morgan, K., and Wilding, G. (1997). Phase I clinical and pharmacokinetic trial of flavopiridol. Proceedings of the 88th Annual Meeting of the American Association of Cancer Research. San Diego, California.

Toledo, L., Lydon, N. B., and Elbaum, D. (1999). Structure-based design of ATP-site directed protein kinase inhibitors. *Curr. Med. Chem.* **6,** 775–805.

Vesely, J., Havlicek, L., Strnad, M., Blow, J. J., Donella-Deana, A., Pinna, L., Letham, D. S., Kato, J. Y., Détivaud, L., Leclerc, S., and Meijer, L. (1994). Inhibition of cyclin-dependent kinases by purine derivatives. *Eur. J. Biochem.* **224,** 771–86.

Vogt, P. K., and Reed S. I. (1998). Cyclin dependent kinase (CDK) inhibitors. "Current Topics in Microbiology and Immunology". Springer-Verlag, Berlin and New York.

Wright, J., Blatner, G. L., and Cheson, B. D. (1998). Clinical trials referral resource. Clinical trials of flavopiridol. *Oncology* **12,** 1018–1023.

Yatabe, Y., Suzuki, R., Tobinai, K., Matsuno, Y., Ichinohasama, R., Okanoto, M., Yamaguchi, M., Tamaru, J., Uike, N., Hashimoto, Y., Morishima, Y., Suchi, T., Seto, M., and Nakamura, S. (2000). Significance of cyclin D1 overexpression for the diagnosis of mantle cell lymphoma: A clinicopathologic comparison of cyclin D1-positive MCL and cyclin D1-negative MCL-like B-cell lymphoma. *Blood* **95,** 2253–2261.

Zaharevitz, D., Gussio, R., Leost, M., Senderowicz, A. M., Lahusen, T., Kunick, C., Meijer, L., and Sausville, E. A. (1999). Discovery and initial characterization of the paullones, a novel class of small-molecule inhibitors of cyclin-dependent kinases. *Cancer Res.* **59,** 2566–2569.

Zhang, Z. N., Liu, E. K., and Zheng, T. L. (1985). Treatment of chronic myelocytic leukemia (CML) by traditional Chinese medicine and Western medicine alternatively. *J. Tradit. Chin. Med.* **5,** 246–248.

Zimmermann, J. (1995). Pharmacologically active pyrimidine derivatives and processes for the preparation thereof. PCT Ciba-Geigy, WO 95/09853.

NF1 and Other RAS-Binding Peptides

Hiroshi Maruta
Ludwig Institute for Cancer Research
Royal Melbourne Hospital
Parkville/Melbourne 3050, Australia

I. RAS MOLECULES: NORMAL VERSUS ONCOGENIC MUTANTS

Oncogenic mutations of RAS family GTPases (at least Ki-RAS, Ha-RAS, N-RAS, and perhaps M-RAS) contribute to the development of more than 30% of all human cancers, notably more than 90% pancreatic carcinomas and 50% of colon carcinomas (Bos, 1989). To understand the molecular mechanism underlying RAS-induced malignant transformation, and to explore potential therapeutics for the treatment of these RAS-associated cancers, RAS molecules of around 21 kDa have been the most extensively studied among proto-oncoproteins or oncoproteins, both biochemically and genetically, since some of these RAS genes were first cloned in early 1980s. The majority of oncogenic mutations take place at residues 12, 13, 59, and 61 of each RAS molecule (Bos, 1989; Maruta and Burgess, 1994).

Like many other G proteins or signaling GTPases, each normal RAS cycles between the active GTP-bound form and the inactive GDP-bound form. The GTP-bound RAS is slowly converted to GDP-bound form by its intrinsic GTPase activity. However, several distinct RAS-specific GTPase activating proteins (GAPs) in the cytosol greatly stimulate the GTP hydrolysis by each normal RAS, and quickly attenuate its signaling activity (Maruta and Burgess, 1994; Maruta, 1998). Once each normal RAS is in GDP-form, it is reactivated to GTP-bound form by a few distinct activators called GDP-dissociation stimulators (GDSs) or GDP–GTP exchange factors (GEFs) such as SOS1 and SOS2 through the adaptor protein Grb-2 and growth factor receptor Tyr kinases in response to the mitogenic stimulation by several distinct growth factors such as epidermal growth factor (EGF) family ligands and platelet-derived growth factor (PDGF) (Maruta and Burgess, 1994; Maruta, 1998).

The most important difference between normal RAS and its oncogenic mutants is in their response to RAS GAPs: the intrinsic GTPase activity of normal RAS is highly stimulated by these GAPs, whereas that of oncogenic RAS mutants is not at all, although each oncogenic RAS mutant binds all these RAS GAPs, including NF1 and p120 GAP (Trahey and McCormick, 1987; Maruta, 1998). As a consequence, normal RAS is activated only transiently in response to extracellular growth factors, whereas oncogenic RAS mutants are constitutively activated (remaining in the GTP-bound form) without any external mitogenic stimulation.

II. SUPER GAP?

Once each RAS is activated, it would activate several distinct effectors such as a Ser/Thr kinase called Raf, PI-3 kinase, and Ral GDS, in addition to binding the RAS-specific GAPs (Maruta, 1996; Downward, 1998; Maruta, 1998). Thus, in principle, there would be at least three distinct ways to block the oncogenic RAS signaling. One approach is to create/screen for a mutant of GAPs or chemical compound that is able to stimulate the intrinsic GTPase activity of oncogenic RAS mutants, thereby converting them to the inactive GDP-bound form. Here such a hypothetical GAP is called Super GAP. Although such a Super GAP has not been discovered or invented as yet, a unique GTP analog called DABP-GTP (for the chemical structure, see Figure 1) was developed which, unlike GTP, is rapidly hydrolyzed by oncogenic RAS mutants, thereby converting them to the inactive GDP-bound form *in vitro* (Ahmadian *et al.*, 1999). A serious problem associated with this compound in any *in vivo* application, including the clinical use for the chemotherapy of RAS-associated cancers, is the poor cell permeability of this GTP analog,

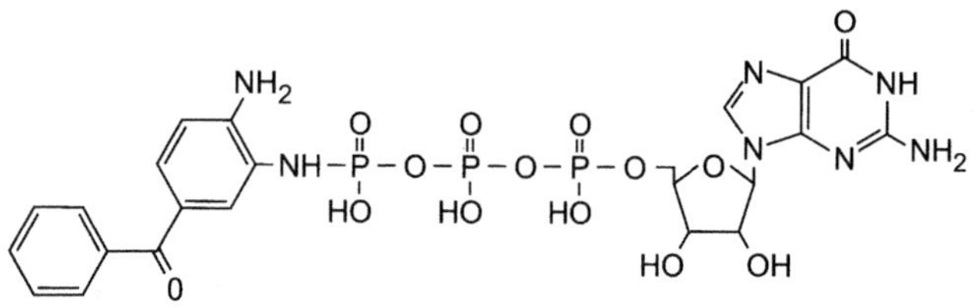

Figure 1 Chemical structure of DABP-GTP.

mainly due to its high negative charge. Conjugation of this analog with a positively charged side chain or liposomes such as lipofectamine would make it more cell-permeable.

A second approach is to block the plasma membrane localization of oncogenic RAS mutants by inhibiting the farnesylation of their C-terminal CAAX box, as their farnesylation-dependent membrane localization is essential or important for RAS transformation. Several distinct RAS farnesyltransferase inhibitors (FTIs) have been developed and are currently in clinical trials (Koblan and Kohl, 1998). This FTIs approach will be discussed in detail in Chapter 15. A third approach is to block the interaction of oncogenic RAS mutants with one of their downstream effectors by the miminal RAS-binding motifs of GAPs or effectors.

III. RAS-BINDING FRAGMENT OF NF1

The minimal GAP domain of bovine p120 GAP/GAP1, that is still able to stimulate the intrinsic GTPase activity of each normal RAS, turned out to be a relatively large fragment consisting of the C-terminal half (GAP1C, residues 720–1044) (Maruta *et al.*, 1991). NF1 (neurofibromatosis type 1 gene product) is a gigantic protein of 2818 amino acids (Nur-E-Kamal *et al.*, 1993). Based on the amino acid sequence similarity to the minimal GAP domain of GAP1 (Vogel *et al.*, 1988; Xu *et al.*, 1990a; Martin *et al.*, 1990), the GAP-related domain (GRD) of human NF1 was identified in the residues 1175–1531 (Martin *et al.*, 1990; Xu *et al.*, 1990b). Our further deletion analysis of the GRD has revealed that the minimal GAP domain of NF1 consists of only 78 amino acids (residues 1441–1518) called NF78 (Nur-E-Kamal *et al.*, 1993).

However, both GAP1C and NF78 fail to stimulate the intrinsic GTPase activity of oncogenic RAS mutants. Thus, the generation of a Super GAP molecule would need a further extensive mutational analysis of NF78 and its three-dimensional (3D) structure analysis which allow us to identify each critical residue that is essential for the GAP activity. The deletion of the C-terminal 22 amino acids from NF78 abolishes completely its GAP activity (Nur-E-Kamal *et al.*, 1993), suggesting that residues 1497–1518 are essential for the GAP activity, and would be the major targets for further mutational analysis to convert the NF78 to a Super GAP. The remaining domain of 56 amino acids called NF56 (residues 1441–1496) still binds at least oncogenic RAS mutants such as v-Ha-RAS (Fridman *et al.*, 1994). Overexpression of NF56 strongly suppresses RAS transformation, as does NF78, by blocking the interaction of RAS with any effectors such as Raf family kinases (Fridman *et al.*, 1994).

IV. c-RAF-1

There are at least three distinct members of Raf family kinases called c-Raf-1, B-Raf, and A-Raf in mammalian cells, and each Raf binds both normal and oncogenic RAS mutants in their GTP-bound form (Fridman *et al.*, 1994; Shimizu *et al.*, 1996). c-Raf-1 is the best biochemically characterized molecule among the RAS effectors. It is a Ser/Thr kinase of 648 amino acids (see Figure 2), and its kinase domain is in the C-terminal half (residues 305–648) which phosphorylates and activates the downstream effector MEK, a kinase that in turn activates Erk family kinases (Maruta, 1996).

Its N-terminal half contains two distinct RAS-binding sites, sites I/RBD (residues 51–131) and site II/CRD (residues 152–184), in addition to a so-called negatively regulating domain (NEG, residues 274–305). The full-length c-Raf-1 is a latent form of kinase which fails to activates its downstream kinase MEK, and is not oncogenic (Maruta, 1996). However, once the NEG domain is deleted, it becomes a constitutively active kinase and highly oncogenic (Maruta, 1996). In fact the oncogene v-Raf encodes only the C-terminal kinase domain of c-Raf-1. These findings strongly suggest that the NEG domain normally suppresses the catalytic function of the C-terminal kinase domain.

The site I (residues 51–131) has a much higher affinity for RAS than the site II (residues 152–184) (Maruta, 1996). Furthermore, the binding of RAS

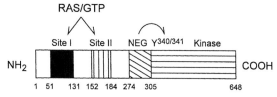

Figure 2 Functional domains of c-Raf-1. Sites I and II, RAS-binding domains; NEG, negatively regulating domain; Y340/341, phosphorylatable Tyr residues.

to site II appears to require the C-terminal CAAX motif in which farnesylation and a few other posttranslational modifications take place for the translocation of RAS to the plasma membranes (Maruta, 1996). The binding of RAS to the site I of c-Raf-1 is thought to play a role in the translocation of the cytosolic c-Raf-1 to the plasma membranes where an additional posttranslational modification including the phosphorylation by c-Src at Tyr-340 and Tyr-341 or PAK at Ser-338 takes place for the activation of c-Raf-1 (Fabian et al., 1993; Chaudhary et al., 2000).

The deletion analysis of c-Raf-1 by us and others has revealed that the 81 amino acid fragment (Raf 81) corresponding to the site I shows a high affinity for RAS only in the active GTP-bound form (Chuang et al., 1994; Fridman et al., 1994). Overexpression of Raf 81 strongly suppresses v-Ha-RAS-induced transformation as does NF56 (Fridman et al., 1994), indicating that Raf 81 alone is able to interfere with the RAS–Raf interaction in cells to block the oncogenic RAS pathways. Since the efficacy of Raf 81 to suppress RAS transformation should depend on the affinity for RAS, we have created a few mutants of Raf 81 which form a much more stable complex with v-Ha-RAS than the wild-type Raf 81 (Maruta, 1996; Fridman et al., 2000). Replacement of either Gln-64 by Lys, Val-70 by Arg, Ala-85 by Lys, or Val-88 by Arg improves the binding of Raf 81 to the oncogenic RAS. These Raf 81 mutants should be useful for the gene therapy of RAS-associated cancer in the future.

V. PI-3 KINASE

PI-3 kinase, a complex of the catalytic subunit (p110) and the regulatory subunit (p85), phosphorylates the D3 position of phosphatidyl inositol and activates several downstream effectors through its end-product (PIP3) (Downward, 1998). Downward (1998) and colleagues found that RAS activates PI-3 kinase by binding p110. The RAS-binding domain of p110 is within the N-terminal half (RBD, residues 150–314) (Downward, 1998). Thus, in theory, this RBD of p110 should suppress RAS transformation when it is overexpressed. However, no one has examined its potential anti-RAS effect as yet.

VI. RAL GDS

Ral GDS is an activator of the G protein Ral and is essential for RAS transformation (Albright et al., 1993). Kikuchi and colleagues identified the minimal RAS-binding domain (RID, residues 602–768) of Ral GDS (Kikuchi et al., 1994). Overexpression of the RID strongly suppresses RAS transformation, but not v-Raf transformation (Okazaki et al., 1996). Interestingly, RID blocks not only the RAS–Ral GDS interaction, but also the RAS–Raf interaction and inhibits RAS induced Raf-dependent activation of the c-Fos gene as does Raf 81 (Okazaki *et al.*, 1996). These observations suggest that the 3D structure of RID may be similar to that of Raf 81, although these two RAS-binding fragments share little primary sequence homology.

How does Ral GDS or Ral contribute to RAS-induced malignant transformation? To answer such a question, Kikuchi and colleagues examined the effect of Ral mutants on the endocytosis of the EGF receptor, and found that Ral is responsible for blocking the internalization of the EGF receptor (Nakashima *et al.*, 1999). Since the internalization of the EGF receptor (ErbB1) keeps the effect of continuously existing EGF or other related ligands which bind ErbB1 transient, this observation suggests that Ral (and its activator Ral GDS) might contribute to RAS transformation by keeping ErbB1 continuously activated.

Why should ErbB1 be kept continuously activated for oncogenic RAS to transform normal cells? The answer appears to be the following. Both ErbB1 and ErbB2 are essential for oncogenic RAS to activate PAK family kinases which are absolutely required for RAS transformation (He *et al.*, 2000). In normal cells EGF can activate PAKs only transiently, whereas in RAS transformants PAKs are constitutively activated by EGF family ligands which are upregulated by oncogenic RAS mutants (He *et al.*, 2000). Thus, it is quite conceivable that a chemical inhibitor specific for either Ral or Ral GDS could suppress RAS transformation efficiently by allowing ErbB1 to be internalized.

References

Ahmadian, M., Zor, T., Vogt, D., *et al.* (1999). GTPase stimulation of oncogenic RAS mutants. *Proc. Natl. Acad. Sci. U.S.A.* **96,** 7065–7070.

Albright, C., Giddings, B., Liu, J., *et al.* (1993). Characterization of a GDP dissociation stimulator for Ras-related GTPase. *EMBO J.* **12,** 339–347.

Bos, J. (1989). Ras oncogenes in human cancer. *Cancer Res.* **49,** 4682–4689.

Chaudhary, A., King, W., Mattaliano, M., *et al.* (2000). PI-3 kinase regulates Raf-1 through PAK phosphorylation of Ser 338. *Curr. Biol.* **10,** 551–554.

Chuang, E., Barnard, D., Hettich, L., *et al.* (1994). Critical binding and regulatory interactions between Ras and Raf occur through a small, stable N-terminal domain of Raf and specific Ras effector residues. *Mol. Cell. Biol.* **14,** 5318–5325.

Downward, J. (1998). Oncogenic RAS signaling network. *In* "G Proteins, Cytoskeleton and Cancer" (H. Maruta and K. Kohama, eds.), pp. 172–183. Landes Biosciences, Austin, Texas.

Fabian, J., Daar, I. and Morrison, D. (1993). Critical Tyr residues regulate the enzymatic and biological activity of Raf-1 kinase. *Mol. Cell. Biol.* **13,** 7170–7179.

Fridman, M., Tikoo, A., Varga, M., *et al.* (1994). The minimal fragment of c-Raf-1 and NF1 that can suppress v-Ha-RAS-induced malignant phenotype. *J. Biol. Chem.* **269,** 30105–30108.

Fridman, M., Maruta, H., Gonez, J., *et al.* (2000). Point mutants of c-Raf-1 RBD with elevated binding to v-Ha-RAS. *J. Biol. Chem.* **275,** 30363–30371.

He, H., Hirokawa, Y., Manser, E., *et al.* (2001). Signal therapy of RAS-induced cancers in combination of AG 879 and PP1, specific inhibitors for ErbB2 and Src family kinases, that block PAK activation. *Cancer J.* **7,** 191–202.

Kikuchi, A., Demo, S., Ye, Z. H., *et al.* (1994). Ral GDS family members interact with the effector loop of RAS. *Mol. Cell. Biol.* **14,** 7483–7491.

Koblan, K. and Kohl, N. (1998). Farnesyltransferase inhbitors: Agents for the treatment of human cancer. *In* "G Proteins, Cytoskeleton and Cancer" (H. Maruta and K. Kohama, eds.), pp. 291–302. Landes Biosciences, Austin, Texas.

Martin, G., Viskochil, D., Bollag, G., *et al.* (1990). The GAP-related domain of the neurofibromatosis type 1 gene product interacts with Ras. *Cell* **63,** 843–849.

Maruta, H. (1996). "The Regulation of RAS Signaling Networks" (H. Maruta and A. W. Burgess, eds.), pp. 139–180. Springer-Verlag/R. G. Landes, Heidelberg, Germany/Austin, Texas.

Maruta, H. (1998). Regulators of Ras/Rho family GTPases: GAPs, GDSs & GDIs. *In* "G Proteins, Cytoskeleton & Cancer" (H. Maruta and K. Kohama, eds.), pp. 151–170. Landes Biosciences, Austin, Texas.

Maruta, H., and Burgess, A. W. (1994). Regulation of the RAS signaling network. *BioEssays* **16,** 489–496.

Maruta, H., Holden, J., Sizeland, A., *et al.* (1991). The residues of Ras and Rap proteins that determine their GAP specificities. *J. Biol. Chem.* **266,** 11661–11668.

Nakashima, S., Morinaka, K., Koyama, S., *et al.* (1999). Samll G protein Ral and its downstream molecules regulate endocytosis of EGF and insulin receptors. *EMBO J.* **18,** 3629–3642.

Nur-E-Kamal, M. S. A., Varga, M., and Maruta, H. (1993). The GTPase activating NF1 fragment of 91 amino acids reverses v-Ha-RAS-induced malignant phenotype. *J. Biol. Chem.* **268,** 22331–22337.

Okazaki, M., Kishida, S., Murai, H., *et al.* (1996). RAS-interacting domain of Ral GDS like (RGL) reverses v-RAS-induced transformation and Raf-1 activation in NIH 3T3 cells. *Cancer Res.* **56,** 2387–2392.

Shimizu, K., Ohtsuka, T., and Takai, Y. (1996). From RAS to MAPK: Cell-free assay system for RAS-and Rap1-dependent B-Raf activation. *In* "the Regulation of RAS Signaling Network" (H. Maruta and A. W. Burgess, eds.), pp. 181–200. Springer-Verlag/R. G. Landes, Heidelberg, Germany/Austin, Texas.

Trahey, M., and McCormick, F. (1987). A cytoplasmic protein stimulates normal Ras GTPase, but does not affect oncogenic mutants. *Science* **238,** 542–545.

Vogel, U., Dixon, R., Schaber, M., *et al.* (1988). Cloning of bovine GAP and its interaction with oncogenic Ras. *Nature* **335,** 90–93.

Xu, G., O' Connell, P., Viskochil, D., *et al.* (1990a). The neurofibromatosis type 1 gene encodes a protein related to GAP. *Cell* **62,** 599–608.

Xu, G., Lin, B., Tanaka, K., *et al.* (1990b). The catalytic domain of the neurofibromatosis type 1 gene product stimulates GTPase and complements Ira mutants of *S. cerevisiae. Cell* **63,** 835–841.

9

Cytoskeletal Tumor Suppressor Genes

Noboru Kuzumaki* and Hiroshi Maruta[†,1]

*Institute for Genetic Medicine
Hokkaido University
Sapporo 060-0815, Japan

†Ludwig Institute for Cancer Research
Royal Melbourne Hospital
Parkville/Melbourne 3050, Australia

I. Introduction (Historical Background)
II. Type I Cytoskeletal Tumor Suppressors
III. Type II Cytoskeletal Tumor Suppressors
References

I. INTRODUCTION (HISTORICAL BACKGROUND)

Actin-cytoskeleton, consisting of actin (thin) filaments, myosin (thick) filaments, and various other actin-binding proteins, is present in all eukaryotes from yeast to human, and plays an important role not only in muscle contraction, but also in a variety of nonmuscle cell motilities such as cytokinesis, phagocytosis, pinocytosis, cell migration, and cell adhesion to the substratum. The interactions of actin filaments with several distinct myosin isoforms and other actin-binding proteins that cap, severe, or bundle actin filaments are regulated by their phosphorylation, and interaction with PIP2/PIP3 or calcium (Maruta and Korn, 1977; Maruta et al., 1984; Yin, 1987).

[1] To whom correspondence should be addressed.

Tumor-Suppressing Viruses, Genes, and Drugs
Innovative Cancer Therapy Approaches
Copyright © 2002 by Academic Press.
All rights of reproduction in any form reserved.

Each actin filament has a functional polarity, being demonstrated in a so-called arrowhead structure by the decoration with myosin ATPase heads or S1 fragments. The barbed end corresponds to the plus (or fast-growing) end of actin filament where a rapid actin polymerization takes place, due to the higher affinity for actin monomers. The pointed end corresponds to the minus (or slow-growing) end of actin filament where depolymerization of actin mainly takes place, due to the lower affinity for actin monomers, in a steady state. Thus, when the plus end is capped by several distinct proteins called plus-end F-actin cappers such as gelsolin and tensin, the rapid actin polymerization is blocked at that end, and consequently the capped actin filament gets shorter, due to depolymerization at the opposite minus end (Maruta et al., 1984).

Interestingly, malignant transformation of normal fibroblasts such as NIH 3T3 cells or epithelial cells caused by SV40 viruses or oncogenic mutants of RAS such as v-Ha-RAS and v-Ki-RAS is accompanied with disruption of actin stress fibers and induction of both actin-based membrane ruffling and microspike formation (Weber et al., 1974; Bar-Sagi and Feramisco, 1986; Maruta et al., 1999), suggesting that these oncogenic proteins affect the interactions of actin filaments with some of these actin binding proteins by regulating either their expression level or posttranslational (covalent or noncovalent) modification.

The biochemical change in these actin-binding proteins that was first found in accompany with SV40/RAS-induced malignant transformation was the downregulation of expression of several distinct actin-binding proteins such as gelsolin, α-actinin, vinculin, tropomyosin, myosin light chain, and α-actin (Vandekerckhove et al., 1990). This critical finding took place just a decade ago, and eventually led to the identification of at least a few distinct actin-binding proteins as tumor suppressors (Rodriguez-Fernandez et al., 1992; Glueck et al., 1993; Muellauer et al., 1993). The first two of them were vinculin and α-actinin whose overexpression suppresses SV40-induced malignant transformation of NIH 3T3 cells (Rodriguez-Fernandez et al., 1992; Glueck et al., 1993). A third one was a mutant of gelsolin whose overexpression suppresses v-Ha-RAS-induced transformation of the same fibroblasts (Muellauer et al., 1993). Interestingly, the genes encoding these three proteins are suppressed by either SV40 or oncogenic RAS mutants. Here a group of these three tumor suppressors shall be called type I cytoskeletal tumor suppressors whose expression is subject to the downregulation by oncogenes. A big mystery is that overexpression of only one of these suppressive genes is sufficient for the complete suppression of malignancy.

The exact molecular mechanism underlying the suppression of malignancy by these actin-binding proteins still remains to be clarified. Shortly after the finding of these three actin-binding tumor suppressors in the early 1990s, we found a unique actin-binding protein called NF2/Merlin (neurofibromatosis type 2 gene product) that also suppresses RAS transformation, but its expression is not suppressed by oncogenic RAS at all (Tikoo et al., 1994). Since then we identified several other tumor suppressing actin-binding proteins (cytoskeletal tumor suppressors) such as HS1, tensin, and cofilin whose expression is not affected by RAS transformation (He et al., 1998; Maruta, 1998; Tikoo et al., 1999). Here a group of these RAS-independent tumor suppressors including NF2, HS1, tensin, and cofilin shall be called type II cytoskeletal tumor suppressors. These findings suggest that the simple restoration of their suppressed expression is not the mechanism underlying their anti-RAS action. More interestingly, a mutant (Gln-112/114) of cofilin which is capable of suppressing RAS transformation lacks its binding to actin filaments (Maruta, 1998), suggesting that the actin binding is not absolutely essential for the suppression of RAS transformation.

Is there any common property, other than actin binding, that these cytoskeletal tumor suppressors share? Yes, there is one. All these proteins bind PIP2/PIP3 invariably. Moreover, it is now well known that oncogenic RAS leads to overproduction of PIP2/PIP3 through PI-3 kinase, RAC/CDC42, and PI-4/ PI-5 kinases, and PIP2/PIP3 uncaps the plus end of actin filaments and induces a rapid actin polymerization which is essential for both membrane ruffling and microspike formation (He et al., 1998; Maruta, 1998). Furthermore, another tumor suppressor called PTEN is a phosphatase that dephosphorylates PIP3 at the D3 position and suppresses RAS transformation (Li et al., 1997; Steck et al., 1997; Maehama and Dixon, 1998; Tolkacheva and Chan, 2000). Thus, if PTEN and these actin-binding proteins exert their anti-RAS action through a common mechanism, it is most likely that their sequestering/destruction of PIP3 is responsible for their suppression of RAS transformation. In this chapter, we shall discuss the biochemical properties of the cytoskeletal tumor suppressors (type I and II) that are capable of binding both actin filaments and PIP2/PIP3.

II. TYPE I CYTOSKELETAL TUMOR SUPPRESSORS

The expression of the type I cytoskeletal tumor suppressors is downregulated by either oncogenic RAS mutant or SV40 virus (Vandekerckhove et al., 1990), as is that of several other noncytoskeletal tumor suppressors such as fibronectin, the homeodomain protein Hox 1.4, and PTEN (Maruta, 1996; Zhu et al., 2000). Interestingly, the RAS-induced suppression of gelsolin, JunD, and a CDK inhibitor called WAF-1/p21/CIP-1 is reversed by a few distinct inhibitors specific for histone deacetyltransferase (HDA) such as TSA (Trichostatin A), FR901228, and Oxamflatin as well as a nuclear exportin inhibitor (p53 reactivator) called Leptomycin B (Yoshida and Horiuchi, 1998, 1999; Kim et al.,

1999), suggesting that their gene expression is activated by histone acetylation in an SP1-dependent manner but through a p53-independent pathway. Although the detailed mechanism underlying the RAS-induced suppression of their expression still remains to be clarified at the molecular level, the downregulation of PTEN absolutely requires PAKs, and appears to involve the Raf–MEK–Erks pathway at least in part (Zhu *et al.*, 2000).

A. Gelsolin

Gelsolin is an actin/PIP2-binding protein of 80 kDa which severs actin filaments as well as caps actin filaments at the plus ends in a Ca^{2+}-dependent manner, thereby shortening actin filaments only in the presence of Ca^{2+} (Liu *et al.*, 1998). Interestingly, PIP2 inhibits both its F-actin severing and capping activities by competing actin for binding to gelsolin (Liu *et al.*, 1998).

Gelsolin is a ubiquitous protein expressed in almost all mammalian tissues. However, the level of expression varies depending on the type of tissues and their differentiation or malignancy. Such a variation in gelsolin expression is closely associated with a marked difference in the intracellular organization of actin-cytoskeleton. For example, when L929 cells are exposed to the glucocorticoid dexamethasone, synthesis of gelsolin is induced and the cells acquire a flat morphology (Lanks and Kasambalides, 1983). Levels of gelsolin mRNA and protein also increase during the differentiation of embryonal carcinoma cells (Dieffenbach *et al.*, 1989). An increase in gelsolin mRNA and protein levels has been found to accompany the tetradecanoyl-phorbolacetate (TPA)-induced differentiation of human myelogenic leukemia cell lines such as HL60 into macrophage-like cells (Kwiatkowski, 1988). In fact gelsolin was first found in macrophages (Yin and Stossel, 1980). These observations suggest that an increase in gelsolin is strongly correlated to the differentiation of embryonal and myeloid cells, and may contribute to the differentiation process.

Gelsolin is hardly detectable in many malignant cells such as RAS-transformed NIH 3T3 fibroblasts and undifferentiated embryonal carcinoma cell lines (Dieffenbach *et al.*, 1989: Muellauer *et al.*, 1990). Gelsolin is one of most prominently downregulated proteins in several SV40-transformed cell lines (Vandekerckhove *et al.*, 1990). Diminished gelsolin expression has been documented in human stomach (Moriya *et al.*, 1994), colon (Furuuchi *et al.*, 1996), breast (Asch *et al.*, 1996), bladder (Tanaka *et al.*, 1995), lung (Dosaka-Akita *et al.*, 1998), prostate (Lee *et al.*, 1999), endometrium, and ovarium (Afify and Werness, 1998) cancer cells compared with normal epithelium. So far no point mutation has been found in any regions of gelsolin in human breast cancer cells (Mielnicki *et al.*, 1999).

Treatment with Trichostatin A (TSA), an inhibitor of histone deacetylase, induced a dramatic upregulation of gelsolin RNA and protein levels, and supports a role for epigenetic changes in chromatin structure leading to downregulation of gelsolin expression in human breast cancer (Mielnicki *et al.*, 1999). Overexpression of wild-type gelsolin in a human colon or bladder cancer cell line greatly reduces both its colony-forming ability in soft agar and tumorigenicity *in vivo* (Furuuchi *et al.*, 1996, Tanaka *et al.*, 1995). Pretreatment of human urinary bladder cancer cells with retroviral packaging cells that produce a retroviral vector expressing gelsolin at high levels resulted in marked inhibition of their tumor growth in nude mice and prolonged survival time in the majority of animals tested (Tanaka *et al.*, 1999). These findings suggest that partial or total loss of gelsolin expression is involved in the development of various cancers as one of the early events that occurs during carcinogenesis, and that the tumor suppressor gelsolin could be used for the gene therapy of human cancers.

Treatment of EJ-NIH 3T3 cells, transformed with the human activated c-Ha-Ras gene, with the mutagen ethyl methanesulfonate generates a flat revertant cell called R1, which carries a mutant (*His321*) of gelsolin (Kuzumaki *et al.*, 1989). In this mutant, an A replaces the second base C of codon 321 in the gelsolin gene, thereby causing the Pro to His mutation (Muellauer *et al.*, 1993). EJ-NIH 3T3 cells overexpressing either the *His321* mutant or the wild-type gelsolin lost or reduced tumorigenicity in syngeneic mice. The *His321* mutant suppresses more strongly the growth of EJ-NIH 3T3 cells than the wild-type gelsolin (Muellauer *et al.*, 1993). The recombinant *His321* gelsolin shows a decreased F-actin-severing activity and increased nucleating activity than the wild-type gelsolin *in vitro* (Fujita *et al.*, 1995), clearly indicating that at least its severing activity is not essential for its tumor suppressor activity.

Furthermore, both actin-nucleation and F-actin severing activities of the *His321* mutant are much more sensitive to the inhibition by the phosphoinositides (PIs): phosphatidylinositol 4-phosphate or 4,5-bisphosphate (PIP, PIP2), compared to those of the wild-type gelsolin, suggesting that the mutation increases the affinity of gelsolin for PIs. The sequence similar to the PIs-binding motif of gelsolin was found in several other PIs-binding proteins, including a highly conserved region of the phospholipase C (PLC) family (Yu *et al.*, 1992). PLC-γ 1 hydrolyzes PIP2, and modulates the intracellular levels of the second messenger molecules: Ca^{2+} and diacylglycerol (DAG) (Kamat *et al.*, 1997). A wide variety of growth factor receptor tyrosine kinases phosphorylate and activate PLC-γ 1, and play a biological role in signal transduction pathways of mitogenesis. For example, Western and Northern blot analyses of PI-specific PLCs revealed elevated expression of PLC-γ at both the protein and mRNA levels in most colorectal carcinomas when compared with paired adjacent normal mucosa samples (Noh *et al.*, 1994). *His321* gelsolin inhibits PIP2 hydrolysis by PLC-γ 1 more strongly than wild-type gelsolin *in vitro* because of its higher binding capacity for phosphoinositol lipids (Fujita et al., 1995). These results suggest that the wild-type gelsolin may exert the tumor-suppressor activity by inhibiting PIP2 hydrolysis, and the His321 mutation may strongly enhance this activity. The activity of PLC-β is also affected by gelsolin, and only the PIP2-binding sites of gelsolin are effective. Gelsolin also had biphasic effects on tyrosine kinase-phosphorylated PLC-γ (Sun et al., 1997). Phospholipase D (PLD) is also activated in mammalian cells in response to a variety of growth factors and may play a role in cell proliferation (Daniel et al., 1999). PLD activities require the lipid cofactor PIP2 (Liscovitch, 1996). It was proposed that DAG formation by PLC, PLD, and phosphatidate phosphohydrolase resulted in long-term stimulation of protein kinase C (PKC) (Carsberg et al., 1995). Actin rearrangements and PLD signaling are coordinately regulated through the physical association between PLD and gelsolin, and this interaction may also serve to amplify both PLD signaling and actin reorganization (Steed et al., 1996).

Overexpression of gelsolin suppresses bradykinin-induced activation of PLC and PLD (Banno et al., 1999). Translocation of PKC-α and PKC-β 1 was reduced in the gelsolin-overexpressed cells. PKC family kinases have been widely implicated in the regulation of cell growth/cell cycle progression and differentiation (Black, 2000). Human urinary bladder cancer cells overexpressing gelsolin after ultraviolet (UV) irradiation demonstrated very low increase of DAG indicating a reduced function of PLD, an accumulation of cells in G2, and/or a protracted delay in G2 phase as compared neotransfected cells (Sakai et al., 1999). The activity of cdk1 (histone 1) kinase in the neotransfectants decreased after 20 hr, while that of the gelsolin transfected cell remained high during the G2 delay. More recent studies indicate that lysophosphatidic acid (LPA) is a bioactive metabolite potentially generated as a result of PLD activation, while LPA activates PLD and stimulates proliferation in cancer cells (Qi et al., 1998). Gelsolin also binds LPA (Goetzl et al., 2000). A PIP2-binding peptide counteracted the effects mediated by LPA, suggesting that LPA binds to the same target region in actin-binding proteins (Meerschaert et al., 1998). These results altogether indicate that gelsolin could inhibit not only PLC but also PLD through phosphoinositol lipid metabolism, leading to the growth suppression of various human tumors (Figure 1).

To conclude this section, we should point out that in fact nobody knows exactly which function of gelsolin is essential or sufficient for suppressing RAS transformation. For nobody has deliberately created a specific mutant of gelsolin which either only caps F-actin, only severs F-actin, only binds PIP2/PIP3, or only nucleates actin polymerization. However, it would be reasonable to assume that at least both its PIP2/PIP3-binding and F-actin capping activities are responsible. The reasons are as follow: (1) a mutant of cofilin which no longer binds actin but binds PIP2/PIP3 can suppresses RAS transformation, as mentioned later in this chapter (Section III, C), and (2) a cytochalasin called CK which caps F-actin,

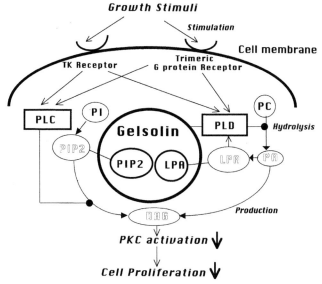

Figure 1 A possible role of the gelsolin–phospholipids interaction in signal transduction for cell proliferation. Gelsolin binds and sequesters both PIP2 and LPA. As a result, signal transduction pathways involving PIP2 and LPA are blocked, reducing production of a PKC activator DAG, and leading to inhibition of cell proliferation. TK, Tyrosine kinase; PLC, phospholipase C; PI, phosphatidylinositol; PIP2, phosphatidylinositol 4,5-bisphosphate; LPA, lysophosphatidic acid; PLD, phospholipase D; PC, phosphatidylcholine; PA, phosphatidic acid; DAG, diacylglycerol; PKC, protein kinase C.

but fails to both sever F-actin and bind PIP2/PIP3 also can suppress RAS transformation, as discussed in Chapter 15.

B. α-Actinin and vinculin

Alpha-actinin is an F-actin cross-linking protein of 200 kDa, and is a homodimer of two subunits. Each subunit has a single actin-binding site at the C-terminus. The actin binding of nonmuscle α-actinin is calcium-sensitive, whereas the muscle-type isoform is calcium-insensitive. The expression of α-actinin-1 is reduced in SV40-transformed BALB/c 3T3 (SVT2) cells. SVT2 cells overexpressing human α-actinin display a flatter phenotype, a decreased ability to grow in suspension culture (in soft agar), and a marked reduction in their ability to form tumors in syngeneic BALB/c mice and in athymic nude mice (Gluck *et al.*, 1993). On the contrary, α-actinin-1-deficient 3T3 cells that overexpress an antisense α-actinin construct still form tumors in nude mice (Gluck and Ben-Ze'ev, 1994).

A second nonmuscle α-actinin isoform called ACTN4 (actinin-4) is abundant in normal, substrate-adherent human neuroblastoma cell variants but absent or only weakly expressed in malignant, poorly substrate-adherent neuroblasts (Nikolopoulos et al., 2000). Although the deduced amino acid sequences suggest a possible difference between these two isoforms (actinin-1 and actinin-4) in both their biological function and calcium sensitivity, ACTN4 also exhibits tumor suppressor activity. Overexpression of ACTIN4 in highly malignant neuroblastoma stem cells causes a decrease in their anchorage-independent growth, loss of the tumorigenicity in nude mice, and downregulation of the proto-oncogene N-myc expression (Nikolopoulos et al., 2000).

Vinculin is an F-actin bundling protein of 130 kDa which is located in the focal adhesion plaques (Geiger, 1979; Jockusch and Isenberg, 1981), and was identified as the first substrate for the oncogenic Tyr kinase v-Src (Sefton et al., 1981), although the effect of its phosphorylation on biological function still remains to be determined. SV40-transformation of normal 3T3 cells reduces the expression of vinculin to only 25% of the control level, and a rat adenocarcinoma has no detectable vinculin (Rodriguez-Fernandez et al., 1992). Overexpression of vinculin in these tumor cell lines strongly suppresses their malignancy, resulting in an apparent increase in substratum adhesion, a decrease in the ability to grow in soft agar, and suppression of their capacity to develop tumors in syngeneic hosts or nude mice (Rodriguez-Fernandez et al., 1992). These findings clearly indicate that, like α-actinin, vinculin is a tumor suppressor.

By means of an antibody specific for PIP2, α-actinin and vinculin were identified as two major PIP2-bound proteins (Fukami et al., 1994). The levels of PIP2 bound to these two proteins decrease in response to platelet-derived growth factor (PDGF). By immunofluorescent staining, PIP2 was found to be present densely in the central areas around nuclei, microfilament bundles, and focal contacts, where α-actinin and vinculin are distributed. PDGF stimulation decreases the intensity of PIP2 staining in these areas. These findings suggest that tyrosine kinase-activated PLC hydrolyzes PIP2 bound to α-actinin and vinculin, leading to the simultaneous generation of second messengers and reorganization of the actin-cytoskeleton (Fukami et al., 1994). Phosphatidylinositol 3-kinase (PI 3-kinase) has been shown to play an important role in the signal transduction of cell growth (Kaplan et al., 1987). Alpha-actinin copurifies with PI 3-kinase from bovine thymus (Shibasaki et al., 1994). These results suggest that PI 3-kinase forms a complex with α-actinin and regulates actin-cytoskeleton.

The SH3 domain of PLC-γ 1 binds to vinculin, and the SH2 domains may mediate the binding of PLC-γ 1 to receptor tyrosine kinase (Park et al., 1999). Vinculin plays a role in the establishment or regulation of the cadherin-based cell adhesion complex by direct interaction with β-catenin (Hazan et al.,

1997). Inactivation of the adenomatous polyposis coli (APC) gene is a critical event in the development of human colorectal cancers. The β-catenin-binding domain in the central third of APC is sufficient for its tumor suppressor activity (Shih *et al.*, 2000). These results suggest that α-actinin and vinculin suppress tumorigenicity through regulation of phospholipid metabolism as gelsolin does or through catenin–APC-associated signal transduction pathways.

C. Tropomyosins

Tropomyosin (TM) family of α-helical proteins of 30–40 kDa bind the side of actin filaments and stabilize the filaments. Each TM molecule binds seven actin monomers through seven so-called α-repeats (actin-binding motifs), each consisting of 21 amino acids (Hitchcock-DeGregori and Varnell, 1990). It dimerizes to form a two-stranded coiled coil that is orientated head to tail on the actin filament. There are larger muscle–type and smaller nonmuscle–type isoforms of TM. In vertebrates multiple isoforms of nonmuscle TM are expressed which are generated from four different genes by a combination of alternative promoter activities and splicing (Lin *et al.*, 1997). They may be required for the regulation of actin filament stability, intracellular granule movement, cell shape determination, and cytokinesis.

The differential expression of TM isoforms found in cell transformation and differentiation, as well as the differential localization of TM isoforms in some types of culture cells, suggests a differential isoform function *in vivo* (Lin *et al.*, 1997). Synthesis of certain members of the TM family is suppressed in fibroblasts neoplastically transformed by a number of retroviral oncogenes, by transforming growth factor α, and by chemical mutagens (Lin *et al.*, 1997). The levels of high molecular weight isoforms of TM are markedly reduced in Ras-transformed cells (Takenaga *et al.*, 1988; Prasad *et al.*, 1993). Synthesis of TM2 is significantly suppressed in high-metastatic Lewis lung carcinoma cells and v-Ha-Ras-transformed NIH 3T3 cells compared with that in low-metastatic cells or normal NIH 3T3 cells, suggesting that the suppression of TM2 synthesis is involved in the expression of the metastatic phenotype (Takenaga *et al.*, 1988). Abnormalities in TM expression in neoplastic cells are not limited to fibroblasts, but are observed among human breast cancer cell lines, raising the possibility that such abnormalities in expression of TM may play a role in human neoplasia (Bhattacharya *et al.*, 1990).

Ras-transformed NIH 3T3 cells expressing the exogenous TM1 cDNA showed partial restoration of microfilament bundle formation together with increased cytoplasmic spreading (Prasad *et al.*, 1993; Braverman *et al.*, 1996). They lost anchorage-independent growth capability, and the onset of tumor growth in athymic mice was delayed. Expression of TM2 protein has no effect on the transformed phenotype of the Ras-transformants (Braverman *et al.*,

1996). However, enhanced expression of both TM1 and TM2 results in the formation of well-organized microfilaments, a morphology that resembles normal fibroblasts, and suppression of tumorigenicity (Shah *et al.*, 1998). These data show that TM1 is a suppressor of the transformed phenotype, while TM2 cooperates with TM1 in the reorganization of microfilaments. v-Src transformed fibroblasts express decreased TM1 (Prasad *et al.*, 1999). The transformed cells overexpressing TM1 grew at a lower rate in monolayer, exhibited well spread, flat morphology, and failed to grow in soft agar. These data demonstrate that TM1 is as an antagonist of functionally diverse oncogenes (Prasad *et al.*, 1999).

So far no evidence has been reported that TM binds phospholipids. Thus, although the mechanism underlying TM-induced tumor suppression still remains to be clarified, it may differ from that involving other PIP2-binding type I cytoskeletal tumor suppressors. Since F-actin cross-linkers such as HS1 and MKT-077 suppresses RAS transformation (for detail, see the following Section III, A, and Chapter 15), it is conceivable that TM exerts its tumor suppressor activity by cross-linking actin filaments.

III. TYPE II CYTOSKELETAL TUMOR SUPPRESSORS

The expression of the type II tumor suppressors is not affected by either RAS or SV40. Instead, the biological function (actin-binding) of these proteins appears to be inhibited by oncogenic RAS mutants at least, through either their phosphorylation or PIP2-binding which provides their actin-binding sites with negative charges.

A. HS1

Cortactin, HS1, myosin I, and a few other related proteins form a small family of F-actin bundling SH3 domain proteins (He *et al.*, 1998). Most members of this family, except for spectrin, contain their SH3 domains at the C-termini (the last 51 amino acids), whereas spectrin contains the SH3 domain in the center of the molecule. In the SH3 domain, HS1 shares 80 and 50% sequence identity with cortactin and myosin I, respectively. HS1 and cortactin have the unique actin/PIP2-binding motifs of 37 amino acids in the N-terminal half (see Figure 2). In these actin-binding motifs, three motifs (residues 82–192) of HS1 share 60 to 70% sequence identity with the six corresponding motifs (residues 83–304) of cortactin. However, only the fourth motif (residues 194–230) of cortactin binds actin, whereas all three motifs of HS1 bind actin (Maruta and He, 1995; He *et al.*, 1998). Thus, although HS1 monomer alone is able to bundle actin filaments, only cortactin dimer (not monomer) can bundle actin filaments (Huang *et al.*, 1997; He *et al.*, 1998).

			* * *			
HS1-1	82-118	GYGG	*RFGVERDR*	MDKSAVGH	EYVA EVEKHS	SQTDAAK
HS1-2	119-155	GFGG	*KYGVERDR*	ADKSAVGF	DYKGEVEKHT	SQKDYSR
HS1-3	156-192	GFGG	*RYGVERDK*	WDKAALGY	DYKGETEKHE	SQRDYAK

			* * *			
EMS1-1	83-119	GYGG	KFGVEQDR	MDKSAVGH	EYQS KLSKHC	SQVDSVR
EMS1-2	120-156	GFGG	KFGVQMDR	VDQSAVGF	EYQGKTEKHA	SQKDYSS
EMS1-3	157-193	GFGG	KYGVQADR	VDKSAVGF	DYQGKTEKHE	SQRDYSK
EMS1-4	194-230	GFGG	*KYG I D KDK*	VDKSAVGF	EYQGKTEKHE	SQKDYVK
EMS1-5	231-267	GFGG	KFGVQTDR	QDKCALGW	DHQEKLQLHE	SQKDYKT
EMS1-6	268-304	GFGG	KFGVQ SER	QDSAAVGF	DYKEKLALHE	SQQDYSK

Figure 2 F-actin/PIP2-binding motifs of HS1 and EMS1/cortactin. The PIP2-binding motifs (in boxes) are found only in the actin-binding repeats of 37 amino acids (underlined) of HS1 and EMS1.

Interestingly, only within these three actin-binding motifs of HS1 and the fourth motif of cortactin, are there PIP2/PIP3-binding motifs of 8 amino acids (see Figure 2). As expected, the affinity of HS1 for PIP2 is much higher than that of cortactin (He et al., 1998). The binding of PIP2 to these motifs abolishes the ability of both HS1 and cortactin to bundle actin filaments, simply because the negatively charged actin and PIP2 compete for their binding to these positively charged motifs. One of the significant differences between HS1 and cortactin is that HS1 is expressed only in hematopoietic cell lineages, whereas cortactin is expressed ubiquitously in almost all cells (Kitamura et al., 1989; Wu et al., 1991).

We found that overexpression of HS1 suppresses RAS-induced malignant transformation of NIH 3T3 fibroblasts (He et al., 1998). During our further study to understand the molecular mechanism underlying the HS1-induced suppression of RAS transformation, we found that in normal fibroblasts, cortactin forms a stable complex with actomyosin II (actin–myosin II complex), whereas in RAS-transformants, this cortactin–actomyosin II complex disassembles (He et al., 1998). No change in the expression of either actin, myosin II, or cortactin is detected during RAS transformation. Interestingly, when RAS transformation is suppressed by HS1 expression, the cortactin–actomyosin II complex is restored. How does oncogenic RAS signal disrupt the cortactin–actomyosin II complex?

There is a drug called SCH51344 that suppresses RAS transformation by blocking Rac-induced membrane ruffling (Walsh et al., 1997). Although it still remains to be clarified what the direct target molecule of this drug is, it appears that this drug downregulates the expression of a protein(s) essential for membrane ruffling or upregulates the expression of a protein(s) that blocks membrane ruffling, most likely at transcription levels. We found that SCH51344 also restores the cortactin–actomyosin II complex during suppression of RAS transformation (He *et al.*, 1998), suggesting that the RAS-induced disruption of this complex depends on a Rac-induced pathway leading to membrane ruffling. Rac is activated by RAS through PI-3 kinase, and its end-product. Rac then induces the production of PIP2 by activating PI-4/PI-5 kinases (Hartwig *et al.*, 1995). As a consequence, PIP2 binds/inactivates a few distinct F-actin capping proteins/cappers such as tensin, gelsolin, and Cap39/CapG that normally cap the plus ends of actin filaments to block a rapid growth (elongation) of actin filaments at this end (Maruta *et al.*, 1999). The inactivation of these F-actin cappers by PIP2 would induce a rapid actin polymerization at the plus end, eventually leading to membrane ruffling.

PIP2 also binds cortactin and blocks its ability to bundle actin filaments, and therefore would disrupt the cortactin–actomyosin II complex. Thus, it is quite conceivable that SCH51344 restores the cortactin–actomyosin II complex, perhaps by blocking the Rac-induced PIP2 production somehow. How does HS1 restore the cortactin–actomyosin II complex? As mentioned before, HS1 has a much higher affinity for PIP2 than cortactin. Thus, it is possible that overexpressed HS1 sequesters PIP2 overproduced in RAS transformants, and restores the ability of cortactin to bundle actin filaments. In fact, PIP2-induced disruption of the cortactin-induced actin bundles can be restored by a tiny amount of HS1 *in vitro* (He *et al.*, 1998), suggesting that the HS1-induced suppression of RAS transformation could be due to its PIP2-sequestering action, at least in part. Moreover, the deletion of all three actin/PIP2-binding motifs of 37 amino acids (residues 82–192) from HS1 completely abolishes the anti-RAS tumor suppressing activity (He *et al.*, 1998). However, it still remains to be determined whether either its actin-binding or PIP2-binding (or both) is essential for the HS1-induced suppression of RAS transformation.

In addition to the N-terminal actin/PIP2-binding motifs, HS1 requires at least one more domains and two phosphorylatable Tyr residues for its anti-RAS action. One is the C-terminal SH3 domain whose deletion causes the translocation of HS1 from cytoplasm to the nucleus, due to the exposure of the nuclear location signal (NLS), and this SH3-minus mutant is no longer able to suppress RAS transformation (He *et al.*, 1998). Furthermore, Tyr-378 and Tyr-397 of HS1 are required for the apoptosis of B lymphocytes, and when they are phosphorylated by the kinases called Syk and Lin, then the phosphorylated HS1 is translocated into the nucleus (Yamanashi *et al.*, 1997). To assess the role of these Tyr residues in the anti-RAS action of HS1, we replaced these two critical Tyr residues by Glu residues so that the resultant HS1 mutant is no longer phosphorylatable. This Glu-378/397 mutant is no longer able to suppress RAS transformation, although it still stays in the cytoplasm (He *et al.*, 1998), suggesting that phosphorylation of these residues is essential for suppression of RAS-transformation.

Interestingly, the Tyr kinase Syk, that phosphorylates these Tyr residues of HS1, is also a tumor suppressor whose expression inhibits the growth of invasive, non-Syk-expressing, breast cancer cells (Coopman *et al.*, 2000). These observations suggest that Syk suppresses malignant transformation probably by phosphorylating HS1 or a related, but more ubiquitous protein(s) such as cortactin.

B. Tensin

Among the F-actin cappers that cap the plus ends of actin filaments, thereby shortening the filaments, tensin is so far the largest (consisting of 1744 amino acids), and remains to be unique in having an SH2 domain (Davis *et al.*, 1991). This SH2 domain (residues 1471–1580) alone is sufficient to localize tensin in focal adhesion plaque (Lo *et al.*, 1994). Its targets are Tyr-phosphorylated proteins localized in the focal adhesion plaque (FAP) such as Src and paxillin (Lo *et al.*, 1994). Among other functional domains, tensin has an F-actin capping domain around the center of the molecule. However, the precise location of the F-actin capping domain has not been settled. According to the Chen group (Lo *et al.*, 1994), it is reportedly localized between residues 888–989, whereas the Lin group (Chuang *et al.*, 1995) insisted that the F-actin capping domain is localized further downstream (residues 1013–1097). Nevertheless, it is clear that tensin caps the plus end of actin filaments (Davis *et al.*, 1991; Lo *et al.*, 1994; Chuang *et al.*, 1995).

As mentioned in the preceding HS1 section, these plus-end cappers are inactivated by PIP2 whose production is induced by RAS through PI-3 kinase and Rac/CDC42. As a consequence, RAS can uncap the plus end of actin filament and induces a rapid actin polymerization which eventually leads to membrane ruffling. In fact, cytochalasins, a family of chemical compounds that cap the plus end of actin filament, block the RAS/Rac-induced membrane ruffling (Yahara *et al.*, 1982). Furthermore, a plus-end capper called gelsolin, in particular its mutant in which Pro-321 is replaced by His, suppresses RAS transformation (Muellauer *et al.*, 1993). These observations suggest that capping the plus end of the actin filament might be sufficient for suppressing oncogenic RAS signaling. However, gelsolin has another biological function, that is to say, severing actin filaments in a Ca^{2+}-dependent manner (Yin, 1987). Thus, it still remains to be clarified whether capping or severing actin filaments is responsible for the suppression of RAS transformation.

Because tensin has no severing activity, we have tested whether tensin is able to suppress RAS transformation. In support of our notion, overexpression of full-length tensin in RAS transformants completely suppresses the malignant phenotype (Tikoo *et al.*, 1999), indicating that the F-actin severing activity of gelsolin is not absolutely essential for its anti-RAS action. These observations strengthen the idea that all plus-end F-actin cappers are anti-RAS tumor suppressors. As described in Chapter 16, this notion has prompted us to screen for the plus-end F-actin capping chemicals such as cytochalasins as potential therapeutics for the treatment of RAS-associated cancers.

C. Cofilin

Like gelsolin, cofilin and its related proteins called ADF (actin-depolymerizing factor) or destrin are able to severe actin filaments (Maruta *et al.*, 1999). However, unlike gelsolin whose severing activity depends on calcium, the severing activity of the cofilin/ADF family proteins is independent of calcium, but depends on protons or pH (Matsuzaki *et al.*, 1988). Human cofilin consists of 166 amino acids (Suzuki *et al.*, 1995). Serine-3 is the critical phosphorylation site whose phosphorylation markedly reduces the affinity of cofilin for actin (Ressad *et al.*, 1998), and therefore almost abolishes its severing activity (Moriyama *et al.*, 1996). The kinase responsible for the phosphorylation of Ser-3 in cofilin is called LIM-kinase (Arber *et al.*, 1998; Yang *et al.*, 1998). This kinase is activated by at least two distinct kinases, PAK and Rock, which phosphorylate Thr-508 of LIM-kinase (Edwards *et al.*, 1999; Maekawa *et al.*, 1999). Since PAK is activated by RAS through PI-3 kinase and Rac/CDC42, it is quite conceivable that the phosphorylation of cofilin at Ser-3, and the subsequent loss of its severing activity, is caused by RAS.

Interestingly, cofilin also binds PIP2 (Moriyama *et al.*, 1992). However, its binding to PIP2 is not affected by the Ser-3 phosphorylation (Moriyama *et al.*, 1996). Thus, in RAS transformants, cofilin is expected to bind PIP2, but no longer severe actin filaments. A mutant (K112/114Q) of cofilin, in which both Lys-112 and Lys-114 are replaced by Gln, no longer binds actin, but is still able to bind PIP2 (Moriyama *et al.*, 1992). We have found that overexpression of this mutant is able to suppress RAS transformation (Maruta *et al.*, 1999), suggesting that PIP2-binding/sequestering alone is sufficient for the suppression of RAS transformation. In general, PIP2-binding proteins bind PIP3 as well, and a PIP3 phosphatase called PTEN is a tumor suppressor that converts PIP3 to PIP2, antagonizing PI-3 kinase (Maehama and Dixon, 1998).

D. NF2/Merlin (neurofibromatosis type 2 gene product)

NF2 is a protein of 595 amino acids, and deletion of its gene or dysfunction of this protein is closely associated with the development of tumors in the central nervous system called neurofibromatosis type 2 such as schwannomas and meningiomas (Trofatter *et al.*, 1993; Rouleau *et al.*, 1993). NF2 belongs to ERM family proteins including ezrin, radixin, and moesin that bind actin filaments (Tikoo *et al.*, 1994). However, unlike other ERM family proteins which bind actin filaments through their C-terminal tail, NF2 binds actin filaments through its N-terminal half (Tikoo *et al.*, 1994; Xu and Gutmann, 1998). We demonstrated that overexpression of full-length NF2 or its N-terminal half (NF2N, residues 1–359) suppresses RAS transformation, clearly indicating that NF2 is an anti-RAS tumor suppressor (Tikoo *et al.*, 1994).

The C-terminal half of NF2 contains a Pro-rich motif which is expected to bind some yet unknown SH3 domain proteins. Interestingly, a human lymphoblast cell line called GUS5722 which was derived from an NF2 patient contains an NF2 mutant that lacks 78 amino acids (residues 447–524) including this Pro-rich motif (Trofatter *et al.*, 1993). We found more recently that overexpression of the NF2 C-terminal half (NF2C, residues 354–595) alone is able to suppress RAS transformation, but not the mutant of NF2C that lacks the 78 amino acids (Tikoo *et al.*, 2000), indicating that some of the missing 78 amino acids are essential for the anti-RAS action (see Figure 3).

NF2 binds actin filaments and several other proteins such as microtubules, βII-spectrin, a cofactor of Na^+/H^+ exchange ATPase called NHE-RF, and SCHIP-1 as well as PIP2, but none of them binds NF2C (Scoles *et al.*, 1998; Murthy *et al.*, 1998; Goutebroze *et al.*, 2000; Tikoo *et al.*, 2000). Thus, we have searched a specific NF2C-binding protein by a yeast two-hybrid screening system, and found that another tumor suppressor protein called GPS2 binds NF2C (Tikoo *et al.*, 2000). Interestingly, GPS2 has been shown to suppress RAS transformation by inhibiting the RAS/Rac-induced activation of JNKs, c-Jun N-terminal kinases (Spain *et al.*, 1996; Jin *et al.*, 1997), suggesting the possibility that GPS2 controls the oncogenic JNKs cascade through its association with NF2. In fact, overexpression of full-length NF2 or NF2C alone strongly reduces the kinase activity of JNK in RAS transformants (Tikoo *et al.*, 2000),

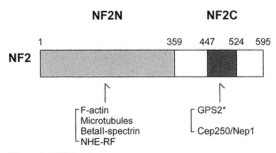

Figure 3 NF2-binding partners.

indicating that NF2 suppresses RAS transformation in part by blocking JNKs cascade, as does GPS2. In other words, a chemical inhibitor specific for JNKs would be potentially useful for the chemotherapy of RAS-associated cancers. However, so far no specific chemical compound has been found that directly inhibits JNKs.

The antibiotic K252a is a furanosylated indolocarbazole compound that inhibits Trk, a NGF receptor Tyr kinase, in a relatively specific manner (Tapley *et al.*, 1992). Although K252a itself does not affect the activation of JNKs, a derivative of K252a called CEP-1347 or K-7515 was more recently shown to inhibit the activation of JNKs, but indirectly (Maroney *et al.*, 1998). PAKs are responsible for the Rac/CDC42-dependent activation of JNKs (Bagrodia *et al.*, 1995; Minden *et al.*, 1995). Furthermore, MLK3/SPRK activates another kinase called MKK4 that in turn activates JNKs (Vacratsis and Gallo, 2000). Thus, we have suspected that the direct targets CEP-1347 might be the CDC42/Rac-dependent Ser/Thr kinases such as PAKs and MLK3. In fact we more recently found that this drug directly inhibits PAKs (for detail, see Chapter 18). Thus, this PAK-inhibitor would be potentially useful for the treatment of RAS-associated cancers, as tumor suppressors such as NF2 and GSP2, that suppress RAS transformation by blocking JNKs pathway. So far CEP-1347 has been tested in clinical trials only for the treatment of neuronal degeneration diseases such as Alzheimer's disease and Parkinson's disease, as it strongly blocks the JNK-mediated apoptosis of neuronal cells.

References

Afify, A. M., and Werness, B. A. (1998). Decreased expression of the actin-binding protein gelsolin in endometrial and ovarian adenocarcinomas. *Appl. Immuno-histochem.* **6**, 30–34.

Arber, S., Barbayannis, F., Hanser, H., *et al.* (1998). Regulation of actin dynamics through phosphorylation of cofilin by LIM-kinase. *Nature* **393**, 805–809.

Asch, H. L., Head, K., Dong, Y., Natoli, F., Winston, J. S., Connolly, J. L., and Asch, B. B. (1996). Widespread loss of gelsolin in breast cancers of humans, mice, and rats. *Cancer Res.* **56**, 4841–4845.

Bagrodia, S., Derijard, B., Davis, R., *et al* (1995). CDC42 and PAK-mediated signaling leads to JNK and p38 MAP kinase activation. *J. Biol. Chem.* **270**, 27995–27998.

Banno, Y., Fujita, H., Ono, Y., Nakashima, S., Ito, Y., Kuzumaki, N., and Nozawa, Y. (1999). Differential phospholipase D activation by bradykinin and sphingosine 1-phosphate in NIH 3T3 fibroblasts overexpressing gelsolin. *J. Biol. Chem.* **274**, 27385–27391.

Bar-Sagi, D., and Feramisco, J. (1986). Induction of membrane ruffling and fluid-phase pinocytosis in quiescent fibroblasts by Ras proteins. *Science* **233**, 1061–1065.

Bhattacharya, B., Prasad, G. L., Valverius, E. M., Salomon, D. S., and Cooper, H. L. (1990). Tropomyosins of human mammary epithelial cells: Consistent defects of expression in mammary carcinoma cell lines. *Cancer Res.* **50**, 2105–2112.

Black, J. D. (2000). Protein kinase C-mediated regulation of the cell cycle. *Front. Biosci.* **5**, D406–D423.

Braverman, R. H., Cooper, H. L., Lee, H. S., and Prasad, G. L. (1996). Anti-oncogenic effects of tropomyosin: Isoform specificity and importance of protein coding sequences. *Oncogene* **13,** 537–545.

Carsberg, C. J., Ohanian, J., and Friedmann, P. S. (1995). Ultraviolet radiation stimulates a biphasic pattern of 1,2-diacylglycerol formation in cultured human melanocytes and keratinocytes by activation of phospholipases C and D. *Biochem. J.* **305,** 471–477.

Chuang, J. Z., Lin, D., and Lin, S. (1995). Molecular cloning, expression, and mapping of the high-affinity actin-capping domain of chicken cardiac tensin. *J. Cell Biol.* **128,** 1095–1109.

Coopman, P., Do, M., Barth, M., *et al.* (2000). The Syk Tyr kinase suppresses malignant growth of human breast cancer cells. *Nature* **406,** 742–747.

Daniel, L. W., Sciorra, V. A., and Ghosh, S. (1999). Phospholipase D, tumor promoters, proliferation and prostaglandins. *Biochim. Biophy. Acta* **30,** 265–276.

Davis, S., Lu, M. L., Lo, S. H., *et al* (1991). Presence of an SH2 domain in the actin-binding protein tensin. *Science* **252,** 712–715.

Dieffenbach, C. W., SenGupta, D. N., Krause, D., Sawzak, D., and Silverman, R. H. (1989). Cloning of murine gelsolin and its regulation during differentiation of embryonal carcinoma cells. *J. Biol. Chem.* **264,** 13281–13288.

Dosaka-Akita, H., Hommura, F., Fujita, H., Kinoshita, I., Nishi, M., Morikawa, T., Katoh, H., Kawakami, Y., and Kuzumaki, N. (1998). Frequent loss of gelsolin expression in non-small cell lung cancers of heavy smokers. *Cancer Res.* **58,** 322–327.

Edwards, D., Sanders, L., Bokoch, G., *et al.* (1999). Activation of LIM-kinase by PAK couples Rac/CDC42 GTPase signaling to actin cytoskeletal dynamics. *Nat. Cell Biol.* **1,** 253–259.

Fujita, H., Laham, L. E., Janmey, P. A., Kwiatkowski, D. J., Stossel, T. P., Banno, Y., Nozawa, Y., Mullauer, L., Ishizaki, A., and Kuzumaki, N. (1995). Functions of [His321] gelsolin isolated from a flat revertant of Ras-transformed cells. *Eur. J. Biochem.* **229,** 615–620.

Fukami, K., Endo, T., Imamura, M., and Takenawa, T. (1994). α-actinin and vinculin are PIP2-binding proteins involved in signaling by tyrosine kinase. *J. Biol. Chem.* **269,** 1518–1522.

Furuuchi, K., Fujita, H., Tanaka, M., Shitinohe, T., Senmaru, N., Ogiso, Y., Moriya, S., Hamada, M., Kato, H., and Kuzumaki, N. (1996). Gelsolin as a suppressor of malignant phenotype in human colon cancer. *Tumor Targeting* **2,** 277–283.

Geiger, B. (1979). A 130K protein from chicken gizzard: Its localization at the termini of microfilament bundle in cultured chicken cells. *Cell* **18,** 193–205.

Gluck, U., and Ben-Ze ev, A. (1994). Modulation of alpha-actinin levels affects cell motility and confers tumorigenicity on 3T3 cells. *J. Cell Sci.* **107,** 1773–1782.

Gluck, U., Kwiatkowski, D. J., and Ben-Ze'ev, A. (1993). Suppression of tumorigenicity in simian virus 40-transformed 3T3 cells transfected with α-actinin cDNA. *Proc. Natl. Acad. Sci. U.S.A.* **90,** 383–387.

Goetzl, E. J., Lee, H., Azuma, T., Stossel, T. P., Turck, C. W., and Karliner, J. S. (2000). Gelsolin binding and cellular presentation of lysophosphatidic acid. *J. Biol. Chem.* **275,** 14573–14578.

Goutebroze, L., Brault, E., Muchardt, C., *et al.* (2000). Cloning and characterization of SCHIP-1, a novel protein interacting specifically with spliced isoforms and naturally occurring mutant NF2 proteins. *Mol. Cell. Biol.* **20,** 1699–1712.

Hartwig, J., Bokoch, G., Carpenter, C., *et al.* (1995). Thrombin receptor ligation and activated Rac uncap actin filament barbed ends through PIP2 synthesis. *Cell* **82,** 643–653.

Hazan, R. B., Kang, L., Roe, S., Borgen, P. I., and Rimm, D. L. (1997). Vinculin is associated with the e-cadherin adhesion complex. *J. Biol. Chem.* **272,** 32448–32453.

He, H., Watanabe, T., Zhan, X., *et al.* (1998). Role of PIP2 in RAS/Rac-induced disruption of the cortactin–actomyosin II complex and malignant transformation. *Mol. Cell. Biol.* **18,** 3829–3837.

Hitchcock-DeGregori, S., and Varnell, T. (1990). Tropomyosin has discrete actin-binding sites with seven-fold and fourteen-fold periodicities. *J. Mol. Biol.* **214,** 885–896.

Huang, C., Ni, Y., Wang, T., et al. (1997). Down-regulation of F-actin cross-linking activity of cortactin/EMS1 by Src-mediated Tyr phosphorylation. J. Biol. Chem. 272, 13911–13915.

Jin, D. Y., Teramoto, H., Giam, C. Z., et al. (1997). A human suppressor of c-Jun N-terminal kinase 1 activation by tumor necrosis factor alpha. J. Biol. Chem. 272, 25816–25823.

Jockusch, B., and Isenberg, G. (1981). Interaction of α-actinin and vinculin with actin: Opposite effects on filament network formation. Proc. Natl. Acad. Sci. U.S.A. 78, 3005–3009.

Kamat, A., and Carpenter, G. (1997). Phospholipase C-γ1: Regulation of enzyme function and role in growth factor-dependent signal transduction. Cytokine and Growth Factor Rev., 8, 109–117.

Kaplan, D. R., Whitman, M., Schaffhausen, B., Pallas, D. C., White, M., Cantley, L., and Roberts, T. M. (1987). Common elements in growth factor stimulation and oncogenic transformation: 85 kd phosphoprotein and phosphatidylinositol kinase activity. Cell 50, 1021–1029.

Kim, Y. B., Lee, K. H., Sugita, K., et al. (1999). Oxamflatin is a novel anti-tumor compound that inhibits mammalian histone deacetylase. Oncogene 18, 2461–2470.

Kitamura, D., Kaneko, H., Miyagoe, Y., et al. (1989). Isolation and characterization of a novel human gene expressed specifically in the cell of hematopoietic lineage. Nucleic Acids Res. 17, 9367–9379.

Kuzumaki, N., Ogiso, Y., Oda, A., Fujita, H., Suzuki, H., Sato, C., and Mullauer, L. (1989). Resistance to oncogenic transformation in revertant R1 of human Ras-transformed NIH 3T3 cells. Mol. Cell. Biol. 9, 2258–2263.

Kwiatkowski, D. (1988). Predominant induction of gelsolin and actin-binding protein during myeloid differentiation. J. Biol. Chem. 263, 13857–13862.

Lanks, K. W., and Kasambalides, E. J. (1983). Dexamethasone induces gelsolin synthesis and altered morphology in L929 cells. J. Cell Biol. 96, 577–581.

Lee, H. K., Driscoll, D., Asch, H., Asch, B., and Zhang, P. J. (1999). Down-regulated gelsolin expression in hyperplastic and neoplastic lesions of the prostate. Prostate 40, 14–19.

Li, J., Yen, C., Liaw, D., et al. (1997). PTEN, a putative protein Tyr phosphatase gene mutated in human brain, breast, and prostate cancer. Science 275, 1943–1947.

Lin, J. J., Warren, K. S., Wamboldt, D. D., Wang, T., and Lin, J. L. (1997). Tropomyosin isoforms in non-muscle cells. Int. Rev. Cytol. 170, 1–38.

Liscovitch, M. (1996). Phospholipase D: Role in signal transduction and membrane traffic. J. Lipid Mediators and Cell Signalling 14, 215–221.

Liu, Y. T., Rozelle, A., and Yin, H. L. (1998). The gelsolin family of actin filament severers and cappers. In "G Proteins, Cytoskeleton and Cancer" (H. Maruta and K. Kohama, eds.), pp. 19–35. Landes Biosciences, Austin, Texas.

Lo, S. H., Weisberg, E., and Chen, L. B. (1994). Tensin: A potential link between the cytoskeleton and signal transduction. BioEssays 16, 817–823.

Maehama, T., and Dixon, J. (1998). The tumor suppressor, PTEN, dephosphorylates the lipid second messenger, PIP3. J. Biol. Chem. 273, 13375–13378.

Maekawa, M., Ishizaki, T., Boku, S., et al. (1999). Signaling from Rho to the actin cytoskeleton through protein kinases ROCK and LIM-kinase. Science 285, 895–898.

Maroney, A., Glicksman, M., Basma, A., et al. (1998). Motoneuron apoptosis is blocked by CEP-1347 (KT 7515), a novel inhibitor of the JNK signaling pathway. J. Neurosci. 18, 104–111.

Maruta, H. (1996). Downstream of RAS and actin-cytoskeleton. In "Regulation of the RAS Signaling Networks" (H. Maruta and A. W. Burgess, eds.), pp. 139–180. Springer-Verlag and R. G. Landes, Heldelberg, Germany/Austin, Texas.

Maruta, H. (1998). PIP2: A proto-oncogenic phopsholipid. In "G Proteins, Cytoskeleton and Cancer" (H. Maruta and K. Kohama, eds.), pp. 133–147. Landes Biosciences, Austin, Texas.

Maruta, H., and He, H. (1995). Cytoskeletal SH3 proteins. Biochemistry (Tokyo) 67, 1210–1217.

Maruta, H., and Korn, E. D. (1977). Acanthamoeba cofactor protein is a heavy chain kinase required for actin-activation of the Mg^{2+}-ATPase activity of Acanthamoeba myosin I. J. Biol. Chem. 252, 8329–8332.

Maruta, H., Knoerzer, W., Hinssen, H., *et al.* (1984). Regulation of actin polymerization by non-polymerizable actin-like proteins. *Nature* **312**, 424–428.

Maruta, H., He, H., Tikoo, A., *et al.* (1999). G proteins, PIP2, and actin-cytoskeleton in the control of cancer growth. *Microsc. Res. Tech.* **47**, 61–66.

Matsuzaki, F., Matsumoto, S., Yahara, I., *et al.* (1988). Cloning and characterization of porcine brain cifilin cDNA. *J. Biol. Chem.* **263**, 11564–11568.

Meerschaert, K., De Corte, V., De Ville, Y., Vandekerckhove, J., and Gettemans, J. (1998). Gelsolin and functionally similar actin-binding proteins are regulated by lysophosphatidic acid. *EMBO J.* **17**, 5923–5932.

Mielnicki, L. M., Ying, A. M., Head, K. L., Asch, H. L., and Asch, B. B. (1999). Epigenetic regulation of gelsolin expression in human breast cancer cells. *Exp. Cell Res.* **249**, 161–176.

Minden, A., Lin, A., Claret, F. X., *et al.* (1995). Selective activation of the JNK signaling cascade and c-Jun transcritional activity by Rac and CDC42. *Cell* **81**, 1147–1157.

Moriya, S., Yanagihara, K., Fujita, H., and Kuzumaki, N. (1994). Differential expression of HSP90, gelsolin and GST-π in human gastric carcinoma cell lines. *Int. J. Oncol.* **5**, 1347–1351.

Moriyama, K., Yonezawa, N., Sakai, H., *et al.* (1992). Mutational analysis of an actin-binding site of cofilin and characterization of cofilin/destrin chimeras. *J. Biol. Chem.* **267**, 7240–7244.

Moriyama, K., Iida, K., and Yahara, I. (1996). Phosphorylation of Ser 3 of cofilin regulates its essential function on actin. *Genes Cells* **1**, 73–86.

Muellauer, L., Fujita, H., Suzuki, H., Katabami, M., Hitomi, Y., Ogiso, Y., and Kuzumaki, N. (1990). Elevated gelsolin and α-actin expression in a flat revertant R1 of Ha-Ras oncogene-transformed NIH/3T3 cells. *Biochem. Biophy. Res. Commun.* **171**, 852–859.

Muellauer, L., Fujita, H., Ishizaki, A., and Kuzumaki, N. (1993). Tumor-suppressive function of mutated gelsolin in Ras-transformed cells. *Oncogene* **8**, 2531–2536.

Murthy, A., Gonzalez-Agosti, C., Cordero, E., *et al.* (1998). NHE-RF, a regulatory cofactor for Na$^+$/H$^+$ exchange, is a common interactor for Merlin and ERM proteins. *J. Biol. Chem.* **273**, 1273–1276.

Nikolopoulos, S. N., Spengler, B. A., Kisselbach, K., Evans, A. E., Biedler, J. L., and Ross, R. A. (2000). The human non-muscle α-actinin protein encoded by the *ACTN4* gene suppresses tumorigenicity of human neuroblastoma cells. *Oncogene* **19**, 380–386.

Noh, D. Y., Lee, Y. H., Kim, S. S., Kim, Y. I., Ryu, S. H., Suh, P. G., and Park, J. G. (1994). Elevated content of phospholipase C-γ 1 in colorectal cancer tissues. *Cancer* **73**, 36–41.

Park, S. Y., Barron, E., Suh, P. G., Ryu, S. H., and Kay, E. P. (1999). FGF-2 facilitates binding of SH3 domain of PLC-γ 1 to vinculin and SH2 domains to FGF receptor in corneal endothelial cells. *Mol. Vision* **5**, 18.

Prasad, G. L., Fuldner, R. A., and Cooper, H. L. (1993). Expression of transduced tropomyosin 1 cDNA suppresses neoplastic growth of cells transformed by the Ras oncogene. *Proc. Natl Acad. Sci. U.S.A.* **90**, 7039–7043.

Prasad, G. L., Masuelli, L., Raj, M. H., and Harindranath, N. (1999). Suppression of src-induced transformed phenotype by expression of tropomyosin-1. *Oncogene* **18**, 2027–2031.

Qi, C., Park, J. H., Gibbs, T. C., Shirley, D. W., Bradshaw, C. D., Ella, K. M., and Meier, K. E. (1998). Lysophosphatidic acid stimulates phospholipase D activity and cell proliferation in PC-3 human prostate cancer cells. *J. Cell. Physiol.* **174**, 261–272.

Ressad, F., Didry, D., Xia, G. X., *et al.* (1998). Kinetic analysis of the interaction of actin-depolymerizing factor ADF/cofilin with G-and F-actins: Comparison of plant and human ADFs and effect of phosphorylation. *J. Biol. Chem.* **273**, 20894–20902.

Rodriguez-Fernandez, J. L., Geiger, B., Salomon, D., Sabanay, I., Zoller, M., and Ben-Ze ev, A. (1992). Suppression of tumorigenicity in transformed cells after transfection with vinculin cDNA. *J. Cell Biol.* **119**, 427–438.

Rouleau, G., Merel, P., Lutchman, M., *et al.* (1993). Alteration in a new gene encoding a putative membrane-organizing protein causes NF2. *Nature* **363**, 515–521.

Sakai, N., Ohtsu, M., Fujita, H., Koike, T., and Kuzumaki, N. (1999). Enhancement of G2 checkpoint function by gelsolin transfection in human cancer cells. *Exp. Cell Res.* **251**, 224–233.

Scoles, D., Huynh, D., Morcos, P., *et al.* (1998). Neurofibromatosis 2 tumor suppressor schwannomin interacts with betaII-spectrin. *Nat. Genet.* **16**, 354–359.

Sefton, B. M., Hunter, T., Ball, E. H., and Singer, S. J. (1981). Vinculin: A cytoskeletal target of the transforming protein of Rous sarcoma virus. *Cell* **24**, 165–174.

Shah, V., Braverman, R., and Prasad, G. L. (1998). Suppression of neoplastic transformation and regulation of cytoskeleton by tropomyosins. *Somat. Cell Mol. Genet.* **24**, 273–280.

Shibasaki, F., Fukami, K., Fukui, Y., and Takenawa, T. (1994). Phosphatidylinositol 3-kinase binds to α-actinin through the p85 subunit. *Biochem. J.* **302**, 551–557.

Shih, I. M., Yu, J., He, T. C., Vogelstein, B., and Kinzler, K. W. (2000). The β-catenin binding domain of adenomatous polyposis coli is sufficient for tumor suppression. *Cancer Res.* **60**, 1671–1676.

Spain, B., Bowdish, K., Pacal, A., *et al.* (1996). Two human cDNAs, including a homolog of *Arabidopsis* FUS6 (COP11), suppress G protein-MAP kinase mediated signal transduction in yeast and mammalian cells. *Mol. Cell. Biol.* **16**, 6698–6706.

Steck, P., Pershouse, M., Jasser, S., *et al.* (1997). Identification of a candidate tumour suppressor gene, MMAC1, at chromosome 10q23.3 that is mutated in multiple advanced cancers. *Nat. Genet.* **15**, 356–362.

Steed, P. M., Nagar, S., and Wennogle, L. P. (1996). Phospholipase D regulation by a physical interaction with the actin-binding protein gelsolin. *Biochemistry* **35**, 5229–5237.

Sun, H., Lin, K., and Yin, H. L. (1997). Gelsolin modulates phospholipase C activity *in vivo* through phospholipid binding. *J. Cell Biol.* **138**, 811–820.

Suzuki, K., Yamaguchi, T., Tanaka, T., *et al.* (1995). Activation induces dephosphorylation of cofilin and its translocation to plasma membranes in neutrophil-like differentiated HL-60 cells. *J. Biol. Chem.* **270**, 19551–19556.

Takenaga, K., Nakamura, Y., and Sakiyama, S. (1988). Suppression of synthesis of tropomyosin isoform 2 in metastatic v-Ha-ras-transformed NIH3T3 cells. *Biochem. Biophy. Res. Commu.* **157**, 1111–1116.

Tanaka, M., Muellauer, L., Ogiso, Y., Fujita, H., Moriya, S., Furuuchi, K., Harabayashi, T., Shinohara, N., Koyanagi, T., and Kuzumaki, N. (1995). Gelsolin: A candidate for suppressor of human bladder cancer. *Cancer Res.* **55**, 3228–3232.

Tanaka, M., Sazawa, A., Shinohara, N., Kobayashi, Y., Fujioka, Y., Koyanagi, T., and Kuzumaki, N. (1999). Gelsolin gene therapy by retrovirus producer cells for human bladder cancer in nude mice. *Cancer Gene Ther.* **6**, 482–487.

Tapley, P., Lamballe, F., and Barbacid, M. (1992). K252a is a selective inhibitor of the Tyr kinase activity of the Trk family of oncogenes and neurotrophin receptors. *Oncogene* **7**, 371–381.

Tikoo, A., Varga, M., Ramesh, V., *et al.* (1994). An anti-RAS function of neurofibromatosis type 2 gene product (NF2/Merlin). *J. Biol. Chem.* **269**, 23387–23390.

Tikoo, A., Cutler, H., Lo, S. H., *et al.* (1999). Treatment of RAS-induced cancer by the F-actin cappers tensin and chaetoglobosin k (CK), in combination with the caspase-1 inhibitor N1445. *Cancer J. Sci. Am.* **5**, 293–300.

Tikoo, A., Renner, C., Wadle, A., *et al.* (2000). Wild-type NF2, but not a mutant from NF2 patient, suppresses RAS transformation through the C-terminal half that blocks JNKs. Submitted for publication.

Tolkacheva, T., and Chan, A. (2000). Inhibition of H-RAS transformation by the PTEN tumor suppressor gene. *Oncogene* **19**, 680–689.

Trofatter, J., MacCollin, M., Rutter, I., *et al.* (1993). A novel moesin/ezrin/radixin-like protein is a candidate for NF2 tumor suppressor. *Cell* **72**, 791–800.

Vacratsis, P., and Gallo, K. (2000). Zipper-mediated oligomerization of the MLK3 is not required for its activation by CDC42 but is necessary for its activation of the JNK pathway. *J. Biol. Chem.* **275,** 27893–27900.

Vandekerckhove, J., Bauw, G., Vancompernolle, K., Honore, B., and Celis, J. (1990). Comparative two-dimensional gel analysis and microsequencing identifies gelsolin as one of the most prominent downregulated markers of transformed human fibroblast and epithelial cells. *J. Cell Biol.* **111,** 95–102.

Walsh, A., Dhanasekaran, M., Bar-Sagi, D., *et al.* (1997). SCH51344-induced reversal of RAS-transformation is accompanied by the specific inhibition of the RAS/Rac-induced membrane ruffling. *Oncogene* **15,** 2553–2560.

Weber, K., Lazarides, E., Goldman, R., *et al.* (1974). Localization and distribution of actin fibers in normal, transformed and revertant cells. *Cold Spring Harbor Symp. Quant. Biol.* **39,** 363–369.

Wu, H., Reynolds, A., Kanner, S., *et al.* (1991). Identification and characterization of a novel cytoskeleton-associated Src substrate. *Mol. Cell. Biol.* **11,** 5113–5124.

Xu, H. M., and Gutmann, D. (1998). Merlin differentially associates with the microtubule and actin cytoskeleton. *J. Neurosci. Res.* **51,** 403–415.

Yahara, I., Harada, F., Sekita, S., *et al.* (1982). Correlation between effects of 24 different cytochalasins on cellular structures and events and those on actin *in vitro. J. Cell Biol.* **92,** 69–78.

Yamanashi, Y., Fukuda, T., Nishizumi, H., *et al.* (1997). Role of Tyr phosphorylation of HS1 in B-cell antigen receptor-mediated apoptosis. *J. Exp. Med.* **185,** 1387–1392.

Yang, N., Higuchi, O., Ohashi, K., *et al.* (1998). Cofilin phosphorylation by LIM-kinase and its role in Rac-mediated actin reorganization. *Nature* **393,** 809–812.

Yin, H. L. (1987). Gelsolin: Calcium and PIP2–regulated actin-modulating protein. *BioEssays* **7,** 176–179.

Yin, H. L., and Stossel, T. (1980). Control of cytoplasmic actin gel-sol transformation by gelsolin, a calcium-dependent protein. *Nature* **218,** 583–586.

Yoshida, M., and Horiuchi, S. (1998). Histone deacetylase inhibitors: Possible anti-cancer therapeutics. *In* "G Proteins, Cytoskeleton and Cancer" (H. Maruta and K. Kohama, eds.), pp. 327–342. Landes Biosciences, Austin, Texas.

Yoshida, M., and Horiuchi, S. (1999). Trichostatin and leptomycin. *Ann. N.Y. Acad. Sci.* **886,** 23–36.

Yu, F. X., Sun, H. Q., Janmey, P. A., and Yin, H. L. (1992). Identification of a polyphosphoinositide-binding sequence in an actin monomer-binding domain of gelsolin. *J. Biol. Chem.* **267,** 14616–14621.

Zhu, H., Iaria, J., Nheu, T., *et al.* (2000). Down-regulation of PTEN expression by RAS through the PAKs–Raf–Erks pathway. Manuscript in preparation.

10

TGF-β Signaling and Carcinogenesis

Masahiro Kawabata

Department of Biochemistry
The Cancer Institute of the Japanese Foundation for Cancer Research (JFCR)
Toshima-ku, Tokyo 170-8455, Japan

I. INTRODUCTION

Transforming growth factor-β (TGF-β) is a potent inhibitor of cell proliferation, and disruption of the TGF-β signaling pathway leads to tumorigenesis. TGF-β, on the other hand, promotes tumor progression through pleiotropic activities under certain conditions. To understand the role of TGF-β in carcinogenesis, it is important to study the molecular functions of the signaling components. TGF-β binds to a heteromeric complex of transmembrane serine–threonine kinase receptors that phosphorylate and activate Smad proteins. Smads propagate signals from the cell membrane to the nucleus, and regulate the expression of target genes. Mutations of the receptors and Smads have been identified in cancers of various origins such as pancreas and colon, providing *in vivo* evidence that the TGF-β signaling pathway plays a role in suppressing tumorigenesis. This chapter summarizes more recent findings on TGF-β signaling and discusses the effects of genetic alterations of signaling components of the TGF-β pathway.

Tumor-Suppressing Viruses, Genes, and Drugs
Innovative Cancer Therapy Approaches
Copyright © 2002 by Academic Press.

II. DUAL ROLE OF TGF-β IN CARCINOGENESIS

Transforming growth factor-β (TGF-β) is the prototype of a large family of secreted polypeptides that regulate a wide variety of biological activities including cell growth, differentiation, apoptosis, extracellular matrix production, and immune function. The TGF-β superfamily includes TGF-βs, bone morphogenetic proteins (BMPs), and activins/inhibins. Over 40 members including other related ligands have been identified from invertebrates to vertebrates (Derynck and Feng, 1997). TGF-β was originally isolated as a factor that induces anchorage-independent growth of normal cells; however, it was soon found that the direct effect of TGF-β on cell growth is inhibitory. Thus, TGF-β can act as a safeguard against deregulated cell growth. TGF-β promotes cell growth under certain conditions, but is thought to do so indirectly, for example, by induction of secretion of other growth factors such as platelet-derived growth factor (PDGF).

It has been reported that a variety of cancer cells are rendered insensitive to TGF-β, suggesting that components of the TGF-β signaling pathway may act as tumor suppressors (Markowitz and Roberts, 1996). In accord with this, a variety of experimental findings including those for TGF-β1 transgenic mice (Pierce et al., 1995) and genetic deletion of the TGF-β1 gene (Glick et al., 1994; Tang et al., 1998) suggests that TGF-β counteracts oncogenic stimulus. TGF-β, on the other hand, may promote tumor progression in different situations (Blobe et al., 2000). Once cancer cells have lost responsiveness to TGF-β, TGF-β secreted from such cancer cells or surrounding cells may promote tumor invasion by enhancing neovascularization and cell adhesion. TGF-β also interferes with immune surveillance that prevents tumor progression. TGF-β thus plays a dual role in carcinogenesis depending on its stage. It is thus important to examine the molecular basis of the switch in cellular responsiveness to TGF-β in order to understand the mechanism of carcinogenesis.

III. TGF-β SUPERFAMILY SIGNALING

TGF-β binds to a heteromeric set of two types of serine–threonine kinase receptors (Derynck and Feng, 1997). The type II receptor (TβR-II) is a constitutively active kinase and can bind TGF-β on its own. The type I receptor (TβR-I) recognizes TGF-β only in the presence of TβR-II. On ligand binding, TβR-II transphosphorylates TβR-I at its juxtamembrane region denoted by the GS domain rich in glycines and serines. The activated TβR-I in turn phosphorylates downstream substrates. Intracellular signaling of TGF-β is mediated by more recently identified Smad proteins (Heldin et al., 1997; Massagué, 1998). Smads

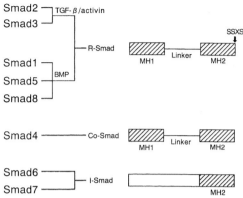

Figure 1 Classification and structure of mammalian Smads. Smad2 and Smad3 are receptor-regulated Smads (R-Smads) activated by TGF-β or activin, whereas Smads1, 5, and 8 are BMP-regulated R-Smads. Smad4 is the only mammalian common partner Smad (Co-Smad). Smad6 and Smad7 are inhibitory Smads (I-Smads). Two domains, MH1 and MH2, are conserved among Smads, although I-Smads lack the MH1 domain. R-Smads have the SSXS motif that is phosphorylated by type I serine–threonine kinase receptors.

were originally identified through genetic approaches using invertebrates. The molecular functions of the mammalian Smads have since been extensively investigated. Eight mammalian Smads have been identified and are classified into three groups depending on their structures and functions (Figure 1). Receptor-regulated Smads (R-Smads) are directly phosphorylated by type I receptors. Smad2 and Smad3 are the substrates of TGF-β and activin type I receptors, whereas Smad1, 5, and 8 are phosphorylated by BMP type I receptors. Smad4 is the only mammalian common partner Smad (Co-Smad) that acts downstream of TGF-β, activin, and BMP. In Xenopus, another Co-Smad, Smad4β, was isolated (Masuyama et al., 1999; Howell et al., 1999). Smad4β has functions similar to those of Smad4, but is expressed only in the very early stage of development of Xenopus. In the Drosophila genome as known at present, Medea is the only Co-Smad, and it is not known whether mammals have another Co-Smad. Inhibitory Smads (I-Smads) include Smad6 and Smad7, which antagonize R-Smads. Whereas Smad6 preferentially inhibits BMP signaling, Smad7 inhibits both TGF-β and BMP signaling.

Smads share two conserved domains: the N-terminal MH1 domain and the C-terminal MH2 domain (Figure 1). The two domains are connected by a linker region with variable length and amino acid sequence. Both R-Smads and Co-Smads contain the MH1 and MH2 domains, but I-Smads share only the MH2 domain, and their N-terminal regions diverge significantly from the conserved MH1 domain. R-Smads have a serine-serine-X-serine (SSXS) motif at the C-terminal end which is the direct phosphorylation site by type I receptors. In the unphosphorylated form, the MH1 and MH2 domains interact intramolecularly and inhibit activation of each other (Hata et al., 1997). Phosphorylation of the SSXS motif somehow induces a conformational change that activates the protein. Once R-Smads are phosphorylated at the SSXS motif, they interact with Co-Smads, and the heteromeric complexes translocate from the cytoplasm to the nucleus where they regulate transcription of target genes (Figure 2). The MH1 domain has DNA binding ability, whereas the MH2 domain has intrinsic transactivation activity.

The transient interaction of R-Smads and type I receptors is regulated by a protein called SARA (Smad anchor for receptor activation) (Tsukazaki et al., 1998). SARA has a FYVE domain that attaches the protein to the cell membrane through binding phosphatidylinositol-3-phosphate. SARA binds an unphosphorylated form of Smad2 and Smad3 but not BMP-regulated R-Smads. SARA also interacts with the TGF-β receptor complex. On ligand binding, Smad2/3 is phosphorylated and dissociates from SARA and forms heteromeric complexes with Smad4. SARA thus recruits Smad2/3 to the TGF-β receptor complex. A SARA-like protein in BMP signaling has not been identified. The crystal structure of the Smad2 MH2 domain, together with the Smad binding domain (SBD) of SARA, has been determined (Wu et al., 2000). The SBD of SARA exhibits an extended conformation that consists of a rigid coil, an α helix, a proline-rich turn, and a β strand. The interaction of the rigid coil with Smad2 determines the binding specificity, whereas the β strand contributes to binding affinity. Smads have more recently been reported to associate with microtubules (Dong et al., 2000). Smads bind to microtubules in the absence of TGF-β, and ligand stimulation induces dissociation of Smads from microtubules. Interestingly, destabilization of microtubules with various reagents such as nocodazole increases TGF-β responsiveness, suggesting that microtubules serve as a sequestration compartment for Smads. The relationship of microtubules to SARA in TGF-β signaling has not been clarified.

Two distinct mechanisms of nuclear translocation of Smad proteins have been proposed. Xiao et al. (2000a) identified a lysine-rich sequence (Lys-Lys-Leu-Lys-Lys) in the MH1 domain of Smad3 as a nuclear localizing signal (NLS). The NLS sequence is conserved among all R-Smads. Mutations in the NLS interfere with TGF-β-dependent nuclear translocation of Smad3. Xiao et al. (2000b) also showed that importin β interacts with the Smad3 MH1

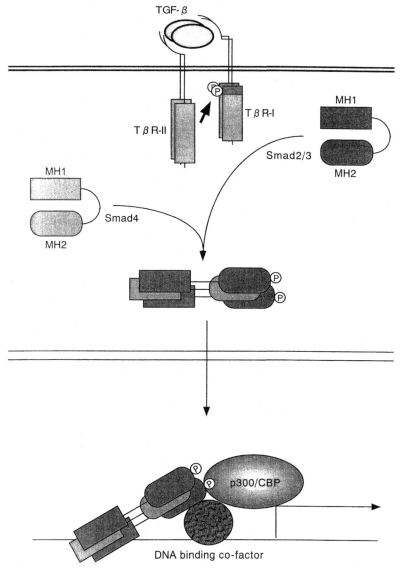

Figure 2 TGF-β signaling by Smads. On TGF-β binding, the type II receptor (TβR-II) transphosphorylates the type I receptor (TβR-I). The activated TβR-I then phosphorylates Smad2/3. Phosphorylated Smad2/3 associates with Smad4 and forms heteromeric complexes that translocate into the nucleus. Smads interact with other DNA binding proteins and p300/CBP, a transcriptional coactivator. Smads recruit transcriptional corepressors instead of coactivators depending on cellular conditions. Smads thus regulate transcription of target genes.

domain, and that mutations in the NLS inhibited this interaction. In contrast, Importin α did not interact with the MH1 domain. However, Xu et al. (2000) showed using an in vitro system that the MH2 domain but not MH1 is required for the nuclear translocation of Smad2. Interestingly, even the unphosphory-lated form of Smad2 was present in the nucleus, suggesting that the MH2 domain has intrinsic nuclear import activity. Xu et al. showed that SARA anchors Smad2 in the cytoplasm in vivo, and that receptor-induced release of Smad2 from SARA triggers accumulation of Smad2 in the nucleus. They further showed that importin β and Smad2 may compete for components in the nuclear translocation machinery. It will be necessary to resolve the discrepancy between these two observations in future studies.

Once Smads enter the nucleus, the R-Smad–Co-Smad complexes bind to DNA. However, compared with other DNA binding proteins, Smads are unique in that the DNA binding affinity of the Smad complexes per se is not high, and their binding specificity is not strict (Derynck et al., 1998; Massagué and Wotton, 2000). Thus, Smads interact with a DNA-binding partner to achieve stable and sequence-specific DNA binding. The first evidence that Smads interact with a transcription factor on DNA was presented by Chen et al. (1996). Activin induces the expression of the Mix.2 gene in Xenopus. They identified the activin-responsive element (ARE) in the Mix.2 promoter as well as the ARE binding complex, activin-responsive factor (ARF). ARF was shown to contain FAST-1, a winged-helix transcription factor with a forkhead domain, and Smad2. Smad2 itself cannot bind directly to DNA because of an intervening insert in its MH1 domain (Yagi et al., 1999; Dennler et al., 1999). Smad4 is also incorporated in ARF, and promotes binding of ARF to ARE (Liu et al., 1997). In fact, FAST-1 and Smad4 bind to distinct but adjacent sites in ARE (Yeo et al., 1999). Likewise, Smads interact with a variety of transcription factors such as AP1 (Zhang et al., 1998; Liberati et al., 1999), Gli3 (Liu et al., 1998), TFE3 (Hua et al., 1999), PEBP2/AML/CBF (Hanai et al., 1999; Pardali et al., 2000; Zhang and Derynck, 2000), Hoxc-8 (Shi et al., 1999; Verschueren et al., 1999) ATF2 (Sano et al., 1999), vitamin D receptor (Yanagisawa et al., 1999), gluco-corticoid receptor (Song et al., 1999), and OAZ (Hata et al., 2000). The outcome of such interactions appears to vary from one case to another. For example, Hoxc-8 binds to the promoter of the osteopontin gene, and represses its expression. On BMP stimulation, Smad1 binds to Hoxc-8 and dislocates the protein from the promoter, resulting in induction of expression of the osteopontin gene.

Smads also interact with other nuclear proteins including transcriptional coactivators and corepressors. Evi-1 is an oncoprotein that may be involved in leukemogenesis. Evi-1 interacts with Smad3 and antagonizes the growth-inhibitory effect of TGF-β, which could contribute to oncogenesis by Evi-1 (Kurokawa et al., 1998). p300 and CBP are transcriptional coactivators that acetylate histones and neutralize the positive electric charge of the proteins.

p300/CBP then loosens the tight chromatin structure, thereby activating tran-scription. Smads have been shown by several groups to recruit p300/CBP (Wrana, 2000). E1A is an adenoviral oncoprotein known to inhibit TGF-β signaling. More recently, E1A was reported to interact with R-Smads and to interfere with their recruitment of p300 (Nishihara et al., 1999). Smad4 cooperates with another coactivator, MSG1, in transactivation (Shioda et al., 1998; Yahata et al., 2000). Smads recruit not only transcriptional coactivators but also repressors. TGIF was identified as a protein that interacts with Smad2 (Wotton et al., 1999). TGIF inhibits transactivation by TGF-β. TGF-β induced interaction of Smad2 with a histone deacetylase, HDAC1, in the presence of TGIF. HDAC1 has an ef-fect opposite that of p300/CBP on transcription. TGIF also competes with p300 in binding to Smad2. c-Ski is a protooncogene product and forms a protein family with Sno. c-Ski and Sno have been shown by several groups to interact with Smads (Massagué and Wotton, 2000). c-Ski recruits HDAC1 and potently inhibits transactivation by TGF-β. c-Ski also suppresses the growth-inhibitory effect of TGF-β. Thus, one of the general mechanisms of carcinogenesis by nuclear oncoproteins may be the inhibition of TGF-β signaling through associa-tion with Smad proteins.

PCR-based screening of random sequences identified palindromic GTCTAGAC as a Smad binding element (SBE) (Zawel et al., 1998). Close examination of several TGF-β-inducible genes such as plasminogen activator inhibitor-1 (PAI-1) (Dennler et al., 1998) and JunB (Jonk et al., 1998) has revealed that short sequences containing CAGACA (CAGA box) serve as Smad binding sequences. Similar sequences have been identified in the TGF-β-responsive Igα promoter (Lin and Stavnezer, 1992), the activin-responsive Mix.2 promoter (Yeo et al., 1999), and other TGF-β inducible promoters (Dennler et al., 1998). Reporter genes with multiple repeats of this motif responded to TGF-β. Intrigu-ingly, GTCTAGAC and CAGACA share AGAC or its complementary GTCT, suggesting that this four-base sequence is critical for DNA binding of Smads. A reporter with the JunB CAGA box also responded to BMP, whereas a reporter from the PAI-1 gene was not activated by BMP stimulation. These results suggest that other factors such as an adjacent sequence may be important for response specificity. The crystal structure of the MH1 domain of Smad3 complexed with the GTCT motif has been determined (Shi et al., 1998): the β-hairpin consisting of β strands 2 and 3 is the major DNA binding domain of Smad3. The β-hairpin protrudes from the globular MH1 domain and interacts with the major groove of the GTCT motif. The amino acid sequence of the β-hairpin is conserved between Smad2/3 and Smad1/5/8, raising the possibility that both classes of R-Smads use the same principle of DNA binding. However, Smad1 did not bind to the CAGA box of the PAI-1 gene (Dennler et al., 1998). Thus, unknown factors may con-tribute to the DNA binding specificity in vivo. The first evidence that Smads can bind to DNA was obtained in Drosophila (Kim et al., 1997). Mad is a Drosophila

R-Smad homologous to Smad1/5/8. Mad mediates signaling of Dpp, a BMP homolog in flies. Mad binds to GC-rich sequences (GCCG box) of various Dpp-responsive genes. Medea, a Smad4 homolog in *Drosophila*, also binds to the GCCG box (Xu *et al.*, 1998). Smad4 was shown to bind to GC-rich sequences in the *goosecoid* gene (Labbé *et al.*, 1998). Smad1 binds to the GCCG box and activates reporter genes with multiple copies of this motif (Kusanagi *et al.*, 2000). However, these reporters did not respond to TGF-β or activin stimulation. Again, factors other than the β-hairpin–DNA interaction appear to determine signaling specificity.

Smad6 and Smad7 are I-Smads that interfere with signaling of the TGF-β superfamily. On ligand binding, R-Smads are recruited to activated receptors, phosphorylated, immediately released from the receptors, and accumulate in the nucleus. In contrast, I-Smads stably bind to activated receptors and inhibit receptor-mediated phosphorylation of R-Smads (Hayashi *et al.*, 1997; Imamura *et al.*, 1997; Nakao *et al.*, 1997). Smad6 also competes with Smad4 in binding to phosphorylated Smad1 (Hata *et al.*, 1998). One important characteristic of I-Smads is induction of their expression by ligands. Dad, a *Drosophila* I-Smad, was identified as a molecule the expression of which is induced by Dpp (Tsuneizumi *et al.*, 1997). Expression of Smad6 and Smad7 is also induced by the TGF-β superfamily members (Nakao *et al.*, 1997; Takase *et al.*, 1998; Afrakhte *et al.*, 1998). I-Smads thus form a negative feedback loop in signaling by the TGF-β superfamily. Examination of murine and human Smad7 promoter identified SBE as a responsive element for TGF-β and activin (Nagarajan *et al.*, 1999; von Gersdorff *et al.*, 2000; Denissova *et al.*, 2000). Mouse Smad6 promoter was shown to contain a GC-rich BMP responsive element with four overlapping GCCG box like motifs (Ishida *et al.*, 2000). Moreover, γ-interferon antagonizes TGF-β signaling by inducing the expression of Smad7 (Ulloa *et al.*, 1999). Tumor necrosis factor-α inhibits TGF-β signaling through induction of Smad7 as well (Bitzer *et al.*, 2000). I-Smads may have other functions as well. Although steady laminar shear stress of blood induces the expression of I-Smads in vascular endothelial cells, its physiological significance remains unknown (Topper *et al.*, 1997). Smad7 was found to exist predominantly in the nucleus in the absence of TGF-β, and to accumulate in the cytoplasm on ligand stimulation (Itoh *et al.*, 1998). I-Smads may thus play roles in the nucleus. Smad6 interacts with Hoxc-8, and the protein complex that results inhibits the interaction of Smad1 with Hoxc-8 and transactivation of the *osteopontin* gene by Smad1 (Bai *et al.*, 2000). Antisense experiments have shown that TGF-β induces apoptosis of prostatic carcinoma cells through induction of Smad7 (Landstrom *et al.*, 2000).

Signaling by the TGF-β superfamily is also negatively regulated by protein degradation. Smurf1 (Smad ubiquitination regulatory factor-1) was identified as a ubiquitin ligase for Smad1 in *Xenopus* (Zhu *et al.*, 1999). Smurf1 belongs to the Hect family of E3 ubiquitin ligases and contains two WW domains. The WW domains recognize a PPXY sequence that exists in the linker region of Smad1. Although R-Smads contain a PY motif in the linker region, Smurf1 specifically targets Smad1/5. The interaction of Smurf1 and Smad1 is ligand-independent, suggesting that Smurf1 regulates the basal level of Smad1. TGF-β induces degradation of Smad2 through ubiquitination (Lo and Massagué, 1999). It was shown that nuclear translocation of Smad2 is sufficient for the degradation of Smad2 because Smad2 attached with an artificial NLS underwent degradation without receptor-induced phosphorylation. UbcH5b/c is probably the E2 ubiquitin conjugating enzyme in this process, although the E3 ligase participating in it has not been identified.

Smads constitute a major signaling pathway of the TGF-β superfamily, but a non-Smad pathway also exists. A number of reports have described the activation of the MAP kinase cascade by TGF-β. At present three MAP kinases are known: Erk, SAPK/JNK, and p38. TGF-β activates each of these kinases depending on the cellular context (Mulder, 2000; Miyazono et al., 2000). TGF-β induces the expression of extracellular matrix components including fibronectin. The induction of fibronectin is mediated by the activation of JNK by TGF-β in a human fibrosarcoma cell line (Hocevar et al., 1999). Dominant-negative JNK inhibited the induction of fibronectin by TGF-β. Fibronectin was produced on stimulation by TGF-β even in cells lacking intact Smad4. These results indicate that JNK but not Smad4 is required for the induction of fibronectin by TGF-β. TAK1 was identified as a MAP kinase kinase kinase that is activated by TGF-β (Yamaguchi et al., 1995). It was shown that TAK1 induces the phosphorylation of ATF-2 via p38 and that ATF-2 associates with the Smad3-Smad4 heterooligomer, forming an active transcription complex (Sano et al., 1999; Hanafusa et al., 1999).

IV. PERTURBATION OF TGF-β SIGNALING IN CANCER CELLS

Perturbation of TGF-β signaling has been identified in a variety of cancers. In particular, high incidences of mutations affecting components of the TGF-β signaling pathway have been reported in pancreatic and colon cancers (Blobe et al., 2000). Various cancer cells have lost receptors for TGF-β (Markowitz and Roberts, 1996). One distinctive mechanism of the inactivation of TβR-II has been found in hereditary nonpolyposis colorectal cancer (HNPCC) (Markowitz et al., 1995). A major population of individuals with HNPCC exhibits defects in DNA mismatch repair, and are also referred to as replication error (RER)-positive. The coding region of the extracellular domain of TβR-II has a consecutive stretch of 10 adenines that are prone to RER. Insertion or deletion in this poly A stretch results in truncation of TβR-II, which disrupts TGF-β signaling.

A case of HNPCC without a mismatch repair defect was shown to have a point mutation in the $T\beta R$-II gene which inactivates the normal function of TβR-II (Lu et al., 1998). These results suggest that TβR-II acts as a tumor suppressor in colorectal carcinogenesis. Decrease or loss of TβR-II expression correlated with grade of malignancy in gastric cancer, T-cell lymphomas, and head and neck squamous carcinomas (Blobe et al., 2000; Muro-Cacho et al., 1999). Deregulation of TβR-I expression has also been reported in cases such as a prostate carcinoma cell line (Kim et al., 1996) and chronic lymphocytic leukemia (DeCoteau et al., 1997). Homozygous carriers of a mutant TβR-I lacking alanines at the N-terminus are significantly more susceptible to various types of cancers than normal individuals (Pasche et al., 1998; Chen et al., 1999).

Smad4, originally denoted DPC4 (homozygously deleted in pancreatic carcinoma, locus 4), was identified as a tumor suppressor in pancreatic carcinomas (Hahn et al., 1996). Mutations of Smad4 have been found in approximately 50% of pancreatic cancers, 20% of colorectal cancers, and less frequently in lung, breast, ovarian, prostate, and other cancers (Duff and Clarke, 1998). Smad2 has been found to be mutated in colorectal and lung cancers (Blobe et al., 2000). Although Smad3 also mediates TGF-β signaling, no genetic alterations of Smad3 have been reported in human cancers, the reason for which is unclear. Smad4 is also mutated in a subset of individuals with familial juvenile polyposis (FJP) (Howe et al., 1998). Patients with this disease are predisposed to hamartomatous polyposis and gastrointestinal and pancreatic cancers. A four-base deletion in exon 9 in the Smad4 gene was detected at relatively high frequency in FJP patients (Friedl et al., 1999).

Most of the Smad4 and Smad2 tumorigenic mutations map to the MH2 domain. This finding appears reasonable, because the MH2 domain has multiple functions including various protein–protein interactions, and may be more susceptible to subtle structural alterations than the MH1 domain. The effects of some MH2 mutations have been predicted from structural study of the MH2 domain. The crystal structure of the MH2 domain of Smad4 has been determined (Shi et al., 1997), and results revealed that the Smad4 MH2 domain forms a homotrimeric complex in vitro, although Smad4 appears to exist as a monomer in the absence of ligand in vivo (Kawabata et al., 1998). Ligand stimulation induces the formation of Smad oligomers. In the MH2 crystal structure, most of the MH2 mutations are located at the interface between monomers, and these mutations interfere with oligomer formation. This finding indicates that oligomer formation is a critical step in Smad signaling. These mutations include D351H, R361C, V370D, D493H, and Smad2(D450E) corresponding to D537 in Smad4 (Figure 3A). The D450E mutation in Smad2 may have an additional effect on Smad2 function. Smad2(D450E) is not phosphorylated by TβR-I (Eppert et al., 1996) and stably interacts with TβR-I, as do I-Smads (Lo et al., 1998). Smad3 with the corresponding mutation thus acted as a dominant-negative Smad in mediating both gene induction and growth inhibition by TGF-β (Goto et al., 1998). R441P, Smad2(L440R) and Smad2(P445H) corresponding to I527 and A532 in Smad4, respectively, are structural residues, and these mutations may disrupt the conformational integrity of the MH2 domain. In other studies, the Smad2(L440R) protein was shown to be unstable and to undergo ubiquitination (Eppert et al., 1996; Xu and Attisano, 2000). Smad2(P445H) was not phosphorylated by TβR-I (Eppert et al., 1996). The effect of the R420H mutation cannot be readily explained from the crystal structure of Smad4. Highly conserved G508 is mutated in *Drosophila Mad* (G409S) (Sekelsky et al., 1995), *Medea* (G701D) (Wisotzkey et al., 1998), and *C. elegans sma-2* (G372D) (Savage et al., 1996). Introduction of the corresponding mutation to Smad1, Smad2, and Smad5 abrogated receptor-dependent phosphorylation of these R-Smads (Hoodless et al., 1996; Lo et al., 1998; Nishimura et al., 1998). Smad4(G408S) failed to heterooligomerize with Smad2 (Shi et al., 1997).

No missense mutations have been reported in the linker region, and only a few missense mutations have been found in the MH1 domain (Takagi et al., 1996; Shi et al., 1998). Interestingly, two independent mutations, Smad4(R100T) and Smad2(R133C), affect the equivalent amino acid in the MH1 domain (Figure 3B). Several different interpretations exist regarding the pathological significance of this mutation. Hata et al. (1997) reported that this mutation enhances the intramolecular interaction between the MH1 and MH2 domains, thereby inhibiting the activation of Smad. Introduction of the MH1 mutation to the corresponding amino acid in Mad abolished DNA binding of Mad (Kim et al., 1997). Smad4(R100T) did not bind to DNA either (Le Dai et al., 1998; Song et al., 1998; Stroschein et al., 1999). Most recently, it has been shown that the arginine mutation targets Smad2 and Smad4 to the ubiquitin–proteasome pathway and causes instability of both proteins (Xu and Attisano, 2000).

The Smad signaling pathway exhibits cross talk with other signaling pathways. Erk, a member of the MAP kinase family, phosphorylates PXSP motifs in the linker region of Smad1 and inhibits its nuclear translocation (Kretzschmar et al., 1997), although the mechanism by which it does so is still not clearly known. The growth of epithelial cells is inhibited by TGF-β under normal conditions. However, once cells acquire an oncogenic mutation in the *ras* gene, they often lose responsiveness to the antiproliferative effect of TGF-β. Oncogenic Ras induced phosphorylation of the linker region of Smad2 or Smad3 through activation of the Erk kinases, resulting in inhibition of nuclear translocation of Smad2/3 (Kretzschmar et al., 1999). Smad3 with mutated Erk phosphorylation sites rescued the growth inhibition by TGF-β in Ras-transformed cells. Thus, oncogenic mutation of Ras blocks TGF-β signaling in cancer cells. SW480.7 colon carcinoma cells exhibit at least two genetic alterations (Calonge and Massagué, 1999): they are defective in Smad4 function, and their *Ki-ras* gene is

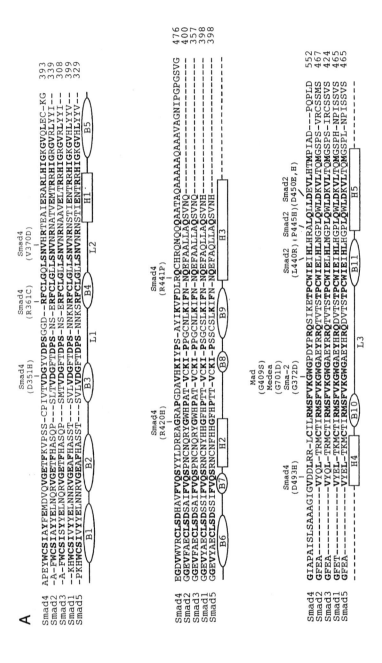

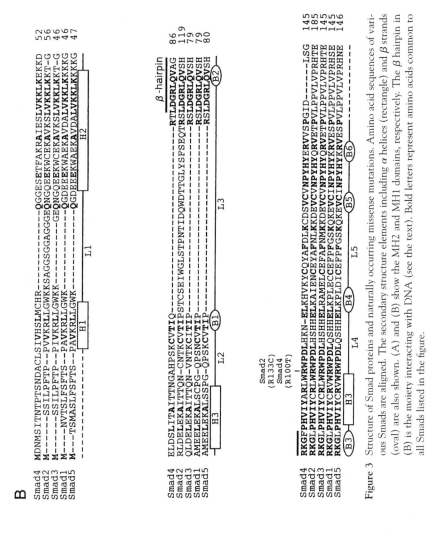

Figure 3 Structure of Smad proteins and naturally occurring missense mutations. Amino acid sequences of various Smads are aligned. The secondary structure elements including α helices (rectangle) and β strands (oval) are also shown. (A) and (B) show the MH2 and MH1 domains, respectively. The β hairpin in (B) is the moiety interacting with DNA (see the text). Bold letters represent amino acids common to all Smads listed in the figure.

activated. SW480.7 cells are thus resistant to growth inhibition by TGF-β. Introduction of Ras phosphorylation-resistant but not wild-type Smad3 together with Smad4 restored responsiveness to TGF-β.

The role of Smads in carcinogenesis has been investigated using gene targeting. Homozygous deletion of *Smad2* or *Smad4* causes early embryonal leathality in mice (Waldrip *et al.*, 1998; Nomura and Li, 1998; Weinstein *et al.*, 1998; Sirard *et al.*, 1998; Yang *et al.*, 1998). Although these results emphasize the importance of Smad2 and Smad4 in early development, the early lethality hampered analysis of the role of these Smads in carcinogenesis. Concomitant heterozygous loss of *Smad4* and *Apc*, the tumor suppressor genes identified in familial adenomatous polyposis coli, caused mice to develop invasive intestinal tumors (Takaku *et al.*, 1998). Furthermore, mice with heterozygous loss of *Smad4* developed gastric and duodenal polyps, and wild-type *Smad4* was lost in polyps (Takaku *et al.*, 1999). These results provide *in vivo* evidence that Smad4 is a tumor suppressor. The *Smad3* gene has been targeted by several groups. One of them reported that the mutant mice developed metastatic colorectal adeno-carcinomas (Zhu *et al.*, 1998), suggesting that Smad3 may be involved in colorectal carcinogenesis.

V. PERSPECTIVES

The discovery of Smad proteins has greatly contributed to the understanding of the signaling mechanism of TGF-β. However, new questions have been raised. Smad2 and Smad3 are almost identical in structure and are activated by TβR-I on ligand binding. Although the two R-Smads are likely to play redundant roles in many biological situations, they have distinct functions under particular conditions. It was biochemically demonstrated that Smad3 can bind directly to DNA, but that Smad2 cannot (Yagi *et al.*, 1999; Dennler *et al.*, 1999). This difference may result in transactivation of distinct sets of genes. In fact, Smad2 activates expression of the mouse *goosecoid* gene, whereas Smad3 inhibits its transcription (Labbé *et al.*, 1998). The phenotypes of *Smad2* and *Smad3* knockout mice differ remarkably. These results might be explained by the difference in pattern of expression of the two genes, but other reasons for them may exist as well. Most intriguingly, mutations of *Smad2* have been found in cancers, but no mutations of *Smad3* have been reported. Conditional gene targeting of *Smad2* may provide clues in answering this question.

The components of the TGF-β signaling pathway are tumor suppressors. It is thus clinically reasonable to restore the TGF-β signaling pathway in neoplastic cells, and this should counteract clonal expansion of malignant cells. As discussed above, however, TGF-β exerts diverse biological effects, and can suppress or promote tumor progression depending on the cellular context. Inhibition of TGF-β signaling may instead be advantageous in preventing tumor invasion or metastasis through various mechanisms under certain conditions. Inhibition of neoplastic cell growth and simultaneous suppression of tumor invasion by modifying specific TGF-β effects would constitute ideal treatment of cancers. It is thus important to develop drugs that target specific activities of TGF-β such as growth control, extracellular matrix production, neovascularization, and immunosuppression. Although the origin of the diversity of TGF-β activities is not yet fully understood, future characterization of the functions of the receptors and Smads will dissect the complex signaling specificity of the ligand.

Acknowledgment

I am grateful to Ms. A. Nishitoh-Sakai for her assistance in the preparation of this chapter.

References

Afrakhte, M., Morén, A., Jossan, S., Itoh, S., Sampath, K., Westermark, B., Heldin, C.-H., Heldin, N.-E., and ten Dijke, P. (1998). Induction of inhibitory Smad6 and Smad7 mRNA by TGF-β family members. *Biochem. Biophys. Res. Commun.* 249, 505–511.

Bai, S., Shi, X., Yang, X., and Cao, X. (2000). Smad6 as a transcriptional corepressor. *J. Biol. Chem.* 275, 8267–8270.

Bitzer, M., von Gersdorff, G., Liang, D., Dominguez-Rosales, A., Beg, A. A., Rojkind, M., and Böttinger, E. P. (2000). A mechanism of suppression of TGF-β/SMAD signaling by NF-κ B/RelA. *Genes Dev.* 14, 187–197.

Blobe, G. C., Schiemann, W. P., and Lodish, H. F. (2000). Role of transforming growth factor β in human disease. *N. Engl. J. Med.* 342, 1350–1358.

Calonge, M. J., and Massagué, J. (1999). Smad4/DPC4 silencing and hyperactive Ras jointly disrupt transforming growth factor-β antiproliferative responses in colon cancer cells. *J. Biol. Chem.* 274, 33637–33643.

Chen, T., de Vries, E. G., Hollema, H., Yegen, H. A., Vellucci, V. F., Strickler, H. D., Hildesheim, A., and Reiss, M. (1999). Structural alterations of transforming growth factor-β receptor genes in human cervical carcinoma. *Int. J. Cancer* 82, 43–51.

Chen, X., Rubock, M. J., and Whitman, M. (1996). A transcriptional partner for MAD proteins in TGF-β signalling. *Nature* 383, 691–696.

DeCoteau, J. F., Knaus, P. I., Yankelev, H., Reis, M. D., Lowsky, R., Lodish, H. F., and Kadin, M. E. (1997). Loss of functional cell surface transforming growth factor β (TGF-β) type 1 receptor correlates with insensitivity to TGF-β in chronic lymphocytic leukemia. *Proc. Natl. Acad. Sci. U.S.A.*, 94, 5877–5881.

Denissova, N. G., Pouponnot, C., Long, J., He, D., and Liu, F. (2000). Transforming growth factor β-inducible independent binding of SMAD to the Smad7 promoter. *Proc. Natl. Acad. Sci. U.S.A.* 97, 6397–6402.

Dennler, S., Itoh, S., Vivien, D., ten Dijke, P., Huet, S., and Gauthier, J. M. (1998). Direct binding of Smad3 and Smad4 to critical TGF β-inducible elements in the promoter of human plasminogen activator inhibitor-type 1 gene. *EMBO J.* 17, 3091–3100.

Dennler, S., Huet, S., and Gauthier, J. M. (1999). A short amino-acid sequence in MH1 domain is responsible for functional differences between Smad2 and Smad3. *Oncogene* **18**, 1643–1648.

Derynck, R., and Feng, X.-H. (1997). TGF-β receptor signaling. *Biochim. Biophys. Acta* **1333**, F105–F150.

Derynck, R., Zhang, Y., and Feng, X.-H. (1998). Smads: Transcriptional activators of TGF-β responses. *Cell* **95**, 737–740.

Dong, C., Li, Z., Alvarez, R., Jr., Feng, X.-H., and Goldschmidt-Clermont, P. J. (2000). Microtubule binding to Smads may regulate TGF-β activity. *Mol. Cell* **5**, 27–34.

Duff, E. K., and Clarke, A. R. (1998). Smad4 (DPC4)—a potent tumour suppressor? *Br. J. Cancer* **78**, 1615–1619.

Eppert, K., Scherer, S. W., Ozcelik, H., Pirone, R., Hoodless, P., Kim, H., Tsui, L. C., Bapat, B., Gallinger, S., Andrulis, I. L., Thomsen, G. H., Wrana, J. L., and Attisano, L. (1996). MADR2 maps to 18q21 and encodes a TGF-β-regulated MAD-related protein that is functionally mutated in colorectal carcinoma. *Cell* **86**, 543–552.

Friedl, W., Kruse, R., Uhlhaas, S., Stolte, M., Schartmann, B., Keller, K. M., Jungck, M., Stern, M., Loff, S., Back, W., Propping, P., and Jenne, D. E. (1999). Frequent 4-bp deletion in exon 9 of the SMAD4/MADH4 gene in familial juvenile polyposis patients. *Genes Chromosomes Cancer* **25**, 403–406.

Glick, A. B., Lee, M. M., Darwiche, N., Kulkarni, A. B., Karlsson, S., and Yuspa, S. H. (1994). Targeted deletion of the TGF-β 1 gene causes rapid progression to squamous cell carcinoma. *Genes Dev.* **8**, 2429–2440.

Goto, D., Yagi, K., Inoue, H., Iwamoto, I., Kawabata, M., Miyazono, K., and Kato, M. (1998). A single missense mutant of Smad3 inhibits activation of both Smad2 and Smad3, and has a dominant negative effect on TGF-β signals. *FEBS Lett.* **430**, 201–204.

Hahn, S. A., Schutte, M., Hoque, A. T., Moskaluk, C. A., da Costa, L. T., Rozenblum, E., Weinstein, C. L., Fischer, A., Yeo, C. J., Hruban, R. H., and Kern, S. E. (1996). DPC4, a candidate tumor suppressor gene at human chromosome 18q21.1. *Science* **271**, 350–353.

Hanafusa, H., Ninomiya-Tsuji, J., Masuyama, N., Nishita, M., Fujisawa, J., Shibuya, H., Matsumoto, K., and Nishida, E. (1999). Involvement of the p38 mitogen-activated protein kinase pathway in transforming growth factor-β-induced gene expression. *J. Biol. Chem.* **274**, 27161–27167.

Hanai, J., Chen, L. F., Kanno, T., Ohtani-Fujita, N., Kim, W. Y., Guo, W. H., Imamura, T., Ishidou, Y., Fukuchi, M., Shi, M. J., Stavnezer, J., Kawabata, M., Miyazono, K., and Ito, Y. (1999). Interaction and functional cooperation of PEBP2/CBF with Smads. Synergistic induction of the immunoglobulin germline Cα promoter. *J. Biol. Chem.* **274**, 31577–31582.

Hata, A., Lo, R. S., Wotton, D., Lagna, G., and Massagué, J. (1997). Mutations increasing autoinhibition inactivate tumour suppressors Smad2 and Smad4. *Nature* **388**, 82–87.

Hata, A., Lagna, G., Massagué, J., and Hemmati-Brivanlou, A. (1998). Smad6 inhibits BMP/Smad1 signaling by specifically competing with the Smad4 tumor suppressor. *Genes Dev.* **12**, 186–197.

Hata, A., Seoane, J., Lagna, G., Montalvo, E., Hemmati-Brivanlou, A., and Massagué, J. (2000). OAZ uses distinct DNA- and protein-binding zinc fingers in separate BMP–Smad and Olf signaling pathways. *Cell* **100**, 229–240.

Hayashi, H., Abdollah, S., Qiu, Y., Cai, J., Xu, Y. Y., Grinnell, B. W., Richardson, M. A., Topper, J.N., Gimbrone, M. A., Jr., Wrana, J. L., and Falb, D. (1997). The MAD-related protein Smad7 associates with the TGFβ receptor and functions as an antagonist of TGFβ signaling. *Cell* **89**, 1165–1173.

Heldin, C.-H., Miyazono, K., and ten Dijke, P. (1997). TGF-β signalling from cell membrane to nucleus through SMAD proteins. *Nature* **390**, 465–471.

Hocevar, B. A., Brown, T. L., and Howe, P. H. (1999). TGF-β induces fibronectin synthesis through a c-Jun N-terminal kinase-dependent, Smad4-independent pathway. *EMBO J.* **18**, 1345–1356.

Hoodless, P. A., Haerry, T., Abdollah, S., Stapleton, M., O'Connor, M. B., Attisano, L., and Wrana, J. L. (1996). MADR1, a MAD-related protein that functions in BMP2 signaling pathways. *Cell* **85,** 489–500.

Howe, J. R., Roth, S., Ringold, J. C., Summers, R. W., Jarvinen, H. J., Sistonen, P., Tomlinson, I. P., Houlston, R. S., Bevan, S., Mitros, F. A., Stone, E. M., and Aaltonen, L. A. (1998). Mutations in the SMAD4/DPC4 gene in juvenile polyposis. *Science* **280,** 1086–1088.

Howell, M., Itoh, F., Pierreux, C. E., Valgeirsdottir, S., Itoh, S., ten Dijke, P., and Hill, C. S. (1999). *Xenopus* Smad4β is the co-Smad component of developmentally regulated transcription factor complexes responsible for induction of early mesodermal genes. *Dev. Biol.* **214,** 354–369.

Hua, X., Miller, Z. A., Wu, G., Shi, Y., and Lodish, H. F. (1999). Specificity in transforming growth factor β-induced transcription of the plasminogen activator inhibitor-1 gene: Interactions of promoter DNA, transcription factor μE3, and Smad proteins. *Proc. Natl. Acad. Sci. U.S.A.* **96,** 13130–13135.

Imamura, T., Takase, M., Nishihara, A., Oeda, E., Hanai, J., Kawabata, M., and Miyazono, K. (1997). Smad6 inhibits signalling by the TGF-β superfamily. *Nature* **389,** 622–626.

Ishida, W., Hamamoto, T., Kusanagi, K., Yagi, K., Kawabata, M., Takehara, K., Sampath, T. K., Kato, M., and Miyazono, K. (2000). Smad6 is a Smad1/5-induced Smad inhibitor. Characterization of bone morphogenetic protein-responsive element in the mouse Smad6 promoter. *J. Biol. Chem.* **275,** 6075–6079.

Itoh, S., Landstrom, M., Hermansson, A., Itoh, F., Heldin, C.-H., Heldin, N.-E., and ten Dijke, P. (1998). Transforming growth factor β1 induces nuclear export of inhibitory Smad7. *J. Biol. Chem.* **273,** 29195–29201.

Jonk, L. J., Itoh, S., Heldin, C.-H., ten Dijke, P., and Kruijer, W. (1998). Identification and functional characterization of a Smad binding element (SBE) in the JunB promoter that acts as a transforming growth factor-β, activin, and bone morphogenetic protein-inducible enhancer. *J. Biol. Chem.* **273,** 21145–21152.

Kawabata, M., Inoue, H., Hanyu, A., Imamura, T., and Miyazono, K. (1998). Smad proteins exist as monomers *in vivo* and undergo homo- and hetero-oligomerization upon activation by serine/threonine kinase receptors. *EMBO J.* **17,** 4056–4065.

Kim, I. Y., Ahn, H. J., Zelner, D. J., Shaw, J. W., Sensibar, J. A., Kim, J. H., Kato, M., and Lee, C. (1996). Genetic change in transforming growth factor β (TGF-β) receptor type I gene correlates with insensitivity to TGF-β 1 in human prostate cancer cells. *Cancer Res.* **56,** 44–48.

Kim, J., Johnson, K., Chen, H. J., Carroll, S., and Laughon, A. (1997). *Drosophila* Mad binds to DNA and directly mediates activation of vestigial by Decapentaplegic. *Nature* **388,** 304–308.

Kretzschmar, M., Doody, J., and Massagué, J. (1997). Opposing BMP and EGF signalling pathways converge on the TGF-β family mediator Smad1. *Nature* **389,** 618–622.

Kretzschmar, M., Doody, J., Timokhina, I., and Massagué, J. (1999). A mechanism of repression of TGFβ/Smad signaling by oncogenic Ras. *Genes Dev.* **13,** 804–816.

Kurokawa, M., Mitani, K., Irie, K., Matsuyama, T., Takahashi, T., Chiba, S., Yazaki, Y., Matsumoto, K., and Hirai, H. (1998). The oncoprotein Evi-1 represses TGF-β signalling by inhibiting Smad3. *Nature* **394,** 92–96.

Kusanagi, K., Inoue, H., Ishidou, Y., Mishima, H. K., Kawabata, M., and Miyazono, K. (2000). Characterization of a bone morphogenetic protein-responsive Smad-binding element. *Mol. Biol. Cell* **11,** 555–565.

Labbé, E., Silvestri, C., Hoodless, P. A., Wrana, J. L., and Attisano, L. (1998). Smad2 and Smad3 positively and negatively regulate TGF β-dependent transcription through the forkhead DNA-binding protein FAST2. *Mol. Cell* **2,** 109–120.

Landstrom, M., Heldin, N.-E., Bu, S., Hermansson, A., Itoh, S., Dijke, P., and Heldin, C.-H. (2000). Smad7 mediates apoptosis induced by transforming growth factor β in prostatic carcinoma cells. *Curr. Biol.* **10**, 535–538.

Le Dai, J., Turnacioglu, K. K., Schutte, M., Sugar, A. Y., and Kern, S. E. (1998). Dpc4 transcriptional activation and dysfunction in cancer cells. *Cancer Res.* **58**, 4592–4597.

Liberati, N. T., Datto, M. B., Frederick, J. P., Shen, X., Wong, C., Rougier-Chapman, E. M., and Wang, X.-F. (1999). Smads bind directly to the Jun family of AP-1 transcription factors. *Proc. Natl. Acad. Sci. U.S.A.* **96**, 4844–4849.

Lin, Y. C., and Stavnezer, J. (1992). Regulation of transcription of the germ-line Ig α constant region gene by an ATF element and by novel transforming factor-β 1-responsive elements. *J. Immunol.* **149**, 2914–2925.

Liu, F., Pouponnot, C., and Massagué, J. (1997). Dual role of the Smad4/DPC4 tumor suppressor in TGFβ- inducible transcriptional complexes. *Genes Dev.* **11**, 3157–3167.

Liu, F., Massagué, J., and Ruizi Altaba, A. (1998). Carboxy-terminally truncated Gli3 proteins associate with Smads. *Nat. Genet.* **20**, 325–326.

Lo, R. S., and Massagué, J. (1999). Ubiquitin-dependent degradation of TGF-β-activated Smad2. *Nat. Cell Biol.* **1**, 472–478.

Lo, R. S., Chen, Y. G., Shi, Y., Pavletich, N. P., and Massagué, J. (1998). The L3 loop: A structural motif determining specific interactions between SMAD proteins and TGF-β receptors. *EMBO J.* **17**, 996–1005.

Lu, S. L., Kawabata, M., Imamura, T., Akiyama, Y., Nomizu, T., Miyazono, K., and Yuasa, Y. (1998). HNPCC associated with germline mutation in the TGF-β type II receptor gene. *Nat. Genet.* **19**, 17–18.

Markowitz, S. D., and Roberts, A. B. (1996). Tumor suppressor activity of the TGF-β pathway in human cancers. *Cytokine Growth Factor Rev.* **7**, 93–102.

Markowitz, S., Wang, J., Myeroff, L., Parsons, R., Sun, L., Lutterbaugh, J., Fan, R. S., Zborowska, E., Kinzler, K. W., Vogelstein, B., Brattain, M., and Willson, K. V. (1995). Inactivation of the type II TGF-β receptor in colon cancer cells with microsatellite instability. *Science* **268**, 1336–1338.

Massagué, J. (1998). TGF-β signal transduction. *Annu. Rev. Biochem.* **67**, 753–791.

Massagué, J., and Wotton, D. (2000). Transcriptional control by the TGF-β/Smad signaling system. *EMBO J.* **19**, 1745–1754.

Masuyama, N., Hanafusa, H., Kusakabe, M., Shibuya, H., and Nishida, E. (1999). Identification of Two Smad4 Proteins in *Xenopus*. Their common and distinct properties. *J. Biol. Chem.* **274**, 12163–12170.

Miyazono, K., ten Duke, P., and Heldin, C.-H. (2000). TGF-β signaling by Smad proteins. *Adv. Immunol.* **75**, 115–157.

Mulder, K. M. (2000). Role of Ras and Mapks in TGFβ signaling. *Cytokine Growth Factor Rev.* **11**, 23–35.

Muro-Cacho, C. A., Anderson, M., Cordero, J., and Munoz-Antonia, T. (1999). Expression of transforming growth factor β type II receptors in head and neck squamous cell carcinoma. *Clin. Cancer Res.* **5**, 1243–1248.

Nagarajan, R. P., Zhang, J., Li, W., and Chen, Y. (1999). Regulation of Smad7 promoter by direct association with Smad3 and Smad4. *J. Biol. Chem.* **274**, 33412–33418.

Nakao, A., Afrakhte, M., Morén, A., Nakayama, T., Christian, J. L., Heuchel, R., Itoh, S., Kawabata, M., Heldin, N.-E., Heldin, C.-H., and ten Dijke, P. (1997). Identification of Smad7, a TGFβ-inducible antagonist of TGF-β signalling. *Nature* **389**, 631–635.

Nishihara, A., Hanai, J., Imamura, T., Miyazono, K., and Kawabata, M. (1999). E1A inhibits transforming growth factor-β signaling through binding to Smad proteins. *J. Biol. Chem.* **274**, 28716–28723.

Nishimura, R., Kato, Y., Chen, D., Harris, S. E., Mundy, G. R., and Yoneda, T. (1998). Smad5 and DPC4 are key molecules in mediating BMP-2-induced osteoblastic differentiation of the pluripotent mesenchymal precursor cell line C2C12. *J. Biol. Chem.* **273**, 1872–1879.

Nomura, M., and Li, E. (1998). Smad2 role in mesoderm formation, left–right patterning and craniofacial development. *Nature* **393**, 786–790.

Pardali, E., Xie, X. Q., Tsapogas, P., Itoh, S., Arvanitidis, K., Heldin, C.-H., ten Dijke, P., Grundstrom, T., and Sideras, P. (2000). Smad and AML proteins synergistically confer transforming growth factor β1 responsiveness to human germ-line IgA genes. *J. Biol. Chem.* **275**, 3552–3560.

Pasche, B., Luo, Y., Rao, P. H., Nimer, S. D., Dmitrovsky, E., Caron, P., Luzzatto, L., Offit, K., Cordon-Cardo, C., Renault, B., Satagopan, J. M., Murty, V. V., and Massagué, J. (1998). Type I transforming growth factor β receptor maps to 9q22 and exhibits a polymorphism and a rare variant within a polyalanine tract. *Cancer Res.* **58**, 2727–2732.

Pierce, D. F., Jr., Gorska, A. E., Chytil, A., Meise, K. S., Page, D. L., Coffey, R. J., Jr., and Moses, H. L. (1995). Mammary tumor suppression by transforming growth factor β 1 transgene expression. *Proc. Natl. Acad. Sci. U.S.A.* **92**, 4254–4258.

Sano, Y., Harada, J., Tashiro, S., Gotoh-Mandeville, R., Maekawa, T., and Ishii, S. (1999). ATF-2 is a common nuclear target of Smad and TAK1 pathways in transforming growth factor-β signaling. *J. Biol. Chem.* **274**, 8949–8957.

Savage, C., Das, P., Finelli, A. L., Townsend, S. R., Sun, C. Y., Baird, S. E., and Padgett, R. W. (1996). *Caenorhabditis elegans* genes sma-2, sma-3, and sma-4 define a conserved family of transforming growth factor β pathway components. *Proc. Natl. Acad. Sci. U.S.A.* **93**, 790–794.

Sekelsky, J. J., Newfeld, S. J., Raftery, L. A., Chartoff, E. H., and Gelbart, W. M. (1995). Genetic characterization and cloning of mothers against dpp, a gene required for decapentaplegic function in *Drosophila melanogaster*. *Genetics* **139**, 1347–1358.

Shi, X., Yang, X., Chen, D., Chang, Z., and Cao, X. (1999). Smad1 interacts with homeobox DNA-binding proteins in bone morphogenetic protein signaling. *J. Biol. Chem.* **274**, 13711–13717.

Shi, Y., Hata, A., Lo, R. S., Massagué, J., and Pavletich, N. P. (1997). A structural basis for mutational inactivation of the tumour suppressor Smad4. *Nature* **388**, 87–93.

Shi, Y., Wang, Y. F., Jayaraman, L., Yang, H., Massagué, J., and Pavletich, N. P. (1998). Crystal structure of a Smad MH1 domain bound to DNA: Insights on DNA binding in TGF-β signaling. *Cell* **94**, 585–594.

Shioda, T., Lechleider, R. J., Dunwoodie, S. L., Li, H., Yahata, T., de Caestecker, M. P., Fenner, M. H., Roberts, A. B., and Isselbacher, K. J. (1998). Transcriptional activating activity of Smad4: Roles of SMAD hetero-oligomerization and enhancement by an associating transactivator. *Proc. Natl. Acad. Sci. U.S.A.* **95**, 9785–9790.

Sirard, C., de la Pompa, J. L., Elia, A., Itie, A., Mirtsos, C., Cheung, A., Hahn, S., Wakeham, A., Schwartz, L., Kern, S. E., Rossant, J., and Mak, T. W. (1998). The tumor suppressor gene Smad4/Dpc4 is required for gastrulation and later for anterior development of the mouse embryo. *Genes Dev.* **12**, 107–119.

Song, C. Z., Siok, T. E., and Gelehrter, T. D. (1998). Smad4/DPC4 and Smad3 mediate transforming growth factor-β (TGF-β) signaling through direct binding to a novel TGF-β-responsive element in the human plasminogen activator inhibitor-1 promoter. *J. Biol. Chem.* **273**, 29287–29290.

Song, C. Z., Tian, X., and Gelehrter, T. D. (1999). Glucocorticoid receptor inhibits transforming growth factor-β signaling by directly targeting the transcriptional activation function of Smad3. *Proc. Natl. Acad. Sci. U.S.A.* **96**, 11776–11781.

Stroschein, S. L., Wang, W., and Luo, K. (1999). Cooperative binding of Smad proteins to two adjacent DNA elements in the plasminogen activator inhibitor-1 promoter mediates transforming growth factor β-induced Smad-dependent transcriptional activation. *J. Biol. Chem.* **274**, 9431–9441.

Takagi, Y., Kohmura, H., Futamura, M., Kida, H., Tanemura, H., Shimokawa, K., and Saji, S. (1996) Somatic alterations of the DPC4 gene in human colorectal cancers in vivo. *Gastroenterology* 111, 1369–1372.

Takaku, K., Oshima, M., Miyoshi, H., Matsui, M., Seldin, M. F., and Taketo, M. M. (1998). Intestinal tumorigenesis in compound mutant mice of both Dpc4 (Smad4) and Apc genes. *Cell* 92, 645–656.

Takaku, K., Miyoshi, H., Matsunaga, A., Oshima, M., Sasaki, N., and Taketo, M. M. (1999). Gastric and duodenal polyps in Smad4 (Dpc4) knockout mice. *Cancer Res.* 59, 6113–6117.

Takase, M., Imamura, T., Sampath, T. K., Takeda, K., Ichijo, H., Miyazono, K., and Kawabata, M. (1998). Induction of Smad6 mRNA by bone morphogenetic proteins. *Biochem. Biophys. Res. Commun.* 244, 26–29.

Tang, B., Böttinger, E. P., Jakowlew, S. B., Bagnall, K. M., Mariano, J., Anver, M. R., Letterio, J. J., and Wakefield, L. M. (1998). Transforming growth factor-β1 is a new form of tumor suppressor with true haploid insufficiency. *Nat. Med.* 4, 802–807.

Topper, J. N., Cai, J., Qiu, Y., Anderson, K. R., Xu, Y. Y., Deeds, J. D., Feeley, R., Gimeno, C. J., Woolf, E. A., Tayber, O., Mays, G. G., Sampson, B. A., Schoen, F. J., Gimbrone, M. A., Jr., and Falb, D. (1997). Vascular MADs: Two novel MAD-related genes selectively inducible by flow in human vascular endothelium. *Proc. Natl. Acad. Sci. U.S.A.* 94, 9314–9319.

Tsukazaki, T., Chiang, T. A., Davison, A. F., Attisano, L., and Wrana, J. L. (1998). SARA, a FYVE domain protein that recruits Smad2 to the TGFβ receptor. *Cell* 95, 779–791.

Tsuneizumi, K., Nakayama, T., Kamoshida, Y., Kornberg, T. B., Christian, J. L., and Tabata, T. (1997). Daughters against dpp modulates dpp organizing activity in *Drosophila* wing development. *Nature* 389, 627–631.

Ulloa, L., Doody, J., and Massagué, J. (1999). Inhibition of transforming growth factor-β/SMAD signalling by the interferon-γ/STAT pathway. *Nature* 397, 710–713.

Verschueren, K., Remacle, J. E., Collart, C., Kraft, H., Baker, B. S., Tylzanowski, P., Nelles, L., Wuytens, G., Su, M. T., Bodmer, R., Smith, J. C., and Huylebroeck, D. (1999). SIP1, a novel zinc finger/homeodomain repressor, interacts with Smad proteins and binds to 5'-CACCT sequences in candidate target genes. *J. Biol. Chem.* 275, 20489–20498.

von Gersdorff, G., Susztak, K., Rezvani, F., Bitzer, M., Liang, D., and Böttinger, E. P. (2000). Smad3 and Smad4 mediate transcriptional activation of the human Smad7 promoter by transforming growth factor β. *J. Biol. Chem.* 275, 11320–11326.

Waldrip, W. R., Bikoff, E. K., Hoodless, P. A., Wrana, J. L., and Robertson, E. J. (1998). Smad2 signaling in extraembryonic tissues determines anterior–posterior polarity of the early mouse embryo. *Cell* 92, 797–808.

Weinstein, M., Yang, X., Li, C., Xu, X., Gotay, J., and Deng, C. X. (1998). Failure of egg cylinder elongation and mesoderm induction in mouse embryos lacking the tumor suppressor Smad2. *Proc. Natl. Acad. Sci. U.S.A.* 95, 9378–9383.

Wisotzkey, R. G., Mehra, A., Sutherland, D. J., Dobens, L. L., Liu, X., Dohrmann, C., Attisano, L., and Raftery, L. A. (1998). Medea is a *Drosophila* Smad4 homolog that is differentially required to potentiate DPP responses. *Development* 125, 1433–1445.

Wotton, D., Lo, R. S., Lee, S., and Massagué, J. (1999). A Smad transcriptional corepressor. *Cell* 97, 29–39.

Wrana, J. L. (2000). Regulation of Smad activity. *Cell* 100, 189–192.

Wu, G., Chen, Y. G., Ozdamar, B., Gyuricza, C. A., Chong, P. A., Wrana, J. L., Massagué, J., and Shi, Y. (2000). Structural basis of Smad2 recognition by the Smad anchor for receptor activation. *Science* 287, 92–97.

Xiao, Z., Liu, X., Henis, Y. I., and Lodish, H. F. (2000a). A distinct nuclear localization signal in the N terminus of Smad 3 determines its ligand-induced nuclear translocation. *Proc. Natl. Acad. Sci. U.S.A.* 97, 7853–7858.

Xiao, Z., Liu, X., and Lodish, H. F. (2000b). Importin β mediates nuclear translocation of Smad3. *J. Biol. Chem.* **275**, 23425–23428.

Xu, J., and Attisano, L. (2000). Mutations in the tumor suppressors Smad2 and Smad4 inactivate transforming growth factor β signaling by targeting Smads to the ubiquitin–proteasome pathway. *Proc. Natl. Acad. Sci. U.S.A.* **97**, 4820–4825.

Xu, L., Chen, Y. G., and Massagué, J. (2000). The nuclear import function of Smad2 is masked by SARA and unmasked by TGFβ-dependent phosphorylation. *Nat. Cell Biol.* **2**, 559–562.

Xu, X., Yin, Z., Hudson, J. B., Ferguson, E. L., and Frasch, M. (1998). Smad proteins act in combination with synergistic and antagonistic regulators to target Dpp responses to the *Drosophila* mesoderm. *Genes Dev.* **12**, 2354–2370.

Yagi, K., Goto, D., Hamamoto, T., Takenoshita, S., Kato, M., and Miyazono, K. (1999). Alternatively spliced variant of Smad2 lacking exon 3. Comparison with wild-type Smad2 and Smad3. *J. Biol. Chem.* **274**, 703–709.

Yahata, T., de Caestecker, M. P., Lechleider, R. J., Andriole, S., Roberts, A. B., Isselbacher, K. J., and Shioda, T. (2000). The MSG1 non-DNA-binding transactivator binds to the p300/CBP coactivators, enhancing their functional link to the Smad transcription factors. *J. Biol. Chem.* **275**, 8825–8834.

Yamaguchi, K., Shirakabe, K., Shibuya, H., Irie, K., Oishi, I., Ueno, N., Taniguchi, T., Nishida, E., and Matsumoto, K. (1995). Identification of a member of the MAPKKK family as a potential mediator of TGF-β signal transduction. *Science* **270**, 2008–2011.

Yanagisawa, J., Yanagi, Y., Masuhiro, Y., Suzawa, M., Watanabe, M., Kashiwagi, K., Toriyabe, T., Kawabata, M., Miyazono, K., and Kato, S. (1999). Convergence of transforming growth factor-β and vitamin D signaling pathways on SMAD transcriptional coactivators. *Science* **283**, 1317–1321.

Yang, X., Li, C., Xu, X., and Deng, C. (1998). The tumor suppressor SMAD4/DPC4 is essential for epiblast proliferation and mesoderm induction in mice. *Proc. Natl. Acad. Sci. U.S.A.* **95**, 3667–3672.

Yeo, C. Y., Chen, X., and Whitman, M. (1999). The role of FAST-1 and Smads in transcriptional regulation by activin during early *Xenopus* embryogenesis. *J. Biol. Chem.* **274**, 26584–26590.

Zawel, L., Dai, J. L., Buckhaults, P., Zhou, S., Kinzler, K. W., Vogelstein, B., and Kern, S. E. (1998). Human Smad3 and Smad4 are sequence-specific transcription activators. *Mol. Cell* **1**, 611–617.

Zhang, Y., and Derynck, R. (2000). Transcriptional regulation of the transforming growth factor-β-inducible mouse germ line Ig α constant region gene by functional cooperation of Smad, CREB, and AML family members. *J. Biol. Chem.* **275**, 16979–16985.

Zhang, Y., Feng, X.-H., and Derynck, R. (1998). Smad3 and Smad4 cooperate with c-*Jun*/c-*Fos* to mediate TGF-β-induced transcription. *Nature* **394**, 909–913.

Zhu, H., Kavsak, P., Abdollah, S., Wrana, J. L., and Thomsen, G. H. (1999). A SMAD ubiquitin ligase targets the BMP pathway and affects embryonic pattern formation. *Nature* **400**, 687–693.

Zhu, Y., Richardson, J. A., Parada, L. F., and Graff, J. M. (1998). Smad3 mutant mice develop metastatic colorectal cancer. *Cell* **94**, 703–714.

11

DAN Gene

Shigeru Sakiyama
Chiba Cancer Center Research Institute
Chuoh-ku, Chiba 260-8717, Japan

I. INTRODUCTION

Cancer is thought to result from accumulated genetic alterations, that is, both activation of proto-oncogenes which prevail over cellular growth regulatory commands, and inactivation of tumor suppressor genes which render cells free of growth control. The alterations of such genes as a result of several types of mutations contribute to the disruption of the harmonized network that regulates normal cell growth and hence oncogenenicity of human cancers (Weinberg, 1991). Many of the suppressor genes were identified and isolated following pedigree and cytogenetic analysis. Meanwhile, several gene transfer and positive selection experiments to clone putative tumor-suppressor (TS) genes have been carried out (Sager, 1989). Such an approach is indispensable for identifying and characterizing as many of the potential TS genes as possible

Tumor-Suppressing Viruses, Genes, and Drugs
Innovative Cancer Therapy Approaches
Copyright © 2002 by Academic Press.
All rights of reproduction in any form reserved.

in order to better understand the diverse phenotypes of cancer. TS genes can be identified in gene populations whose expression is downregulated in response to oncogenic stimuli such as activation of proto-oncogenes. These genes can be classified as class II in which loss of function is cause by downregulation of their expression, whereas class I TS genes comprise those whose function is lost by a variety of mutations including deletion (Lee *et al.*, 1991).

II. CLONING OF DAN cDNA

We initiated our studies to obtain novel gene(s) whose expression is retained in normal rat fibroblasts (3Y1) but blocked or dramatically reduced in those transformed with a Schmidt-Ruppin strain of Rous sarcoma virus (SR-3Y1) (Ozaki and Sakiyama, 1993). Using a differential hybridization screening method of cDNAs prepared from 3Y1 and SR-3Y1, several clones showing relatively higher signals in 3Y1 cells compared with SR-3Y1 were obtained (DAN, which was previously named N03, is among those clones). The downregulation of DAN expression is not specific for Src transformation but can also be observed in Mos- or SV40-transformants (but not in Ras-transformants) (Figure 1). Such a downregulation is also seen in mouse NIH 3T3 cells transformed with various oncogenes (Ozaki *et al.*, 1996). It is intriguing to note that expression of the *DRM* gene, which was isolated using differential display analysis between v-Mos-transformants and their flat revertants, turned out to be a member of the DAN family of genes (see below). DRM is also suppressed in rat embryonal fibroblasts transformed with various viral oncogenes such as *Mos, Ras, Src Raf,* and *Fos* (Topol *et al.*, 1997).

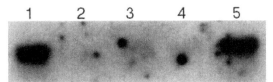

Figure 1 DAN gene expression in 3Y1 cells and those transformed with various oncogenes. Northern blot analysis was done with RNA of 3Y1 cells (1), and those transformed with v-*src* (2), v-*mos* (3), SV40 (4), and v-Ha-*ras* (5). From T. Ozaki and S. Sakiyama (1993). Molecular cloning and characterization of a cDNA showing negative regulation in v-*src*-transformed 3Y1 rat fibroblasts. *Proc. Natl. Acad. Sci. U.S.A.* **90**, 2593–2597. Copyright 1993 National Academy of Sciences, U.S.A.

III. TRANSFECTION OF DAN

It is crucial to test if DAN plays causative role(s) in cell growth control. When DAN is overexpressed in SR-3Y1, several important alterations in SR-3Y1 cells are induced. First, it causes a dramatic morphological reversion (Figure 2). Second, growth rates in culture dishes and in soft agar are significantly reduced. Third, tumor formation of these cells in nude mice is suppressed (Ozaki and Sakiyama, 1994). Furthermore, overexpression of DAN in the normal 3Y1 cells causes a delay in the entry into the S phase (Ozaki et al., 1995). These data collectively suggest that DAN protein might act as a negative regulator in cell growth control, and its loss of expression leads to the transformed phenotypes.

Growth suppressive activity of DAN is not limited to rodent cells. Human osteosarcoma-derived Saos-2 cells lack functional p53 and RB and do not express DAN as judged by a Northern blot. Overexpression of DAN in the Saos-2 cells resulted in suppression of cell growth as well as colony formation, suggesting that DAN does not require either p53 or RB to exert its effect (Hanaoka et al., 2001).

IV. ROLE OF DAN IN NEUROBLASTOMAS

To further explore the possible role of DAN in detail, we analyzed the human DAN gene. Human DAN is mapped to chromosome 1p36.11–p36.13 (Enomoto et al., 1994), which is well known to show highly significant linkage with genesis and/or progression of a variety of human cancers such as neuroblastoma (Schwab et al., 1996). Expression levels of DAN gene in cultured neuroblastoma cell lines are variable, but it is completely suppressed in several cell lines, most of which are accompanied with N-Myc gene amplification (Enomoto et al., 1994, Nakamura et al., 1998). It is possible that there exists the inverse relationship between DAN expression and N-Myc gene amplification in some of neuroblastoma cells. Thus, a loss of DAN function might contribute to acquisition of malignant properties in a subset of human neuroblastoma cell lines. Furthermore, Southern analysis of the DAN gene in neuroblastoma tissues revealed aberrant bands in accompany with N-Myc gene amplification (Figure 3). Hence came the name of the gene DAN, a differential-screening-selected gene aberrative in neuroblastoma (Enomoto et al., 1994). Although it has been shown that the DAN gene locus lies outside of the consensus region of allelic loss of neuroblastoma (White et al., 1995), the role of DAN as a class II TS gene should be pursued further, especially at the molecular level.

Because neuroblastoma cell lines differentiate by extending neurites and forming ganglion-like aggregates in response to retinoic acid (RA) associated with N-Myc downregulation (Peverali et al., 1996), the role of DAN in the

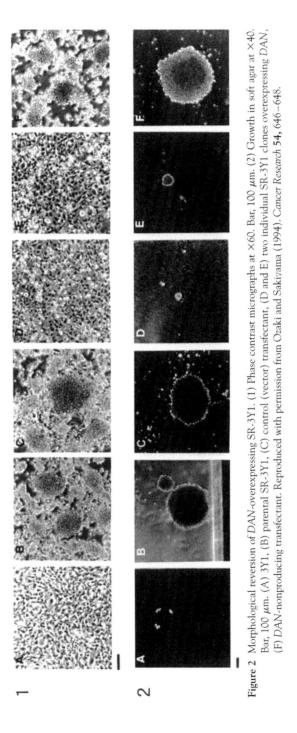

Figure 2 Morphological reversion of *DAN*-overexpressing SR-3Y1. (1) Phase contrast micrographs at ×60. Bar, 100 μm. (2) Growth in soft agar at ×40. Bar, 100 μm. (A) 3Y1, (B) parental SR-3Y1, (C) control (vector) transfectant, (D and E) two individual SR-3Y1 clones overexpressing *DAN*, (F) *DAN*-nonproducing transfectant. Reproduced with permission from Ozaki and Sakiyama (1994). *Cancer Research* **54**, 646–648.

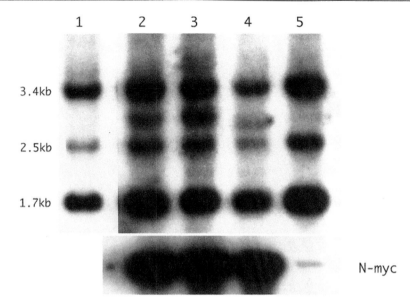

Figure 3 Southern blot analysis of *DAN* gene in neuroblastomas. Lane 1 represents a pattern observed in DNA prepared from 18 unrelated individuals. Lanes 2–4 show DNA patterns of three different neuroblastoma cases, and lane 5 shows the leukocyte DNA pattern of the case shown at lane 4. Lower panel shows Southern analysis of N-myc gene in each case. Reproduced with permission from Enomoto *et al.* (1994). *Oncogene* **9**, 2785–2791.

RA-induced neuronal differentiation was studied. When neuroblastoma cell lines, SH-SY5Y and RTBM-1, were treated with RA, DAN expression was induced 4 to 7 days after the treatment, a period when significant neurite extension was detected (Figure 4). Forced expression of DAN in SY5Y cells further potentiated the onset and extent of neurite extension, that is, the neurite outgrowth was observed as early as 2 days after RA treatment. These results suggested that DAN could play a role(s) in the RA-induced cellular differentiation in neuroblastoma cells (Nakamura *et al.*, 1998). Although the presence of RA responsible element(s) in the promoter region of DAN genes is speculated, so far no typical RA response element has been found within 2 kb upstream of the DAN coding region.

V. STRUCTURAL FEATURES OF THE DAN PROTEIN

DAN differs in only several amino acids between rat and human, and the several putative functional motifs are highly conserved (Enomoto *et al.*, 1994). Based on the deduced amino acid sequence, it is predicted that there are two casein kinase

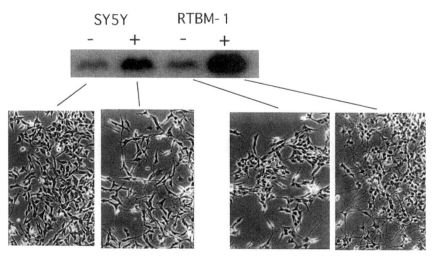

Figure 4 Retinoic acid-induced *DAN* expression and neurite formation in neuroblastoma cells. Induction of *DAN* mRNA and neurite formation 4 days after retinoic acid treatment in each cell line.

II target sequences, one N-glycosylation site, and an $(Hx)_n$ repeat sequence (Kondo *et al.*, 1995) in DAN of both species. In addition, the N-terminal region (amino acid 1–24) of DAN is highly hydrophobic and may possibly comprise the leader peptide (Figure 5). DAN is a protein of 27 kDa localized in both cytoplasm and the culture medium (Figure 6). About 80% of the total DAN protein appears to be secreted from the cells as judged by the SDS–PAGE analysis. Possible N-glycosylation of DAN was confirmed by the peptide *N*-glycosidase F (PNGase F) treatment of both secreted and cytoplasmic forms of DAN. It causes a reduction of the apparent mass of DAN by about 3 kDa (Nakamura *et al.*, 1997).

To examine the role(s) of the secreted form of DAN, the deletion mutant of DAN which lacks the signal sequence was tested for its biological activity in Saos-2 cells. Unlike the wild-type, this DAN mutant fails to inhibit

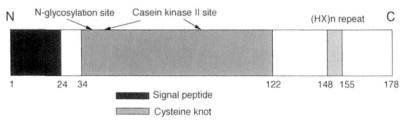

Figure 5 Structural features of rat DAN.

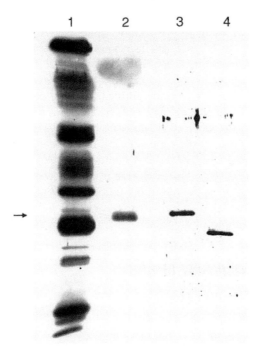

Figure 6 Secretion and glycosylation of DAN. Cell
lysates (lane 1) and the culture medium (lane
2) of DAN-overexpressing 3Y1 cells were analyzed for the presence of DAN by Western blot
analysis. The sample represented in lane 2 was
treated (lane 4) or untreated (lane 3) with
PDGFase followed by Western blot analysis.
Reprinted from *European Journal of Cancer*
33; Y. Nakamura, T. Ozaki, A. Nakagawara,
and S. Sakiyama; A product of DAN, a novel
candidate tumor suppressor gene, is secreted
into the culture medium and suppresses DNA
synthesis; pp. 1986–1990. Copyright 1997,
with permission from Elsevier Science.

their colony formation, suggesting that secretion of DAN is a prerequisite for its
growth suppression (Hanaoka *et al.*, 2001). However, it is also possible that
membrane-association of DAN may be essential for its function as suggested in
the case of DRM (Topol *et al.*, 2000).

It has been suggested that clusters containing Cys and/or His bind
metal ions and contribute to a metal-dependent protein–protein interaction
(Berg, 1986). The significance of an $(HX)_n$ repeat sequence at the C-terminal

region (amino acid 148–155) was examined using DAN mutants truncated from the C-terminus, using an Ni^{2+}-affinity column. DAN does indeed possess a strong affinity for Ni^{2+}. It has also been suggested that this region can be utilized for dimer/oligomer formation of DAN. (Kondo et al., 1995).

To investigate the biological effect(s) of secreted DAN (if any) in a more direct way, DAN was partially purified from the serum-free conditioned medium of DAN-overexpressing 3Y1 cells by ammonium sulfate fractionation. When this fraction was added at a concentration of approximately 100 ng/ml to the SR-3Y1 cells which hardly express DAN, it inhibited their DNA synthesis by 40% (Nakamura et al., 1997). This function of DAN is abolished by heat treatment (100°C, 10 min). The inhibition never exceeded 40% even at higher concentrations, suggesting the possibility that an impurity of the DAN preparation might antagonize DAN, or that the cytoplasmic form of DAN might also contribute to the complete inhibition.

VI. GENOMIC STRUCTURE OF DAN

Structural analysis of the DAN gene revealed that both rat and human DAN genes are composed of four exons (Ozaki et al., 1997; Ozaki and Sakiyama, 1997). The translation initiation and termination codons start at exons 2 and 4, respectively. Putative TATA-like sequences were 31 and 38 bp upstream from the transcription initiation site for rat and human genes, respectively. Using a reporter assay, it has been shown that a TATA-like sequence is involved for proper transcription of the rat DAN gene. In a series of experiments using deletion mutants, the presence of the regions that regulate rat DAN expression positively (-57 to $+118$) and negatively (-1231 to -636) were demonstrated. Furthermore, there exist two dinucleotide repeats, $(CA)_7$ and $(CA)_8$ in the first intron of the human DAN gene. These repeats could be useful to distinguish each allele of the DAN gene.

VII. DAN FAMILY

A Xenopus homolog of DRM was found as a novel antagonist of bone morphogenetic protein (BMP) signaling that acts as a dorsaling factor expressed in the Xenopus neural crest (Hsu et al., 1998). DAN, DRM/Gremlin, and the head-inducing factor Cerberus (Bouwmeester et al., 1996) were suggested to form a novel group called DAN family (Hsu et al., 1998). Dante, CeCan 1 (Pearce et al., 1999), Cer1 (Stanley et al., 1998), PRDC (Minabe-Saegusa et al., 1998) and Caronte (Esteban et al., 1999) were also found to belong to the DAN family. Xenopus DAN has been shown to indeed possess an ability to dorsalize ventral

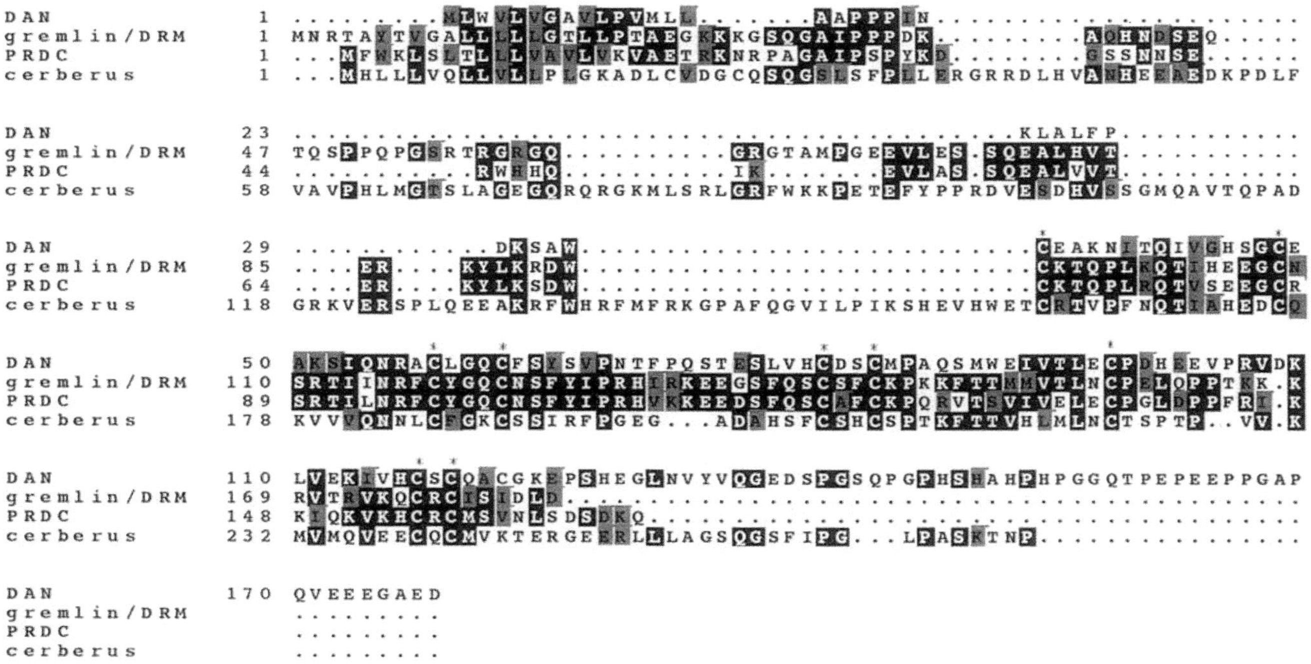

```
DAN           1   · · · · · MEWVLVGAVLPVMLL · · · · · · · · · · AAPPPN · · · · · · · · AQHNDSEQ · · · ·
gremlin/DRM   1   MNRTAYTVGALLLLGTLLPTAEG KKKGSQGAIPPDK · · · · · · · · · · · · · · · · · · · · · · · ·
PRDC          1   · · · MFVKLSLTLLVAVLVKVAET RKNRPAGAIPSPYKD · · · · · · · · GSSNNSE · · · ·
cerberus      1   · · · MHLLLVQLLVLPLGKADLCVDGCQSQGSLFPLLERGRRDLHVANHEEAEDKPDLF

                                                                              KLALFP · · · · ·
DAN          23   · · · · · · · · · · · · · · · · · · · · · · · GRGTAMPGEEVLES · SQEALHVT · · · · · KLALFP
gremlin/DRM  47   TQSPPQPGSRTRGRGQ · · · · · · · · · · · · · IK · · · · · EVLAS · SQEALVT · · · · · · · · · ·
PRDC         44   · · · · · · · RWHHQ · · · · · · · · · · · · IK · · · · · EVLAS · SQEALVT · · · · · · · · · ·
cerberus     58   VAVPHLMGTSLAGEGQRQRGKMLSRLGRFWKKPETEFYPPRDVESDHVSSGMQAVTQPAD

DAN          29   · · · · · · · · · DKSAW · · · · · · · · · · · · · · · · · · · · · · · CEAKNITQIVGHSGCE
gremlin/DRM  85   · · · · · ER · · · KYLKSDW · · · · · · · · · · · · · · · · · · · · CKTQPL · KQTIHEEGCR
PRDC         64   · · · · · ER · · · KYLKSDW · · · · · · · · · · · · · · · · · · · · CKTQPL · RQTVSEEGCR
cerberus    118   GRKVERSPLQEEAKRFWHRFMFRKGPAFQGVILPIKSHEVHWETCRTVPFNQTIAHEDCQ

DAN          50   AKSIQNRACLGQCFSYSVPNTFPQSTESLVHCDSCMPAQSMWEIVTLECPDHEVPRVDK
gremlin/DRM 110   SRTIINRFCYGQCNSFYIPRHIRKEEGSFQSCSFCKPKKFTTMAVTLNCPELQPPTKK
PRDC         89   SRTILNRFCYGQCNSFYIPRHVKKEEDSFQSCAFCKPQRVTSIVELECPGLDPPFRI · KK
cerberus    178   KVVQNNLCFGKCSSIRFPGEG · · ADAHSFCSHCSPTKFTTVHMLNCTSPTP · · VV · K

DAN         110   LVEKIVHCSCQACGKEPSHEGLNVVVQGEDSPGSQPGPHSHAHPHPGGQTPEPEEPPGAP
gremlin/DRM 169   RVTRVKQCRCISIDLD · · · · · · · · · · · · · · · · · · · · · · · · · · · · · · · · ·
PRDC        148   KIQKVKHCRCMSVNLSDSDKQ · · · · · · · · · · · · · · · · · · · · · · · · · · · · · ·
cerberus    232   MVMQVEECQCMVKTERGEERLLLAGSQGSFIPG · · · LPASKTNP · · · · · · · · · · · · ·

DAN         170   QVEEEGAED
gremlin/DRM       · · · · · · · · ·
PRDC              · · · · · · · · ·
cerberus          · · · · · · · · ·
```

Figure 7 Alignment of DAN family members. Black and shaded regions represent identical and conserved amino acids residues, respectively. Numbers denote the amino acid positions in the corresponding proteins. Asterisks indicate the positions of conserved Cys residues in the Cys knot domain.

229

mesoderm and neutralize ectoderm in *Xenopus* embryos. Furthermore, the biological activities of DAN family members are mediated through direct association with BMP2 at a binding domain similar to Noggin (Hsu *et al.*, 1998). These molecules share a conserved Cys knot motif (see Figure 7) that is also found in a number of signaling molecules such as TGF-β-superfamily (McDonald and Hendrickson, 1993).

Although the DAN family is well conserved at the Cys knot motif, members are quite diverse at their sequences outside the Cys knot as well as its position in each molecule. These might imply the unique function of each member in the DAN family. For instance, the IC_{50} of DRM/Gremlin, DAN, and Cer-1 are different for the BMP-mediated activation of the target gene (*Tlx-2*), and DAN has a much weaker activity than DRM (Pearce *et al.*, 1999). Furthermore, DAN and Cerberus are quite distinct from each other in *Xenopus* animal cap assays (Stanley *et al.*, 1998). Induction of ventrolateral mesoderm by BMP4 was completely inhibited by coinjection of Cerberus, but not DAN. BMP4 effectively suppresses the DAN-induced expression of marker genes, suggesting the possibility that the anti-BMP activity of DAN is less stable or less potent against BMP4.

BMPs, members of the TGF-β superfamily, have a variety of critical functions in many biological contexts (Hogan, 1996), and their activities can be extracellularly modulated through the availability of BMP inhibitors (Massagué and Chen, 2000). Because BMPs are expressed in many organs, and assuming that each member of DAN family has a different affinity for BMPs, the number of these combinations would be enormous, and thus could affect the fate of individual tissues and cells in a variety of ways. Although the studies on interaction between DAN and BMP families were done mostly at embryonic development, it is important to note that expression of DAN is ubiquitous among various adult mammalian tissues except liver (Stanley *et al.*, 1998 and S. Sakiyama, unpublished observations, 2000). These data strongly suggest that DAN is playing an important role(s) in different adult tissues. The role(s) of DAN family members as BMP antagonists might not simply explain the phenomenon that DAN and DRM suppress transformed phenotypes, because it has been demonstrated that BMP signaling promotes apoptosis in a variety of adult tissues (Rodriguez-Leon *et al.*, 1999). Therefore, other possible activities for DAN remain to be explored.

Acknowledgments

The author is grateful to Dr. Toshinori Ozaki, Ms. Yohko Nakamura, and Dr. Akira Nakagawara, and also to the staff members of the Division of Biochemistry, Chiba Cancer Center Research Institute, who contributed to the present works on DAN. The works presented here were supported in part by a Grant-in-Aid for a New Comprehensive 10-Year Strategy for Cancer Control from the Ministry of Health and Welfare, Japan.

References

Berg, J. M. (1986). Potential metal-binding domains in nucleic acid binding proteins. *Science* **232**, 485–487.

Bouwmeester, T., Kim, S.-H., Sasai, Y., Lu, B., and Robertis, E. M. (1996). Cerberus is a head-inducing secreted factor expressed in the anterior endoderm of Spemenn's organizer. *Nature* **382**, 595–601.

Enomoto, H., Ozaki, T., Takahashi, E., Nomura, N., Tabata, S., Takahashi, H., Ohnuma, N., Tanabe, M., Iwai, J., Yoshida, H., Matsunaga, T., and Sakiyama, S. (1994). Identification of human *DAN* gene, mapping to the putative neuroblastoma tumor suppressor locus. *Oncogene* **9**, 2785–2791.

Esteban, C. R., Capdevila, J., Economides, A. N., Pascual, J., Oritz, A., and Belmonte, J. C. I. (1999). The novel Cer-like protein Caronte mediates the establishment of embryonic left–right asymmetry. *Nature* **401**, 243–251.

Hanaoka, E., Ozaki, T., Nakamura, Y., Moriya, H., Nakagawara, A., and Sakiyama, S. (2000). Overexpression of DAN causes a growth suppression in p53-deficient SAOS-2 cells. *Biochem. Biophys. Res. Commun.* **278**, 20–26.

Hogan B. L. M. (1996). Bone morphogenetic proteins: Multifunctional regulators of vertebrate development. *Genes Dev.* **10**, 1580–1594.

Hsu, D. R., Economides, A. N., Wang, X., Eimon, P., and Harland, R. M. (1998). The *Xenopus* dorsalizing factor gremlin identifies a novel family of secreted proteins that antagonize BMP activities. *Mol. Cell* **1**, 673–683.

Kondo, K., Ozaki, T., Nakamura, Y., and Sakiyama, S. (1995). DAN gene product has an affinity for Ni^{2+} *Biochem. Biophys. Res. Commun.* **216**, 209–215.

Lee, S. W., Tomasetto, C., and Sager, R. (1991). Positive selection of candidate tumor-suppressor genes by subtraction hybridization. *Proc. Natl. Acad. Sci. U.S.A.* **88**, 2825–2829.

McDonald N. Q., and Hendrickson, W. A. (1993). A structural superfamily of growth factors containing a cysteine knot motif. *Cell* **73**, 421–424.

Massagué, J., and Chen, Y.-G. (2000). Controlling TGF-β signaling. *Genes Dev.* **14**, 627–644.

Minabe-Saegusa, C., Saegusa, H., Tsukahara, M., and Noguchi, S. (1998). Sequence and expression of a novel mouse gene PRDC (protein related to DAN and Cerberus) identified by a gene trap approach. *Dev. Growth Differ.* **40**, 343–353.

Nakamura, Y., Ozaki, T., Nakagawara, A., and Sakiyama, S. (1997). A product of *DAN*, a novel candidate tumor suppressor gene, is secreted into the culture medium and suppresses DNA synthesis. *Eur. J. Cancer* **33**, 1986–1990.

Nakamura, Y., Ozaki, T., Ichimiya, S., Nakagawara, A., and Sakiyama, S. (1998). Ectopic expression of DAN enhances the retinoic acid-induced neuronal differentiation in human neuroblastoma cell lines. *Biochem. Biophys. Res. Commun.* **24**, 722–726.

Ozaki, T., and Sakiyama, S. (1993). Molecular cloning and characterization of a cDNA showing negative regulation in v-src-transformed 3Y1 rat fibroblasts. *Proc. Natl. Acad. Sci. U.S.A.* **90**, 2593–2597.

Ozaki, T., and Sakiyama, S. (1994). Tumor-suppressive activity of N03 gene product in v-src-transformed rat 3Y1 fibroblasts. *Cancer Res.* **54**, 646–648.

Ozaki, T., and Sakiyama, S. (1997). Identification of essential cis-acting regulatory elements for transcription of the rat *DAN* gene. *DNA Cell Biol.* **16**, 779–786.

Ozaki, T., Nakamura, Y., Enomoto, H., Hirose, M., and Sakiyama, S. (1995). Overexpression of DAN gene product in normal rat fibroblasts causes a retardation of the entry into the S phase. *Cancer Res.* **55**, 895–900.

Ozaki, T., Ma, J., Takenaga, K., and Sakiyama, S. (1996). Cloning of mouse DAN cDNA and its down-regulation in transformed cells. *Jpn. J. Cancer Res.* **87**, 58–61.

Ozaki, T., Enomoto, H., Nakamura, Y., Kondo, K., Seki, N., Ohira, M., Nomura, N., Ohki, M., Nakagawara, A., and Sakiyama, S. (1997). The genomic analysis of human *DAN* gene. *DNA Cell Biol.* **16**, 1031–1039.

Pearce, J. J. H., Peny., G., and Rossant. J. (1999). A mouse cerberus/Dan-related gene family. *Dev. Biol.* **209**, 98–110.

Peverali, F. A., Orioli, D., Tonon, L., Ciana, P., Bunone, G., Negri, M., and Della-Valle, G. (1996). Retinoic acid-induced growth arrest and differentiation of neuroblastoma cells are counteracted by N-myc and enhanced max overexpression. *Oncogene* **12**, 457–462.

Rodriguez-Leon, J., Merino, R., Macias, D., Ganan, Y., Santesteban, E., and Hurle, J. M. (1999). Retinoic acid regulates programmed cell death through BMP signaling. *Nat. Cell Biol.* **1**, 125–126.

Sager, R. (1989). Tumor suppressor gene: The puzzle and the promise. *Science* **246**, 1406–1411.

Schwab, M., Praml, C., and Amler, L. C. (1996). Genomic instability in 1p and human malignancies. *Genes Chromosomes Cancer* **16**, 211–229.

Stanley, E., Biben, C., Kotecha, S., Febri, L., Tajbakhsh, S., Wang, C.-C., Hatzistavrou, T., Roberts, B., Drinkwater, C., Lah, M., Buckingham, M., Hilton, D., Nash, A., Mohun, T., and Harvey, R. P. (1998). DAN is a secreted glycoprotein related to *Xenopus cerberus*. *Mech. Dev.* **77**, 173–184.

Topol, L. Z., Marks, M., Laugier, D., Bogdanova, N. N., Boubnov, N. V., Clausen, P. A., Calothy, G., and Blair, D. G. (1997). Identification of *drm*, an novel gene whose expression is suppressed in transformed cells and which can inhibit growth of normal but not transformed cells in culture. *Mol. Cell. Biol.* **17**, 4801–4810.

Topol, L. Z., Bardot, B., Zhang, Q., Resau, J., Huillard, E., Marx, M., Calothy, G., and Blair, D. G. (2000). Biosynthesis, post-translational modification, and functional characterization of Drm/Gremlin. *J. Biol. Chem.* **275**, 8785–8793.

Weinberg, R. A. (1991). Tumor suppressor gene. *Science* **254**, 1138–1146.

White, P. S., Maris, J. M., Beltinger, C., Sulman, E., Marshall, H., Fujimori, M., Kaufman, B. A., Biegel, J. A., Allen, C., Hilliard, C., Valentine, M. B., Look, T., Enomoto, H., Sakiyama, S., and Brodeur, G. M. (1995). A region of consistent deletion in neuroblastoma maps within 1p36.2-.3. *Proc. Natl. Acad. Sci. U.S.A.* **92**, 5520–5524.

Design of Hammerhead Ribozymes and Allosterically Controllable Maxizymes for Cancer Gene Therapy

12

Hiroaki Kawasaki, Tomoko Kuwabara, and Kazunari Taira[1]
Department of Chemistry and Biotechnology
Graduate School of Engineering, The University of Tokyo
Hongo, Tokyo 113-8656, Japan, and
Gene Discovery Research Center
National Institute of Advanced Industrial Science and Technology (AIST)
Tsukuba Science City 305-8562, Japan

I. INTRODUCTION

Hammerhead ribozymes are one of the smallest catalytic RNA molecules and have a high specificity for their target RNAs. The sequence motif, with three duplex stems and a conserved core of two nonhelical segments that are responsible for the self-cleavage reaction (cis action), was first recognized in the

---

[1]Correspondence should be addressed to Prof. Kazunari Taira at the following address: Department of Chemistry and Biotechnology, Graduate School of Engineering, The University of Tokyo, Hongo, Tokyo 113-8656, Japan.

Tumor-Suppressing Viruses, Genes, and Drugs
Innovative Cancer Therapy Approaches

satellite RNAs of certain viruses (Symons, 1992). The ribozymes can act "in trans" (Uhlenbeck, 1987; Haseloff and Gerlach, 1988), and trans-acting hammerhead ribozymes, consisting of an antisense section (stems I and stem III) and a catalytic core with a flanking stem–loop II section (Figure 1A), have been tested as potential therapeutic agents and used extensively in mechanistic studies (Zhou and Taira, 1998; Takagi *et al.*, 2001). The ribozymes can cleave RNAs at specific sites (namely, after the sequence NUX, where N and X can be A, G, C, or U and A, C, or U, respectively, with the most efficient cleavage occurring after a GUC triplet (Ruffner *et al.*, 1990; Shimayama *et al.*, 1995). In the case of trans-acting hammerhead ribozymes, most of the conserved nucleotides that are essential for the cleavage reaction are included in the catalytic core. Therefore, RNA molecules consisting of only about 30 nucleotides can be generated for use as artificial endonucleases that can cleave specific RNA molecules. Because of their small size and considerable potential as inhibitors of gene expression, hammerhead ribozymes are used most frequently in studies directed toward therapeutic applications *in vivo* (Sarver *et al.*, 1990; Krupp and Gaur, 2000; Kuwabara *et al.*, 2000).

A cancer is a multistep process involving sequential genetic activations and inactivations of certain key elements involved in cell proliferation, differentiation, and apoptosis. Oncogenes are often overexpressed or mutated in the signal transduction pathway, leading to uncontrolled cell growth. The majority of ribozyme gene therapy of cancer has focused on inhibition of specific oncogene expression in tumor cells. Several oncogenes such as K-*ras* (Tokunaga *et al.*, 2000; Zhang *et al.*, 2000; Funato *et al.*, 2000; Tsuchida *et al.*, 2000), c-*myc* (Cheng *et al.*, 2000), *neu* (Suzuki *et al.*, 2000), *bcl-2* (Gibson *et al.*, 2000), *MDR1* (Kobayashi *et al.*, 1999; Wang *et al.*, 1999), *hTERT* (Yokoyama *et al.*, 2000), and the RNA component of telomerase (Folini *et al.*, 2000) were made to be a target for the ribozymes, and these ribozymes, were tested in several cancer cell lines.

However, in general, it is still difficult to construct effective ribozymes *in vivo*. Some of the reasons for this ineffectiveness *in vivo* are as follows.

Figure 1 The secondary structures of tRNA^{Val}-ribozymes that were predicted by computer folding. (A) The secondary structure of the parental hammerhead ribozyme consists of an antisense section and a catalytic domain with a flanking stem–loop II section. (B) The secondary structure of the wild-type tRNA^{Val}. (C) The Rz-A to Rz-E are ribozymes targeted against p300 mRNA. The sequence of the hammerhead ribozyme (bold capital letters) was ligated downstream of that of the seven-base-deleted tRNA^{Val} (capital letters) with various linker sequences. The sequences that correspond to the internal promoter of tRNA^{Val}, namely, the A and B boxes, are indicated by shaded boxes. The recognition arms of ribozymes are indicated by underlining.

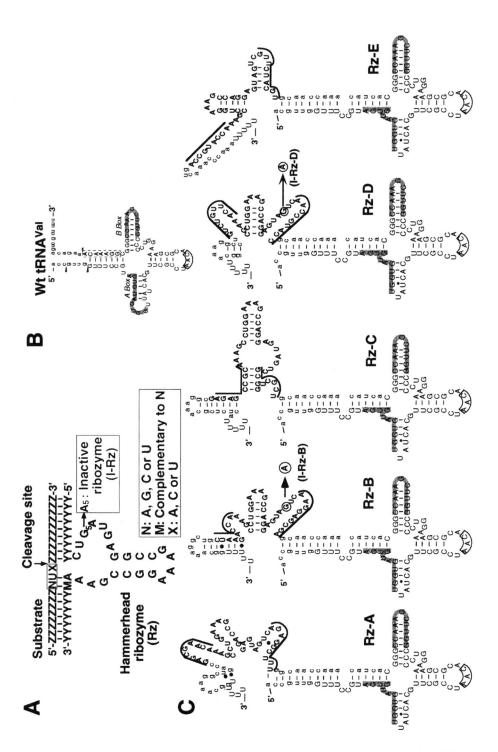

(1) Cellular proteins may inhibit the binding of the ribozyme to its target RNA or may disrupt the active conformation of the ribozyme. (2) The intracellular concentration of the metal ions that are essential for ribozyme-mediated cleavage might not be sufficient for functional activity. (3) Ribozymes are easily attacked by RNases. (4) Transcribed ribozymes remain inactive because of improper folding and/or inability to colocalize with their target. Moreover, studies *in vivo* have suggested that the following factors are important for the effective ribozyme-mediated inactivation of genes: a high level of ribozyme expression (Yu *et al.*, 1993); the intracellular stability of the ribozyme (Rossi and Sarver, 1990; Eckstein and Lilley, 1996); colocalization of the ribozyme and its target RNA in the same cellular compartment (Sullenger *et al.*, 1990; Koseki *et al.*, 1999; Bertrand *et al.*, 1994, 1997; Kato *et al.*, 2001); and the cleavage activity of the transcribed ribozyme (Thompson *et al.*, 1995).

In this chapter, we describe the design of effective hammerhead ribozymes and allostetrically controllable maxizymes for a cancer gene therapy.

II. RIBOZYME EXPRESSION SYSTEM IN CELLS

A. RNA polymerase II dependent expression system

The polymerase II (pol II) system is the system that is normally used for the expression of proteins in cells. This type of expression system was adopted in early experiments for the expression of ribozymes. In attempts to express functionally active ribozymes, strong and constitutive promoters, such as the cytomegalovirus (CMV) promoter, the simian virus 40 (SV40) promoter, and the promoter of the human gene for β-actin have been used successfully. However, the pol II promoter is not suitable for production of short RNAs. In the pol II system, at least several hundred nucleotides are necessary between the promoter and the terminator for effective transcription and termination at the correct site. Thus, either extra sequences must be added at both ends of the ribozyme sequence or, alternatively, the ribozyme sequence must be inserted into the noncoding region of a stable mRNA. Cameron and Jennings (1989) demonstrated that the ribozyme sequence targeted to the mRNA for chloramphenicol acetyl transferase (CAT) was introduced into the 3' noncoding region of the mRNA for luciferase, and the transcription was controlled by the SV40 early promoter. In this system, the expression of CAT was suppressed to 30% of the control level. By contrast, Sarver *et al.* (1990) used the promoter of the human gene for β-actin for expression of a ribozyme targeted to HIV-1 RNA. They achieved a considerable reduction in the rate of viral replication in cultured cells.

Although strong and constitutive promoters of the type mentioned above are effective in cultured cells, they may not always be useful under conditions when analysis depends on, for example, the stage of development of germ cells or the phase of the cell cycle. Under such conditions, a temporally controlled promoter is required. Thus, for example, Zhao and Pick (1993) used the promoter of a heat-shock gene to control expression of a ribozyme targeted to the transcript of the *fushi tarazu* gene. They generated transgenic flies that carried a heat-shock-inducible ribozyme-expression cassette, and they were able to examine the function of the *fushi tarazu* gene at each stage of development by appropriately timed induction of the ribozyme. Efrat *et al.* (1994) used the promoter of a gene for insulin, namely, a tissue-specific promoter that operates specifically in pancreatic cells, to regulate the expression of a ribozyme targeted to the transcript of a gene for glucokinase. However, such successful examples are rare. Because the efficiency of transcription from a conditionally controlled promoter is lower than that from a constitutive-type promoter, such as the CMV or SV40 promoter, it is often difficult to obtain an effective intracellular concentration of the ribozyme of interest.

B. RNA polymerase III dependent expression system

The polymerase III (pol III) system is used for transcription of small RNA molecules, such as tRNAs, U6 snRNA, and adenoviral-VA 1 RNA. Two important features of this system are that (1) minimal numbers of extra nucleotides are required at the 3′ and 5′ ends for production of properly initiated and terminated transcripts and (2) the level of expression is at least one order of magnitude higher than that obtained with the pol II system (Cotten and Birnstiel, 1989; Ohkawa and Taira, 2000). Indeed, in our studies, pol III-driven ribozymes but not pol II-driven ribozymes could be detected by Northern blotting analysis, providing evidence of the higher level of the former transcripts (Kuwabara *et al.*, 2000; Ohkawa and Taira, 2000). Thus, the pol III system appears to be particularly useful for the expression of ribozymes. In addition, in the case of the tRNA promoter and the VA 1 RNA promoter, because each expression unit is compact (less than 200 bp including the ribozyme sequence), the promoters are suitable for the expression of multiple ribozymes from a single vector. However, in many cases, the expected effects of ribozymes have not been achieved, in spite of the apparently favorable features of the pol III system.

Among the pol III-type promoters, the promoters of genes for tRNAs (tRNA promoters) have been well studied as controllers of the expression of ribozymes. Cotten and Birnstiel (1989) used a tRNA^{Met} promoter from

Xenopus, and they inserted the coding sequence of a ribozyme, targeted to U7 snRNA, into the anticodon loop of the tRNA. They coinjected radioisotopically labeled U7 snRNA and the ribozyme-expression plasmid into the nucleus of *Xenopus* oocytes. Twenty hours after the injection, the target RNA had completely vanished. The transcribed ribozyme had remained localized in the nucleus for the most part, whereas the target RNA had been rapidly exported to the cytoplasm. In this case, large amounts of ribozymes, as compared to the amount of target RNA (about 500- to 1000-fold molar excess of ribozyme over target RNA), were needed for cleavage of the target RNA. The Wong-Staal group used another strategy that exploited a tRNA promoter (Yu *et al.*, 1993; Yamada *et al.*, 1994). They used the human tRNA[Val] promoter and replaced the amino-acceptor stem sequence with a hairpin ribozyme sequence targeted against an HIV-1. Replication of various strains of HIV-1 was efficiently inhibited by a single ribozyme in cell culture. The Bertrand and Rossi groups used a human tRNA[Met] promoter for expression of a hammerhead ribozyme directed against HIV-1 (the target site was identical to that in the study by the Wong-Staal group). However, the effective inhibition of viral replication was not observed. The ribozymes had remained mainly in the cell nuclei (Bertrand *et al.*, 1997). Immature mRNA in the nucleus is decorated with splicesomes and other RNA-binding proteins, and mature mRNA is exported rapidly to the cytoplasm by the protein factors that serve as an export system. Therefore, a ribozyme might not be able to find its target site within an mRNA or even the target mRNA itself in the nucleus. Furthermore, the structure of mRNA that has already been exported to the cytoplasm can be loosened by the helicase activity of eIF4A (eukaryotic initiation factor) prior to binding of ribosomes. Thus, we might expect more efficient suppression of a specific gene if a ribozyme could be designed to attack its target mRNA in the cytoplasm (Koseki *et al.*, 1999; Kato *et al.*, 2001; Kuwabara *et al.*, 2001).

We also used a human tRNA[Val] promoter for expression of a hammerhead ribozyme (Kawasaki *et al.*, 1996, 1998a; Koseki *et al.*, 1999; Kuwabara *et al.*, 1998a,b, 1999, 2000, 2001; Tanabe *et al.*, 2000a,b). The ribozyme sequence was connected to the tRNA[Val] promoter sequence with a short intervening linker sequence (Figure 1B,C). We found that the putative structural differences influenced the cleavage activity *in vitro*, as well as the stability of the transcripts in cultured cells. The steady-state concentration of the ribozyme differed over a 30-fold range (Koseki *et al.*, 1999). Moreover, tRNA[Val]-ribozymes were exported to the cytoplasm (Kawasaki *et al.*, 1998a; Kuwabara *et al.*, 1998b; Koseki *et al.*, 1999). Thus, the ribozymes and their target were present within the same cellular compartment. Under these conditions, our cytoplasmic tRNA[Val]-ribozymes had significant activities in cultured cells (Kawasaki *et al.*, 1998a; Kuwabara *et al.*, 1998b, 1999, 2001; Koseki *et al.*, 1999; Kato *et al.*, 2001).

The importance of the cytoplasmic localization of ribozymes is also supported by the results of several studies in which the promoter of the gene for U6 snRNA (an RNA that is known to remain in the nucleus) was used for expression of ribozymes (Beck and Nassal, 1995; Koseki *et al.*, 1999; Ohkawa and Taira, 2000). In all cases examined, even though large amounts of the U6 snRNA-ribozymes were present in cells, no inhibition of expression of target genes occurred. The failure of the ribozymes to cleave their targets in cells might have been due to the propensity for nuclear localization of the U6 snRNA portion of the ribozyme transcript and, indeed, it was confirmed that the U6 snRNA-ribozyme remained localized mainly in the nucleus (Bertrand *et al.*, 1997; Ohkawa and Taira, 2000; Kato *et al.*, 2001).

A protein, designated Exportin-t (Xpo-t), which transports tRNA from the nucleus to the cytoplasm has been identified (Arts *et al.*, 1998, 1999; Kutay *et al.*, 1998). Xpo-t binds RanGTP in the absence of tRNA but it does not bind tRNA in the absence of RanGTP. Therefore, a model for the transport of tRNAs was proposed wherein Xpo-t first associates with RanGTP in the nucleus, and then the complex binds a mature tRNA molecule. The entire complex is then translocated through a nuclear pore to the cytoplasm, where the Ran-bound GTP is hydrolyzed, releasing the tRNA into the cytoplasm and allowing Xpo-t to be recycled to the nucleus. The minimal sequence or structure within a tRNA that can be recognized by Xpo-t is being clarified (Arts *et al.*, 1999; Lund and Dahlberg, 1999; Lipowsky *et al.*, 1999; Simos and Hurt, 1999). The 3'-terminal CCA region of the tRNA appears to be important for recognition by Xpo-t. With regard to the export of the tRNA^{Val}-ribozymes transcribed from our expression system, we cannot exclude the possibility that these tRNA^{Val}-ribozymes might be recognized by an export receptor similar to Xpo-t (Arts *et al.*, 1998, 1999; Kutay *et al.*, 1998; Lund and Dahlberg, 1999; Lipowsky *et al.*, 1999; Simos and Hurt, 1999) since all tRNA^{Val}-ribozymes that had a cloverleaf tRNA-like structure were found in the cytoplasm. Studies with *Xenopus* oocytes indicate that only tRNAs with mature 5' and 3' ends are recognized by Xpo-t and can be exported to the cytoplasm by Xpo-t (Arts *et al.*, 1999; Lund and Dahlberg, 1999; Lipowsky *et al.*, 1999). Clearly, our tRNA^{Val}-ribozymes have a ribozyme sequence at their 3' ends, but nevertheless, it appears that they can be exported to the cytoplasm in mammalian cells as long as their secondary structures retain a cloverleaf tRNA-like configuration (Kawasaki *et al.*, 1996, 1998a; Koseki *et al.*, 1999; Kuwabara *et al.*, 1998b, 1999, 2000, 2001). At present, we are trying to identify the export receptor that recognizes tRNA^{Val}-ribozymes in mammalian cells. The results from our laboratory indicate that a tRNA-like structure is important for the production of intracellularly active tRNA^{Val}-ribozymes (Kawasaki *et al.*, 1996, 1998a; Koseki *et al.*, 1999; Kuwabara *et al.*, 1998b, 1999, 2000, 2001; Tanabe *et al.*, 2000a,b).

III. DESIGN OF THE tRNA^{Val}-DRIVEN RIBOZYME THAT IS TRANSCRIBED BY pol III

A. Transcriptional coactivator p300 as a tumor suppressor protein

Transcriptional coactivators p300 and the related protein CBP (CREB binding protein) are functionally conserved proteins that act in concert with other factors to regulate a transcription (Giles *et al.*, 1998; Goodman and Smolik, 2000). These proteins participate in several tumor-suppressor pathways via interaction with tumor suppressor proteins such as the p53 (Avantaggiati *et al.*, 1997; Gu *et al.*, 1997; Lill *et al.*, 1997) or the Smad (Janknecht *et al.*, 1998; Feng *et al.*, 1998). In addition, monoallelic mutation of the CBP locus causes Rubinstein-Taybi syndrome (RTS) which is accompanied with a cancer (Petrij *et al.*, 1995). And biallelic inactivating somatic mutations of the *p300* gene have been observed in gastric and colon cancers (Muraoka *et al.*, 1996).

Embryonal carcinoma (EC) F9 cells can be induced to differentiate into three distinct extraembryonic types of cells on exposure to a retinoic acid (RA) that is known as an antitumor agent (Bernstine and Ephrussi, 1976; Strickland and Mahdavi, 1978; Strickland *et al.*, 1980). Concomitantly with the induction of the differentiated phenotype, the RA can also trigger an apoptosis (programmed cell death) and a cell cycle arrest in F9 cells (Atencia *et al.*, 1994; Clifford *et al.*, 1996). It was reported previously that p300 is involved in the regulated transcription of the *c-jun* gene during the RA-mediated differentiation of F9 cells (Kitabayashi *et al.*, 1995; Kawasaki *et al.*, 1998b, 2000). However, the functional significance of p300 protein in the effect of RA was still unknown. Thus, we decided to regulate the expression of *p300* gene by tRNA^{Val}-driven hammerhead ribozymes.

B. Construction of a test plasmid (p300-luc), for the selection of the best target site of the tRNA^{Val}-ribozymes, that encoded the amino-terminal region of p300 and luciferase

For the functional analysis of an unknown gene, in general, appropriate target sites have to be chosen first since some sites (e.g., those sites embedded within a stem structure) are inaccessible to a ribozyme. In order to suppress the expression of p300, we at first tried to identify the best target site within the p300 transcript for ribozyme-mediated cleavage. Because the phenotype of suppressed p300 clones cannot be used for the quantitation of the efficiency of ribozyme-mediated cleavage and, moreover, because the cleavage products, in general, cannot easily be detected *in vivo*, we constructed a rapid assay system, for the selection of the best target site, by fusing the amino-terminal region of the gene for p300 in frame

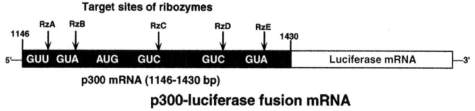

Figure 2 The location of the target sites relative to the gene for p300. Positions of target sites included the triplets of each ribozyme designated Rz-A to Rz-E, are indicated by arrows. Initiation codon (AUG) is indicated at nucleotide 1200.

with the gene for luciferase (Figure 2). The fused gene product was produced under the control of the SV40 promoter and its enhancer (Kawasaki *et al.*, 1996).

C. Construction of tRNA^{Val}-ribozyme expression plasmids

We once examined the generality of NUX rule, and we demonstrated from kinetic studies *in vitro* that, in reactions catalyzed by trans-acting hammerhead ribozymes, mutant substrates that contained the GUA or GUU triplet were cleaved much less efficiently than the wild-type substrate with the GUC triplet [GUG could not be cleaved (Shimayama *et al.*, 1995)]. In choosing potential target sites for the examination of the efficiency of their ribozyme-mediated cleavage *in vivo*, we made sure that each of the selected conserved sites contained one of all possible GUX triplets.

The five selected target sites are indicated in Figure 2, and the corresponding ribozymes are designated Rz-A to Rz-E in the upstream to the downstream direction (Figure 1C). In our pol III-driven ribozyme-expression cassette, in order to block 3'-end processing of the transcript without any negative effects on transcription, we first removed the last seven nucleotides from the mature tRNA^{Val} (Wt tRNA^{Val} in Figure 1B; removed nucleotides are indicated by capital letters in dashed box). The removed nucleotides were replaced by a linker (lowercase letters in Figure 1C). The overall secondary structures were adjusted to maintain the cloverleaf tRNA-like configuration by the selection of an appropriate linker sequence. The selected target sequence might also be expected to influence the overall structure of the transcripts and, naturally, it is important that the 5' and 3' substrate-recognition arms of the ribozyme be available to the substrate so that the ribozyme and substrate RNA together can form the stem structures that ensure subsequent cleavage of the substrate. Thus, to ensure the accessibility of the recognition arms of the ribozyme (recognition arms are under-lined), adjustments were made, when necessary, by properly choosing appropriate linker sequences. Each ribozyme had the same 24-nucleotide-long catalytic domain, and each was equipped with nine bases on both substrate-binding arms (recognition arms are underlined) that were targeted to the relatively well-conserved sequences of p300 mRNA. The flexibility of the substrate-recognition arm of ribozymes differed depending on the specific target site of each ribozyme.

The inactive ribozyme (designated I-Rz) differed from the active ribozyme by a single G_5-to-A_5 mutation in the catalytic core [the numbering system follows the rule for hammerhead ribozymes (Hertel *et al.*, 1992)]. This single-base change should diminish the cleavage activity while the antisense effect, if any, should be unaffected (Ruffner *et al.*, 1990; Inokuchi *et al.*, 1994).

D. Suppression of expression of the p300-luc fusion gene: Results of transient transfection assays

The effects of tRNA^{Val}-ribozymes on the specific target mRNA, namely, the transcript of the p300-luc fusion gene, were examined by measuring the luciferase activity of HeLa S3 cells that had been cotransfected with the plas-mids that encoded tRNA^{Val}-ribozymes and the p300-luciferase. To determine the effectiveness of ribozyme-mediated inhibition of expression of the p300-luc fusion gene in the transient expression assay, a control plasmid, namely, pUC-tRVP from which the ribozyme sequence had been deleted, was used to allow generation of the luciferase activity that was designated 100%. As shown in Figure 3, all the active ribozymes were capable of decreasing the luciferase activity *in vivo*. The extent of inhibition by the ribozyme expressed from the pol III promoter varied from 47 to 96% when the molar ratio of template DNAs for the target and the ribozyme, was only 1:4. However, the inactive ribozyme-coding plasmid did not have any inhibitory effect (average inhibition was 4%).

The order of efficiency of cleavage was as follows: Rz-B (GUA) > Rz-E (GUA) > Rz-A (GUU) > Rz-D (GUC) > Rz-C (GUC), where the target triplet is indicated in parenthesis. These results would not have been predicted from the cleavage activity of NUX triplets *in vitro* (Shimayama *et al.*, 1995), and they sug-gest, in turn, that the accessibility of the target site and/or the structure of the transcripts govern the effectiveness of the ribozyme *in vivo* (Kato *et al.*, 2001; Warashina *et al.*, 2001). With regard to the flexibility of the substrate-recognition arm of the ribozymes, the most effective Rz-B in this assay has high flexibility in the substrate-recognition arms. However, the least effective Rz-C in this assay

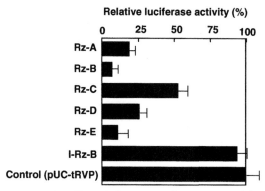

Figure 3 The effects of ribozymes on the expression of the p300-luc fusion gene in HeLa S3 cells cotransfected with p300-Rz encoding plasmids at 4 μg and p300-luc at 2 μg per dish. The results shown are averages of results from five sets of experiments and are given as percentages relative to the control value of 100% (pUC-tRVP).

formed a stem structure between the 3′ and 5′ substrate-recognition arms. Importantly, a system for the rapid evaluation of ribozyme activity *in vivo*, such as the one described herein, is of obvious value.

E. Role of tumor suppressor p300 in the effect of RA in F9 cells

Since we obtained tRNA^{Val}-ribozymes, targeted to p300 mRNA, which have high activity, we next investigated the function of tumor supressor p300 in F9 cells. Among the several active tRNA^{Val}-ribozymes tested (Figure 3), we chose Rz-D to generate tRNA^{Val}-ribozyme expressing stable lines of F9 cells because this specific ribozyme can be used not only in human cells but also in murine cells including the F9 cells (all ribozymes listed in Figure 1C were targeted to human p300 mRNA). The concentrations of p300 transcripts from these cell lines were measured by competitive reverse transcription–polymerase chain reaction (RT–PCR). The level of p300 mRNA was drastically reduced from 1×10^{-2} to less than 1×10^{-4} attomole/μl in F9 cells expressing an active p300-ribozyme, Rz-G (Kawasaki et al., 1998a). In the same cells, the level of homologous CBP transcripts was unchanged. An inactive ribozyme, I-Rz-D, targeted to p300 mRNA did not alter mRNA levels for p300 or CBP. These data indicate that the employed tRNA^{Val}-ribozymes acted specifically on p300 transcripts despite the high homology of p300 and CBP mRNAs within conserved regions. Compared

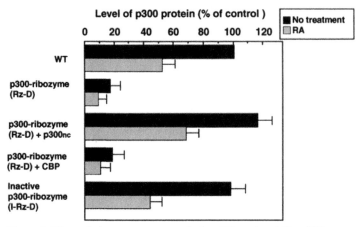

Figure 4 Effects of ribozyme vectors specific for p300 on the p300 or CBP protein
levels. Relative levels of p300 proteins were compared by an amplified
sandwich ELISA as described elsewhere (Kawasaki *et al.*, 1998a). Nor-
malized levels of p300 in WT F9 cells were taken arbitrarily as 100%
(A450 nm for p300 was 0.605). All values are the average of results from
at least four experiments, and the standard deviation for each value is
indicated.

to the inactive tRNA^{Val}-ribozyme expressing F9 cell lines, p300 mRNAs were
reduced about 100-fold in the active tRNA^{Val}-ribozyme expressing cell lines.

We next examined the levels of p300 protein. In F9 cells expressing
the active p300-ribozyme, the protein levels of p300 were at least sixfold lower
than that in F9 cells harboring the inactive p300-ribozyme, or in WT F9 cells
(Figure 4).

We then asked whether F9 cells lacking p300 can still respond to treatment with RA. When challenged with RA, F9 cells harboring the p300-ribozyme
failed to show morphological signs of differentiation. These cells continued to
express high levels of the cell-surface antigen SSEA-1 (Solter and Knowles, 1978),
which is characteristic for undifferentiated F9 cells, and did not induce synthesis of
collagen IV and laminin B1, two proteins specific for differentiated cells (Strickland
and Mahdavi, 1978; Strickland *et al.*, 1980). WT F9 cells and F9 cells containing
the inactive p300-ribozyme downregulated expression of SSEA-I after RA treatment, and upregulated collagen IV. Prolonged incubation with RA of F9 cells
expressing the p300-ribozyme (up to 10 days) did not result in enhanced differentiation (Kawasaki *et al.*, 1998a). These results demonstrate that p300 plays a critical
role in mediating RA-induced differentiation of F9 cells.

Apart from differentiating, a fraction of F9 cells treated with RA dies by
apoptosis (Atencia *et al.*, 1994; Clifford *et al.*, 1996). To explore the dependence

of this process on p300, we evaluated the occurrence of nicked DNA [by the TUNEL assay (Gavrieli *et al.*, 1992)], as the hallmarks of apoptosis. F9 cells harboring the p300-ribozyme were much less sensitive to RA-induced nicking of DNA than were WT cells or those containing the inactive ribozyme (Kawasaki *et al.*, 1998a). These results demonstrate that RA-induced apoptosis requires intact p300 expression.

Treatment of F9 cells with RA decreases the rate of proliferation of F9 cells (Dean *et al.*, 1986). The consequence of decreasing p300 protein levels on the rate of cell proliferation was assessed by BrdU incorporation. In the absence of RA, the basal rate of proliferation of F9 cells expressing the p300-ribozyme was almost the same as that of WT F9 cells. Surprisingly, exposure to RA failed to reduce proliferation of F9 cells expressing the p300-ribozyme (Kawasaki *et al.*, 1998a).

To gain insights into the molecular mechanisms underlying the failure of cells harboring ribozymes to reduce proliferation rates, the expression of inhibitors of cyclin-dependent kinases (cdk), such as $p21^{Cip1}$, was analyzed. The $p21^{Cip1}$ expression was induced during differentiation of many cell types, including F9 cells (WT), and certain events triggering cell-cycle arrest (Sherr and Roberts, 1995). Consistent with the failure of the p300-ribozyme containing cells to differentiate and decrease proliferation, $p21^{Cip1}$ levels were not upregulated in these cells after RA treatment, whereas expression of another cdk inhibitor $p27^{Kip1}$ was normal in those expressing the p300-ribozyme (Kawasaki *et al.*, 1998a). These results indicate that the induced levels of $p27^{Kip1}$ appear to be insufficient for triggering cell-cycle arrest. A differential sensitivity of $p21^{Cip1}$ and $p27^{Kip1}$ expression to reduced p300 protein levels was also observed in p300-ribozyme expressing F9 cells. Unlike with the above two cdk inhibitors, the level of $p15^{INK4b}$ protein (cdk inhibitor) did not vary among the various F9 cell lines. Thus, expression of $p15^{INK4b}$ appears to be independent of p300.

We demonstrated in this section, by regulating the expression of p300 by $tRNA^{Val}$-ribozymes, that p300 has important functions during RA-induced differentiation and cessation of cell proliferation. An unexpected finding was that reduced levels of p300 do no longer support differentiation of F9 cells. This observation suggests that p300 may control the expression of target genes which are necessary for RA-induced differentiation. The identical conclusion was made based on analysis of mice lacking p300 (Yao *et al.*, 1998). In support of this view, the cdk inhibitors $p21^{Cip1}$ and $p27^{Kip1}$ were differently affected in F9 cells expressing the p300-ribozyme. Induction of $p21^{Cip1}$ expression was dependent on intact p300 levels, whereas elevated levels of $p27^{Kip1}$ were neither necessary nor sufficient for differentiation: induction of $p21^{Cip1}$ correlated with F9 cell differentiation. Previous studies have shown similar correlations (Deng *et al.*, 1995; Sherr and Roberts, 1995). Because mice lacking $p21^{Cip1}$ exhibited normal differentiation

(Deng *et al.*, 1995), the inability of F9 cells with reduced p300 levels to differentiate is likely due to altered expression of additional target genes which depend on an intact p300 protein level.

For apoptosis and full G1 arrest, p300 is required. Since tumor suppressor p53 is often involved in these two processes, the finding that p300 and CBP are directly linked to the activity of p53 (Avantaggiati *et al.*, 1997; Gu *et al.*, 1997; Lill *et al.*, 1997) is very intriguing. In particular, a dominant negative p300 was shown to significantly inhibit both p53-mediated apoptosis and irradiation-induced cell-cycle arrest (Avantaggiati *et al.*, 1997). Thus, the failure to correctly activate p53 may explain in part the resistance of p300-ribozyme containing cells to the apoptosis and the full G1 arrest.

In summary, these results demonstrate that in the context of regulation of cell growth and differentiation, p300 has an important role in the effect of RA. Therefore, properly designed tRNA^{Val}-ribozymes appeared to be very useful as tools in molecular biology, with potential utility for a cancer gene therapy as well.

IV. DESIGN OF ALLOSTERICALLY CONTROLLED MAXIZYMES

A. Discovery of maxizymes that act as a dimeric form of short ribozymes

For the development of chemically synthesized ribozymes as potential therapeutic agents, it would certainly be advantageous to remove any surplus nucleotides that are not essential for catalytic activity. Removal of surplus nucleotides would obviously reduce the cost of synthesis, increase the overall yield of the desired polymer, and simplify purification. These considerations led to the production of short ribozymes (minizymes), namely, conventional hammerhead ribozymes with a deleted stem−loop II region (McCall *et al.*, 1992; Tuschl and Eckstein, 1993; Fu *et al.*, 1994; Long and Uhlenbeck, 1994). However, the activities of the minizymes turned out to be two to three orders of magnitude lower than those of the parental hammerhead ribozymes, a result that led to the suggestion that minizymes might not be suitable as gene-inactivating agents (Long and Uhlenbeck, 1994).

We tried to create variants of hammerhead ribozymes with deletions in the stem−loop II region (Figure 5A) and, fortunately, we found that some shortened forms of hammerhead ribozymes had high cleavage activity that was similar to that of the wild-type parental hammerhead ribozyme (Amontov and Taira, 1996). Moreover, the active species appeared to form dimeric structures with a common stem II (Figure 5B,C). In the active short ribozymes, the linker sequences that replaced the stem−loop II region were palindromic so that two

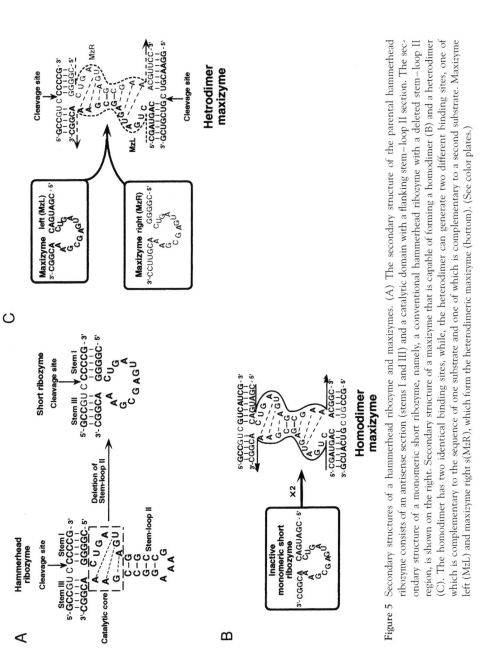

Figure 5 Secondary structures of a hammerhead ribozyme and maxizymes. (A) The secondary structure of the parental hammerhead ribozyme consists of an antisense section (stems I and III) and a catalytic domain with a flanking stem–loop II section. The secondary structure of a monomeric short ribozyme, namely, a conventional hammerhead ribozyme with a deleted stem–loop II region, is shown on the right. Secondary structure of a maxizyme that is capable of forming a homodimer (B) and a heterodimer (C). The homodimer has two identical binding sites, while, the heterodimer can generate two different binding sites, one of which is complementary to one substrate and one of which is complementary to a second substrate. Maxizyme left (MzL) and maxizyme right s(MzR), which form the heterodimeric maxizyme (bottom). (See color plates.)

247

short ribozymes were capable of forming a dimeric structure with a common stem II. We named it the *maxizyme* that was initially chosen as an abbreviation of *m*inimized, *a*ctive, *x*-shaped (heterodimeric), and *i*ntelligent (allosterically controllable) ribo*zyme* (Kuwabara *et al.*, 1998b, 1999, 2000; Tanabe *et al.*, 2000a,b). The activity of the homodimeric maxizyme (a dimer with two identical binding sequences) depended on Mg^{2+} and, in addition, interactions with the substrates also stabilized the dimeric structures. In addition, we tried to obtain more direct evidence for the existence of homodimeric maxizymes by NMR spectroscopy (Kuwabara *et al.*, 1998a). The observations suggested that the major population of maxizymes was in a dimeric form even in the absence of the substrate.

We next synthesized heterodimeric maxizymes, in which one maxizyme left (MzL) and one maxizyme right (MzR) were the monomers that together formed one heterodimeric maxizyme, as shown in Figure 5C. Such a heterodimeric maxizyme has two independent substrate-recognition arms and can cleave its substrates only when the individual maxizymes form a heterodimer (Figure 5C). Increases in the length of the common stem II region were associated with increases in the activity of the heterodimeric maxizymes *in vitro* (Kuwabara *et al.* 1996; Nakayama *et al.*, 1999). Moreover, we found that the cleavage rate of one of the substrates by the heterodimeric maxizyme increased linearly with increases in the concentration of the other substrate, a clear demonstration that the formation of the dimer was essential for cleavage of the substrate and, moreover, that the dimeric structure could be stabilized in the presence of the substrates.

B. Construction of tRNA^{Val}-embedded maxizymes

For the application of maxizymes to gene therapy, they must be expressed constitutively *in vivo* under the control of a strong promoter. As described in previous sections, we succeeded in establishing an effective ribozyme-expression system that was based on the pol III promoter. High-level expression under control of the pol III promoter would be advantageous for maxizymes and enhance the likelihood of their dimerization. Therefore, we chose the expression system with the promoter of a human gene for tRNA^{Val}, which we had used successfully in the suppression of target genes by conventional ribozymes, as discussed above. We embedded maxizymes similarly in the 3′-modified side of the tRNA^{Val} portion of the human gene.

In the case of the tRNA^{Val}-embedded maxizymes, we worried initially about the possibility that the tRNA^{Val} portion might cause severe steric hindrance that would inhibit dimerization, with resultant production of monomers with extremely low activity. To our surprise, the tRNA^{Val}-embedded maxizymes still had significant cleavage activity *in vitro* (Kuwabara *et al.*, 1998a). We performed modeling studies to confirm that the formation of dimeric structures by tRNA^{Val}-

embedded homodimeric maxizymes was theoretically feasible (Kuwabara *et al.*, 1998a). The resultant model structure appeared feasible, and steric hindrance by the two tRNA moieties seemed not to be a problem. More direct evidence for the formation of dimers was provided by gel-shift analysis in the absence of the substrate.

C. A novel allosterically trans-activated ribozyme (maxizyme) that acts as an exceptionally specific inhibitor of gene expression

1. Chronic myelogenous leukemia and the potential for ribozyme therapy

In some cytogenetic abnormalities, such as certain leukemias, chimeric fusion mRNAs connecting strange exons result from reciprocal chromosomal translocations and cause abnormalities (Nowell and Hungerford, 1960). For the design of ribozymes that can disrupt such chimeric RNAs, it is necessary to target the junction sequence. Otherwise, normal mRNAs that share part of the chimeric RNA sequence will also be cleaved by the ribozyme, with resultant damage to the host cells.

Chronic myelogenous leukemia (CML) is a clonal myeloproliferative disorder of hematopoietic stem cells that is associated with the Philadelphia chromosome. The reciprocal chromosomal translocation t(9; 22) (q34; q11) (K28 translocations and L6 translocations) results in the formation of the *BCR–ABL* fusion genes that encode two types of mRNA: b3a2 (consisting of *BCR* exon 3 and *ABL* exon 2) and b2a2 (consisting of *BCR* exon 2 and *ABL* exon 2). Both of these mRNAs are translated into a protein of 210 kDa ($p210^{BCR–ABL}$) which is unique to the malignant phenotype (Konopka *et al.*, 1984). In the case of the b2a2 sequence, there are no triplet sequences that are potentially cleavable by hammerhead ribozymes within two or three nucleotides of the junction in question. GUC triplets are generally the most susceptible to cleavage by a hammerhead ribozyme, and one such triplet is located 45 nucleotides from the junction. If this GUC triplet were cleaved by a ribozyme, normal *ABL* mRNA that shares part of the sequence of the abnormal *BCR–ABL* mRNA would also be cleaved by the ribozyme, with resultant damage to host cells. In designing ribozymes that might cleave b2a2 mRNA, we must be sure to avoid cleavage of normal *ABL* mRNA.

Previous attempts at the cleavage of *BCR–ABL* (b2a2) mRNA have involved a combination of a long antisense arm and a ribozyme sequence for generation of specificity (Pachuk *et al.*, 1994; James *et al.*, 1996). However, we demonstrated that the antisense-type of ribozyme cleaved normal *ABL* mRNA nonspecifically both *in vitro* and *in vivo* (Kuwabara et al., 1998b), most probably because hammerhead ribozymes can cleave their substrates even when the binding arms are as little as three nucleotides in length (Hertel *et al.*, 1996; Birikh *et al.*, 1997).

2. Design of an allosterically controllable maxizyme and its specificity of the cleavage of the chimeric $BCR-ABL$ substrate *in vitro*

In extending our studies of dimeric maxizymes, we tried to create an allosterically controllable ribozyme, based on the heterodimeric maxizyme, that would have cleavage activity only when the target sequence of interest was present. Such a ribozyme should be controlled such that it is active only in the presence of an abnormal RNA sequence, for example, the junction sequence of a chimeric mRNA. We first thought about using one of the two substrate-binding regions of the heterodimeric maxizyme as sensor arms. Then, since one domain of the maxizyme was to be used solely for recognition of the target sequence of interest, we deleted its catalytic core completely to generate an even smaller monomeric unit (Figure 6A, right). Our goal was to control the activity of maxizymes allosterically by introducing sensor arms so that, only in the presence of the correct target sequence of interest, would the maxizyme be able to create a cavity for the capture of catalytically indispensable Mg^{2+}. We wanted to ensure that the active structure of the maxizyme with a Mg^{2+}-binding pocket would be formed only in the presence of the sequence of interest.

In order to achieve high substrate-specificity, we wanted our maxizyme to adopt an active conformation exclusively in the presence of the abnormal $BCR-ABL$ junction (Figure 6B), with the maxizyme remaining in its inactive conformations in the presence of normal ABL mRNA and in the absence of the abnormal $BCR-ABL$ junction (Figure 6C). The specifically designed sequences, which are shown in Figure 6B (note that the lengths and sequences of sensor arms and those of common stem II are the variables), should ensure such conformations in the presence and in the absence, respectively, of the abnormal b2a2

Figure 6 Design of the novel allosteric maxizyme. (A) Schematic representation for the specific cleavage of chimeric mRNA by the tRNA^{Val}-driven maxizyme. The heterodimer [MzL (maxizyme left) and MzR (maxizyme right)] can generate two different binding sites: one is complementary to the sequence of a substrate, the other is complementary to a second substrate (left structure). One of the catalytic cores of the heterodimer can be deleted completely to yield the even smaller maxizyme (right structure) in that the substrate-recognition sequences recognize the abnormal chimeric junction, acting as sensor arms. In order to achieve high substrate-specificity, the tRNA^{Val}-driven maxizyme should be in an active conformation only in the presence of the abnormal $BCR-ABL$ junction (B), while the conformation should remain inactive in the presence of normal ABL mRNA or in the absence of the $BCR-ABL$ junction (C). (See color plates.)

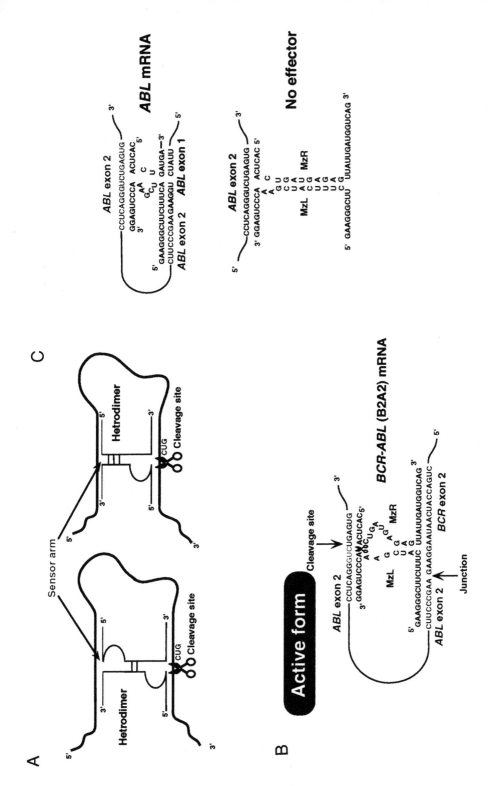

mRNA. This phenomenon would resemble the changes in conformation of allosteric proteinaceous enzymes that occur in response to their effector molecules. Later, the term of maxizyme was used simply to distinguish monomeric forms of conventional minizymes that have extremely low activity from the novel dimeric short ribozyme with high-level activity.

In order to prove *in vitro* that changes in the conformation of our heterodimeric maxizyme depended on the presence or absence of the abnormal b2a2 mRNA, specificity was examined by incubating the maxizyme, which had been transcribed *in vitro*, with the 5'-^{32}P-labeled short 16-nucleotide substrate (S16) that corresponded to the target (cleavage) site in the presence and in the absence of either a 20-nucleotide normal *ABL* effector molecule or a 28-nucleotide *BCR–ABL* effector molecule. Indeed, no products of cleavage of substrate S16 were detected in the absence of the *BCR–ABL* junction or in the presence of the normal *ABL* sequence (effector molecule), demonstrating the expected high substrate-specificity of the maxizyme. These results proved that the maxizyme was subjected to complete allosteric control *in vitro*, in accord with the predicted conformational changes that should occur in response to the effector molecule (the *BCR–ABL* junction) that was added in trans.

3. The intracellular activity and specificity of the maxizyme against an endogenous *BCR–ABL* cellular target in mammalian cells

We next examined the activity of the maxizyme against an endogenous *BCR–ABL* (b2a2 mRNA) target in mammalian cells. For this purpose, we established a line of murine cells, BaF3/p210$^{BCR–ABL}$, that expressed human b2a2 mRNA constitutively, by integrating into the murine genome a plasmid construct that expressed p210$^{BCR–ABL}$ (p210$^{BCR–ABL}$ was generated from human b2a2 mRNA). Although the parental BaF3 cell line is an interleukin-3-dependent (IL-3-dependent) hematopoietic cell line, our transformed BaF3/p210$^{BCR–ABL}$ cells were IL-3-independent because of the tyrosine kinase activity of p210$^{BCR–ABL}$ and, thus, the latter transformed cells were able to grow in the absence of IL-3. However, if the expression of p210$^{BCR–ABL}$ were to be inhibited, BaF3/p210$^{BCR–ABL}$ cells should become IL-3-dependent and, in the absence of IL-3, they should undergo apoptosis. In order to examine the specificity of the maxizyme, we also used H9 cells, which originated from human T cells and expressed normal *ABL* mRNA, as control cells.

These cell lines led us to examine whether the maxizyme could play a role in the regulation of apoptosis. We transfected BaF3 cells that stably expressed *BCR–ABL* (b2a2) mRNA with plasmids that encoded the wild-type

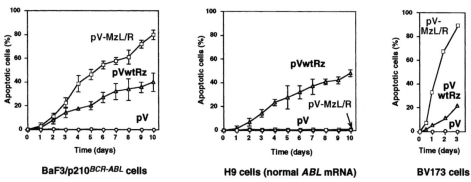

Figure 7 Efficiency of cleavage by the maxizyme of the endogenous *BCR–ABL* mRNA target. Measurements of viability of tRNA^{Val}-enzyme-transduced BaF3/p210$^{BCR-ABL}$ cells and H9 cells. The viability of BV173 cells, that were derived from a patient with a Philadelphia chromosome and that were transiently expressing tRNA^{Val}-enzymes, was also shown.

ribozyme (pVwtRz), the maxizyme (pV-MzL/R), or the parental vector (pV), and we selected cells by exposing them to puromycin 24 hr after transfection. After incubation for 60 hr in the presence of puromycin, dead cells were removed, and puromycin-resistant cells were cultured for various times in medium without IL-3. The viability of cells was assessed in terms of their ability to exclude trypan blue dye. As shown in the left panel of Figure 7, BaF3/p210$^{BCR-ABL}$ cells that expressed the maxizyme died rapidly, whereas the control-transfected BaF3/p210$^{BCR-ABL}$ (pV) cells were still alive 10 days after withdrawal of IL-3. Moreover, the maxizyme did not kill any H9 cells that expressed normal *ABL* mRNA, a result that demonstrates the high specificity of the maxizyme for its target, the chimeric *BCR–ABL* gene. By contrast, in the presence of the conventional hammerhead ribozyme wtRz, apoptosis was induced in both BaF3/p210$^{BCR-ABL}$ and H9 cells. This result was consistent with the observation that wtRz can target the transcripts of both the *BCR–ABL* gene and the normal *ABL* gene *in vitro*. Furthermore, the maxizyme also killed many more BV173 cells, derived from a leukemic patient with a Philadelphia chromosome, than did the wild-type ribozyme or the parental vector (Figure 7, the right panel).

    To examine the potential utility of the maxizyme in the treatment of CML, we assessed the antitumor effect of the maxizyme in murine models of CML (Tanabe *et al.*, 2000b). All mice, without exception, that had been injected with cells from a patient with CML died of diffuse leukemia within 13 weeks. In marked contrast, when maxizyme-treated CML cells were injected, all mice remained disease-free. Thus, the maxizyme apparently functioned not only in cultured cells but also in an established animal system with exceptional efficacy. To our knowledge, ours is the first successful demonstration of the action of an artificial, allosterically controllable enzyme in animals. These results suggest that maxizyme technology might provide a useful approach to the treatment of CML.

Our novel maxizymes were more effective than similarly transcribed standard ribozymes in cells. To the best of our knowledge, our novel maxizyme is superior to other nucleic acid-based drugs reported to date because of its extremely high substrate-specificity and high cleavage activity. Novel maxizymes, whose activity can be controlled allosterically by sensor arms that recognize abnormal mRNAs specifically, should be powerful tools for the disruption of abnormal chimeric targets and might provide the basis for future gene therapy for the treatment of CML and other diseases.

D. Generality of the maxizyme application

There have been many attempts to construct artificial allosteric enzymes, but none have been demonstrated *in vivo* (Tanabe *et al.*, 2000a). The experimental results described above represent the first successful construction of an allosteric enzyme that functions *in vivo*. The maxizyme is indeed novel in that it imparts a sensor function to the short ribozymes that function as a dimer. By using this sensor function, it has become possible to specifically cleave abnormal chimeric mRNA without affecting the normal mRNA. This cannot be achieved with the conventional hammerhead ribozymes. The length and sequences of the sensor arms and of the common stem II of the maxizyme can easily be adjusted to attack specific targets. Indeed, it has been demonstrated that maxizymes can be designed for specific cleavage of several sequences of interest. In one case, maxizymes disrupted the expression of an abnormal chimeric gene that lacks a NUX cleavage site at the junction; this type of target, which was generated from reciprocal chromosomal translocation, was observed frequently among several leukemia diseases. For example, acute lymphoblastic leukemia (ALL), acute promyelocytic leukemia (APL), and CML include this type of abnormal gene, and maxizymes have successfully cleaved only those abnormal targets without causing any damage to normal genes (Figure 8A). In a second case, the maxizyme was used to investigate α and β genes with unknown functions (Figure 8B). If the difference between the two genes consisted of the insertion of one extra sequence, it would be easy to disrupt the α transcript using the wild-type ribozyme, as long as this region contained a cleavable NUX site. The β transcript would present a more difficult target unless this transcript contained a NUX site at the exon X–Y junction. The maxizyme allows us the cleavage of the respective transcript with high specificity.

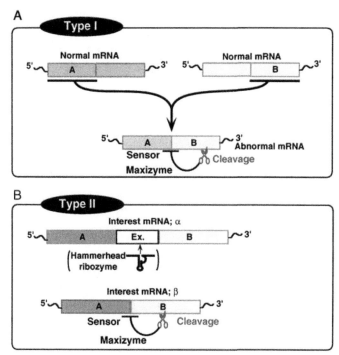

Figure 8 Different types of transcripts that can be disrupted by a maxizyme. (A) A chimeric transcript can be recognized by sensor arms at the junction and cleaved at a distant site by the maxizyme (the cleavage site is indicated by scissors); (B) the maxizyme can specifically cleave the hypothetical β mRNA. In this case, the α mRNA can be disrupted by the wild-type ribozyme, whereas the β mRNA must be recognized at the junction and cleaved by a maxizyme.

V. CONCLUSION

A ribozyme is a potentially useful tool for the suppression of the expression of specific genes since it can be engineered to act on other RNA molecules with high specificity. Although many trials have been successful, it remains difficult to design an effective ribozyme-expression system that can be used *in vivo*. One major challenge related to the use of ribozymes and antisense RNAs as therapeutic or genetic agents is the development of suitable expression vectors. We used the pol III system and the promoter of a human gene for tRNA^{Val} for transcription of ribozymes. This promoter is not only suitable for transcription of small RNAs, but its use also facilitates prediction of secondary structure by computer folding. More importantly, if properly designed, it allows export of transcribed tRNA^Val-ribozymes from the nucleus to the cytoplasm so that the tRNA^Val-ribozymes can find their mRNA targets. The tRNA^Val-vector may be useful for expression of functional RNAs other than ribozymes whose target molecules are localized in the cytoplasm. Although colocalization in the cytoplasm cannot by itself guarantee effectiveness, we can clearly increase the probability of success. In our hands, tRNA^Val-ribozymes have consistently high activities, at least in cultured cells.

Moreover, the maxizyme is an allosterically controllable ribozyme with powerful biosensor capacity that appears to function even in mice. Its biosensor functions allow specific inhibition of expression of a gene of interest only, without any effect on the normal mRNA. The maxizyme provides the first example of successful allosteric control of the activity of an artificially created allosteric enzyme, not only *in vitro* but also in animals, and its potential utility in medicine cannot be ignored. By modulating the sequences of sensor arms, we can easily adjust the activity of the maxizyme. Thus, maxizymes are powerful gene-inactivating agents with allosteric functions that can cleave any type of chimeric mRNA and/or any RNA by recognizing a specific sequence of interest for a cancer gene therapy.

References

Amontov, S., and Taira, K. (1996). Hammerhead minizymes with high cleavage activity: A dimeric structure as the active conformation of minizymes. *J. Am. Chem. Soc.* **118**, 1624–1628.

Arts, G.-J., *et al.* (1998). Identification of a nuclear export receptor for tRNA. *Curr. Biol.* **6**, 305–314.

Arts, G.-J., *et al.* (1999). The role of exportin-t in selective nuclear export of mature tRNAs. *EMBO J.* **17**, 7430–7441.

Atencia, R., *et al.* (1994). Apoptosis during retinoic acid-induced differentiation of F9 embryonal carcinoma cells. *Exp. Cell Res.* **214**, 663–667.

Avantaggiati, M. L., *et al.* (1997). Recruitment of p300/CBP in p53-dependent signal pathways. *Cell* **89**, 1175–1184.

Beck, J., and Nassal, M. (1995). Efficient hammerhead ribozyme-mediated cleavage of the structured hepatitis B virus encapsidation signal *in vitro* and in cell extracts, but not in intact cells. *Nucleic Acids Res.* **23**, 4954–4962.

Bernstine, E. G., and Ephrussi, B. (1976). Alkaline phosphatase activity in embryonal carcinoma and its hybrids with neuroblastoma. *In* "Teratomas and Differentiation." (M. I. Sherman and D. Solter, eds.), pp. 271–287. Academic Press, New York.

Bertrand, E., *et al.* (1994). Can hammerhead ribozymes be efficient tools for inactivating gene function? *Nucleic Acids Res.* **22**, 293–300.

Bertrand, E., *et al.* (1997). The expression cassette determines the functional activity of ribozymes in mammalian cells by controlling their intracellular localization. *RNA* **3**, 75–88.

Birikh, K. R., *et al.* (1997). The hammerhead ribozyme—structure, function and application. *Eur. J. Biochem.* **245**, 1–16.

Cameron, F. H., and Jennings, P. A. (1989). Specific gene suppression by engineered ribozymes in monkey cells. *Proc. Natl. Acad. Sci. U.S.A.* **86**, 9139–9143.

Cheng, J., *et al.* (2000). Inhibition of cell proliferation in HCC-9204 hepatoma cells by a c-myc specific ribozyme. *Cancer Gene Ther.* **7**, 407–412.

Clifford, J., et al. (1996). RXRα-null F9 embryonal carcinoma cells are resistant to the differentiation, anti-proliferative and apoptotic effects of retinoids. *EMBO J.* **15**, 4142–4155.

Cotten, M., and Birnstiel, M. (1989). Ribozyme mediated destruction of RNA *in vivo*. *EMBO J.* **8**, 3861–3866.

Dean, M., et al. (1986). c-myc regulation during retinoic acid-induced differentiation of F9 cells is posttranscriptional and associated with growth arrest. *Mol. Cell. Biol.* **6**, 518–524.

Deng, C., et al. (1995). Mice lacking p21$^{Cip1/WAF1}$ undergo normal development, but are defective in G1 checkpoint control. *Cell* **82**, 675–684.

Eckstein, F., and Lilley, D. M. J. (eds.) (1996). Catalytic RNA. *In* "Nucleic Acids and Molecular Biology," Vol. 10. Springer-Verlag, Berlin.

Efrat, S., et al. (1994). Ribozyme-mediated attenuation of pancreatic beta-cell glucokinase expression in transgenic mice results in impaired glucose-induced insulin secretion. *Proc. Natl. Acad. Sci. U.S.A.* **91**, 2051–2055.

Feng, X. H., et al. (1998). The tumor suppressor Smad4/DPC4 and transcriptional adaptor CBP/p300 are coactivators for smad3 in TGF-beta-induced transcriptional activation. *Genes Dev.* **12**, 2153–2163.

Folini, M., et al. (2000). Inhibition of telomerase activity by a hammerhead ribozyme targeting the RNA component of telomerase in human melanoma cells. *J. Invest. Dermatol.* **114**, 259–267.

Fu, D. J., et al. (1994). Hammerhead ribozymes containing non-nucleoside linkers are active RNA catalysts. *J. Am. Chem. Soc.* **116**, 4591–4598.

Funato, T., et al. (2000). Anti-K-ras ribozyme induces growth inhibition and increased chemosensitivity in human colon cancer cells. *Cancer Gene Ther.* **7**, 495–500.

Gavrieli, Y., et al. (1992). Identification of programmed cell death *in situ* via specific labeling of nuclear DNA fragmentation. *J. Cell Biol.* **119**, 493–501.

Gibson, S. A., et al. (2000). Induction of apoptosis in oral cancer cells by an anti-bcl-2 ribozyme delivered by an adenovirus vector. *Clin. Cancer. Res.* **6**, 213–222.

Giles, R. H., et al. (1998). Conjunction dysfunction: CBP/p300 in human disease. *Trends Genet.* **14**, 178–183.

Goodman, R. H., and Smolik, S. (2000). CBP/p300 in cell growth, transformation, and development. *Genes Dev.* **14**, 1553–1577.

Gu, W., et al. (1997). Synergistic activation by CBP and p53. *Nature* **387**, 819–823.

Haseloff, J., and Gerlach, W. L. (1988). Simple RNA enzymes with new and highly specific endonuclease activities. *Nature* **334**, 585–591.

Hertel, K. J., et al. (1992). Numbering system for the hammerhead. *Nucleic Acids Res.* **20**, 3252.

Hertel, K. J., et al. (1996). Specificity of hammerhead ribozyme cleavage. *EMBO J.* **15**, 3751–3757.

Inokuchi, Y., et al. (1994). A hammerhead ribozyme inhibits the proliferation of an RNA coliphage SP in *E. coli. J. Biol. Chem.* **269**, 11361–11366.

James, H., et al. (1996). Investigating and improving the specificity of ribozymes directed against the bcr–abl translocation. *Leukemia* **10**, 1054–1064.

Janknecht, R., et al. (1998). TGF-beta-stimulated cooperation of smad proteins with the coactivators CBP/p300. *Genes Dev.* **12**, 2114–2119.

Kato, Y., et al. (2001). Relationships between the activities *in vitro* and *in vivo* of various kinds of ribozyme and their intracellular localization in mammalian cells. *J. Biol. Chem.* **276**, 15378–15385.

Kawasaki, H., et al. (1996). Selection of the best target site for ribozyme-mediated cleavage within a fusion gene for adenovirus E1A-associated 300 kDa protein (p300) and luciferase. *Nucleic Acids Res.* **24**, 3010–3016.

Kawasaki, H., et al. (1998a). Distinct roles of the co-activators p300 and CBP in retinoic-acid-induced F9-cell differentiation. *Nature* **393**, 284–289.

Kawasaki, H., *et al.* (1998b). p300 and ATF-2 are components of the DRF complex, which regulates retinoic acid- and E1A-mediated transcription of the c-jun gene in F9 cells. *Genes Dev.* **12**, 233–245.

Kawasaki, H., *et al.* (2000). ATF-2 has intrinsic histone acetyltransferase activity which is modulated by phosphorylation. *Nature* **405**, 195–200.

Kitabayashi, I., *et al.* (1995). Phosphorylation of the adenovirus E1A-associated 300-kDa protein in response to retinoic acid and E1A during differentiation of F9 cells. *EMBO. J.* **14**, 3496–3509.

Kobayashi, H., *et al.* (1999). Retrovirus-mediated transfer of anti-MDR1 hammerhead ribozymes into multidrug-resistant human leukemia cells: Screening for effective target sites. *Int. J. Cancer* **81**, 944–950.

Konopka, J. B., *et al.* (1984). An alteration of the human c-abl protein in K562 leukemia cells unmasks associated tyrosine kinase activity. *Cell* **37**, 1035–1042.

Koseki, S., *et al.* (1999). Factors governing the activity *in vivo* of ribozymes transcribed by RNA polymerase III. *J. Virol.* **73**, 1868–1877.

Krupp, G., and Gaur, R. K. (2000). "Ribozyme, Biochemistry and Biotechnology." Eaton, Natick, Massachusetts.

Kutay, U., *et al.* (1998). Identification of a tRNA-specific nuclear export receptor. *Mol. Cell* **1**, 359–369.

Kuwabara, T., *et al.* (1996). Characterization of several kinds of dimer minizyme: Simultaneous cleavage at two sites in HIV-1 tat mRNA by dimer minizymes. *Nucleic Acids Res.* **24**, 2302–2310.

Kuwabara, T., *et al.* (1998a). Formation of a catalytically active dimer by tRNA^{Val}-driven short ribozymes. *Nat. Biotechnol.* **16**, 961–965.

Kuwabara, T., *et al.* (1998b). A novel allosterically trans-activated ribozyme, the maxizyme, with exceptional specificity *in vitro* and *in vivo*. *Mol. Cell* **2**, 617–627.

Kuwabara, T., *et al.* (1999). Novel tRNA^{Val}-heterodimeric maxizymes with high potential as gene-inactivating agents: Simultaneous cleavage at two sites in HIV-1 tat mRNA in cultured cells. *Proc. Natl. Acad. Sci. U.S.A.* **96**, 1886–1891.

Kuwabara, T., Warashina, M., and Taira, K. (2000). Allosterically controllable maxizymes cleave mRNA with high efficiency and specificity. *Trends Biotechnol.* **18**, 462–468.

Kuwabara, T., *et al.* (2001). Significantly higher activity of a cytoplasmic hammerhead ribozyme than a corresponding nuclear counterpart: Engineered tRNAs with an extended 3′ end can be exported efficiently and specifically to the cytoplasm in mammalian cells. *Nucleic Acids Res.* **29**, 2780–2788.

Lill, N. L., *et al.* (1997). Binding and modulation of p53 by p300/CBP activators. *Nature* **387**, 823–827.

Lipowsky, G., *et al.* (1999). Coordination of tRNA nuclear export with processing of tRNA. *RNA* **5**, 539–549.

Long, D. M., and Uhlenbeck, O. C. (1994). Kinetic characterization of intramolecular and intermolecular hammerhead RNAs with stem II deletions. *Proc. Natl. Acad. Sci. U.S.A.* **91**, 6977–6981.

Lund, E., and Dahlberg, J. E. (1999). Proofreading and aminoacylation of tRNAs before export from the nucleus. *Science* **282**, 2082–2085.

McCall, M. J., *et al.* (1992). Minimal sequence requirements for ribozyme activity. *Proc. Natl. Acad. Sci. U.S.A.* **89**, 5710–5714.

Muraoka, M., *et al.* (1996). p300 gene alterations in colorectal and gastric carcinomas. *Oncogene* **12**, 1565–1569.

Nakayama, A., *et al.* (1999). CTAB-mediated enrichment for active forms of novel dimeric maxizymes. *FEBS Lett.* **448**, 67–74.

Nowell, P. C., and Hungerford, D. A. (1960). A minute chromosome in human chronic granulocytic leukemia. *Science* **132**, 1497–1499.

Ohkawa, J., and Taira, K. (2000). Control of the functional activity of an antisense RNA by a tetracycline-responsive derivative of the human U6 snRNA promoter. *Hum. Gene. Ther.* **11**, 577–585.

Pachuk, C. J., et al. (1994). Selective cleavage of bcr–abl chimeric RNAs by a ribozyme targeted to non-contiguous sequences. Nucleic Acids Res. 22, 301–307.

Petrij, F., et al. (1995). Rubinstein-Taybi syndrome caused by mutations in the transcriptional co-activator CBP. Nature 376, 348–351.

Rossi, J. J., and Sarver, N. (1990). RNA enzymes (ribozymes) as antiviral therapeutic agents. Trends Biotechnol. 8, 179–183.

Ruffner, D. E., et al. (1990). Sequence requirements of the hammerhead RNA self-cleavage reaction. Biochemistry 29, 10695–10702.

Sarver, N., et al. (1990). Ribozymes as potential anti-HIV-1 therapeutic agents. Science 247, 1222–1225.

Sherr, C. J., and Roberts, J. M. (1995). Inhibitors of mammalian G1 cyclin-dependent kinases. Genes Dev. 9, 1149–1163.

Shimayama, T., et al. (1995). Generality of the NUX rule: Kinetic analysis of the results of systematic mutations in the trinucleotide at the cleavage site of hammerhead ribozymes. Biochemistry 34, 3649–3654.

Simos, G., and Hurt, E. (1999). Transfer RNA biogenesis: A visa to leave the nucleus? Curr. Biol. 9, 238–241.

Solter, D., and Knowles, B. (1978). Monoclonal antibody defining a stage-specific mouse embryonic antigen (SSEA-1). Proc. Natl. Acad. Sci. U.S.A. 75, 5565–5569.

Strickland, S., and Mahdavi, V. (1978). The induction of differentiation in teratocarcinoma stem cells by retinoic acid. Cell 15, 393–403.

Strickland, S., et al. (1980). Hormonal induction of differentiation in teratocarcinoma stem cells: Generation of parietal endoderm by retinoic acid and dibutyryl cAMP. Cell 21, 347–355.

Sullenger, B. A., et al. (1990). Expression of chimeric tRNA-driven antisense transcripts renders NIH 3T3 cells highly resistant to Moloney murine leukemia virus replication. Mol. Cell. Biol. 10, 6512–6523.

Suzuki, T., et al. (2000). Adenovirus-mediated ribozyme targeting of HER-2/neu inhibits in vivo growth of breast cancer cells. Gene Ther. 7, 241–248.

Symons, R. H. (1992). Small catalytic RNAs. Annu. Rev. Biochem. 61, 641–671.

Takagi, Y., et al. (2001). Recent advances in the elucidation of the mechanisms of action of ribozymes. Nucleic Acids Res. 29, 1815–1834.

Tanabe, T., et al. (2000a). Maxizymes, novel allosterically controllable ribozymes can be designed to cleave various substrates. Biomacromolecules 1, 108–117.

Tanabe, T., et al. (2000b). Oncogene inactivation in a mouse model: Tissue invasion by leukaemic cells is stalled by loading them with a designer ribozyme. Nature 406, 473–474.

Tokunaga, T., et al. (2000). Ribozyme-mediated inactivation of mutant K-ras oncogene in a colon cancer cell line. Br. J. Cancer 83, 833–839.

Thompson, D. J., et al. (1995). Improved accumulation and activity of ribozymes expressed from a tRNA-based RNA polymerase III promoter. Nucleic Acids Res. 23, 2259–2268.

Tsuchida, T., et al. (2000). Adenovirus-mediated anti-K-ras ribozyme induces apoptosis and growth suppression of human pancreatic carcinoma. Cancer Gene Ther. 7, 373–383.

Tuschl, T., and Eckstein, F. (1993). Hammerhead ribozymes: Importance of stem-loop II activity. Proc. Natl. Acad. Sci. U.S.A. 90, 6991–6994.

Uhlenbeck, O. C. (1987). A small catalytic oligonucleotide. Nature 328, 596–600.

Wang, F. S., et al. (1999). Retrovirus-mediated transfer of anti-MDR1 ribozymes fully restores chemosensitivity of P-glycoprotein-expressing human lymphoma cells. Hum. Gene. Ther. 10, 1185–1195.

Warashina, M., et al. (2001). RNA-protein hybrid ribozymes that efficiently cleave any mRNA independently of the structure of the target RNA. Proc. Natl. Acad. Sci. U.S.A. 98, 5572–5577.

Yamada, O., et al. (1994). Intracellular immunization of human T cells with a hairpin ribozyme against human immunodeficiency virus type 1. Gene Ther. 1, 38–45.

Yao, T.-P., *et al.* (1998). Gene dosage-dependent embryonic development and proliferation defects in mice lacking the transcriptional integrator p300. *Cell* **93,** 361–372.

Yokoyama, Y., *et al.* (2000). The 5'-end of hTERT mRNA is a good target for hammerhead ribozyme to suppress telomerase activity. *Biochem. Biophys. Res. Commun.* **273,** 316–321.

Yu, M., *et al.* (1993). A hairpin ribozyme inhibits expression of diverse strains of human immunodeficiency virus type 1. *Proc. Natl. Acad. Sci. U.S.A.* **90,** 6340–6344.

Zhang, Y. A., *et al.* (2000). Generation of a ribozyme-adenoviral vector against K-ras mutant human lung cancer cells. *Mol. Biotechnol.* **15,** 39–49.

Zhao, J. J., and Pick, L. (1993). Generating loss-of-function phenotypes of the *fushi tarazu* gene with a targeted ribozyme in *Drosophila*. *Nature* **365,** 448–451.

Zhou, D.-M., and Taira, K. (1998). The hydrolysis of RNA: From theoretical calculations to the hammerhead ribozyme-mediated cleavage of RNA. *Chem. Rev.* **98,** 991–1026.

Inhibitors of Angiogenesis

Steven A. Stacker[1] and Marc G. Achen

Ludwig Institute for Cancer Research
Royal Melbourne Hospital
Parkville/Melbourne 3050, Australia

I. INTRODUCTION — ANGIOGENESIS

Angiogenesis, the formation of new blood vessels from preexisting vessels, is a process that occurs in a very restricted manner under normal physiological conditions. In adults, blood vessels are relatively quiescent, having turnover rates in the order of years (Folkman and Shing, 1992). This appears to be due to the effectiveness of natural angiogenesis inhibitors that keep in check the positive effect of angiogenic growth factors. Only in areas such as the placenta, endometrium, and developing embryo is angiogenesis observed under nonpathological conditions. In contrast, certain pathological conditions, including tumor formation, exhibit high levels of neovascularization/angiogenesis. This difference in rates of angiogenesis between vessels in normal adult tissues versus a tumor provides a potential target for therapeutic intervention based on inhibiting angiogenesis. Other advantages to an antiangiogenesis approach are (1) Ready

[1]To whom correspondence should be addressed.

access of drug to target endothelium, (2) Drug resistance is less likely to arise because the target is the proliferating endothelial cell rather than tumor cells with high mutation rates, (3) Capacity to block the generation of metastases that spread via blood vessels.

A. Physiological angiogenesis

During early embryogenesis, blood vessels initially arise from mesoderm that differentiates *in situ* to form angioblasts which, in turn, associate to form primitive embryonic vessels (vasculogenesis) (Risau, 1997). Subsequent expansion of these vessels is by angiogenesis, and includes differentiation along specialized pathways to achieve the tissue-specific blood vessels seen in the various organs throughout the body. Other major sites of active angiogenesis, under normal physiological conditions, are in the endometrium in response to the menstrual cycle, in the placenta, and in the developing embryo. Angiogenesis is also an important part of the wound healing response. Importantly, the proliferation and migration of endothelial cells, a critical feature of angiogenesis, is tightly controlled. This is due in part to an array of inhibitory mechanisms that counterbalance the angiogenic stimuli to which blood vessels are exposed. In normal physiological angiogenesis, downregulation of the inhibitors of angiogenesis and upregulation of angiogenic growth factors allows this process to proceed in a controlled manner. In many pathological conditions these processes are disregulated, and uncontrolled angiogenesis is a consequence.

It should be remembered that the growth of lymphatic vessels (lymphangiogenesis) is an important process in the correct formation of the circulatory system (Achen *et al.*, 2000). Although not as widely studied as angiogenesis, lymphangiogenesis requires our attention when considering inhibitors of vessel formation, as this may be an important step in promoting metastatic spread of tumors (Mandriotta *et al.*, 2001; Skobe *et al.*, 2001; Stacker *et al.*, 2001).

B. Pathological angiogenesis

1. Nontumor angiogenesis

Numerous pathological conditions are characterised by excessive and/or uncontrolled angiogenesis. In diabetes, new blood vessels in the retina invade the vitreous, bleed, and thereby cause blindness (Stitt *et al.*, 1998). In fact, ocular neovascularization is the most common cause of blindness and is a feature not only of diabetic retinopathy but also of age-related macular degeneration and recurrent pterygium. In rheumatoid arthritis, new capillary blood vessels invade the joint and destroy cartilage, leading to joint pain and swelling (Storgard *et al.*, 1999). Angiogenesis is also a feature of atherosclerosis as this condition can lead to angiogenesis of the vasa vasorum within the arterial media and adventicia in response to ischemia following arteriole occlusion (Moulton *et al.*, 1999). This can be dangerous as angiogenesis may cause atheroma to rupture or thrombose. Another condition in which angiogenesis plays a role is psoriasis — psoriatic skin is characterized by increased vascularity and vascular permeability, and molecules such as vascular endothelial growth factor (VEGF) are implicated (Bhushan *et al.*, 1999).

2. Tumor angiogenesis

During the 1990s, the association between angiogenic growth factors and tumors was extensively characterised at a molecular level. The original experiments showed that tumors could not grow beyond a given size ($2-3$ mm^3), limited by diffusion, without the recruitment of a new vasculature (Folkman *et al.*, 1963). These experiments implied that populations of tumor cells remained dormant at this size until they underwent the angiogenic switch, that is, the ability to secrete or to induce other cells to secrete factors that stimulate vascularization of the tumor. The tumor cells are then capable of rapid proliferation and spread via blood or lymphatic vessels. Since these early studies at the cellular level, the molecular mechanisms that control the process of tumor angiogenesis have been partly defined. Knowledge taken from our understanding of angiogensis in embryonic development and normal physiology has underpinned much of what we know about angiogenesis in tumors.

From a microscopic examination of tumor vessels it is clear that they have particular characteristics not usually observed in the blood vessels of normal adult tissues. Tumor vessels are frequently enlarged and more irregular in appearance, and are often poorly associated with supporting cells such as pericytes (Dewhirst *et al.*, 1989; Jain, 1987). Furthermore, the basement membrane composition of tumor vessels is often altered, again suggesting an abnormal association with the matrix (Stetler-Stevenson *et al.*, 1993). Tumor vessels also exhibit increased permeability (Jain, 1987), which may be due to their partially disrupted structure and/or the action of angiogenic growth factors (e.g., VEGF) which can induce vascular hyperpermeability (Senger *et al.*, 1983).

In parallel to normal angiogenesis, tumor angiogenesis involves a number of steps that allow for the dissociation of endothelial cells from existing blood vessels, endothelial cell proliferation and migration to a new site, and assembly of endothelial cells into a new vessel. These steps are summarized in Figure 1. One point to emphasize is that each step involves not only positive regulators but also negative regulators or inhibitors that keep angiogenesis under strict control (although in the case of tumor angiogenesis these processes appear to be uncontrolled). In terms of an antiangiogenic strategy, each of these steps provides the opportunity to create inhibitors or to utilize naturally occurring inhibitors to

264

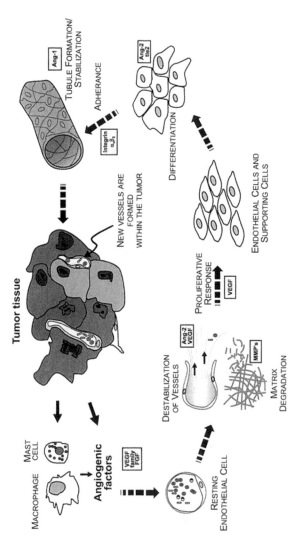

Figure 1 Defined steps in tumor angiogenesis: potential targets for antiangiogenic therapy of tumors. Schematic representation of the multiple processes required for the formation of tumor blood vessels (matrix degradation, cell survival, cell proliferation, cell migration, differentiation/tubule formation). Tumor cells secrete, or induce other cell types such as macrophages and mast cells to secrete, angiogenic factors that act on cells of the blood vessels to stimulate the initial phases of angiogenesis. This involves the dissociation of preexisting vessels, the release of plasma proteins, and the degradation of basement membrane and matrix proteins. Subsequent proliferative signals drive the production of endothelial cells and supporting cells, which undergo differentiation and formation to tubule structures. These new vessels are laid down on new extracellular matrix and basement membrane, although in tumor angiogenesis these are frequently abnormal or absent. The diagram indicates some of the main molecular pathways which mediate these processes. VEGF, Vascular endothelial growth factor; MMP, matrix metaloproteinase; tie2, receptor tyrosine kinase with fibronectin-type III repeats, Ig-like domains, and EGF homology repeats; Ang, angiopoietin; FGF, fibroblast growth factor. (See color plates.)

prevent tumor angiogenesis. In the following sections we have highlighted some of the main drug targets for angiogenesis, they are generally positive regulators to which inhibitory reagents can be made (e.g., VEGF that can be inhibited by an anti-VEGF monoclonal antibody). These have been catagorized into sections depending on which part of the angiogenic process is involved (e.g., matrix association, endothelial cell proliferation, differentiation, or tubule formation). Additionally we have listed antiangiogenic reagents that have been made to such drug targets, and we refer also to naturally occurring inhibitors.

II. ANGIOGENESIS INHIBITORS

A. Matrix, basement membrane, and cell–matrix adhesion molecules

Endothelial cells reside on specialized matrix components that form the basement membrane which helps to maintain blood vessels in a quiescent state. The degradation of this membrane is a prime step in the release of endothelial cells, and the subsequent breakdown of extracellular matrix results in the migration and proliferation of endothelial cells (Klagsburn and D'Amore, 1991). The creation of new extracellular matrix is required for the reassembly of an intact vessel (Klagsburn and D'Amore, 1991). The production of a new basement membrane is required to provide a stable base for the support of the new vessels. The extracellular matrix and basement membrane contain a number of proteins that appear to be critical for tumor angiogenesis and could consequently serve as targets for antiangiogenic therapy. In particular, the matrix metaloproteinases (MMPs), a family of at least 18 zinc-dependent enzymes required for the remodeling and turnover of extracellular matrix (reviewed in Yip *et al.*, 1999), have been implicated in angiogenesis. A number of studies have linked MMP expression in tumor cell lines to the ability to metastasize *in vivo*. Inhibitors of MMPs including Marimastat, Neovastat, and BMS-275291 (see Table 1) are currently in clinical trials as anticancer agents. Another proteolytic system being targeted for development of antiangiogenic agents is the plasminogen activator–plasmin system (see Table 1). This system is thought to be involved in establishing the proteolytic cascade that releases serine proteases and MMPs and activates angiogenic growth factors. A peptide, designated A6, derived from the nonreceptor-binding region of urokinase plasminogen activator (uPA) was more recently shown to inhibit angiogenesis *in vivo*. In combination with the cytotoxic agent cisplatin, A6 inhibits the growth of glioblastomas in mice (Mishima *et al.*, 2000).

The role of integrins in endothelial cell–extracellular matrix interactions in angiogenesis was demonstrated by the finding that disruption of the binding of integrin $\alpha_v\beta_3$ to the extracellular matrix prevented blood vessel

Table 1 Targets for Antiangiogenic Compounds

Target[a]	Antiangiogenic Agent/Mechanism[b]	Clinical Trials Status[c]	References[d]
Matrix, basement membrane, and cell-matrix adhesion molecules			
Matrix metalloproteinases (MMPs)	Marimastat (BB-2516), synthetic MMP inhibitor (MMPI), blocks TNFα convertase	Phase I/II, glioblastoma; phase III, NSC[e] lung	Wojtowicz-Praga et al. (1998), Wojtowicz-Praga (1999)
	BAY 12-9566, nonpeptide bipheynl MMPI	Phase I, solid malignancies	Rowinsky et al. (2000)
	COL-3, synthetic MMPI, tetracycline derivative	Phase I, solid tumors; phase I/II, brain	Fife et al. (2000), Wojtowicz-Praga (1999)
	Neovastat (AE-941), liquid complex from shark cartilage, antiangiogenesis and anti-MMP	Phase III, renal cell (kidney); phase III, NSC lung	Wojtowicz-Praga et al. (1997), Wojtowicz-Praga (1999)
	Ro 113,830, MMPI, inhibits collagenase-3, gelatinase-A and -B		Yip et al. (1999)
	Trocade (Ro-32-3555), MMPI, selective for MMP-1,-8, and-13, i.e., collagenases over the stromelysins and gelatinases		Yip et al. (1999)
	BMS-275291		Heath and Grochow (2000), Wojtowicz-Praga (1999)
	CGS27023A, hydroxamic acid based structure, broad spectrum MMPI	Phase I	Yip et al. (1999)
	Ilomastat (Galardin, GM-6001), first generation MMPI		Galardy et al. (1994)
	Batimastat (BB-94), broad spectrum MMPI, prevents MMP activity		Wylie et al. (1999), Lein et al. (2000), Zhu et al. (2000)
	Prinomastat (AG3340, Agouron), MMPI derived from X-ray crystal structure, MMP-2, 3, and 13 inhibitor	Phase III, NSC lung phase III, prostate; phase II, glioblastoma	Shalinsky et al. (1999)
	Interleukin-10, inhibits MMP-1/2 and 9		Stearns et al. (1999)
	BPHA, inhibits MMP-2, 9, and 14		Maki et al. (1999), Maekawa et al. (1999)
Urokinase plasminogen activator (uPA)	A6, peptide inhibitor derived from uPA that is noncompetitive inhibitor of uPA—uPA receptor interaction		Guo et al. (2000), Mishima et al. (2000)
	Amiloride inhibition of uPA		Jankun and Skrzypczak-Jankun (1999)
	Isogladine maleate, diminishes PA activity in vitro		Ren et al. (1998)
Integrins	Humanized MAb to integrin α$_v$β$_3$ (Vitaxin) —cell adhesion inhibitor		Senger et al. (1997), Coleman et al. (1999), Brooks et al. (1994a,b, 1995, 1998), Rüegg et al. (1998), Friedlander et al. (1995)
	EMD121974, small molecule blocker of α$_v$β$_3$	Phase I, advanced/metastatic cancer	Mitjans et al. (2000), Lode et al. (1999)
	Integrin β, MAb, inhibits endothelial cell migration and tubule formation		Gamble et al. (1993, 1999)
	RGD peptide inhibitor of integrin α$_v$β$_3$		Klotz et al. (2000), Dechantsreiter et al. (1999), Storgard et al. (1999)
	PEX, blocks binding of MMP-2 to integrin α$_v$β$_3$, inhibits MMP-2		Brooks et al. (1998)

266

	Description	Clinical phase/indication	References
Collagens	Collagen type IV fragment, basement membrane collagen, inhibits endothelial cell proliferation, migration, and tubule formation		Maeshima et al. (2000), Colorado et al. (2000)
	Collagenase		Reed et al. (2000)
	Canstatin		Kamphaus et al. (2000)
	U-995, shark-cartilage derived angiogenesis inhibitor, prevents collagenase-induced collagenolysis		Sheu et al. (1998)
Other	Shark cartilage (SC)	Phase I/II, advanced solid tumors	Miller et al. (1998)
	Hyaluronic acid		Tempel et al. (2000)
	Thrombospondin-2, endogenous inhibitor of angiogenesis		Streit et al. (1999), Kyriakides et al. (1998)
Endothelial cell proliferation and differentiation			
VEGFR-2	SU5416, small molecule blocker of VEGFR-2, preventing VEGFR-2 signaling by kinase inhibition	Phase I, head and neck, advanced solid tumors, stage II and IV breast cancer; phase II, AML; phase I/II, Kaposi's sarcoma, glioma, advanced malignancies; phase II, von Hippel Lindau disease; phase II, prostate, colorectal, metastatic melanoma, multiple myeloma, mesothelioma; phase III metastatic colorectal carcinoma	Fong et al. (1999), Mendel et al. (2000)
	SU6668, small molecule blocker of VEGFR-2 (and other RTKs), preventing VEGFR-2 and other RTK signaling by kinase inhibition	Phase I, advanced cancer	Fong et al. (1999), Laird et al. (2000)
	3-[(4,5,6,7-Tetrahydro-1H-indol-2-yl) methylene]-1,3-dihydroindol-2-ones, VEGFR-2 inhibitors		Zhu et al. (1999), Zhu and Witte, (1999), Sun et al. (2000)
	Ribozymes, cleave mRNAs of VEGFRs		Pavco et al. (2000)
	MAb to VEGFR-2 extracellular domain		Witte et al. (1998)
	4-anilinoquinazolines (ZD4190), VEGFR-2 kinase inhibitor		Hennequin et al. (1999), Wedge et al. (2000)
	PTK787/ZK 222584 inhibits kinase activity of VEGFRs		Xu et al. (2000), Wood et al. (2000)
VEGFR-3	Neutralizing monoclonal antibody, blocks ligand binding		Kubo et al. (2000)
VEGF	Anti-VEGF antibody, inhibits the binding of VEGF to VEGFR-2, prevents VEGFR-2 mediated mitogenic effects and VEGF induced angiogenesis	Phase I, refractory solid tumors; phase II, metastatic renal cell cancer	van Bruggen et al. (1999), Gabrilovich et al. (1999), Chen et al. (1999), Kamiya et al. (1999a,b), Schlaeppi et al. (1999), Yuan et al. (1996), Kim et al. (1992, 1993), Gorski et al. (1999), Ryan et al. (1999), Asano et al. (1999), Borgström et al. (1998), Zhu et al. (1999), Zhu and Witte (1999)

(continues)

Table 1 (*Continued*)

Target[a]	Antiangiogenic Agent/Mechanism[b]	Clinical Trials Status[c]	References[d]
	VEGFR-1 soluble extracellular domain		Goldman et al. (1998)
	NX1838, oligonucleotide, inhibits binding of VEGF to endothelial cells		Bell et al. (1999), Siemeister et al. (1998)
	Antisense VEGF RNA and oligonucleotide		Belletti et al. (1999), Oku et al. (1998), Robinson et al. (1996) Cheng et al. (1996), Saleh et al. (1996), Huang et al. (1999), Fachinger et al. (1999)
	CAI, decreases Ca^{2+} signal transduction, decreases VEGF secretion	Phase I, solid tumors; phase II, ovarian cancer, advanced renal cancer; phase III, NSC lung	Bauer et al. (2000)
	Porphyrin analogs, TMPP; blocks VEGF binding to VEGFR-2 (and FGF binding to FGFR)		Aviezer et al. (2000)
	Platelet factor 4, causes impaired downregulation of Cip/Waf-1 and inhibits VEGF activity		Gentilini et al. (1999), Gengrinovitch et al. (1995), Tanaka et al. (1997)
	Prolactin and prolactin-like molecules inhibit VEGF activation of Ras		Duenas et al. (1999), D'Angelo et al. (1999), Struman et al. (1999)
	Octreatide and somatostatin, inhibit endothelial cell proliferation and VEGF secretion		Lawnicka et al. (2000)
VEGF-D	MAb blocks the binding of VEGF-D to both VEGFR-2 and VEGFR-3		Achen et al. (2000), Stacker et al. (2001)
FGFs/FGF receptor	1,3,6-Naphthalenetrisulfonate, FGF inhibitor		Cuevas et al. (1999)
	Heparin		Engelberg (1999)
	TNP-470		Bond et al. (2000)
	Curuminods, inhibit FGF-2 angiogenic response and MMP gelB expression		Mohan et al. (2000)
	Perlecan (coreceptor for FGF-2), antisense treatment		Sharma et al. (1998)
EGF/EGF receptor	EGF has angiogenic activity		Harris et al. (2000)
	Anti-EGFR MAb (C225)		Bruns et al. (2000), Perrotte et al. (1999)
HGF/met receptor	Ribozymes inhibiting the expression of HGF and its receptor c-met		Abounader et al. (1999)
Other growth Factors/receptors	TGF-β2 (inhibits HGF-induced endothelial cell proliferation		Manganini and Maier (2000)
	VEGI—cytokine of TNF family, inhibits endothelial cell mitogenesis		Zhai et al. (1999)
	CXC chemokines		Moore et al. (1998), Addison et al. (2000)
	Endoglin, Anti-human endoglin immunotoxin, involved in signaling with TGF-β receptor complex		Seon et al. (1997)
Other kinases	RTK inhibitors PD166285 and PD173074		Dimitroff et al. (1999)
	Protein kinase A antisense DNA/RNA hybrid inhibitor		Tortora et al. (2000)
	Tranilast, protein kinase C signaling inhibitor		Koyama et al. (1999)
	UCN-01, protein kinase C inhibitor of endothelial cell proliferation		Kruger et al. (1998), Yoshiji et al. (1999)

Others	α-Interferon, downregulates angiogenesis-related genes	Phase II/III	Slaton et al. (1999), see Website for cancer trials[c]
			Fathallah-Shaykh et al. (2000)
	γ-Interferon		Keane et al. (1999)
	γ-Interferon-inducible protein-10 (a CXC chemokine)		
	Interleukin-4		Volpert et al. (1998)
	Interleukin-8		Koch et al. (1993)
	Interleukin-12, inhibits angiogenesis synergistically with indomethacin	Phase I/II, Karposi's sarcoma; phase I/II, IL-12 gene therapy	Golab et al. (2000)
	Interleukin-18		Cao et al. (1999a)
	Retinoids, regulate VEGF expression, inhibit in vivo angiogenesis		Diaz et al. (2000), Liaudet-Coopman et al. (1997)
	Protein kinase C8, inhibits VEGF-mediated cell migration/proliferation		Shizukuda et al. (1999)
	SPARC (BM-40, osteonectin), inhibits VEGF-mediated mitogenesis		Kupprion et al. (1998)
	TNP-470, synthetic analog of fumagillin, inhibits Ets-1		Yamaoka et al. (1993), Zhang et al. (2000)
	Kininogen (kininostatin), inhibits endothelial cell migration/proliferation		Colman et al. (2000)
	Ets-1 transcription factor, inhibitors such as fumagillin block Ets-1 activity, thus blocking angiogenesis		Nakano et al. (2000), Wernert et al. (1999), Sato (1998)
	Peroxisome proliferator-activated receptor γ ligands, nuclear receptors that are transcription factors and inhibit angiogenesis		Xin et al. (1999)
	Apigenin, inhibits endothelial cell proliferation		Trochon et al. (2000)
	Thromboxane A(2) inhibitors		Nie et al. (2000a)
	Eicosanoid inhibitors		Nie et al. (2000b,c)
	Protamine		Arrieta et al. (1998)
	Pigment epithelium-derived factor		Dawson et al. (1999)
Vascular remodeling			
Angiopoietins/tie family of RTKs	Angiopoietins (Ang-1, Ang-2)		Maisonpierre et al. (1997)
	Soluble tie2 extracellular domain		Siemeister et al. (1999)
Endothelial cell migration	Lysophosphatidylcholine inhibits endothelial cell migration/proliferation		Rikitake et al. (2000)
	Anti-CD9 MAb, inhibits endothelial cell migration		Klein-Soyer et al. (2000)
	P38(MAPK)/HSP27 inhibitors, act on smooth muscle cell migration		Hedges et al. (1999)
	Decorin, inhibits endothelial cell migration and fibronectin fibrillogenesis		Kinsella et al. (2000)
Endothelial cell survival	Thrombospondin-1, apoptosis-dependent inhibition of angiogenesis		Jimenez et al. (2000)
	Carboxyamido-triazole, induces apoptosis of endothelial cells		Ge et al. (2000)
	Protein kinase Akt, survival signal for endothelial cells		Hermann et al. (2000)
	p53 target genes		Tokino and Nakamura (2000)

(continues)

Table 1 (*Continued*)

Target[a]	Antiangiogenic Agent/Mechanism[b]	Clinical Trial Status[c]	References[d]
Unknown mechanism	Thalidomide	Phase I/II, melanoma; phase II, head and neck; phase I, gynecologic sarcomas, solid tumors, metastatic colorectal carcinoma; phase III, NSC lung, prostate, kidney, myeloma	Fine et al. (2000), Moreira et al. (1999), D'Amato et al. (1994)
	Squalamine, inhibitor of sodium–hydrogen exchanger NHE3	Phase II, NSC lung, ovarian; phase I, advanced cancers	Schiller and Bittner (1999)
	Combretastatin A-4 prodrug, induces endothelial cell apoptosis and tubulin destabilization	Phase I, solid tumors; phase II, to begin in mid-2000	Parkins et al. (2000)
	Endostatin, 20 kDa C-terminal proteolytic fragment of collagen XVIII, inhibits endothelial cell proliferation	Phase I, solid tumors	Yamaguchi et al. (1999), Kaplan et al. (1997)
	Angiostatin, plasminogen fragment, inhibits endothelial cell migration in angiogenesis		O'Reilly et al. (1994, 1999), Ji et al. (1998)
	Plasminogen kringle 1–3 and kringle 1–5		Joe et al. (1999), Cao et al. (1999b)
	Prothrombin kringle-2		Lee et al. (1998)
	Suramin analogs		Meyers et al. (2000)
	Linomide (for treatment of von Hippel Lindau tumors)		Gross et al. (1999)
	S-Nitrocaptopril crystals, nitric oxide donor		Jia et al. (2000)
	Cyclooxygenase-2 inhibitors, including nonsteroidal anti-inflammatory drugs (NSAIDs)		Fosslien (2000), Sawaoka et al. (1999), Jones et al. (1999)

Emerging targets:

Nimodipine ginko-biloba dipyridamole (Juarez et al., 2000); Paclitaxel (Belotti et al., 1996); soy isoflavones (Zhou et al., 1998) 2-methoxyestradiol (Reiser et al., 1998); cyclosporane (Iurlaro et al., 1998; Lingen et al., 1998); roxithromycin (Yatsunami et al., 1999a,b); sulindac and sulfone metabolite (FGN-1) (Skopinska-Rozewska et al., 1998a,b; 1999); NF-κB inhibitors (Yoshida et al., 1999); angiotensin converting enzyme (Fabre et al., 1999); phytoestrogens (Fotsis et al., 1998); connective tissue growth factor (Shimo et al., 1999); isocoumarin, NM-3 (Nakashima et al., 1999); soybean phytochemicals (Zhou et al., 1999); halofuginone (Elkin et al., 1999a,b); N-acetylcysteine (Cai et al., 1999); arilinophthalazines (Bold et al., 2000); TAC-10 (Murakami et al., 1999); topotecan (Clements et al., 1999); camptothecin analogs (Clements et al., 1999; O'Leary et al., 1999); copper (Rabinovitz, 1999); hypothemycin (Tanaka et al., 1999); docosahexaenoic acid (Rose and Connolly, 1999); prostate-specific antigen (Fortier et al., 1999); vasoactive intestinal polypeptide (VIP) (Ogasawara et al., 1999); sphingosine-1 (Lee et al., 1999); AdmATF (Li et al., 1999); Chrysobal-anus icaco L. extract (Alves et al., 2000); N(G)-nitro-L-arginine methylester (Jadeski et al., 2000; Jadeski and Lala, 1999); farnesyl transferase (Feldkamp et al., 1999; Ashton et al., 1999); lactacystin (Kumeda et al., 1999); FMPA (9-α-fluoromedroxoxyprogesterone acetate) (Yamaji et al., 1999); methotrexate (Joussen et al., 1999); curcumin (Arbiser et al., 1998); selenium (Jiang et al., 1999); small GTPase rho (Uchida et al., 2000); lovastatin and TNFα (Feliesko et al., 1999); CEA-related adhesion molecule-1 (Ergün et al., 2000); β-catenin (Cruz et al., 2000); transcription factor, Sox18 (Pennisi et al., 2000); hypoxia-inducible factor-1 (Zagzag et al., 2000; Zhong et al., 1999); PR39 peptide, inhibited degradation of HIF-1α (Li et al., 2000); NP-1 (VEGF receptor) (Soker et al., 1998).

[a]Targets for antiangiogenic agents have been categorized based on their major role in angiogenesis/endothelial cell biology. Some agents have more than one action, for example, antiproliferative and antisurvival, and have therefore been placed in the category that they are predominantly associated with in the literature. Emerging targets, those for which little data is available, are listed at the bottom of the table.

[b]Antiangiogenic agents have been described using the names that appear most frequently in the research or clinical trials literature. Mechanisms are indicated when these have been well established.

[c]The status of clinical trials is constantly changing. Information on current trials of antiangiogenesis reagents can be found at the Cancer Trials Website of the National Cancer Institute (http://cancertrials.nih.gov/news/angio/) or on the homepage of the various companies sponsoring the trials.

[d]The reference list highlights the most recent work on the antiangiogenic agents.

[e]NSC, non small cell.

formation on the chick chorioallantoic membrane (Friedlander *et al.*, 1995). In fact, integrin $\alpha_v\beta_3$ is preferentially expressed on growing blood vessels (Brooks *et al.*, 1994a). Therefore, inhibitors of these integrins have been actively sought as antiangiogenic therapeutics for cancer treatment (Table 1). For example, a fragment of MMP-2, that comprises the C-terminal hemopexin-like domain, termed PEX, prevents MMP-2 binding to integrin $\alpha_v\beta_3$ and blocks cell surface collagenolytic activity. PEX blocks tumor angiogenesis and tumor growth in animal models (Brooks *et al.*, 1998). A monoclonal antibody against $\alpha_v\beta_3$ designated LM609 (Brooks *et al.*, 1995), a humanized version of which is known as Vitaxin (Coleman *et al.*, 1999) also blocks tumor growth in animal models.

B. Endothelial cell proliferation and differentiation

Although factors such as epidermal growth factor (EGF), transforming growth factor (TGF)-α, TGF-β, tumor necrosis factor (TNF), and interleukin (IL)-8 also have positive effects on endothelial cell mitogenesis (see Table 1), the major proliferative signals for vascular endothelial cells are the members of the VEGF and FGF families of secreted growth factors (Gerwins *et al.*, 2000). Indeed, VEGF and its receptors, in particular VEGFR-2, have become the prime targets for antiangiogenic approaches to tumor therapy (see Table 1). VEGF is an attractive target because VEGF mRNA is upregulated in most human tumors (Ferrara and Davis-Smyth, 1997). Not only do hypoxic tumor cells secrete VEGF but tumor-associated stroma can also be a major site of VEGF synthesis (Fukumura *et al.*, 1998). VEGF expression correlates well with microvessel density in breast and gastric carcinoma and in several other malignancies (Maeda *et al.*, 1996; Gasparini *et al.*, 1997). In addition, VEGF plasma levels are higher in cancer patients than in tumor-free individuals, and high VEGF levels prior to chemotherapy are indicative of a poor outcome (Salven *et al.*, 1998). The first direct evidence that VEGF is required for tumor angiogenesis came from studies using a monoclonal antibody against VEGF demonstrating that a range of human xenografts can be growth-inhibited in nude mice with this reagent (Kim *et al.*, 1993). Such treatment caused reductions in the diameter, tortuosity, and permeability of tumor blood vessels and eventually led to regression of these vessels (Yuan *et al.*, 1996). Vessel regression probably occurred because VEGF in addition to being angiogenic, is a survival factor for the endothelium of immature vessels (Alon *et al.*, 1995).

There are numerous opportunities for inhibition of the VEGF signaling pathway, and many of these targets are being exploited in programs for development of antiangiogenesis drugs. These include the following. (1) The synthesis of VEGF may be inhibited in hypoxic tumor cells using antisense RNA (Saleh *et al.*, 1996) or ribozymes (Pavco *et al.*, 2000). (2) Binding of VEGF to its receptors may be blocked: approaches for blocking these interactions include use of soluble VEGF receptor extracellular domains (Takayama *et al.*, 2000; Gagnon *et al.*, 2000), neutralizing anti-VEGF antibodies (Borgström *et al.*, 1999; Rowe *et al.*, 2000), neutralizing anti-VEGF receptor antibodies (Kubo *et al.*, 2000; Zhu and Witte, 1999; Prewett *et al.*, 1999), and generation of short peptides (Binétruy-Tournaire *et al.*, 2000) or other compounds (Aviezer *et al.*, 2000) that block ligand–receptor interactions. Neutralizing anti-VEGF antibodies are currently in clinical trials as anticancer agents (see Table 1). (3) Blocking the tyrosine kinase activity of VEGF receptors: numerous synthetic antiangiogenic compounds have been developed that inhibit the tyrosine kinase activity of VEGFR-2 (Sun *et al.*, 2000; Vajkoczy *et al.*, 2000; Bold *et al.*, 2000; Wood *et al.*, 2000; Wedge *et al.*, 2000; Xu *et al.*, 2000; Hennequin *et al.*, 1999; Shaheen *et al.*, 1999; Angelov *et al.*, 1999), some of which are in clinical trial (Table 1). (4) Future approaches will target the components of the intracellular signal transduction pathways activated by VEGF receptors (see Table 1, Emerging targets). Targets will include cytoplasmic signaling molecules and transcription factors. However, these approaches will require more precise delineation of the signaling pathways, transcription factors, and downstream genes regulated by the VEGF-regulated transcription factors. Some of this information might be revealed by DNA microarray and proteomic studies ongoing in many laboratories.

Although VEGF has been the best studied of the growth factors as a target for antiangiogenic therapy, there is evidence that inhibition of other VEGF family members may also lead to new antitumor therapeutics. It was more recently shown that signaling via VEGFR-3, a receptor that binds VEGF-C (Joukov *et al.*, 1996) and VEGF-D (Achen *et al.*, 1998) but not VEGF, also plays an important role in tumor angiogenesis. Administration of a monoclonal antibody that blocked the binding of these ligands to VEGFR-3 caused hemorrhage of tumor blood vessels and blocked solid tumor growth in a mouse tumor model (Kubo *et al.*, 2000). Clearly VEGF-C, VEGF-D, and VEGFR-3 are all potential antitumor drug targets. In addition to VEGF family members, bFGF may also be a target for blocking tumor growth as indicated by the finding that antisense oligonucleotides targeting bFGF mRNA blocked AIDS-Kaposi's sarcoma cell growth, tumor angiogenesis, and tumor formation in nude mice (Ensoli *et al.*, 1994).

C. Vascular remodeling

The endothelial cell receptor tyrosine kinase Tie2 and its ligands the angiopoietins (Angs) play important roles in vascular remodeling (Holash *et al.*, 1999). Ang-1 is an activating ligand for Tie2 (Davis *et al.*, 1996), whereas Ang-2 is a natural antagonist (Maisonpierre *et al.*, 1997). Ang-1 is thought to signal vessel stabilization by promoting interactions between endothelial cells and support cells such as smooth muscle cells. Ang-2 destabilizes blood vessels. When VEGF is present, Ang-2 induced vascular destabilization is overcome, and both angiogenesis and vascular sprouting occurs. In contrast, in the absence of VEGF, the endothelial cells apoptose and the vessels regress (Gale and Yancopoulos, 1999). The Tie2 signaling system is amenable for therapeutic manipulation to block angiogenesis (Holash et al., 1999). In experimental mouse tumor models, tumor angiogenesis, growth, and metastasis were inhibited using a soluble version of Tie2 consisting of the extracellular domain known to block activation of the endogenous receptor (Lin et al., 1998a; Lin et al., 1997). This effect was observed when tumor bearing mice were treated with the soluble receptor extracellular domain (Lin et al., 1997) or an adenoviral vector encoding this protein (Lin et al., 1998b).

D. Endothelial cell migration

During angiogenesis, endothelial cells degrade their basement membrane and invade the extravascular matrix that surrounds the vessel (Klagsburn and D'Amore, 1991). These cells eventually proliferate and migrate to the site of the new vessel. The migration of endothelial cells is dependent on interactions with the extracellular matrix. In particular, proteins such as laminin, collagens, proteoglycans, and matrix-associated growth factors are important for this process (see Table 1, Matrix, basement membrane, and cell–matrix adhesion molecules). In addition, the integrin family of adhesion molecules are involved, in particular the $\beta 1$ and $\beta 3$ subfamilies (Senger et al., 1997). Therefore, many of the molecules listed in Section II, A (Matrix, basement membrane, and cell–matrix adhesion molecules) may play important roles in the migration of endothelial cells and supporting cells.

E. Endothelial cell survival

The lack of survival or apoptosis of endothelial cells is a common theme in antiangiogenic therapy. The withdrawl of growth and/or survival signals or the induction of apoptosis by other molecules are mechanisms that lead to the death of neovascular endothelial cells. VEGF is known to behave as an endothelial cell survival factor in certain physiological situations (Gerber et al., 1998a, 1999). The antiapoptotic proteins Bcl-2 and A1 have also been shown to be altered in response to stimulation of endothelial cells with VEGF (Gerber et al., 1998b). Furthermore, thrombospondin-1 is a molecule that inhibits angiogenesis in an apoptosis-dependent manner (Jimenez et al., 2000). Other cytoplasmic signaling molecules such as the Shb adaptor protein are also involved in the apoptotic process and are therefore potential targets for antiangiogenic therapy (Gerber *et al.*, 1998a; Dixelius *et al.*, 2000).

F. Mechanism unknown

Many of the agents or targets being tested for antiangiogenic therapy have mechanisms of action which are uncharacterized (Table 1). Progress is being made in elucidating these processes, and some examples clearly illustrate these advances. Two of the most publicized angiogenesis inhibitors of recent times are angiostatin (O'Reilly *et al.*, 1994) and endostatin (O'Reilly *et al.*, 1997). These molecules were isolated from tumor models in which removal of the primary tumor caused increased growth of metastases, suggesting the presence of angiogenesis inhibitors produced by the primary tumor. Angiostatin and endostatin were isolated and found to be naturally occurring inhibitors of angiogenesis which are derived by proteolytic processing of plasminogen and collagen XVIII, respectively. They act as potent inhibitors of tumor angiogenesis. A number of studies have demonstrated the ability of angiostatin and endostatin to inhibit proliferation and to induce apoptosis of endothelial cells (Claesson-Welsh *et al.*, 1998; Dhanabal *et al.*, 1999), resulting in tumor regression without the development of drug-resistant tumor cell populations (Boehm *et al.*, 1997). Others have suggested that angiostatin may act via circulating endothelial progenitor cells, as these cells are more sensitive to the action of angiostatin than endothelial cells which are already established in the vessels of tissues (Ito *et al.*, 1999).

Some studies have shown that the MMP family is involved in regulating the production of angiostatin (O'Reilly *et al.*, 1999), and others have shown that enzymes such as elastase (Wen *et al.*, 1999) and cathepsin L (Felbor *et al.*, 2000) can cleave endostatin. Despite some controversy over the publicity associated with angiostatain and endostatin (Salmon, 1998), it is generally accepted that these agents have potent biological effects (O'Reilly *et al.*, 1996, 1997) and are excellent candidates for antiangiogenic approaches in human cancer therapy. The results of clinical trials with these two agents are being awaited with great interest.

III. FUTURE DIRECTIONS

The use of strategies to target the growth of blood vessels in both primary and metastatic tumors has provided an exciting new approach for development of cancer therapeutics. This approach has a number of theoretical advantages (1) Selective toxicity—As endothelial cells have a very low turnover under normal physiological conditions, targeting molecules specific for angiogenic endothelial cells in a tumor provides selective toxicity, (2) Drug resistance does not arise—The low mutation rate of endothelial cells versus the genetically unstable tumor cells means that cells resistant to antiangiogenic drugs are much less likely to arise, (3) Access of drug to target—As vessels are in direct contact with blood, therapeutics administered systemically can have ready access to the target endothelial cells, (4) Capacity to block the generation of metastases—The angiogenic switch is required for the spread and subsequent expansion of small populations of metastatic cells.

One area of tumor therapy that has received little attention in comparison to antiangiogenesis strategies is lymphangiogenesis—the growth of lymphatic vessels. The establishment of lymphatics in solid tumors plays a role in the spread of tumor cells. This area has not been thoroughly investigated, and the discovery of specific markers which discriminate lymphatic endothelium from vascular endothelium in tumors will greatly assist the development of this important area.

Another major challenge in the field of antiangiogenic therapeutics is the development of surrogate endpoints for cancer therapy (Schatzkin, 2000; Deplanque and Harris, 2000; Brasch and Turetschek, 2000). We require markers that are capable of indicating, in an early and reliable fashion, the effectiveness of antiangiogenic cancer therapy. This would allow the rapid evaluation of potential antiangiogenic agents without the need for long expensive studies. Currently there are a number of markers, based on the abnormal circulation due to the hyperpermeability of tumor blood vessels, which show potential for use in development of surrogate end points, but their validation in the clinical setting is still required (Deplanque and Harris, 2000; Schatzkin, 2000). The characterization of downstream targets for many of the pathways that regulate endothelial cell proliferation and migration also provides an opportunity to use these as readouts of receptor blocking or inactivation. The use of magnetic resonance imaging and macromolecular contrast media to visualize tumor microvasculature offers the opportunity to assess the efficiency of antiangiogenic agents in a noninvasive fashion (Brasch and Turetschek, 2000).

Discoveries over the past decade have defined molecules that have natural antiangiogenic capacity or act as inhibitors of angiogenesis *in vivo* (see Table 1). In addition, the understanding at the molecular level of how angiogenesis is controlled has led to the development of specific inhibitors of pathways involved in the proliferation, differentiation, migration, and establishment of new blood vessels (see Insight Progress. *Nature* **407**, 219–270, 2000). Many of these compounds are currently undergoing evaluation in clinical trials or are in preclinical evaluation in animal models. The next decade promises to reveal the true worth of this approach.

Acknowledgments

The authors thank Tony Burgess for comments on the manuscript. The work of the authors is supported by the National Health and Medical Research Council of Australia and the Anti-Cancer Council of Victoria. We apologize to those whose work is not cited here due to restrictions of space.

References

Abounader, R., Ranganathan, S., Lal, B., Fielding, K., Book, A., Dietz, H., Burger, P., and Laterra, J. (1999). Reversion of human glioblastoma malignancy by U1 small nuclear RNA/ribozyme targeting of scatter factor/hepatocyte growth factor and c-met expression. *J. Natl. Cancer Inst.* **91**, 1548–1556.

Achen, M. G., Jeltsch, M., Kukk, E., Mäkinen, T., Vitali, A., Wilks, A. F., Alitalo, K., and Stacker, S. A. (1998). Vascular endothelial growth factor D (VEGF-D) is a ligand for the tyrosine kinases VEGF receptor 2 (Flk-1) and VEGF receptor 3 (Flt-4). *Proc. Natl. Acad. Sci. U.S.A.* **95**, 548–553.

Achen, M. G., Roufail, S., Domagala, T., Catimel, B., Nice, E. C., Geleick, D. M., Murphy, R., Scott, A. M., Caesar, C., Makinen, T., Alitalo, K., and Stacker, S. A. (2000). Monoclonal antibodies to vascular endothelial growth factor-D block interactions with both VEGF receptor 2 and VEGF receptor-3. *Eur. J. Biochem.* **267**, 2505–2515.

Addison, C. L., Arenberg, D. A., Morris, S. B., Xue, Y. Y., Burdick, M. D., Mulligan, M. S., Iannettoni, M. D., and Strieter, R. M. (2000). The CXC chemokine, monokine induced by interferon-γ, inhibits non-small cell lung carcinoma tumor growth and metastasis. *Hum. Gene Ther.* **11**, 247–261.

Alon, T., Hemo, I., Itin, A., Pe'er, J., Stone, J., and Keshet, E. (1995). Vascular endothelial growth factor acts as a survival factor for newly formed retinal vessels and has implications for retinopathy of prematurity. *Nat. Med.* **1**, 1024–1028.

Alves, D. P., Teruszkin, B., I, Henriques, S. N., Oliveira, C. R., Coelho Kaplan, M. A., Currie, C. M., and Costa Carvalho, M. G. (2000). Chrysobalanus icaco L. extract for antiangiogenic potential observation. *Int. J. Mol. Med.* **5**, 667–669.

Angelov, L., Salhia, B., Roncari, L., McMahon, G., and Guha, A. (1999). Inhibition of angiogenesis by blocking activation of the vascular endothelial growth factor receptor 2 leads to decreased growth of neurogenic sarcomas. *Cancer Res.* **59**, 5536–5541.

Arbiser, J. L., Klauber, N., Rohan, R., van Leeuwen, R., Huang, M. T., Fisher, C., Flynn, E., and Byers, H. R. (1998). Curcumin is an *in vivo* inhibitor of angiogenesis. *Mol. Med.* **4**, 376–383.

Arrieta, O., Guevara, P., Reyes, S., Ortiz, A., Rembao, D., and Sotelo, J. (1998). Protamine inhibits angiogenesis and growth of C6 rat glioma; a synergistic effect when combined with carmustine. *Eur. J. Cancer* **34**, 2101–2106.

Asano, M., Yukita, A., and Suzuki, H. (1999). Wide spectrum of antitumor activity of a neutralizing monoclonal antibody to human vascular endothelial growth factor. *Jpn. J. Cancer Res.* **90**, 93–100.

Ashton, A. W., Yokota, R., John, G., Zhao, S., Suadicani, S. O., Spray, D. C., and Ware, J. A. (1999). Inhibition of endothelial cell migration, intercellular communication, and vascular tube formation by thromboxane A(2). *J. Biol. Chem.* **274**, 35562–35570.

Aviezer, D., Cotton, S., David, M., Segev, A., Khaselev, N., Galili, N., Gross, Z., and Yayon, A. (2000). Porphyrin analogues as novel antagonists of fibroblast growth factor and vascular endothelial growth factor receptor binding that inhibit endothelial cell proliferation, tumor progression, and metastasis. *Cancer Res.* **60**, 2973–2980.

Bauer, K. S., Cude, K. J., Dixon, S. C., Kruger, E. A., and Figg, W. D. (2000). Carboxyamido-triazole inhibits angiogenesis by blocking the calcium-mediated nitric-oxide synthase–vascular endothelial growth factor pathway. *J. Pharmacol. Exp. Ther.* **292**, 31–37.

Bell, C., Lynam, E., Landfair, D. J., Janjic, N., and Wiles, M. E. (1999). Oligonucleotide NX1838 inhibits VEGF$_{165}$-mediated cellular responses *in vitro*. In *Vitro Cell Dev. Biol. Anim.* **35**, 533–542.

Belletti, B., Ferraro, P., Arra, C., Baldassarre, G., Bruni, P., Staibano, S., Rosa, G. D., Salvatore, G., Fusco, A., Persico, M. G., and Viglietto, G. (1999). Modulation of *in vivo* growth of thyroid tumor-derived cell lines by sense and antisense vascular endothelial growth factor gene. *Oncogene* **18**, 4860–4869.

Belotti, D., Vergani, V., Drudis, T., Borsotti, P., Pitelli, M. R., Viale, G., Giavazzi, R., and Taraboletti, G. (1996). The microtubule-affecting drug paclitaxel has antiangiogenic activity. *Clin. Cancer Res.* **2**, 1843–1849.

Bhushan, M., McLaughlin, B., Weiss, J. B., and Griffiths, C. E. (1999). Levels of endothelial cell stimulating angiogenesis factor and vascular endothelial growth factor are elevated in psoriasis. *Br. J. Dermatol.* **141**, 1054–1060.

Binétruy-Tournaire, R., Demangel, C., Malavaud, B., Vassy, R., Rouyre, S., Kraemer, M., Plouët, J., Derbin, C., Perret, G., and Mazié, J. C. (2000). Identification of a peptide blocking vascular endothelial growth factor (VEGF)-mediated angiogenesis. *EMBO J.* **19**, 1525–1533.

Boehm, T., Folkman, J., Browder, T., and O'Reilly, M. S. (1997). Antiangiogenic therapy of experimental cancer does not induce acquired drug resistance. *Nature* **390**, 404–407.

Bold, G., Altmann, K.-H., Frei, J., Lang, M., Manley, P. W., Traxler, P., Wietfeld, B., Brüggen, J., Buchdunger, E., Cozens, R., Ferrari, S., Furet, P., Hofmann, F., Martiny-Baron, G., Mestan, J., Rösel, J., Sills, M., Stover, D., Acemoglu, F., Boss, E., Emmenegger, R., Lässer, L., Masso, E., Roth, R., Schlachter, C., and Vetterli, W. (2000). New anilinophthalazines as potent and orally well absorbed inhibitors of the VEGF receptor tyrosine kinases useful as antagonists of tumor-driven angiogenesis. *J. Med. Chem.* **43**, 2310–2323.

Bond, S. J., Klein, S. A., Anderson, G. L., and Wittliff, J. L. (2000). Interaction of angiogenesis inhibitor TNP-470 with basic fibroblast growth factor receptors. *J. Surg. Res.* **92**, 18–22.

Borgström, P., Bourdon, M. A., Hillan, K. J., Sriramarao, P., and Ferrara, N. (1998). Neutralizing anti-vascular endothelial growth factor antibody completely inhibits angiogenesis and growth of human prostate carcinoma micro tumors *in vivo*. *Prostate* **35**, 1–10.

Borgström, P., Gold, D. P., Hillan, K. J., and Ferrara, N. (1999). Importance of VEGF for breast cancer angiogenesis *in vivo*: Implications from intravital microscopy of combination treatments with an anti-VEGF neutralizing monoclonal antibody and doxorubicin. *Anticancer Res.* **19**, 4203–4214.

Brasch, R., and Turetschek, K. (2000). MRI characterization of tumors and grading angiogenesis using macromolecular contrast media: Status report. *Eur. J. Radiol.* **34**, 148–155.

Brooks, P. C., Clark, R. A. F., and Cheresh, D. A. (1994a). Requirement of vascular integrin $\alpha_v\beta_3$ for angiogenesis. *Science* **264**, 569–571.

Brooks, P. C., Montgomery, A. M. P., Rosenfeld, M., Reisfeld, R. A., Hu, T., Klier, G., and Cheresh, D. A. (1994b). Integrin $\alpha_v\beta_3$ antagonists promote tumor regression by inducing apoptosis of angiogenic blood vessels. *Cell* **79**, 1157–1164.

Brooks, P. C., Strömblad, S., Klemke, R., Visscher, D., Sarkar, F. H., and Cheresh, D. A. (1995). Antiintegrin $\alpha_v\beta_3$ blocks human breast cancer growth and angiogenesis in human skin. *J. Clin. Invest.* **96**, 1815–1822.

Brooks, P. C., Silletti, S., von Schalscha, T. L., Friedlander, M., and Cheresh, D. A. (1998). Disruption of angiogenesis by PEX, a noncatalytic metalloproteinase fragment with integrin binding activity. *Cell* **92**, 391–400.

Bruns, C. J., Harbison, M. T., Davis, D. W., Portera, C. A., Tsan, R., McConkey, D. J., Evans, D. B., Abbruzzese, J. L., Hicklin, D. J., and Radinsky, R. (2000). Epidermal growth factor receptor blockade with C225 plus gemcitabine results in regression of human pancreatic carcinoma growing orthotopically in nude mice by antiangiogenic mechanisms. *Clin. Cancer Res.* **6**, 1936–1948.

Cai, T., Fassina, G., Morini, M., Aluigi, M. G., Masiello, L., Fontanini, G., D Agostini, F., De Flora, S., Noonan, D. M., and Albini, A. (1999). N-Acetylcysteine inhibits endothelial cell invasion and angiogenesis. *Lab. Invest.* **79,** 1151–1159.

Cao, R., Farnebo, J., Kurimoto, M., and Cao, Y. (1999a). Interleukin-18 acts as an angiogenesis and tumor suppressor. *FASEB J.* **13,** 2195–2202.

Cao, R., Wu, H.-L., Veitonmäki, N., Linden, P., Farnebo, J., Shi, G.-Y., and Cao, Y. (1999b). Suppression of angiogenesis and tumor growth by the inhibitor K1–5 generated by plasmin-mediated proteolysis. *Proc. Natl. Acad. Sci. U.S.A.* **96,** 5728–5733.

Chen, Y., Wiesmann, C., Fuh, G., Li, B., Christinger, H. W., McKay, P., de Vos, A. M., and Lowman, H. B. (1999). Selection and analysis of an optimized anti-VEGF antibody: Crystal structure of an affinity-matured Fab in complex with antigen. *J. Mol. Biol.* **293,** 865–881.

Cheng, S.-Y., Huang, H.-J. S., Nagane, M., Ji, X.-D., Wang, D., Shih, C. C. Y., Arap, W., Huang, C.-M., and Cavenee, W. K. (1996). Suppression of glioblastoma angiogenicity and tumorigenicity by inhibition of endogenous expression of vascular endothelial growth factor. *Proc. Natl. Acad. Sci. U.S.A.* **93,** 8502–8507.

Claesson-Welsh, L., Welsh, M., Ito, N., Anand-Apte, B., Soker, S., Zetter, B., O Reilly, M., and Folkman, J. (1998). Angiostatin induces endothelial cell apoptosis and activation of focal adhesion kinase independently of the integrin-binding motif RGD. *Proc. Natl. Acad. Sci. U.S.A.* **95,** 5579–5583.

Clements, M. K., Jones, C. B., Cumming, M., and Daoud, S. S. (1999). Antiangiogenic potential of camptothecin and topotecan. *Cancer Chemother. Pharmacol.* **44,** 411–416.

Coleman, K. R., Braden, G. A., Willingham, M. C., and Sane, D. C. (1999). Vitaxin, a humanized monoclonal antibody to the vitronectin receptor ($\alpha_v\beta_3$), reduces neointimal hyperplasia and total vessel area after balloon injury in hypercholesterolemic rabbits. *Circ. Res.* **84,** 1268–1276.

Colman, R. W., Jameson, B. A., Lin, Y., Johnson, D., and Mousa, S. A. (2000). Domain 5 of high molecular weight kininogen (kininostatin) down-regulates endothelial cell proliferation and migration and inhibits angiogenesis. *Blood* **95,** 543–550.

Colorado, P. C., Torre, A., Kamphaus, G., Maeshima, Y., Hopfer, H., Takahashi, K., Volk, R., Zamborsky, E. D., Herman, S., Sarkar, P. K., Ericksen, M. B., Dhanabal, M., Simons, M., Post, M., Kufe, D. W., Weichselbaum, R. R., Sukhatme, V. P., and Kalluri, R. (2000). Anti-angiogenic cues from vascular basement membrane collagen. *Cancer Res.* **60,** 2520–2526.

Cruz, A., DeFouw, L. M., and DeFouw, D. O. (2000). Restrictive endothelial barrier function during normal angiogenesis *in vivo*: Partial dependence on tyrosine dephosphorylation of β-catenin. *Microvasc. Res.* **59,** 195–203.

Cuevas, P., Carceller, F., Reimers, D., Cuevas, B., Lozano, R. M., and Gimenez-Gallego, G. (1999). Inhibition of intra-tumoral angiogenesis and glioma growth by the fibroblast growth factor inhibitor 1,3,6-naphthalenetrisulfonate. *Neurol. Res.* **21,** 481–487.

D'Amato, R. J., Loughnan, M. S., Flynn, E., and Folkman, J. (1994). Thalidomide is an inhibitor of angiogenesis. *Proc. Natl. Acad. Sci. U.S.A.* **91,** 4082–4085.

D'Angelo, G., Martini, J. F., Iiri, T., Fantl, W. J., Martial, J., and Weiner, R. I. (1999). 16K human prolactin inhibits vascular endothelial growth factor-induced activation of Ras in capillary endothelial cells. *Mol. Endocrinol.* **13,** 692–704.

Davis, S., Aldrich, T. H., Jones, P. F., Acheson, A., Compton, D. L., Jain, V., Ryan, T. E., Bruno, J., Radziejewski, C., Maisonpierre, P. C., and Yancopoulos, G. D. (1996). Isolation of Angiopoietin-1 a ligand for the TIE2 receptor, by secretion-trap expression cloning. *Cell* **87,** 1161–1169.

Dawson, D. W., Volpert, O. V., Gillis, P., Crawford, S. E., Xu, H.-J., Benedict, W., and Bouck, N. P. (1999). Pigment epithelium-derived factor: A potent inhibitor of angiogenesis. *Science* **285,** 245–248.

Dechantsreiter, M. A., Planker, E., Matha, B., Lohof, E., Holzemann, G., Jonczyk, A., Goodman, S. L., and Kessler, H. (1999). N-Methylated cyclic RGD peptides as highly active and selective $\alpha V \beta 3$ integrin antagonists. *J. Med. Chem.* **42**, 3033–3040.

Deplanque, G., and Harris, A. L. (2000). Anti-angiogenic agents. Clinical trial design and therapies in development. *Eur. J. Cancer* **36**, 1713–1724.

Dewhirst, M. W., Tso, C. Y., Oliver, R., Gustafson, C. S., Secomb, T. W., and Gross, J. F. (1989). Morphologic and hemodynamic comparison of tumor and healing normal tissue microvasculature. *Int. J. Radiat. Oncol. Biol. Phys.* **17**, 91–99.

Dhanabal, M., Ramchandran, R., Waterman, M. J. F., Lu, H., Knebelmann, B., Segal, M., and Sukhatme, V. P. (1999). Endostatin induces endothelial cell apoptosis. *J. Biol. Chem.* **274**, 11721–11726.

Diaz, B. V., Lenoir, M. C., Ladoux, A., Frelin, C., Demarchez, M., and Michel, S. (2000). Regulation of vascular endothelial growth factor expression in human keratinocytes by retinoids. *J. Biol. Chem.* **275**, 642–650.

Dimitroff, C. J., Klohs, W., Sharma, A., Pera, P., Driscoll, D., Veith, J., Steinkampf, R., Schroeder, M., Klutchko, S., Sumlin, A., Henderson, B., Dougherty, T. J., and Bernacki, R. J. (1999). Anti-angiogenic activity of selected receptor tyrosine kinase inhibitors, PD166285 and PD173074: Implications for combination treatment with photodynamic therapy. *Invest. New Drugs* **17**, 121–135.

Dixelius, J., Larsson, H., Sasaki, T., Holmqvist, K., Lu, L., Engstrom, A., Timpl, R., Welsh, M., and Claesson-Welsh, L. (2000). Endostatin-induced tyrosine kinase signaling through the Shb adaptor protein regulates endothelial cell apoptosis. *Blood* **95**, 3403–3411.

Duenas, Z., Torner, L., Corbacho, A. M., Ochoa, A., Gutierrez-Ospina, G., Lopez-Barrera, F., Barrios, F. A., Berger, P., Martinez, D. L. E., and Clapp, C. (1999). Inhibition of rat corneal angiogenesis by 16-kDa prolactin and by endogenous prolactin-like molecules. *Invest. Ophthalmol. Visual Sci.* **40**, 2498–2505.

Elkin, M., Reich, R., Nagler, A., Aingorn, E., Pines, M., de Groot, N., Hochberg, A., and Vlodavsky, I. (1999a). Inhibition of matrix metalloproteinase-2 expression and bladder carcinoma metastasis by halofuginone. *Clin. Cancer Res.* **5**, 1982–1988.

Elkin, M., Ariel, I., Miao, H. Q., Nagler, A., Pines, M., de Groot, N., Hochberg, A., and Vlodavsky, I. (1999b). Inhibition of bladder carcinoma angiogenesis, stromal support, and tumor growth by halofuginone. *Cancer Res.* **59**, 4111–4118.

Engelberg, H. (1999). Actions of heparin that may affect the malignant process. *Cancer* **85**, 257–272.

Ensoli, B., Markham, P., Kao, V., Barillari, G., Fiorelli, V., Gendelman, R., Raffeld, M., Zon, G., and Gallo, R. C. (1994). Block of AIDS-Kaposi's sarcoma (KS) cell growth, angiogenesis, and lesion formation in nude mice by antisense oligonucleotide targeting basic fibroblast growth factor. A novel strategy for the therapy of KS. *J. Clin. Invest.* **94**, 1736–1746.

Ergün, S., Kilic, N., Ziegeler, G., Hansen, A., Nollau, P., Götze, J., Wurmbach, J.-H., Horst, A., Weil, J., Fernando, M., and Wagener, C. (2000). CEA-related cell adhesion molecule 1: A potent angiogenic factor and a major effector of vascular endothelial growth factor. *Mol. Cell* **5**, 311–320.

Fabre, J. E., Rivard, A., Magner, M., Silver, M., and Isner, J. M. (1999). Tissue inhibition of angiotensin-converting enzyme activity stimulates angiogenesis *in vivo*. *Circulation* **99**, 3043–3049.

Fachinger, G., Deutsch, U., and Risau, W. (1999). Functional interaction of vascular endothelial-protein-tyrosine phosphatase with the angiopoietin receptor tie-2. *Oncogene* **18**, 5948–5953.

Fathallah-Shaykh, H. M., Zhao, L.-J., Kafrouni, A. I., Smith, G. M., and Forman, J. (2000). Gene transfer of IFN-γ into established brain tumors represses growth by antiangiogenesis. *J. Immunol.* **164**, 217–222.

Felbor, U., Dreier, L., Bryant, R. A. R., Ploegh, H. L., Olsen, B. R., and Mothes, W. (2000). Secreted cathepsin L generates endostatin from collagen XVIII. *EMBO J.* **19**, 1187–1194.

Feldkamp, M. M., Lau, N., and Guha, A. (1999). Growth inhibition of astrocytoma cells by farnesyl transferase inhibitors is mediated by a combination of anti-proliferative, pro-apoptotic and anti-angiogenic effects. *Oncogene* **18**, 7514–7526.

Feleszko, W., Balkowiec, E. Z., Sieberth, E., Marczak, M., Dabrowska, A., Giermasz, A., Czajka, A., and Jakobisiak, M. (1999). Lovastatin and tumor necrosis factor-α exhibit potentiated antitumor effects against Ha-ras-transformed murine tumor via inhibition of tumor-induced angiogenesis. *Int. J. Cancer* **81**, 560–567.

Ferrara, N., and Davis-Smyth, T. (1997). The biology of vascular endothelial growth factor. *Endocr. Rev.* **18**, 4–25.

Fife, R. S., Sledge, G. W., Jr., Sissons, S., and Zerler, B. (2000). Effects of tetracyclines on angiogenesis *in vitro*. *Cancer Lett.* **153**, 75–78.

Fine, H. A., Figg, W. D., Jaeckle, K., Wen, P. Y., Kyritsis, A. P., Loeffler, J. S., Levin, V. A., Black, P. M., Kaplan, R., Pluda, J. M., and Yung, W. K. (2000). Phase II trial of the antiangiogenic agent thalidomide in patients with recurrent high-grade gliomas. *J. Clin. Oncol.* **18**, 708–715.

Folkman, J., and Shing, Y. (1992). Angiogenesis. *J. Biol. Chem.* **267**, 10931–10934.

Folkman, J., Long, D. M., and Becker, F. F. (1963). Growth and metastasis of tumor in organ culture. *Cancer* **16**, 453–467.

Fong, T. A. T., Shawver, L. K., Sun, L., Tang, C., App, H., Powell, T. J., Kim, Y. H., Schreck, R., Wang, X., Risau, W., Ullrich, A., Hirth, K. P., and McMahon, G. (1999). SU5416 is a potent and selective inhibitor of the vascular endothelial growth factor receptor (Flk-1/KDR) that inhibits tyrosine kinase catalysis, tumor vascularization, and growth of multiple tumor types. *Cancer Res.* **59**, 99–106.

Fortier, A. H., Nelson, B. J., Grella, D. K., and Holaday, J. W. (1999). Antiangiogenic activity of prostate-specific antigen. *J. Natl. Cancer Inst.* **91**, 1635–1640.

Fosslien, E. (2000). Molecular pathology of cyclooxygenase-2 in neoplasia. *Ann. Clin. Lab. Sci.* **30**, 3–21.

Fotsis, T., Pepper, M. S., Montesano, R., Aktas, E., Breit, S., Schweigerer, L., Rasku, S., Wahala, K., and Adlercreutz, H. (1998). Phytoestrogens and inhibition of angiogenesis. *Baillieres Clin. Endocrinol. Metab.* **12**, 649–666.

Friedlander, M., Brooks, P. C., Shaffer, R. W., Kincaid, C. M., Varner, J. A., and Cheresh, D. A. (1995). Definition of two angiogenic pathways by distinct α_v integrins. *Science* **270**, 1500–1502.

Fukumura, D., Xavier, R., Sugiura, T., Chen, Y., Park, E.-C., Lu, N., Selig, M., Nielsen, G., Taksir, T., Jain, R. K., and Seed, B. (1998). Tumor induction of VEGF promoter activity in stromal cells. *Cell* **94**, 715–725.

Gabrilovich, D. I., Ishida, T., Nadaf, S., Ohm, J. E., and Carbone, D. P. (1999). Antibodies to vascular endothelial growth factor enhance the efficacy of cancer immunotherapy by improving endogenous dendritic cell function. *Clin. Cancer Res.* **5**, 2963–2970.

Gagnon, M. L., Bielenberg, D. R., Gechtman, Z., Miao, H.-Q., Takashima, S., Soker, S., and Klagsbrun, M. (2000). Identification of a natural soluble neuropilin-1 that binds vascular endothelial growth factor: *In vivo* expression and antitumor activity. *Proc. Natl. Acad. Sci. U.S.A.* **97**, 2573–2578.

Galardy, R. E., Grobelny, D., Foellmer, H. G., and Fernandez, L. A. (1994). Inhibition of angiogenesis by the matrix metalloprotease inhibitor N-[2R-2-(hydroxamidocarbonymethyl)-4-methylpentanoyl)]-L-tryptophan methylamide. *Cancer Res.* **54**, 4715–4718.

Gale, N. W., and Yancopoulos, G. D. (1999). Growth factors acting via endothelial cell-specific receptor tyrosine kinases: VEGFs, angiopoietins, and ephrins in vascular development. *Genes Dev.* **13**, 1055–1066.

Gamble, J. R., Matthias, L. J., Meyer, G., Kaur, P., Russ, G., Faull, R., Berndt, M. C., and Vadas, M. A. (1993). Regulation of *in vitro* capillary tube formation by anti-integrin antibodies. *J. Cell Biol.* **121**, 931–943.

Gamble, J., Meyer, G., Noack, L., Furze, J., Matthias, L., Kovach, N., Harlant, J., and Vadas, M. (1999). B1 integrin activation inhibits *in vitro* tube formation: Effects on cell migration, vacuole coalescence and lumen formation. *Endothelium* **7**, 23–34.

Gasparini, G., Toi, M., Gion, M., Verderio, P., Dittadi, R., Hanatani, M., Matsubara, I., Vinante, O., Bonoldi, E., Boracchi, P., Gatti, C., Suzuki, H., and Tominaga, T. (1997). Prognostic significance of vascular endothelial growth factor protein in node-negative breast carcinoma. *J. Natl. Cancer Inst.* **89**, 139–147.

Ge, S., Rempel, S. A., Divine, G., and Mikkelsen, T. (2000). Carboxyamido-triazole induces apoptosis in bovine aortic endothelial and human glioma cells. *Clin. Cancer Res.* **6**, 1248–1254.

Gengrinovitch, S., Greenberg, S. M., Cohen, T., Gitay-Goren, H., Rockwell, P., Maione, T. E., Levi, B.-Z., and Neufeld, G. (1995). Platelet factor-4 inhibits the mitogenic activity of $VEGF_{121}$ and $VEGF_{165}$ using several concurrent mechanisms. *J. Biol. Chem.* **270**, 15059–15065.

Gentilini, G., Kirschbaum, N. E., Augustine, J. A., Aster, R. H., and Visentin, G. P. (1999). Inhibition of human umbilical vein endothelial cell proliferation by the CXC chemokine, platelet factor 4 (PF4), is associated with impaired downregulation of p21(Cip1/WAF1). *Blood* **93**, 25–33.

Gerber, H. P., McMurtrey, A., Kowalski, J., Yan, M., Keyt, B. A., Dixit, V., and Ferrara, N. (1998a). Vascular endothelial growth factor regulates endothelial cell survival through the phosphatidylinositol 3′-kinase/Akt signal transduction pathway. Requirement for the flk-1/kdr activation. *J. Biol. Chem.* **273**, 30336–30343.

Gerber, H. P., Dixit, V., and Ferrara, N. (1998b). Vascular endothelial growth factor induces expression of the antiapoptotic proteins Bcl-2 and A1 in vascular endothelial cells. *J. Biol. Chem.* **273**, 13313–13316.

Gerber, H.-P., Hillan, K. J., Ryan, A. M., Kowalski, J., Keller, G.-A., Rangell, L., Wright, B. D., Radtke, F., Aguet, M., and Ferrara, N. (1999). VEGF is required for growth and survival in neonatal mice. *Development* **126**, 1149–1159.

Gerwins, P., Sköldenberg, E., and Claesson-Welsh, L. (2000). Function of fibroblast growth factors and vascular endothelial growth factors and their receptors in angiogenesis. *Crit. Rev. Oncol. Hematol.* **34**, 185–194.

Golab, J., Kozar, K., Kaminski, R., Czajka, A., Marczak, M., Switaj, T., Giermasz, A., Stoklosa, T., Lasek, W., Zagozdzon, R., Mucha, K., and Jakobisiak, M. (2000). Interleukin 12 and indomethacin exert a synergistic, angiogenesis-dependent antitumor activity in mice. *Life Sci.* **66**, 1223–1230.

Goldman, C. K., Kendall, R. L., Cabrera, G., Soroceanu, L., Heike, Y., Gillespie, G. Y., Siegal, G. P., Mao, X., Bett, A. J., Huckle, W. R., Thomas, K. A., and Curiel, D. T. (1998). Paracrine expression of a native soluble vascular endothelial growth factor receptor inhibits tumor growth, metastasis, and mortality rate. *Proc. Natl. Acad. Sci. U.S.A.* **95**, 8795–8800.

Gorski, D. H., Beckett, M. A., Jaskowiak, N. T., Calvin, D. P., Mauceri, H. J., Salloum, R. M., Seetharam, S., Koons, A., Hari, D. M., Kufe, D. W., and Weichselbaum, R. R. (1999). Blockage of the vascular endothelial growth factor stress response increases the antitumor effects of ionizing radiation. *Cancer Res.* **59**, 3374–3378.

Gross, D. J., Reibstein, I., Weiss, L., Slavin, S., Stein, I., Neeman, M., Abramovitch, R., and Benjamin, L. E. (1999). The antiangiogenic agent linomide inhibits the growth rate of von Hippel-Lindau paraganglioma xenografts to mice. *Clin. Cancer Res.* **5**, 3669–3675.

Guo, Y., Higazi, A. A., Arakelian, A., Sachais, B. S., Cines, D., Goldfarb, R. H., Jones, T. R., Kwaan, H., Mazar, A. P., and Rabbani, S. A. (2000). A peptide derived from the nonreceptor binding region of urokinase plasminogen activator (uPA) inhibits tumor progression and angiogenesis and induces tumor cell death *in vivo*. *FASEB J.* **14**, 1400–1410.

Harris, V. K., Coticchia, C. M., Kagan, B. L., Ahmad, S., Wellstein, A., and Riegel, A. T. (2000). Induction of the angiogenic modulator fibroblast growth factor-binding protein by epidermal growth factor is mediated through both MEK/ERK and p38 signal transduction pathways. *J. Biol. Chem.* **275**, 10802–10811.

Heath, E. I., and Grochow, L. B. (2000). Clinical potential of matrix metalloprotease inhibitors in cancer therapy. *Drugs* **59**, 1043–1055.

Hedges, J. C., Dechert, M. A., Yamboliev, I. A., Martin, J. L., Hickey, E., Weber, L. A., and Gerthoffer, W. T. (1999). A role for p38(MAPK)/ HSP27 pathway in smooth muscle cell migration. *J. Biol. Chem.* **274**, 24211–24219.

Hennequin, L. F., Thomas, A. P., Johnstone, C., Stokes, E. S., Ple, P. A., Lohmann, J. J., Ogilvie, D. J., Dukes, M., Wedge, S. R., Curwen, J. O., Kendrew, J., and Lambert-Van Der Brempt, C. (1999). Design and structure–activity relationship of a new class of potent VEGF receptor tyrosine kinase inhibitors. *J. Med. Chem.* **42**, 5369–5389.

Hermann, C., Assmus, B., Urbich, C., Zeiher, A. M., and Dimmeler, S. (2000). Insulin-mediated stimulation of protein kinase Akt: A potent survival signaling cascade for endothelial cells. *Arterioscler. Thromb. Vasc. Biol.* **20**, 402–409.

Holash, J., Wiegand, S. J., and Yancopoulos, G. D. (1999). New model of tumor angiogenesis: Dynamic balance between vessel regression and growth mediated by angiopoietins and VEGF. *Oncogene* **18**, 5356–5362.

Huang, L., Sankar, S., Lin, C., Kontos, C. D., Schroff, A. D., Cha, E. H., Feng, S.-M., Li, S.-F., Yu, Z., Van Etten, R. L., Blanar, M. A., and Peters, K. G. (1999). HCPTPA, a protein tyrosine phosphatase that regulates vascular endothelial growth factor receptor-mediated signal transduction and biological activity. *J. Biol. Chem.* **274**, 38183–38188.

Ito, H., Rovira, I. I., Bloom, M. L., Takeda, K., Ferrans, V. J., Quyyumi, A. A., and Finkel, T. (1999). Endothelial progenitor cells as putative targets for angiostatin. *Cancer Res.* **59**, 5875–5877.

Iurlaro, M., Vacca, A., Minischetti, M., Ribatti, D., Pellegrino, A., Sardanelli, A., Giacchetta, F., and Dammacco, F. (1998). Antiangiogenesis by cyclosporine. *Exp. Hematol.* **26**, 1215–1222.

Jadeski, L. C., and Lala, P. K. (1999). Nitric oxide synthase inhibition by *N*(G)-nitro-L-arginine methyl ester inhibits tumor-induced angiogenesis in mammary tumors. *Am. J. Pathol.* **155**, 1381–1390.

Jadeski, L. C., Hum, K. O., Chakraborty, C., and Lala, P. K. (2000). Nitric oxide promotes murine mammary tumour growth and metastasis by stimulating tumour cell migration, invasiveness and angiogenesis. *Int. J. Cancer* **86**, 30–39.

Jain, R. K. (1987). Transport of molecules across tumor vasculature. *Cancer Metastasis Rev.* **6**, 559–593.

Jankun, J., and Skrzypczak-Jankun, E. (1999). Molecular basis of specific inhibition of urokinase plasminogen activator by amiloride. *Cancer Biochem. Biophys.* **17**, 109–123.

Ji, W. R., Castellino, F. J., Chang, Y., Deford, M. E., Gray, H., Villarreal, X., Kondri, M. E., Marti, D. N., Llinas, M., Schaller, J., Kramer, R. A., and Trail, P. A. (1998). Characterization of kringle domains of angiostatin as antagonists of endothelial cell migration, an important process in angiogenesis. *FASEB J.* **12**, 1731–1738.

Jia, L., Wu, C. C., Guo, W., and Young, X. (2000). Antiangiogenic effects of S-nitrosocaptopril crystals as a nitric oxide donor. *Eur. J. Pharmacol.* **391**, 137–144.

Jiang, C., Jiang, W., Ip, C., Ganther, H., and Lu, J. (1999). Selenium-induced inhibition of angiogenesis in mammary cancer at chemopreventive levels of intake. *Mol. Carcinog.* **26**, 213–225.

Jimenez, B., Volpert, O. V., Crawford, S. E., Febbraio, M., Silverstein, R. L., and Bouck, N. (2000). Signals leading to apoptosis-dependent inhibition of neovascularization by thrombospondin-1. *Nat. Med.* **6**, 41–48.

Joe, Y. A., Hong, Y. K., Chung, D. S., Yang, Y. J., Kang, J. K., Lee, Y. S., Chang, S. I., You, W. K., Lee, H., and Chung, S. I. (1999). Inhibition of human malignant glioma growth *in vivo* by human recombinant plasminogen kringles 1–3. *Int. J. Cancer* **82**, 694–699.

Jones, M. K., Wang, H., Peskar, B. M., Levin, E., Itani, R. M., Sarfeh, I. J., and Tarnawski, A. S. (1999). Inhibition of angiogenesis by nonsteroidal anti-inflammatory drugs: Insight into mechanisms and implications for cancer growth and ulcer healing. *Nat. Med.* **5,** 1418–1423.

Joukov, V., Pajusola, K., Kaipainen, A., Chilov, D., Lahtinen, I., Kukk, E., Saksela, O., Kalkkinen, N., and Alitalo, K. (1996). A novel vascular endothelial growth factor, VEGF-C, is a ligand for the Flt-4 (VEGFR-3) and KDR (VEGFR-2) receptor tyrosine kinases. *EMBO J.* **15,** 290–298.

Joussen, A. M., Kruse, F. E., Volcker, H. E., and Kirchhof, B. (1999). Topical application of methotrexate for inhibition of corneal angiogenesis. *Graefe's Arch. Clin. Exp. Ophthalmol.* **237,** 920–927.

Juarez, C. P., Muino, J. C., Guglielmone, H., Sambuelli, R., Echenique, J. R., Hernandez, M., and Luna, J. D. (2000). Experimental retinopathy of prematurity: Angiostatic inhibition by nimodipine, ginkgo-biloba, and dipyridamole, and response to different growth factors. *Eur. J. Ophthalmol.* **10,** 51–59.

Kamiya, K., Konno, H., Tanaka, T., Baba, M., Matsumoto, K., Sakaguchi, T., Yukita, A., Asano, M., Suzuki, H., Arai, T., and Nakamura, S. (1999a). Antitumor effect on human gastric cancer and induction of apoptosis by vascular endothelial growth factor neutralizing antibody. *Jpn. J. Cancer Res.* **90,** 794–800.

Kamiya, K., Konno, H., Tanaka, T., Baba, M., Matsumoto, K., Sakaguchi, T., Yukita, A., Asano, M., Suzuki, H., Arai, T., and Nakamura, S. (1999b). Antitumor effect on human gastric cancer and induction of apoptosis by vascular endothelial growth factor neutralizing antibody. *Jpn. J. Cancer Res.* **90,** 794–800.

Kamphaus, G. D., Colorado, P. C., Panka, D. J., Hopfer, H., Ramchandran, R., Torre, A., Maeshima, Y., Mier, J. W., Sukhatme, V. P., and Kalluri, R. (2000). Canstatin, a novel matrix-derived inhibitor of angiogenesis and tumor growth. *J. Biol. Chem.* **275,** 1209–1215.

Kaplan, J. B., Sridharan, L., Zaccardi, J. A., Dougher-Vermazen, M., and Terman, B. I. (1997). Characterization of a soluble vascular endothelial growth factor receptor-immunoglobulin chimera. *Growth Factors.* **14,** 243–256.

Keane, M. P., Belperio, J. A., Arenberg, D. A., Burdick, M. D., Xu, Z. J., Xue, Y. Y., and Strieter, R. M. (1999). IFN-γ-inducible protein-10 attenuates bleomycin-induced pulmonary fibrosis via inhibition of angiogenesis. *J. Immunol.* **163,** 5686–5692.

Kim, K. J., Li, B., Houck, K., Winer, J., and Ferrara, N. (1992). The vascular endothelial growth factor proteins: Identification of biologically relevant regions by neutralizing monoclonal antibodies. *Growth Factors* **7,** 53–64.

Kim, K. J., Li, B., Winer, J., Armanini, M., Gillett, N., Phillips, H. S., and Ferrara, N. (1993). Inhibition of vascular endothelial growth factor-induced angiogenesis suppresses tumour growth *in vivo.* *Nature* **362,** 841–844.

Kinsella, M. G., Fischer, J. W., Mason, D. P., and Wight, T. N. (2000). Retrovirally mediated expression of decorin by macrovascular endothelial cells. Effects on cellular migration and fibronectin fibrillogenesis *in vitro.* *J. Biol. Chem.* **275,** 13924–13932.

Klagsburn, M., and D'Amore, P. A. (1991). Regulators of angiogenesis. *Annu. Rev. Physiol.* **53,** 217–239.

Klein-Soyer, C., Azorsa, D. O., Cazenave, J. P., and Lanza, F. (2000). CD9 participates in endothelial cell migration during *in vitro* wound repair. *Arterioscler. Thromb. Vasc. Biol.* **20,** 360–369.

Klotz, O., Park, J. K., Pleyer, U., Hartmann, C., and Baatz, H. (2000). Inhibition of corneal neovascularization by α(v)-integrin antagonists in the rat. *Graefe's Arch. Clin. Exp. Ophthalmol.* **238,** 88–93.

Kobayashi, S., Kimura, I., Fukuta, M., Kontani, H., Inaba, K., Niwa, M., Mita, S., and Kimura, M. (1999). Inhibitory effects of tetrandrine and related synthetic compounds on angiogenesis in streptozotocin-diabetic rodents. *Biol. Pharm. Bull.* **22,** 360–365.

Koch, A. E., Polverini, P. J., Kunkel, S. L., Harlow, L. A., DiPietro, L. A., Elner, V. M., Elner, S. G., and Strieter, R. M. (1993). Interleukin-8 as a macrophage-derived mediator of angiogenesis. *Science* **258**, 1798–1801.

Koyama, S., Takagi, H., Otani, A., Suzuma, K., Nishimura, K., and Honda, Y. (1999). Tranilast inhibits protein kinase C-dependent signalling pathway linked to angiogenic activities and gene expression of retinal microcapillary endothelial cells. *Br. J. Pharmacol.* **127**, 537–545.

Kruger, E. A., Blagosklonny, M. V., Dixon, S. C., and Figg, W. D. (1998). UCN-01, a protein kinase C inhibitor, inhibits endothelial cell proliferation and angiogenic hypoxic response. *Invasion Metastasis* **18**, 209–218.

Kubo, H., Fujiwara, T., Jussila, L., Hashi, H., Ogawa, M., Shimizu, K., Awane, M., Sakai, Y., Takabayashi, A., Alitalo, K., Yamaoka, Y., and Nishikawa, S.-I. (2000). Involvement of vascular endothelial growth factor receptor-3 in maintenance of integrity of endothelial cell lining during tumor angiogenesis. *Blood* **96**, 546–553.

Kumeda, S. I., Deguchi, A., Toi, M., Omura, S., and Umezawa, K. (1999). Induction of G1 arrest and selective growth inhibition by lactacystin in human umbilical vein endothelial cells. *Anticancer Res.* **19**, 3961–3968.

Kupprion, C., Motamed, K., and Sage, E. H. (1998). SPARC (BM-40, osteonectin) inhibits the mitogenic effect of vascular endothelial growth factor on microvascular endothelial cells. *J. Biol. Chem.* **273**, 29635–29640.

Kyriakides, T. R., Zhu, Y. H., Smith, L. T., Bain, S. D., Yang, Z., Lin, M. T., Danielson, K. G., Iozzo, R. V., LaMarca, M., McKinney, C. E., Ginns, E. I., and Bornstein, P. (1998). Mice that lack thrombospondin 2 display connective tissue abnormalities that are associated with disordered collagen fibrillogenesis, an increased vascular density, and a bleeding diathesis. *J. Cell Biol.* **140**, 419–430.

Laird, A. D., Vajkoczy, P., Shawver, L. K., Thurnher, A., Liang, C., Mohammadi, M., Schlessinger, J., Ullrich, A., Hubbard, S. R., Blake, R. A., Fong, T. A., Strawn, L. M., Sun, L., Tang, C., Hawtin, R., Tang, F., Shenoy, N., Hirth, K. P., McMahon, G., and Cherrington. (2000). SU6668 is a potent antiangiogenic and antitumor agent that induces regression of established tumors. *Cancer Res.* **60**, 4152–4160.

Lawnicka, H., Stepien, H., Wyczolkowska, J., Kolago, B., Kunert-Radek, J., and Komorowski, J. (2000). Effect of somatostatin and octreotide on proliferation and vascular endothelial growth factor secretion from murine endothelial cell line (HECa10) culture. *Biochem. Biophys. Res. Commun.* **268**, 567–571.

Lee, M. J., Thangada, S., Claffey, K. P., Ancellin, N., Liu, C. H., Kluk, M., Volpi, M., Sha afi, R. I., and Hla, T. (1999). Vascular endothelial cell adherens junction assembly and morphogenesis induced by sphingosine-1-phosphate. *Cell* **99**, 301–312.

Lee, T. H., Rhim, T., and Kim, S. S. (1998). Prothrombin kringle-2 domain has a growth inhibitory activity against basic fibroblast growth factor-stimulated capillary endothelial cells. *J. Biol. Chem.* **273**, 28805–28812.

Lein, M., Jung, K., Le, D. K., Hasan, T., Ortel, B., Borchert, D., Winkelmann, B., Schnorr, D., and Loenings, S. A. (2000). Synthetic inhibitor of matrix metalloproteinases (batimastat) reduces prostate cancer growth in an orthotopic rat model. *Prostate* **43**, 77–82.

Li, H., Griscelli, F., Lindenmeyer, F., Opolon, P., Sun, L. Q., Connault, E., Soria, J., Soria, C., Perricaudet, M., Yeh, P., and Lu, H. (1999). Systemic delivery of antiangiogenic adenovirus AdmATF induces liver resistance to metastasis and prolongs survival of mice. *Hum. Gene Ther.* **10**, 3045–3053.

Li, J., Post, M., Volk, R., Gao, Y., Li, M., Metais, C., Sato, K., Tsai, J., Aird, W., Rosenberg, R. D., Hampton, T. G., Sellke, F., Carmeliet, P., and Simons, M. (2000). PR39, a peptide regulator of angiogenesis. *Nat. Med.* **6**, 49–55.

Liaudet-Coopman, E. D., Berchem, G. J., and Wellstein, A. (1997). In vivo inhibition of angiogenesis and induction of apoptosis by retinoic acid in squamous cell carcinoma. *Clin. Cancer Res.* **3**, 179–184.

Lin, P., Polverini, P., Dewhirst, M., Shan, S., Rao, P. S., and Peters, K. (1997). Inhibition of tumor angiogenesis using a soluble receptor establishes a role for Tie2 in pathologic vascular growth. *J. Clin. Invest.* **100**, 2072–2078.

Lin, P., Buxton, J. A., Acheson, A., Radziejewski, C., Maisonpierre, P. C., Yancopoulos, G. D., Channon, K. M., Hale, L. P., Dewhirst, M. W., George, S. E., and Peters, K. G. (1998a). Antiangiogenic gene therapy targeting the endothelium-specific receptor tyrosine kinase Tie2. . *Proc. Natl. Acad. Sci. U.S.A.* **95**, 8829–8834.

Lin, P., Buxton, J. A., Acheson, A., Radziejewski, C., Maisonpierre, P. C., Yancopoulos, G. D., Channon, K. M., Hale, L. P., Dewhirst, M. W., George, S. E., and Peters, K. G. (1998b). Antiangiogenic gene therapy targeting the endothelium-specific receptor tyrosine kinase Tie2. *Proc. Natl. Acad. Sci. U.S.A.* **95**, 8829–8834.

Lingen, M. W., Polverini, P. J., and Bouck, N. P. (1998). Retinoic acid and interferon alpha act synergistically as antiangiogenic and antitumor agents against human head and neck squamous cell carcinoma. *Cancer Res.* **58**, 5551–5558.

Lode, H. N., Moehler, T., Xiang, R., Jonczyk, A., Gillies, S. D., Cheresh, D. A., and Reisfeld, R. A. (1999). Synergy between an antiangiogenic integrin αv antagonist and an antibody–cytokine fusion protein eradicates spontaneous tumor metastases. *Proc. Natl. Acad. Sci. U.S.A.* **96**, 1591–1596.

Maeda, K., Chung, Y. S., Ogawa, Y., Takatsuka, S., Kang, S. M., Ogawa, M., Sawada, T., and Sowa, M. (1996). Prognostic value of vascular endothelial growth factor expression in gastric carcinoma. *Cancer* **77**, 858–863.

Maekawa, R., Maki, H., Yoshida, H., Hojo, K., Tanaka, H., Wada, T., Uchida, N., Takeda, Y., Kasai, H., Okamoto, H., Tsuzuki, H., Kambayashi, Y., Watanabe, F., Kawada, K., Toda, K., Ohtani, M., Sugita, K., and Yoshioka, T. (1999). Correlation of antiangiogenic and antitumor efficacy of N-biphenyl sulfonyl-phenylalanine hydroxiamic acid (BPHA), an orally-active, selective matrix metalloproteinase inhibitor. *Cancer Res.* **59**, 1231–1235.

Maeshima, Y., Colorado, P. C., Torre, A., Holthaus, K. A., Grunkemeyer, J. A., Ericksen, M. B., Hopfer, H., Xiao, Y., Stillman, I. E., and Kalluri, R. (2000). Distinct antitumor properties of a type IV collagen domain derived from basement membrane. *J. Biol. Chem.* **275**, 21340–21348.

Maisonpierre, P. C., Suri, C., Jones, P. F., Bartunkova, S., Wiegand, S. J., Radziejewski, C., Compton, D., McClain, J., Aldrich, T. H., Papadopoulos, N., Daly, T. J., Davis, S., Sato, T. N., and Yancopoulos, G. D. (1997). Angiopoietin-2, a natural antagonist for Tie2 that disrupts *in vivo* angiogenesis. *Science* **277**, 55–60.

Maki, H., Maekawa, R., Yoshida, H., Hojo, K., Uchida, N., Wada, T., Tanaka, H., Banerji, S., Takeda, Y., Matsumoto, M., Yamada, H., Nishitani, Y., Shono, K., Kasai, H., Sato, S., Okamoto, H., Hayashi, R., Tamura, Y., Tsuzuki, H., Watanabe, F., Sugita, K., and Yoshioka, T. (1999). Antiangiogenic and antitumor effect of BPHA, an orally-active matrix metalloproteinase inhibitor, in *in vivo* murine and human tumor model. *Gan To Kagaku Ryoho* **26**, 1599–1606.

Mandriota, S. J., Jussila, L., Jeltsch, M., Compagni, A., Baetens, D., Prevo, R., Banerji, S., Huarte, J., Montesano, R., Jackson, D. G., Orci, L., Alitalo, K., Christofori, G., and Pepper, M. S. (2001). Vascular endothelial growth factor-C-mediated lymphangiogenesis promotes tumour metastasis. *EMBO J.* **20**, 672–682.

Manganini, M., and Maier, J. A. (2000). Transforming growth factor $\beta 2$ inhibition of hepatocyte growth factor-induced endothelial proliferation and migration. *Oncogene* **19**, 124–133.

Mendel, D. B., Laird, A. D., Smolich, B. D., Blake, R. A., Liang, C., Hannah, A. L., Shaheen, R. M., Ellis, L. M., Weitman, S., Shawver, L. K., and Cherrington, J. M. (2000). Development of SU5416, a selective small molecule inhibitor of VEGF receptor tyrosine kinase activity, as an anti-angiogenesis agent. *Anticancer Drug Design* **15**, 29–41.

Meyers, M. O., Gagliardi, A. R., Flattmann, G. J., Su, J. L., Wang, Y. Z., and Woltering, E. A. (2000). Suramin analogs inhibit human angiogenesis *in vitro*. *J. Surg. Res.* **91**, 130–134.

Miller, D. R., Anderson, G. T., Stark, J. J., Granick, J. L., and Richardson, D. (1998). Phase I/II trial of the safety and efficacy of shark cartilage in the treatment of advanced cancer. *J. Clin. Oncol.* **16**, 3649–3655.

Mishima, K., Mazar, A. P., Gown, A., Skelly, M., Ji, X.-D., Wang, X.-D., Jones, T. R., Cavenee, W. K., and Huang, H.-J. S. (2000). A peptide derived from the non-receptor-binding region of urokinase plasminogen activator inhibits glioblastoma growth and angiogenesis *in vivo* in combination with cisplatin. *Proc. Natl. Acad. Sci. U.S.A.* **97**, 8484–8489.

Mitjans, F., Meyer, T., Fittschen, C., Goodman, S., Jonczyk, A., Marshall, J. F., Reyes, G., and Piulats, J. (2000). *In vivo* therapy of malignant melanoma by means of antagonists of αv integrins. *Int. J. Cancer* **87**, 716–723.

Mohan, R., Sivak, J., Ashton, P., Russo, L. A., Pham, B. Q., Kasahara, N., Raizman, M. B., and Fini, M. E. (2000). Curcuminoids inhibit the angiogenic response stimulated by fibroblast growth factor-2, including expression of matrix metalloproteinase gelatinase B. *J. Biol. Chem.* **275**, 10405–10412.

Moore, B. B., Arenberg, D. A., Addison, C. L., Keane, M. P., and Strieter, R. M. (1998). Tumor angiogenesis is regulated by CXC chemokines. *J. Lab. Clin. Med.* **132**, 97–103.

Moreira, A. L., Friedlander, D. R., Shif, B., Kaplan, G., and Zagzag, D. (1999). Thalidomide and a thalidomide analogue inhibit endothelial cell proliferation *in vitro*. *J. Neurooncol.* **43**, 109–114.

Moulton, K. S., Heller, E., Konerding, M. A., Flynn, E., Palinski, W., and Folkman, J. (1999). Angiogenesis inhibitors endostatin or TNP-470 reduce intimal neovascularization and plaque growth in apolipoprotein E-deficient mice [see comments]. *Circulation* **99**, 1726–1732.

Murakami, K., Sakukawa, R., Sano, M., Hashimoto, A., Shibata, J., Yamada, Y., and Saiki, I. (1999). Inhibition of angiogenesis and intrahepatic growth of colon cancer by TAC-101. *Clin. Cancer Res.* **5**, 2304–2310.

Nakano, T., Abe, M., Tanaka, K., Shineha, R., Satomi, S., and Sato, Y. (2000). Angiogenesis inhibition by transdominant mutant ets-1. *J. Cell. Physiol.* **184**, 255–262.

Nakashima, T., Hirano, S., Agata, N., Kumagai, H., Isshiki, K., Yoshioka, T., Ishizuka, M., Maeda, K., and Takeuchi, T. (1999). Inhibition of angiogenesis by a new isocoumarin, NM-3. *J. Antibiot. (Tokyo)* **52**, 426–428.

Nie, D., Lamberti, M., Zacharek, A., Li, L., Szekeres, K., Tang, K., Chen, Y., and Honn, K. V. (2000a). Thromboxane A(2) regulation of endothelial cell migration, angiogenesis, and tumor metastasis. *Biochem. Biophys. Res. Commun.* **267**, 245–251.

Nie, D., Tang, K., Szekeres, K., Li, L., and Honn, K. V. (2000b). Eicosanoid regulation of angiogenesis in human prostate carcinoma and its therapeutic implications. *Ann. N.Y. Acad. Sci.* **905**, 165–176.

Nie, D., Tang, K., Diglio, C., and Honn, K. V. (2000c). Eicosanoid regulation of angiogenesis: Role of endothelial arachidonate 12-lipoxygenase. *Blood* **95**, 2304–2311.

O'Leary, J. J., Shapiro, R. L., Ren, C. J., Chuang, N., Cohen, H. W., and Potmesil, M. (1999). Antiangiogenic effects of camptothecin analogues 9-amino-20(S)-camptothecin, topotecan, and CPT-11 studied in the mouse cornea model. *Clin. Cancer Res.* **5**, 181–187.

O'Reilly, M. S., Holmgren, L., Shing, Y., Chen, C., Rosenthal, R. A., Moses, M., Lane, W. S., Cao, Y., Sage, E. H., and Folkman, J. (1994). Angiostatin: A novel angiogenesis inhibitor that mediates the suppression of metastases by a Lewis lung carcinoma. *Cell* **79**, 315–328.

O'Reilly, M. S., Holmgren, L., Chen, C., and Folkman, J. (1996). Angiostatin induces and sustains dormancy of human primary tumours in mice. *Nat. Med.* **2**, 689–692.

O'Reilly, M. S., Boehm, T., Shing, Y., Fukai, N., Vasios, G., Lane, W. S., Flynn, E., Birkhead, J. R., Olsen, B. R., and Folkman, J. (1997). Endostatin: An endogenous inhibitor of angiogenesis and tumor growth. *Cell* **88**, 277–285.

O'Reilly, M. S., Wiederschain, D., Stetler-Stevenson, W. G., Folkman, J., and Moses, M. A. (1999). Regulation of angiostatin production by matrix metalloproteinase-2 in a model of concomitant resistance. *J. Biol. Chem.* **274**, 29568–29571.

Ogasawara, M., Murata, J., Kamitani, Y., Hayashi, K., and Saiki, I. (1999). Inhibition by vasoactive intestinal polypeptide (VIP) of angiogenesis induced by murine Colon 26-L5 carcinoma cells metastasized in liver. *Clin. Exp. Metastasis* **17**, 283–291.

Oku, T., Tjuvajev, J. G., Miyagawa, T., Sasajima, T., Joshi, A., Joshi, R., Finn, R., Claffey, K. P., and Blasberg, R. G. (1998). Tumor growth modulation by sense and antisense vascular endothelial growth factor gene expression: Effects on angiogenesis, vascular permeability, blood volume, blood flow, fluorodeoxyglucose uptake, and proliferation of human melanoma intracerebral xenografts. *Cancer Res.* **58**, 4185–4192.

Parkins, C. S., Holder, A. L., Hill, S. A., Chaplin, D. J., and Tozer, G. M. (2000). Determinants of anti-vascular action by combretastatin A-4 phosphate: Role of nitric oxide. *Br. J. Cancer* **83**, 811–816.

Pavco, P. A., Bouhana, K. S., Gallegos, A. M., Agrawal, A., Blanchard, K. S., Grimm, S. L., Jensen, K. L., Andrews, L. E., Wincott, F. E., Pitot, P. A., Tressler, R. J., Cushman, C., Reynolds, M. A., and Parry, T. J. (2000). Antitumor and antimetastatic activity of ribozymes targeting the messenger RNA of vascular endothelial growth factor receptors. *Clin. Cancer Res.* **6**, 2094–2103.

Pennisi, D., Gardner, J., Chambers, D., Hosking, B., Peters, J., Muscat, G., Abbott, C., and Koopman, P. (2000). Mutations in *Sox18* underlie cardiovascular and hair follicle defects in ragged mice. *Nat. Genet.* **24**, 434–437.

Perrotte, P., Matsumoto, T., Inoue, K., Kuniyasu, H., Eve, B. Y., Hicklin, D. J., Radinsky, R., and Dinney, C. P. (1999). Anti-epidermal growth factor receptor antibody C225 inhibits angiogenesis in human transitional cell carcinoma growing orthotopically in nude mice. *Clin. Cancer Res.* **5**, 257–265.

Prewett, M., Huber, J., Li, Y., Santiago, A., O Connor, W., King, K., Overholser, J., Hooper, A., Pytowski, B., Witte, L., Bohlen, P., and Hicklin, D. J. (1999). Antivascular endothelial growth factor receptor (fetal liver kinase 1) monoclonal antibody inhibits tumor angiogenesis and growth of several mouse and human tumors. *Cancer Res.* **59**, 5209–5218.

Rabinovitz, M. (1999). Angiogenesis and its inhibition: The copper connection. *J. Natl. Cancer Inst.* **91**, 1689–1690.

Reed, M. J., Corsa, A. C., Kudravi, S. A., McCormick, R. S., and Arthur, W. T. (2000). A deficit in collagenase activity contributes to impaired migration of aged microvascular endothelial cells. *J. Cell. Biochem.* **77**, 116–126.

Reiser, F., Way, D., Bernas, M., Witte, M., and Witte, C. (1998). Inhibition of normal and experimental angiotumor endothelial cell proliferation and cell cycle progression by 2-methoxyestradiol. *Proc. Soc. Exp. Biol. Med.* **219**, 211–216.

Ren, C. J., Ueda, F., Roses, D. F., Harris, M. N., Mignatti, P., Rifkin, D. B., and Shapiro, R. L. (1998). Irsogladine maleate inhibits angiogenesis in wild-type and plasminogen activator-deficient mice. *J. Surg. Res.* **77**, 126–131.

Rikitake, Y., Kawashima, S., Yamashita, T., Ueyama, T., Ishido, S., Hotta, H., Hirata, K., and Yokoyama, M. (2000). Lysophosphatidylcholine inhibits endothelial cell migration and proliferation via inhibition of the extracellular signal-regulated kinase pathway. *Arterioscler. Thromb. Vasc. Biol.* **20**, 1006–1012.

Risau, W. (1997). Mechanisms of angiogenesis. *Nature* **386**, 671–674.

Robinson, G. S., Pierce, E. A., Rook, S. L., Foley, E., Webb, R., and Smith, L. E. H. (1996). Oligodeoxynucleotides inhibit retinal neovascularization in a murine model of proliferative retinopathy. *Proc. Natl. Acad. Sci. U.S.A.* **93**, 4851–4856.

Rose, D. P., and Connolly, J. M. (1999). Antiangiogenicity of docosahexaenoic acid and its role in the suppression of breast cancer cell growth in nude mice. *Int. J. Oncol.* **15**, 1011–1015.

Rowe, D. H., Huang, J., Kayton, M. L., Thompson, R., Troxel, A., O Toole, K. M., Yamashiro, D., Stolar, C. J., and Kandel, J. J. (2000). Anti-VEGF antibody suppresses primary tumor growth and metastasis in an experimental model of Wilms tumor. *J. Pediatr. Surg.* **35**, 30–32.

Rowinsky, E. K., Humphrey, R., Hammond, L. A., Aylesworth, C., Smetzer, L., Hidalgo, M., Morrow, M., Smith, L., Garner, A., Sorensen, J. M., Von Hoff, D. D., and Eckhardt, S. G. (2000). Phase I and pharmacologic study of the specific matrix metalloproteinase inhibitor BAY 12-9566 on a protracted oral daily dosing schedule in patients with solid malignancies. *J. Clin. Oncol.* **18,** 178–186.

Rüegg, C., Yilmaz, A., Bieler, G., Bamat, J., Chaubert, P., and Lejeune, F. J. (1998). Evidence for the involvement of endothelial cell integrin $\alpha V \beta 3$ in the disruption of the tumor vasculature induced by TNF and IFN-γ. *Nat. Med.* **4,** 408–414.

Ryan, A. M., Eppler, D. B., Hagler, K. E., Bruner, R. H., Thomford, P. J., Hall, R. L., Shopp, G. M., and O'Neill, C. A. (1999). Preclinical safety evaluation of rhuMAbVEGF, an antiangiogenic humanized monoclonal antibody. *Toxicol. Pathol.* **27,** 78–86.

Saleh, M., Stacker, S. A., and Wilks, A. F. (1996). Inhibition of growth of C6 glioma cells *in vivo* by expression of antisense vascular endothelial growth factor sequence. *Cancer Res.* **56,** 393–401.

Salmon, S. (1998). Scientists, media should not overstate 'breakthroughs'. *Ann. Oncol.* **9,** 794–795.

Salven, P., Ruotsalainen, T., Mattson, K., and Joensuu, H. (1998). High pre-treatment serum level of vascular endothelial growth factor (VEGF) is associated with poor outcome in small-cell lung cancer. *Int. J. Cancer* **79,** 144–146.

Sato, Y. (1998). Transcription factor ETS-1 as a molecular target for angiogenesis inhibition. *Hum. Cell* **11,** 207–214.

Sawaoka, H., Tsuji, S., Tsujii, M., Gunawan, E. S., Sasaki, Y., Kawano, S., and Hori, M. (1999). Cyclooxygenase inhibitors suppress angiogenesis and reduce tumor growth *in vivo*. *Lab. Invest.* **79,** 1469–1477.

Schatzkin, A. (2000). Intermediate markers as surrogate endpoints in cancer research. *Hematol. Oncol. Clin. North Am.* **14,** 887–905.

Schiller, J. H., and Bittner, G. (1999). Potentiation of platinum antitumor effects in human lung tumor xenografts by the angiogenesis inhibitor squalamine: Effects on tumor neovascularization. *Clin. Cancer Res.* **5,** 4287–4294.

Schlaeppi, J. M., Siemeister, G., Weindel, K., Schnell, C., and Wood, J. (1999). Characterization of a new potent, *in vivo* neutralizing monoclonal antibody to human vascular endothelial growth factor. *J. Cancer Res. Clin. Oncol.* **125,** 336–342.

Senger, D. R., Galli, S. J., Dvorak, A. M., Perruzzi, C. A., Harvey, V. S., and Dvorak, H. F. (1983). Tumour cells secrete a vascular permeability factor that promotes accumulation of ascites fluid. *Science* **219,** 983–985.

Senger, D. R., Claffey, K. P., Benes, J. E., Perruzzi, C. A., Sergiou, A. P., and Detmar, M. (1997). Angiogenesis promoted by vascular endothelial growth factor: Regulation through $\alpha_1 \beta_1$ and $\alpha_2 \beta_1$ integrins. *Proc. Natl. Acad. Sci. U.S.A.* **94,** 13612–13617.

Seon, B. K., Matsuno, F., Haruta, Y., Kondo, M., and Barcos, M. (1997). Long-lasting complete inhibition of human solid tumors in SCID mice by targeting endothelial cells of tumor vasculature with antihuman endoglin immunotoxin. *Clin. Cancer Res.* **3,** 1031–1044.

Shaheen, R. M., Davis, D. W., Liu, W., Zebrowski, B. K., Wilson, M. R., Bucana, C. D., McConkey, D. J., McMahon, G., and Ellis, L. M. (1999). Antiangiogenic therapy targeting the tyrosine kinase receptor for vascular endothelial growth factor receptor inhibits the growth of colon cancer liver metastasis and induces tumor and endothelial cell apoptosis. *Cancer Res.* **59,** 5412–5416.

Shalinsky, D. R., Brekken, J., Zou, H., McDermott, C. D., Forsyth, P., Edwards, D., Margosiak, S., Bender, S., Truitt, G., Wood, A., Varki, N. M., and Appelt, K. (1999). Broad antitumor and antiangiogenic activities of AG3340, a potent and selective MMP inhibitor undergoing advanced oncology clinical trials. *Ann. N.Y. Acad. Sci.* **878,** 236–270.

Sharma, B., Handler, M., Eichstetter, I., Whitelock, J. M., Nugent, M. A., and Iozzo, R. V. (1998). Antisense targeting of perlecan blocks tumor growth and angiogenesis *in vivo*. *J. Clin. Invest.* **102,** 1599–1608.

Sheu, J. R., Fu, C. C., Tsai, M. L., and Chung, W. J. (1998). Effect of U-995, a potent shark carti-
lage-derived angiogenesis inhibitor, on anti-angiogenesis and anti-tumor activities. *Anticancer
Res.* **18,** 4435–4441.

Shimo, T., Nakanishi, T., Nishida, T., Asano, M., Kanyama, M., Kuboki, T., Tamatani, T., Tezuka,
K., Takemura, M., Matsumura, T., and Takigawa, M. (1999). Connective tissue growth factor
induces the proliferation, migration, and tube formation of vascular endothelial cells *in vitro*, and
angiogenesis *in vivo*. *J. Biochem. (Tokyo)* **126,** 137–145.

Shizukuda, Y., Tang, S., Yokota, R., and Ware, J. A. (1999). Vascular endothelial growth factor-
induced endothelial cell migration and proliferation depend on a nitric oxide-mediated decrease
in protein kinase Cδ activity. *Circ. Res.* **85,** 247–256.

Siemeister, G., Schirner, M., Reusch, P., Barleon, B., Marmé, D., and Martiny-Baron, G. (1998). An
antagonistic vascular endothelial growth factor (VEGF) variant inhibits VEGF-stimulated receptor autophosphorylation and proliferation of human endothelial cells. *Proc. Natl. Acad. Sci.
U.S.A.* **95,** 4625–4629.

Siemeister, G., Schirner, M., Weindel, K., Reusch, P., Menrad, A., Marme, D., and Martiny-Baron,
G. (1999). Two independent mechanisms essential for tumor angiogenesis: Inhibition of human
melanoma xenograft growth by interfering with either the vascular endothelial growth factor
receptor pathway or the Tie-2 pathway. *Cancer Res.* **59,** 3185–3191.

Skobe, M., Hawighorst, T., Jackson, D. G., Prevo, R., Janes, L., Velasco, P., Riccardi, L., Alitalo, K.,
Claffey, K., and Detmar, M. (2001). Induction of tumor lymphangiogenesis by VEGF-C promotes
breast cancer metastasis. *Nat. Med.* **7,** 192–198.

Skopinska-Rozewska, E., Piazza, G. A., Sommer, E., Pamukcu, R., Barcz, E., Filewska, M., Kupis, W.,
Caban, R., Rudzinski, P., Bogdan, J., Mlekodaj, S., and Sikorska, E. (1998a). Inhibition of angiogenesis by sulindac and its sulfone metabolite (FGN-1): A potential mechanism for their antineoplastic properties. *Int. J. Tissue React.* **20,** 85–89.

Skopinska-Rozewska, E., Janik, P., Przybyszewska, M., Sommer, E., and Bialas-Chromiec, B. (1998b).
Inhibitory effect of theobromine on induction of angiogenesis and VEGF mRNA expression in
v-raf transfectants of human urothelial cells HCV-29. *Int. J. Mol. Med.* **2,** 649–652.

Skopinska-Rozewska, E., Krotkiewski, M., Sommer, E., Rogala, E., Filewska, M., Bialas-Chromiec,
B., Pastewka, K., and Skurzak, H. (1999). Inhibitory effect of shark liver oil on cutaneous angiogenesis induced in Balb/c mice by syngeneic sarcoma L-1, human urinary bladder and human
kidney tumour cells. *Oncol. Rep.* **6,** 1341–1344.

Slaton, J. W., Perrotte, P., Inoue, K., Dinney, C. P., and Fidler, I. J. (1999). Interferon-α-mediated
downregulation of angiogenesis-related genes and therapy of bladder cancer are dependent on
optimization of biological dose and schedule. *Clin. Cancer Res.* **5,** 2726–2734.

Soker, S., Takashima, S., Miao, H. Q., Neufeld, G., and Klagsbrun, M. (1998). Neuropilin-1 is
expressed by endothelial and tumor cells as an isoform-specific receptor for vascular endothelial
growth factor. *Cell* **92,** 735–745.

Stacker, S. A., Caesar, C., Baldwin, M. E., Thornton, G. E., Williams, R. A., Prevo, R., Jackson,
D. G., Nishikawa, S., Kubo, H., and Achen, M. G. (2001). VEGF-D promotes the metastatic
spread of tumor cells via the lymphatics. *Nat. Med.* **7,** 186–191.

Stearns, M. E., Garcia, F. U., Fudge, K., Rhim, J., and Wang, M. (1999). Role of interleukin 10 and
transforming growth factor β1 in the angiogenesis and metastasis of human prostate primary
tumor lines from orthotopic implants in severe combined immunodeficiency mice. *Clin. Cancer
Res.* **5,** 711–720.

Stetler-Stevenson, W. G., Aznavoorian, S., and Liotta, L. A. (1993). Tumor cell interactions with
the extracellular matrix during invasion and metastasis. *Annu. Rev. Cell Biol.* **9,** 541–573.

Stitt, A. W., Simpson, D. A., Boocock, C., Gardiner, T. A., Murphy, G. M., and Archer, D. B.
(1998). Expression of vascular endothelial growth factor (VEGF) and its receptors is regulated in
eyes with intraocular tumours. *J. Pathol.* **186,** 306–312.

Storgard, C. M., Stupack, D. G., Jonczyk, A., Goodman, S. L., Fox, R. I., and Cheresh, D. A. (1999). Decreased angiogenesis and arthritic disease in rabbits treated with an $\alpha_v\beta_3$ antagonist. *J. Clin. Invest.* **103,** 47–54.

Streit, M., Riccardi, L., Velasco, P., Brown, L. F., Hawighorst, T., Bornstein, P., and Detmar, M. (1999). Thrombospondin-2: A potent endogenous inhibitor of tumor growth and angiogenesis. *Proc. Natl. Acad. Sci. U.S.A.* **96,** 14888–14893.

Struman, I., Bentzien, F., Lee, H., Mainfroid, V., D'Angelo, G., Goffin, V., Weiner, R. I., and Martial, J. A. (1999). Opposing actions of intact and N-terminal fragments of the human prolactin/growth hormone family members on angiogenesis: An efficient mechanism for the regulation of angiogenesis. *Proc. Natl. Acad. Sci. U.S.A.* **96,** 1246–1251.

Sun, L., Tran, N., Liang, C., Hubbard, S., Tang, F., Lipson, K., Schreck, R., Zhou, Y., McMahon, G., and Tang, C. (2000). Identification of substituted 3-[(4,5,6,7-tetrahydro-1H-indol-2-yl)methylene]-1, 3-dihydroindol-2-ones as growth factor receptor inhibitors for VEGF-R2 (Flk-1/KDR), FGF-R1, and PDGF-Rβ tyrosine kinases. *J. Med. Chem.* **43,** 2655–2663.

Takayama, K., Ueno, H., Nakanishi, Y., Sakamoto, T., Inoue, K., Shimizu, K., Oohashi, H., and Hara, N. (2000). Suppression of tumor angiogenesis and growth by gene transfer of a soluble form of vascular endothelial growth factor receptor into a remote organ. *Cancer Res.* **60,** 2169–2177.

Tanaka, H., Nishida, K., Sugita, K., and Yoshioka, T. (1999). Antitumor efficacy of hypothemycin, a new Ras-signaling inhibitor. *Jpn. J. Cancer Res.* **90,** 1139–1145.

Tanaka, T., Manome, Y., Wen, P., Kufe, D. W., and Fine, H. A. (1997). Viral vector-mediated transduction of a modified platelet factor 4 cDNA inhibits angiogenesis and tumor growth. *Nat. Med.* **3,** 437–442.

Tempel, C., Gilead, A., and Neeman, M. (2000). Hyaluronic acid as an anti-angiogenic shield in the preovulatory rat follicle. *Biol. Reprod.* **63,** 134–140.

Tokino, T., and Nakamura, Y. (2000). The role of p53-target genes in human cancer. *Crit. Rev. Oncol. Hematol.* **33,** 1–6.

Tortora, G., Bianco, R., Damiano, V., Fontanini, G., De Placido, S., Bianco, A. R., and Ciardiello, F. (2000). Oral antisense that targets protein kinase A cooperates with taxol and inhibits tumor growth, angiogenesis, and growth factor production. *Clin. Cancer Res.* **6,** 2506–2512.

Trochon, V., Blot, E., Cymbalista, F., Engelmann, C., Tang, R. P., Thomaidis, A., Vasse, M., Soria, J., Lu, H., and Soria, C. (2000). Apigenin inhibits endothelial-cell proliferation in G(2)/M phase whereas it stimulates smooth-muscle cells by inhibiting P21 and P27 expression. *Int. J. Cancer* **85,** 691–696.

Uchida, S., Watanabe, G., Shimada, Y., Maeda, M., Kawabe, A., Mori, A., Arii, S., Uehata, M., Kishimoto, T., Oikawa, T., and Imamura, M. (2000). The suppression of small GTPase rho signal transduction pathway inhibits angiogenesis *in vitro* and *in vivo. Biochem. Biophys. Res. Commun.* **269,** 633–640.

Vajkoczy, P., Menger, M. D., Goldbrunner, R., Ge, S., Fong, T. A., Vollmar, B., Schilling, L., Ullrich, A., Hirth, K. P., Tonn, J. C., Schmiedek, P., and Rempel, S. A. (2000). Targeting angiogenesis inhibits tumor infiltration and expression of the pro-invasive protein SPARC. *Int. J. Cancer* **87,** 261–268.

van Bruggen, N., Thibodeaux, H., Palmer, J. T., Lee, W. P., Fu, L., Cairns, B., Tumas, D., Gerlai, R., Williams, S.-P., Campagne, M. V. L., and Ferrara, N. (1999). VEGF antagonism reduces edema formation and tissue damage after ischemia/reperfusion injury in the mouse brain. *J. Clin. Invest.* **104,** 1613–1620.

Volpert, O. V., Fong, T., Koch, A. E., Peterson, J. D., Waltenbaugh, C., Tepper, R. I., and Bouck, N. P. (1998). Inhibition of angiogenesis by interleukin 4. *J. Exp. Med.* **188,** 1039–1046.

Wedge, S. R., Ogilvie, D. J., Dukes, M., Kendrew, J., Curwen, J. O., Hennequin, L. F., Thomas, A. P., Stokes, E. S., Curry, B., Richmond, G. H., and Wadsworth, P. F. (2000). ZD4190: An orally active inhibitor of vascular endothelial growth factor signaling with broad-spectrum anti-tumor efficacy. *Cancer Res.* **60,** 970–975.

Wen, W., Moses, M. A., Wiederschain, D., Arbiser, J. L., and Folkman, J. (1999). The generation of endostatin is mediated by elastase. *Cancer Res.* **59,** 6052–6056.

Wernert, N., Stanjek, A., Kiriakidis, S., Hugel, A., Jha, H. C., Mazitschek, R., and Giannis, A. (1999). Inhibition of angiogenesis *in vivo* by ets-1 antisense oligonucleotides-inhibition of Ets-1 transcription factor expression by the antibiotic fumagillin. *Angew. Chem. Int. Ed. Engl.* **38,** 3228–3231.

Witte, L., Hicklin, D. J., Zhu, Z., Pytowski, B., Kotanides, H., Rockwell, P., and Bohlen, P. (1998). Monoclonal antibodies targeting the VEGF receptor-2 (Flk1/KDR) as an anti-angiogenic therapeutic strategy. *Cancer Metastasis Rev.* **17,** 155–161.

Wojtowicz-Praga, S. (1999). Clinical potential of matrix metalloprotease inhibitors. *Drugs R. D.* **1,** 117–129.

Wojtowicz-Praga, S. M., Dickson, R. B., and Hawkins, M. J. (1997). Matrix metalloproteinase inhibitors. *Invest. New Drugs* **15,** 61–75.

Wojtowicz-Praga, S., Torri, J., Johnson, M., Steen, V., Marshall, J., Ness, E., Dickson, R., Sale, M., Rasmussen, H. S., Chiodo, T. A., and Hawkins, M. J. (1998). Phase I trial of Marimastat, a novel matrix metalloproteinase inhibitor, administered orally to patients with advanced lung cancer. *J. Clin. Oncol.* **16,** 2150–2156.

Wood, J. M., Bold, G., Buchdunger, E., Cozens, R., Ferrari, S., Frei, J., Hofmann, F., Mestan, J., Mett, H., O Reilly, T., Persohn, E., Rösel, J., Schnell, C., Stover, D., Theuer, A., Towbin, H., Wenger, F., Woods-Cook, K., Menrad, A., Siemeister, G., Schirner, M., Thierauch, K.-H., Schneider, M. R., Drevs, J., Martiny-Baron, G., Totzke, F., and Marmé, D. (2000). PTK787/ZK 222584, a novel and potent inhibitor of vascular endothelial growth factor receptor tyrosine kinases, impairs vascular endothelial growth factor-induced responses and tumor growth after oral administration. *Cancer Res.* **60,** 2178–2189.

Wylie, S., MacDonald, I. C., Varghese, H. J., Schmidt, E. E., Morris, V. L., Groom, A. C., and Chambers, A. F. (1999). The matrix metalloproteinase inhibitor batimastat inhibits angiogenesis in liver metastases of B16F1 melanoma cells. *Clin. Exp. Metastasis* **17,** 111–117.

Xin, X., Yang, S., Kowalski, J., and Gerritsen, M. E. (1999). Peroxisome proliferator-activated receptor γ ligands are potent inhibitors of angiogenesis *in vitro* and *in vivo*. *J. Biol. Chem.* **274,** 9116–9121.

Xu, L., Yoneda, J., Herrera, C., Wood, J., Killion, J. J., and Fidler, I. J. (2000). Inhibition of malignant ascites and growth of human ovarian carcinoma by oral administration of a potent inhibitor of the vascular endothelial growth factor receptor tyrosine kinases. *Int. J. Oncol.* **16,** 445–454.

Yamaguchi, N., Anand-Apte, B., Lee, M., Sasaki, T., Fukai, N., Shapiro, R., Que, I., Lowik, C., Timpl, R., and Olsen, B. R. (1999). Endostatin inhibits VEGF-induced endothelial cell migration and tumor growth independently of zinc binding. *EMBO J.* **18,** 4414–4423.

Yamaji, T., Tsuboi, H., Murata, N., Uchida, M., Kohno, T., Sugino, E., Hibino, S., Shimamura, M., and Oikawa, T. (1999). Anti-angiogenic activity of a novel synthetic agent, 9 α-fluoromedroxyprogesterone acetate. *Cancer Lett.* **145,** 107–114.

Yamaoka, M., Yamamoto, T., Masaki, T., Ikeyama, S., Sudo, K., and Fujita, T. (1993). Inhibition of tumour growth and metastasis of rodent tumours by the angiogenesis inhibitor o-(chloroacetyl-carbamyl) fumagillol (TNP-470; AGM-1470). *Cancer Res.* **53,** 4262–4267.

Yatsunami, J., Tsuruta, N., Fukuno, Y., Kawashima, M., Taniguchi, S., and Hayashi, S. (1999a). Inhibitory effects of roxithromycin on tumor angiogenesis, growth and metastasis of mouse B16 melanoma cells. *Clin. Exp. Metastasis* **17,** 119–124.

Yatsunami, J., Fukuno, Y., Nagata, M., Tsuruta, N., Aoki, S., Tominaga, M., Kawashima, M., Taniguchi, S., and Hayashi, S. (1999b). Roxithromycin and clarithromycin, 14-membered ring macrolides, potentiate the antitumor activity of cytotoxic agents against mouse B16 melanoma cells. *Cancer Lett.* **147,** 17–24.

Yip, D., Ahmad, A., Karapetis, C. S., Hawkins, C. A., and Harper, P. G. (1999). Matrix metallopro-teinase inhibitors: Applications in oncology. *Invest. New Drugs* **17,** 387–399.

Yoshida, A., Yoshida, S., Ishibashi, T., Kuwano, M., and Inomata, H. (1999). Suppression of retinal neovascularization by the NF-κB inhibitor pyrrolidine dithiocarbamate in mice. *Invest. Ophthalmol. Visual Sci.* **40,** 1624–1629.

Yoshiji, H., Kuriyama, S., Ways, D. K., Yoshii, J., Miyamoto, Y., Kawata, M., Ikenaka, Y., Tsujinoue, H., Nakatani, T., Shibuya, M., and Fukui, H. (1999). Protein kinase C lies on the signaling pathway for vascular endothelial growth factor-mediated tumor development and angiogenesis. *Cancer Res.* **59,** 4413–4418.

Yuan, F., Chen, Y., Dellian, M., Safabakhsh, N., Ferrara, N., and Jain, R. K. (1996). Time-dependent vascular regression and permeability changes in established human tumor xenografts induced by an antivascular endothelial growth factor/vascular permeability factor antibody. *Proc. Natl. Acad. Sci. U.S.A.* **93,** 14765–14770.

Zagzag, D., Zhong, H., Scalzitti, J. M., Laughner, E., Simons, J. W., and Semenza, G. L. (2000). Expression of hypoxia-inducible factor 1α in brain tumors: Association with angiogenesis, invasion, and progression. *Cancer* **88,** 2606–2618.

Zhai, Y., Ni, J., Jiang, G. W., Lu, J., Xing, L., Lincoln, C., Carter, K. C., Janat, F., Kozak, D., Xu, S., Rojas, L., Aggarwal, B. B., Ruben, S., Li, L. Y., Gentz, R., and Yu, G. L. (1999). VEGI, a novel cytokine of the tumor necrosis factor family, is an angiogenesis inhibitor that suppresses the growth of colon carcinomas *in vivo. FASEB J.* **13,** 181–189.

Zhang, Y., Griffith, E. C., Sage, J., Jacks, T., and Liu, J. O. (2000). Cell cycle inhibition by the anti-angiogenic agent TNP-470 is mediated by p53 and p21[WAF1/CIP1]. *Proc. Natl. Acad. Sci. U.S.A.* **97,** 6427–6432.

Zhong, H., De Marzo, A. M., Laughner, E., Lim, M., Hilton, D. A., Zagzag, D., Buechler, P., Isaacs, W. B., Semenza, G. L., and Simons, J. W. (1999). Overexpression of hypoxia-inducible factor 1α in common human cancers and their metastases. *Cancer Res.* **59,** 5830–5835.

Zhou, J. R., Mukherjee, P., Gugger, E. T., Tanaka, T., Blackburn, G. L., and Clinton, S. K. (1998). Inhibition of murine bladder tumorigenesis by soy isoflavones via alterations in the cell cycle, apoptosis, and angiogenesis. *Cancer Res.* **58,** 5231–5238.

Zhou, J. R., Gugger, E. T., Tanaka, T., Guo, Y., Blackburn, G. L., and Clinton, S. K. (1999). Soybean phytochemicals inhibit the growth of transplantable human prostate carcinoma and tumor angiogenesis in mice. *J. Nutr.* **129,** 1628–1635.

Zhu, W. H., Guo, X., Villaschi, S., and Francesco, N. R. (2000). Regulation of vascular growth and regression by matrix metalloproteinases in the rat aorta model of angiogenesis. *Lab. Invest.* **80,** 545–555.

Zhu, Z., and Witte, L. (1999). Inhibition of tumor growth and metastasis by targeting tumor-associated angiogenesis with antagonists to the receptors of vascular endothelial growth factor. *Invest. New Drugs* **17,** 195–212.

Zhu, Z., Lu, D., Kotanides, H., Santiago, A., Jimenez, X., Simcox, T., Hicklin, D. J., Bohlen, P., and Witte, L. (1999). Inhibition of vascular endothelial growth factor induced mitogenesis of human endothelial cells by a chimeric anti-kinase insert domain-containing receptor antibody. *Cancer Lett.* **136,** 203–213.

Geranylgeranylated RhoB Mediates the Apoptotic and Antineoplastic Effects of Farnesyltransferase Inhibitors: New Insights into Cancer Cell Suicide

14

George C. Prendergast

The DuPont Pharmaceuticals Company
Glenolden Laboratory
Glenolden, Pennsylvania 19036, and
The Wistar Institute
Tumor Biology Group
Philadelphia, Pennsylvania 19104

I. INTRODUCTION

Farnesyltransferase inhibitors (FTIs) are a novel class of cancer chemotherapeutics currently in phase II human trials. The development of FTIs was predicated on the discovery that Ras must be farnesylated to drive malignant transformation. However, mechanistic studies have indicated that Ras may not be a crucial target in mediating the antineoplastic effects of FTIs. One important target that has emerged is RhoB, a cytoskeletal actin regulator with a specialized role in receptor trafficking. An unexpected twist in the drug mechanism is the gain-of-function of geranylgeranylated RhoB (RhoB-GG). In this chapter, we survey the evidence for RhoB-GG in FTI action, summarize recent advances concerning how FTIs drive cancer cell suicide, and discuss how these advances impact clinical applications. We suggest that FTIs may be more suited to addressing Rho-driven malignancies, such as melanoma, breast cancer, and possibly prostate cancer, rather than Ras-driven malignancies, such as pancreatic, lung, and colon cancers.

II. DO FARNESYLTRANSFERASE INHIBITORS TARGET A UNIQUE ASPECT OF NEOPLASTIC PATHOPHYSIOLOGY?

Initial cell biological studies performed in H-Ras-transformed rat fibroblasts showed that FTIs could suppress anchorage-independent growth potential in the absence of nonspecific toxicity (Garcia et al., 1993; James et al., 1993; Kohl et al., 1993). Interestingly, this effect was associated with phenotypic reversion and acquisition of morphological and growth regulatory characteristics of non-transformed cells (James et al., 1994; Prendergast et al., 1994). Cells were similarly affected if they were transformed by v-Src, which requires Ras, but not by v-Raf, which acts independently of Ras (Kohl et al., 1993; Prendergast et al., 1994). FTIs also inhibited the anchorage-independent growth of a variety of human tumor cell lines (Sepp-Lorenzino et al., 1995). Furthermore, they blunted or reversed tumor formation in a variety of xenograft or transgenic mouse models (Kohl et al., 1994, 1995; Liu et al., 1998; Mangues et al., 1998; Nagasu et al., 1995; Omer et al., 2000; Sun et al., 1995, 1998). Notably, FTIs were neither cytotoxic nor cytostatic at doses that completely blocked H-Ras farnesylation in vitro or in vivo and that were associated with antitumor properties (James et al., 1994; Kohl et al., 1995; Nagasu et al., 1995; Prendergast et al., 1994). Thus, it appeared that FTIs could target some unique feature of neoplastic pathophysiology.

III. Ras IS NOT A CRUCIAL TARGET OF FARNESYLTRANSFERASE INHIBITORS

Our early investigations of the unusual biological properties of FTIs prompted us to question the causal relationship with effects on Ras. First, as mentioned earlier, it was surprising that FTIs displayed little if any effect on the proliferation of normal cells *in vitro* or *in vivo,* because of the evidence that Ras is required for division in normal cells (e.g., Mulcahy *et al.,* 1985). Second, close characterization of the phenotypic reversion in H-Ras-transformed cells revealed that cell biological changes preceded significant reduction in the steady-state levels of H-Ras protein (Prendergast *et al.,* 1994). FTIs block the function of newly synthesized H-Ras, not preexisting H-Ras, the function of which is attenuated in drug-treated cells only by protein turnover. Strikingly, although fully processed H-Ras turns over with a half-life of ~24 hr, phenotypic reversion was almost complete within the same period (Prendergast *et al.,* 1994). This phenomenon could not be explained by the production of soluble mutant Ras species with dominant inhibitory effects: unfarnesylated Ras is unstable, so it does not accumulate to significant steady-state levels (Prendergast *et al.,* 1994), and in any case, only the Ras L61 allele rather than the Ras V12 mutant allele used in the majority of experimental models has significant dominant inhibitory activity (Gibbs *et al.,* 1989).

Additional evidence unlinking FTI biology from the effects on Ras includes the following. First, FTIs block the anchorage-independent growth of cells transformed with N-myristylated Ras, which does not require farnesylation to function (Lebowitz *et al.,* 1995). Second, the susceptibility of human tumor cells to growth inhibition by FTIs was uncorrelated with Ras status (Sepp-Lorenzino *et al.,* 1995). Third, FTIs suppressed the proliferation of K-Ras-transformed cells, even though K-Ras remained prenylated and functional in drug-treated cells owing to its alternate modification by geranylgeranyltransferase-I (James *et al.,* 1995; Lerner *et al.,* 1997; Omer *et al.,* 2000; Rowell *et al.,* 1997; Song *et al.,* 2000; Whyte *et al.,* 1997). This finding is especially important to therapeutic strategy, because K-Ras is the Ras family member most frequently mutated in human cancer. It should be noted that FTIs do not inhibit transformation-associated processes driven by K-Ras as readily as those driven by H-Ras. However, since FTIs do not inhibit K-Ras prenylation, it is telling that K-Ras biology is susceptible at all. Taken together, the evidence unlinks FTI biology from inhibition of Ras prenylation, implying that FTIs mediate effects by altering farnesylated protein(s) other than Ras.

IV. RhoB IS A CRUCIAL TARGET OF FARNESYLTRANSFERASE INHIBITORS

We hypothesized that Rho alteration formed the basis for FTI biology because of our observations that FTIs rapidly stimulated the formation of actin stress fibers in both normal and H-Ras-transformed cells (Prendergast *et al.,* 1994). Rho proteins are small GTPases that like Ras must be isoprenylated to function. Rho proteins regulate cytoskeletal actin organization, adhesion, survival, motility, and proliferation (reviewed in Symons, 1996; Tapon and Hall, 1997; Van Aelst and D'Souza-Schorey, 1997). For several reasons, we investigated and corroborated RhoB as a crucial FTI target (Lebowitz and Prendergast, 1998; Prendergast, 2000a,b). RhoB is unique among Rho proteins in that it is regulated by growth and stress stimuli (Fritz and Kaina, 1997; Fritz et al., 1995; Jahner and Hunter, 1991a,b; Zalcman et al., 1995). Most Rho proteins are geranylgeranylated, but RhoB is either farnesylated or geranylgeranylated in cells (Adamson et al., 1992). Figure 1 summarizes the effects of FTI treatment on the prenylation and associated cellular properties of RhoB, H-Ras, and K-Ras. FTI treatment causes a rapid depletion of the farnesylated species of RhoB (RhoB-F) in cells because of its short half-life (Lebowitz et al., 1995). Geranylgeranylated forms of RhoB (RhoB-GG) accumulate in drug-treated cells owing to the action of geranylgeranyltransferase-I, which is unaffected by FTI treatment (Lebowitz et al., 1997a). Significantly, elevation of RhoB-GG is sufficient and necessary to mediate most of the cellular effects of FTIs, a finding

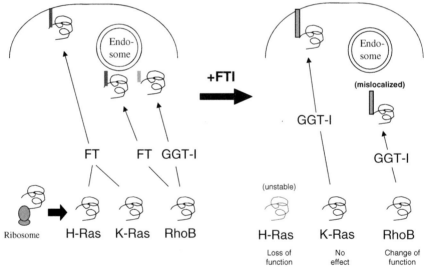

Figure 1 Effects of FTI treatment on RhoB, H-Ras, and K-Ras. FTIs elicit three different types of effects on farnesylated proteins in the cell as illustrated here. (1) The function of newly synthesized H-Ras is abolished by inhibition of its farnesylation. (2) The function of K-Ras is not appreciably affected because newly synthesized K-Ras can be geranylgeranylated by GGT-I in the absence of FT activity. (3) The function of RhoB is changed by a shift in its prenylation pattern and an associated mislocalization from endosomes. This mislocalization is likely critical to change of function.

discussed later (Du *et al.*, 1999a; Liu *et al.*, 2000). Although the role of differential prenylation of this protein is not exactly clear, it appears that the change in the ratio of RhoB-F:RhoB-GG is important, because cells can be rendered FTI resistant by ectopic expression of an engineered form of RhoB that is prenylation-independent and active due to N-myristylation (Lebowitz *et al.*, 1995). In summary, the results support the conclusion that RhoB is a crucial target of FTI treatment in malignant cells.

V. FARNESYLTRANSFERASE INHIBITORS ACT THROUGH A GAIN OF FUNCTION MECHANISM INVOLVING RhoB-GG

Farnesyltransferase inhibitors were initially conceived as a method to induce a loss of function in farnesylated target proteins, in particular Ras proteins. However, investigations of the effects of FTIs on RhoB have established that the drugs act in large part by inducing gain of function. The rapid increase of RhoB-GG elicited in cells by FTI treatment has been shown to be (a) sufficient to mediate growth inhibition and (b) sufficient and necessary to mediate apoptosis (Du *et al.*, 1999b; Liu *et al.*, 2000). These properties have been confirmed in human carcinoma cells that are susceptible to FTI treatment (Du and Prendergast, 1999). The requirement for RhoB-GG to drive apoptosis is an especially important component of FTI action, since loss of this effect significantly compromises antineoplastic effects *in vivo* (Liu *et al.*, 2000). RhoB-GG elevation has no discernible effect on normal cells, similar to FTI treatment (within a pharmacologically relevant range). Thus, the identification of RhoB-GG as a key mediator of the cancer-selective properties of FTIs may lead to a greater understanding of the basis for this selectivity. As discussed later, cell adhesion status appears to be one determinant of apoptotic susceptibility to FTI treatment (Lebowitz *et al.*, 1997b). Rho proteins are important regulators and mediators of cell adhesion signals, and studies of RhoB function in knockout mice suggest a role in controlling $\beta 1$ integrin-dependent cell motility (A. X. Liu, N. Rane, and G. C. Prendergast, unpublished observations, 2000). Thus, we favor the notion that a key feature of neoplastic pathophysiology targeted by FTIs is the radical difference in cell adhesion signaling requirements displayed by malignantly transformed cells (Du *et al.*, 1999b; Lebowitz *et al.*, 1997b; Liu *et al.*, 2000).

The realization that RhoB-GG levels are elevated following FTI treatment addresses one conundrum that was raised by the FTI–Rho hypothesis. Rho proteins induce stress fiber formation, so loss of RhoB function in transformed cells might be expected to lead to the disappearance rather than the appearance of actin stress fibers. However, FTIs elevate levels of RhoB-GG, which in terms of prenylation and cell localization resembles RhoA (Lebowitz *et al.*, 1995). Therefore, one aspect of FTI treatment is to increase a RhoA-like function in cells, which could explain why FTIs promote actin stress fiber formation. This explanation is consistent with observed effects of FTIs in both normal and transformed cells (Prendergast *et al.*, 1994). It is tempting to speculate that the gain of RhoB-GG associated with stress fiber formation is compatible with anchorage-dependent but not with anchorage-independent growth. Normal cells exist in a state where stress fibers dominate, so RhoB-GG may have little effect in this setting. In contrast, transformed cells exist in a disorganized state characterized by a loss of polarity and disordered actin cytoskeleton, such that the reenforcement of a stress fiber-based structure might radically alter transformed cell structure and function.

VI. RhoB-GG IS REQUIRED TO MEDIATE APOPTOSIS BY FARNESYLTRANSFERASE INHIBITORS

A. Substratum adhesion status: Why FTIs do not kill normal cells

Perhaps the most dramatic demonstration of the anticancer potential of FTIs was illustrated by experiments in transgenic Ras "oncomice" that get mammary carcinomas. In these animals, FTI treatment led to dramatic, nearly complete, regression of the carcinoma (Barrington *et al.*, 1998; Kohl *et al.*, 1995; Mangues *et al.*, 1998). Regression was associated with significantly increased rates of apoptosis in the tumor (Barrington *et al.*, 1998). However, it was unclear why regression was possible *in vivo*, because FTIs were not cytotoxic *in vitro* (Prendergast *et al.*, 1994; Sepp-Lorenzino *et al.*, 1995). This issue was complicated by the fact that against human tumor cell xenografts FTIs were also only cytostatic (Ito *et al.*, 1996; Kohl *et al.*, 1994; Nagasu *et al.*, 1995). Since cytotoxicity is preferred in a cancer chemotherapeutic, the factors influencing the ability of FTIs to kill cells and to cause tumor regression continue to be of great interest.

Two other issues were raised by mice experiments. First, resistance to tumor regression was observed in some animals, a phenomenon that could be selected for by repeated cycles of drug exposure and withdrawal (Kohl *et al.*, 1995). Although this phenomenon has not been examined extensively in animals, *in vitro* selection for FTI resistance is not associated with amplification of the multiple drug resistance gene MDR-1 (Prendergast *et al.*, 1996). Second, even though tumors regressed completely, cessation of FTI treatment led to a rapid relapse (Kohl *et al.*, 1995). Thus, some malignant cells persisted during long-term drug treatment. Tumor persistence would require continuous FTI treatment that could increase the likelihood of side effects and drug resistance. Therefore, understanding the basis of tumor resistance and persistence, as well as determining the basis for cytotoxic versus cytostatic effects, are important to maximize clinical efficacy.

The elucidation of RhoB as a key target offers a new viewpoint and explanative power regarding these issues. Rho proteins have been implicated in focal adhesion formation and integrin signaling (Symons, 1996). Therefore, we compared the effects of FTIs under conditions where cell–cell or cell–matrix attachment of transformed cells was favored. These experiments revealed that attachment parameters dictated the cellular response to FTIs (Lebowitz et al., 1997b). For example, if H-Ras-transformed cells were cultured in suspension, where cell–cell but not cell–substratum attachment is possible, then FTIs induced a massive apoptosis instead of reversion. Although this response might be predicted by the ability of Ras to block anoikis (adhesion deprival-induced apoptosis) (Rak et al., 1995b), the death mechanism used by FTIs was not based on effects on Ras but instead on RhoB (Lebowitz et al., 1997b). FTI-induced death was little affected by p53 mutation but was inhibited by Bcl-X$_L$ (Lebowitz et al., 1997a), a member of the Bcl-2 family of apoptosis regulators that broadly influence cell death responses. These findings pointed to Rho-dependent integrin signaling pathways as a realm for understanding the antitransforming and antitumor properties of the drugs.

The identification of a link between FTI action and cell adhesion capacity suggested a way to explain how nontoxic tumor regression was possible and provided a starting point to address the basis for drug resistance. Lack of appropriate substratum attachment might facilitate anoikis of tumor cells, thereby causing tumor regression. Integrin-dependent attachment and survival signals, which involve Rho, are probably universally subverted in cancer. Thus, if by altering RhoB FTIs induce a physiological attachment response, then cancer cells might react much differently than normal cells, since the latter already exist in a physiologically attached state (Ruoslahti, 1996). Tumor cells at privileged locations, perhaps in the periphery of the tumor where normal anchorage cues may exist, might be able to revert to a normal attached phenotype and survive, explaining the tumor cell persistence phenomenon in oncomice. Genetic alterations that block the apoptotic response, such as Bcl-X$_L$ overexpression, might arise in a percentage of nonprivileged tumor cells, allowing them to survive FTI-induced anoikis and contribute to relapse. The recurrent tumor might then be resistant to tumor regression when FTI treatment is reimplemented. This explanation is compatible with the fact that FTIs are not cytotoxic to human tumor cells, since such cells must have generally evolved significant anoikis resistance during malignant progression. The ability of Bcl-X$_L$ to defeat FTI-induced cell death raises questions about whether the drugs will cause regression of invasive and metastatic cancers in the clinic. For example, a significant proportion of carcinomas overexpress Bcl-X$_L$ or its related family member Bcl-2 (Krajewska et al., 1996). As primary therapeutics, FTIs may be more useful in irradicating premalignant lesions or localized tumors that retain some normal adhesion requirements and thus anoikis susceptibility. In this regard, FTIs may also prove useful as cancer preventative agents, especially given their relative lack of normal cell toxicity.

B. PI3′K−Akt pathway modulates apoptosis

Two lines of investigation led us to hypothesize a role for the phosphatidylinositol 3′-kinase (PI3′K)−Akt cell survival pathway in modulating the response to FTI treatment. First, as discussed earlier, substratum adhesion was identified as a critical determinant of apoptotic susceptibility in transformed cells (Lebowitz et al., 1997b). Attached cells have elevated levels of PI3′K−Akt activity relative to unattached cells, due to the effects of integrin engagement. Second, levels of serum, in particular the level of insulin-like growth factor-1 (IGF-1), was another determinant of apoptotic susceptibility (Du et al., 1999b; Suzuki et al., 1998). IGF1 is also a strong inducer of PI3′K−Akt activity. We, therefore, tested the hypothesis that PI3′K−Akt activity dictated growth inhibitory versus apoptotic responses of transformed cells to FTI treatment. Inhibition of PI3′K by cell treatment with the specific inhibitor LY294002 switched the response of attached cells cultured in the presence of IGF1 from reversion and growth inhibition to apoptosis. Similarly, ectopic activation of Akt blocked the ability of FTIs to induce apoptosis under any conditions (Du et al., 1999b). We concluded that Akt activity was one factor that influenced the manifestation of transformation-specific apoptosis by FTIs.

Do FTIs themselves influence Akt activity? There is some evidence to support the notion that this is the case in certain epithelial cell settings. First, Cheng and colleagues have reported that the elevated levels of Akt-2 activity which occur in ovarian cancer cells can be suppressed by FTI treatment in a manner that correlates with induction of apoptosis (Jiang et al., 2000). These findings are provocative, insofar as they suggest FTIs may be efficacious in ovarian cancer, but they need to be corroborated because FTIs generally have not been found to be proapoptotic against human tumor cells. However, work on the proapoptotic effector mechanism used by RhoB-GG to kill cells supports the notion that FTIs may suppress Akt activity in certain settings. Specifically, RhoB-GG is sufficient to mediate the suppressive effects of FTIs on Akt-1 in COS or MCF7 breast epithelial cells, a mechanism that does not operate in transformed fibroblast models (Liu and Prendergast, 2000). The biochemical linkage between RhoB-GG and Akt is unclear. However, it is tempting to speculate that it involves the protein kinase C-like kinase PRK, a crucial RhoB effector (Gampel et al., 1999; Mellor et al., 1998), and PDK1, the Akt regulatory kinase that is reported to functionally interact with PRK (Balendran et al., 1999). There is some support for the notion that PRK may modulate cell survival, but the evidence is indirect [e.g., cleavage and activation in apoptotic cells (Takahashi *et al.*, 1998)]. A direct connection between RhoB and PDK1 is suggested by the work of Parker and colleagues, who reported that RhoB interacts with and regulates PRK-PDK1 in a ternary complex (Flynn *et al.*, 2000). Thus, it is conceivable that PDK1-mediated regulation of Akt is influenced by RhoB-GG through PRK. In summary, existing evidence suggests that Akt may antagonize FTI-induced cell death, but that in certain epithelial cancer cells FTIs may directly overcome this effect by an RhoB-GG-dependent mechanism. In any case, the PI3′K–Akt pathway is clearly one key determinant of the apoptotic prowess of FTIs.

C. Role for Rb inhibition suggested by CDKI synergy

A recent report indicates that the cell division kinase inhibitors (CDKIs) roscovitine and olomoucine synergize effectively with FTIs in promoting apoptosis (Edamatsu *et al.*, 2000). Roscovitine is a more potent derivative of olomoucine, each of which effectively inhibits CDK1 and CDK2 in cells in the micromolar range. The proapoptotic effects of FTI plus CDKI treatment were observed in HL60 and CEM human leukemia cells and also in LNCaP prostate cancer cells. By themselves, neither treatment elicited much apoptosis, but when added together a rapid and dramatic increase in apoptosis occurred. The effect appeared to be mechanistically distinct from the combination of FTI plus PI3′K inhibition, which also synergized to drive apoptosis in this study, insofar as the CDKIs acted without affecting Akt activity. Under the conditions tested, the CDKIs partially inhibited steady-state levels of phosphorylated Rb consistent with effects on CDK2. By itself, FTI treatment had similar effects. In combination, Rb phosphorylation was fully inhibited. Although undefined, the basis for the collaborative effect may reflect the ability of FTIs to increase levels of the cellular CDK inhibitor p21WAF1, through a RhoB-GG-dependent mechanism (Du *et al.*, 1999a; Sepp-Lorenzino and Rosen, 1998). One interpretation of the results is that cell death resulted from a cell cycle signaling conflict brought into play by superactivation of Rb. Although Rb activation is generally linked to arrest and antiapoptosis, some reports have shown that Rb can activate apoptosis. Therefore, another interpretation is that apoptosis results from a direct effect of Rb activation. Lastly, since roscovitine and olomoucine probably affect kinases other than CDKs (a common problem with kinase inhibitors), it is possible that the synergy with FTIs is off-target. It should be possible to rule out the latter interpretation in experiments using a dominant positive mutant of Rb, which should mimic the effects of FTI plus CDKI treatment. In addition, if the synergy is Rb-dependent, then it should be abolished by Rb deletion or neutralization (e.g., in HPV-positive cervical cancer cells). Such information may be clinically useful based on the ability to judge Rb status in tumors.

VII. RhoB-GG AND THE ANTIANGIOGENIC PROPERTIES OF FARNESYLTRANSFERASE INHIBITORS

Two universal features of cancer are alterations in cell adhesion properties and a dependence on angiogenesis. We presented earlier the evidence that FTIs kill transformed cells by interfering with some aspect of Rho-dependent cell adhesion processes. However, the ability of FTIs to cause nontoxic tumor regression in the H-Ras oncomouse model (Kohl *et al.*, 1995) also raised the question whether some of this activity was also due to antiangiogenic effects. Kerbel and colleagues have provided evidence that Ras activates VEGF secretion and that FTIs suppress this effect (Rak *et al.*, 1995a). More recent investigations corroborate this effect and, interestingly, extend the antiangiogenic effects to the tumor vasculature as well as the tumor (Charvat *et al.*, 1999; Gu *et al.*, 1999). Whether RhoB-GG is involved in these effects is under study. In the transformed cell, a role for RhoB in FTI regulation of VEGF levels would be consistent with evidence that Ras can upregulate VEGF by a PI3′K-independent mechanism (Arbiser *et al.*, 1997). We have observed that FTI treatment reverses mutant Ras-dependent upregulation of VEGF message levels with rapid kinetics that are consistent with a RhoB-GG mechanism. However, RhoB-GG is not sufficient to suppress VEGF in Ras-transformed cells, and dominant inhibitory mutants of RhoB do not block the ability of mutant Ras to upregulate VEGF message levels (P. Lebowitz, W. Du, and G. C. Prendergast, unpublished results, 1999). Thus, FTIs may suppress VEGF expression in the transformed cell through an RhoB-GG-independent mechanism. We are examining this issue and the question of FTI effects on tumor vasculature in knockout systems where RhoB-GG cannot be elicited by drug treatment.

VIII. CLINICAL IMPLICATIONS

A. Utility in Rho-driven tumors that retain death susceptibility factors

The ability of FTIs to selectively induce apoptosis in neoplastically transformed cells is their most interesting and desirable feature with regard to cancer applications. This feature was perhaps most strikingly manifested in the H-Ras oncomouse model. However, the promise offered by these early results dwindled considerably because of two general problems. First, a flawed understanding of the mechanisms through which FTIs manifested their remarkable biological properties. Second, which factors made nontoxic tumor regression possible. Progress has been made on both problems.

Genetic and biochemical evidence has established a crucial role for RhoB-GG in mediating apoptosis. More work is required to understand exactly how RhoB-GG acts. However, its identification as a key player prompts one to reassess clinical strategies to exploit FTIs. If RhoB-GG exerts a dominant inhibitory effect on Rho signaling, then Rho-driven tumors may be more likely to respond to FTI treatment than Ras-driven tumors. Examples of the former include breast cancer, especially the aggressive inflammatory breast cancers (10% of all breast malignancies), where RhoC drives the malignant phenotype (van Golen et al., 1999, 2000). Similarly, RhoC has been reported to be a key driver of malignant melanoma (Clark et al., 2000). These tumors generally lack Ras mutations but may respond favorably to FTIs, which through RhoB-GG may interfere with Rho signaling.

Genes that influence the apoptotic susceptibility of neoplastic cells to FTIs have been identified, including Akt, Bcl-2/Bcl-X$_L$, and possibly Rb. PTEN status is also important since this gene is responsible for governing Akt activity. Other factors that may impact regressions include the substratum status of tumor cells and angiogenic requirements. Given that two oncogenes and possibly at least two tumor suppressors can influence the apoptotic response to FTIs, it is unsurprising that FTIs act so poorly against frank human cancer cells, which have typically sustained alterations in one or more of these genes. With regard to basic research, one existing challenge is to trace the suicide signal transduction pathway downstream of where FTIs act (i.e., downstream of RhoB-GG), to try to manipulate it at a point downstream of the various suppressors. The remarkable regressions displayed in the H-Ras oncomouse model probably reflect the competence of this pathway, and the disappointing lack of regressions in xenograft assays may reflect the result of its suppression and/or inactivation. With regard to clinical applications, one would predict that the human tumors most susceptible to FTI-induced regression would be those sustaining expression of wild-type RhoB, PTEN, and Rb, and lacking activated Akt or Bcl-2/Bcl-X$_L$. Since these genes are altered rather frequently in late stage cancers, FTIs as single therapy agents may be rather impotent in these settings, but they may be more useful against less advanced tumors or when used as antimetastatics (to prolong remission through cytostatic activity). More realistically, combinatorial strategies would seem to offer the best chance to enhance efficacy.

B. Combinatorial strategies to potentiate cell death

A number of preclinical reports suggest that FTIs may synergize with taxanes, DNA damaging agents, CDKIs, and PI3'K–Akt pathway inhibitors. As mentioned previously, PI3'K inhibitors enhance FTI-induced apoptosis in Ras-transformed cells without significantly affecting the survival of normal cells (Du et al., 1999b). These results suggest that inhibitors of the PI3'K survival pathway, including Akt kinase inhibitors, may improve FTI efficacy. Tumors in which RhoB-GG can target Akt activity directly for inhibition, including perhaps certain breast or ovarian cancers (Jiang *et al.*, 2000; Liu and Prendergast, 2000), may be more susceptible to killing by FTIs. Additionally, FTIs have been reported to sensitize Ras-transformed cells or frank carcinoma cells to cell death by irradiation (Bernhard *et al.*, 1996, 1998) or to treatment with taxol (Moasser *et al.*, 1998). Recent results indicate that RhoB-GG mediates this effect (Liu *et al.*, 2001), a role consistent with previous work implicating RhoB in the response to DNA damage (Fritz and Kaina, 1997, 2000; Fritz *et al.*, 1995). Early clinical experience supports the notion that combinatorial applications using DNA damaging agents achieve efficacy in humans. Methods to apply FTIs as radiosensitizers or chemoenhancers in Rho-driven tumors may prove especially worthwhile.

IX. SUMMARY

Recent work suggests that the antineoplastic effects of FTIs are mediated by RhoB-GG. In particular, RhoB-GG is required to mediate proapoptotic effects that are most desirable in the clinic. Exactly how RhoB-GG signals cell death is not yet clear, but a working hypothesis is that it interferes with Rho signaling pathways that are important for the neoplastic phenotype. Several factors influencing the killing prowess of FTIs have been identified. These include the important cell survival kinase Akt (and by inference the tumor suppressor PTEN) and the tumor suppressor Rb. The shift in view about FTI mechanism toward RhoB-GG prompts a reassessment of how to test FTIs in the clinic. Rho proteins are not mutated in cancer. However, there is evidence that RhoA and RhoC are frequently overexpressed in cancer cells and that these events are associated with increased invasive capability and malignant status (Clark *et al.*, 2000; Fritz *et al.*, 1999; van Golen *et al.*, 2000; Yoshioka *et al.*, 1999). We suggest that Rho-driven rather than Ras-driven tumors may be more appropriate to test efficacy in the clinic (Prendergast, 2000b). As a single agent therapy, FTIs may be most efficacious against premalignant lesions and localized tumors that have not yet evolved significant resistance to apoptosis. In addition, the cytostatic properties of FTIs may be useful to block the growth of micrometastases following primary therapy. Combinatorial applications using FTIs as radiosensitizers and chemoenhancing agents appear promising. Whatever their clinical fate, FTIs have been used to pinpoint RhoB-GG as a mediator of a cell death principle that is largely specific to neoplastically transformed cells. It is hoped that mechanistic dissection of this pathway may provide deeper insights into neoplastic pathophysiology as well as an opportunity to attack cancer at its roots.

References

Adamson, P., Marshall, C. J., Hall, A., and Tilbrook, P. A. (1992). Post-translational modification of p21rho proteins. *J. Biol. Chem.* **267,** 20033–20038.

Arbiser, J. L., Moses, M. A., Fernandez, C. A., Ghiso, N., Cao, Y., Klauber, N., Frank, D., Brownlee, M., Flynn, E., Parangi, S., Byers, H. R., and Folkman, J. (1997). Oncogenic H-ras stimulates tumor angiogenesis by two distinct pathways. *Proc. Natl. Acad. Sci. U.S.A.* **94,** 861–866.

Balendran, A., Casamayor, A., Deak, M., Paterson, A., Gaffney, P., Currie, R., Downes, C. P., and Alessi, D. R. (1999). PDK1 acquires PDK2 activity in the presence of a synthetic peptide derived from the carboxyl terminus of PRK2. *Curr. Biol.* **9,** 393–404.

Barrington, R. E., Subler, M. A., Rands, E., Omer, C. A., Miller, P. J., Hundley, J. E., Koester, S. K., Troyer, D. A., Bearss, D. J., Conner, M. W., Gibbs, J. B., Hamilton, K., Koblan, K. S., Mosser, S. D., O'Neill, T. J., Schaber, M. D., Senderak, E. T., Windle, J. J., Oliff, A., and Kohl, N. E. (1998). A farnesyltransferase inhibitor induces tumor regression in transgenic mice harboring multiple oncogenic mutations by mediating alterations in both cell cycle control and apoptosis. *Mol. Cell. Biol.* **18,** 85–92.

Bernhard, E. J., Kao, G., Cox, A. D., Sebti, S. M., Hamilton, A. D., Muschel, R. J., and McKenna, W. G. (1996). The farnesyltransferase inhibitor FTI-277 radiosensitizes H-ras-transformed rat embryo fibroblasts. *Cancer Res.* **56,** 1727–1730.

Bernhard, E. J., McKenna, W. G., Hamilton, A. D., Sebti, S. M., Qian, Y., Wu, J. M., and Muschel, R. J. (1998). Inhibiting Ras prenylation increases the radiosensitivity of human tumor cell lines with activating mutations of ras oncogenes. *Cancer Res.* **58,** 1754–1761.

Charvat, S., Duchesne, M., Parvaz, P., Chignol, M.-C., Schmitt, D., and Serres, M. (1999). The up-regulation vascular endothelial growth factor in mutated Ha-ras HaCaT cell lines is reduced by a farnesyl transferase inhibitor. *Anticancer Res.* **19,** 557–562.

Clark, E. A., Golub, T. R., Lander, E. S., and Hunes, R. O. (2000). Genomic analysis of metastasis reveals an essential role for RhoC. *Nature* **406,** 532–535.

Du, W., and Prendergast, G. C. (1999). Geranylgeranylated RhoB mediates inhibition of human tumor cell growth by farnesyltransferase inhibitors. *Cancer Res.* **59,** 5924–5928.

Du, W., Lebowitz, P., and Prendergast, G. C. (1999a). Cell growth inhibition by farnesyltransferase inhibitors is mediated by gain of geranylgeranylated RhoB. *Mol. Cell. Biol.* **19,** 1831–1840.

Du, W., Liu, A., and Prendergast, G. C. (1999b). Activation of the PI3′K–AKT pathway masks the proapoptotic effect of farnesyltransferase inhibitors. *Cancer Res.* **59,** 4808–4812.

Edamatsu, H., Gau, C.-L., Nemoto, T., Guo, L., and Tamanoi, F. (2000). Cdk inhibitors, roscovitine and olomoucine, synergize with farnesyltransferase inhibitor (FTI) to induce efficient apoptosis of human cancer cell lines. *Oncogene* **19,** 3059–3068.

Flynn, P., Mellor, H., Casamassima, A., and Parker, P. J. (2000). Rho GTPase control of protein kinase C-related protein kinase activation by 3-phosphoinositide-dependent protein kinase. *J. Biol. Chem.* **275,** 11064–11070.

Fritz, G., and Kaina, B. (1997). rhoB encoding a UV-inducible ras-related small GTP-binding protein is regulated by GTPases of the rho family and independent of JNK, ERK, and p38 MAP kinase. *J. Biol. Chem.* **272,** 30637–30644.

Fritz, G., and Kaina, B. (2000). Ras-related GTPase RhoB forces alkylation-induced apoptotic cell death. *Biochem. Biophys. Res. Commun.* **268,** 784–789.

Fritz, G., Kaina, B., and Aktories, K. (1995). The ras-related small GTP-binding protein RhoB is immediate-early inducible by DNA damaging treatments. *J. Biol. Chem.* **270,** 25172–25177.

Gampel, A., Parker, P. J., and Mellor, H. (1999). Regulation of epidermal growth factor receptor traffic by the small GTPase RhoB. *Curr. Biol.* **9,** 955–958.

Garcia, A. M., Rowell, C., Ackermann, K., Kowalczyk, J. J., and Lewis, M. D. (1993). Peptidomimetic inhibitors of Ras farnesylation and function in whole cells. *J. Biol. Chem.* **268,** 18415–18418.

Gibbs, J. B., Schaber, M. D., Schofield, T. L., Scolnick, E. M., and Sigal, I. S. (1989). *Xenopus* oocyte germinal-vesicle breakdown induced by [Val^{12}]Ras is inhibited by a cytosol-localized Ras mutant. *Proc. Natl. Acad. Sci. U.S.A.* **86,** 6630–6634.

Gu, W.-Z., Tahir, S. K., Wang, Y.-C., Zhang, H.-C., Cherian, S. P., O Connor, S., Leal, J. A., Rosenberg, S. H., and Ng, S.-C. (1999). Effect of novel CAAX peptidomimetic farnesyltransferase inhibitor on angiogenesis *in vitro* and *in vivo. Eur. J. Cancer* **35,** 1394–1401.

Ito, T., Kawata, S., Tamura, S., Igura, T., Nagase, T., Miyagawa, J. I., Yamazaki, E., Ishiguro, H., and Matasuzawa, Y. (1996). Suppression of human pancreatic cancer growth in BALB/c nude mice by manumycin, a farnesyl: Protein transferase inhibitor. *Jpn. J. Cancer Res.* **87,** 113–116.

Jahner, D., and Hunter, T. (1991a). The *ras*-related gene *rhoB* is an immediate-early gene inducible by v-Fps, epidermal growth factor, and platelet-derived growth factor in rat fibroblasts. *Mol. Cell. Biol.* **11,** 3682–3690.

Jahner, D., and Hunter, T. (1991b). The stimulation of quiescent rat fibroblasts by v-*src* and v-*fps* oncogenic protein–tyrosine kinases leads to the induction of a subset of immediate early genes. *Oncogene* **6,** 1259–1268.

James, G. L., Goldstein, J. L., Brown, M. S., Rawson, T. E., Somers, T. C., McDowell, R. S., Crowley, C. W., Lucas, B. K., Levinson, A. D., and Marsters, J. C. (1993). Benzodiazepine peptidomimetics: Potent inhibitors of Ras farnesylation in animal cells. *Science* **260,** 1937–1942.

James, G. L., Brown, M. S., Cobb, M. H., and Goldstein, J. L. (1994). Benzodiazepine peptidomimetic BZA-5B interrupts the MAP kinase activation pathway in H-Ras-transformed Rat-1 cells, but not in untransformed cells. *J. Biol. Chem.* **269,** 27705–27714.

James, G. L., Goldstein, J. L., and Brown, M. S. (1995). Polylysine and CVIM sequences of K-RasB dictate specificity of prenylation and confer resistance to benzodiazepine peptidomimetic *in vitro. J. Biol. Chem.* **270,** 6221–6226.

Jiang, K., Coppola, D., Crespo, N. C., Nicosia, S. V., Hamilton, A. D., Sebti, S. M., and Cheng, J. Q. (2000). The phosphoinositide 3-OH kinase/AKT2 pathway as a critical target for farnesyltransferase inhibitor-induced apoptosis. *Mol. Cell. Biol.* **20,** 139–148.

Kohl, N. E., Mosser, S. D., deSolms, S. J., Giuliani, E. A., Pompliano, D. L., Graham, S. L., Smith, R. L., Scolnick, E. M., Oliff, A., and Gibbs, J. B. (1993). Selective inhibition of *ras*-dependent transformation by a farnesyltransferase inhibitor. *Science* **260,** 1934–1937.

Kohl, N. E., Redner, F., Mosser, S., Guiliani, E. A., deSolms, S. J., Conner, M. W., Anthony, N. J., Holtz, W. J., Gomez, R. P., Lee, T.-J., Smith, R. L., Graham, S. L., Hartman, G. D., Gibbs, J., and Oliff, A. (1994). Protein farnesyltransferase inhibitors block the growth of ras-dependent tumors in nude mice. *Proc. Natl. Acad. Sci. U.S.A.* **91,** 9141–9145.

Kohl, N. E., Omer, C. A., Conner, M. W., Anthony, N. J., Davide, J. P., deSolms, S. J., Giuliani, E. A., Gomez, R. P., Graham, S. L., Hamilton, K., Handt, L. K., Hartman, G. D., Koblan, K. S., Kral, A. M., Miller, P. J., Mosser, S. D., O'Neill, T. J., Rands, E., Schaber, M. D., Gibbs, J. B., and Oliff, A. (1995). Inhibition of farnesyltransferase induces regression of mammary and salivary carcinomas in *ras* transgenic mice. *Nat. Med.* **1,** 792–797.

Krajewska, M., Moss, S. F., Krajewski, S., Song, K., Holt, P. R., and Reed, J. C. (1996). Elevated expression of Bcl-x and reduced Bak in primary colorectal adenocarcinomas. *Cancer Res.* **56,** 2422–2427.

Lebowitz, P. F., and Prendergast, G. C. (1998). Non-Ras targets for farnesyltransferase inhibitors: Focus on Rho. *Oncogene* **17,** 1439–1447.

Lebowitz, P. F., Davide, J. P., and Prendergast, G. C. (1995). Evidence that farnesyl transferase inhibitors suppress Ras transformation by interfering with Rho activity. *Mol. Cell. Biol.* **15,** 6613–6622.

Lebowitz, P., Casey, P. J., Prendergast, G. C., and Thissen, J. (1997a). Farnesyltransferase inhibitors alter the prenylation and growth-stimulating function of RhoB. *J. Biol. Chem.* **272,** 15591–15594.

Lebowitz, P. F., Sakamuro, D., and Prendergast, G. C. (1997b). Farnesyltransferase inhibitors induce apoptosis in Ras-transformed cells denied substratum attachment. *Cancer Res.* **57,** 708–713.

Lerner, E. C., Zhang, T. T., Knowles, D. B., Qian, Y. M., Hamilton, A. D., and Sebti, S. M. (1997). Inhibition of the prenylation of K-Ras, but not H- or N-Ras, is highly resistant to CAAX peptidomimetics and requires both a farnesyltransferase and a geranylgeranyltransferase-I inhibitor in human tumor cell lines. *Oncogene* **15,** 1283–1288.

Liu, A.-X., and Prendergast, G. C. (2000). Geranylgeranylated RhoB is sufficient to mediate tissue-specific suppression of Akt kinase activity by farnesyltransferase inhibitors. *FEBS Lett.* **481,** 205–208.

Liu, A.-X., Du, W., Liu, J.-P., Jessell, T. M., and Prendergast, G. C. (2000). RhoB alteration is required for the apoptotic and antineoplastic responses to farnesyltransferase inhibitors. *Mol. Cell. Biol.* **20,** 6105–6113.

Liu, M., Bryant, M. S., Chen, J., Lee, S., Yaremko, B., Lipari, P., Malkowski, M., Ferrari, E., Nielsen, L., Prioli, N., Dell, J., Sinha, D., Syed, J., Dorfmacher, W. A., Nomeir, A. A., Lin, C.-C., Wang, L., Taveras, A. G., Doll, R. J., Njoroge, F. G., Mallams, A. K., Remiszewski, S., Catino, J. J., Girijavallabhan, V. M., Kirschmeier, P., and Bishop, W. R. (1998). Antitumor activity of SCH 6636, an orally bioavailable tricyclic inhibitor of farnesyl protein transferase, in human tumor xenograft models and wap-ras transgenic mice. *Cancer Res.* **58,** 4947–4956.

Liu, A. X., Cerniglia, G. J., Bernhard, E. J., and Prendergast, G. C. (2001). RhoB is required to mediate apoptosis in neoplastically transformed cells after DNA damage. *Proc. Natl. Acad. Sci. U.S.A.* **98,** 6192–6197.

Mangues, R., Corral, T., Kohl, N. E., Symmans, W. F., Lu, S., Malumbres, M., Gibbs, J. B., Oliff, A., and Pellicer, A. (1998). Antitumor effect of a farnesyl protein transferase inhibitor in mammary and lymphoid tumors overexpressing N-ras in transgenic mice. *Cancer Res.* **58,** 1253–1259.

Mellor, J., Flynn, P., Nobes, C. D., Hall, A., and Parker, P. J. (1998). PRK1 is targeted to endosomes by the small GTPase, RhoB. *J. Biol. Chem.* **273,** 4811–4814.

Moasser, M. M., Sepp-Lorenzino, L., Kohl, N. E., Oliff, A., Balog, A., Su, D. S., Danishefsky, S. J., and Rosen, N. (1998). Farnesyl transferase inhibitors cause enhanced mitotic sensitivity to taxol and epothilones. *Proc. Natl. Acad. Sci. U.S.A.* **95,** 1369–1374.

Mulcahy, L. S., Smith, M. R., and Stacey, D. W. (1985). Requirement for ras proto-oncogene function during serum-stimulated growth of NIH3T3 cells. *Nature* **313,** 241–243.

Nagasu, T., Yoshimatsu, K., Rowell, C., Lewis, M. D., and Garcia, A. M. (1995). Inhibition of human tumor xenograft growth by treatment with the farnesyltransferase inhibitor B956. *Cancer Res.* **55,** 5310–5314.

Omer, C. A., Chen, Z., Diehl, R. E., Conner, M. W., Chen, H. Y., Trumbauer, M. E., Gopal-Truter, S., Seeburger, G., Bhimnathwala, H., Abrams, M. T., Davide, J. P., Ellis, M. S., Gibbs, J. B., Greenberg, I., Hamilton, K., Koblan, K. S., Kral, A. M., Liu, D., Lobell, R. B., Miller, P. J., Mosser, S. D., O'Neill, T. J., Rands, E., Schaber, M. D., Senderak, E. T., and Kohl, N. E. (2000). Mouse mammary tumor virus-Ki-RasB transgenic mice devlop mammary carcinomas that can be growth-inhibited by a farnesyl:protein transferase inhibitor. *Cancer Res.* **60,** 2680–2688.

Prendergast, G. C. (2000a). Farnesyltransferase inhibitors: Antineoplastic mechanism and clinical prospects. *Curr. Opin. Cell Biol.* **12,** 166–173.

Prendergast, G. C. (2000b). Mode of action of farnesyltransferase inhibitors. *Lancet Oncol.* **1,** 73.

Prendergast, G. C., Davide, J. P., deSolms, S. J., Giuliani, E., Graham, S., Gibbs, J. B., Oliff, A., and Kohl, N. E. (1994). Farnesyltransferase inhibition causes morphological reversion of *ras*-transformed cells by a complex mechanism that involves regulation of the actin cytoskeleton. *Mol. Cell. Biol.* **14**, 4193–4202.

Prendergast, G. C., Davide, J. P., Lebowitz, P. F., Wechsler-Reya, R., and Kohl, N. E. (1996). Resistance of a variant ras-transformed cell line to phenotypic reversion by farnesyl transferase inhibitors. *Cancer Res.* **56**, 2626–2632.

Rak, J., Mitsuhashi, Y., Bayko, L., Filmus, J., Shirasawa, T., Sasazuki, T., and Kerbel, R. S. (1995a). Mutant ras oncogenes upregulate VEGF/vPF expression: Implications for induction and inhibition of tumor angiogenesis. *Cancer Res.* **55**, 4575–4580.

Rak, J., Mitsuhashi, Y., Erdos, V., Huang, S.-N., Filmus, J., and Kerbel, R. S. (1995b). Massive programmed cell death in intestinal epithelial cells induced by three-dimensional growth conditions: Suppression by mutant c-H-ras oncogene expression. *J. Cell Biol.* **131**, 1587–1598.

Rowell, C. A., Kowalczyk, J. J., Lewis, M. D., and Garcia, A. M. (1997). Direct demonstration of geranylgeranylation and farnesylation of Ki-Ras *in vivo*. *J. Biol. Chem.* **272**, 14093–14097.

Ruoslahti, E. (1996). Integrin signaling and matrix assembly. *Tumour Biol.* **17**, 117–124.

Sepp-Lorenzino, L., and Rosen, N. (1998). A farnesyl:protein transferase inhibitor induces p21 expression and G1 block in p53 wild type tumor cells. *J. Biol. Chem.* **273**, 20243–20251.

Sepp-Lorenzino, L., Ma, Z., Rands, E., Kohl, N. E., Gibbs, J. B., Oliff, A., and Rosen, N. (1995). A peptidomimetic inhibitor of farnesyl: Protein transferase blocks the anchorage-dependent and -independent growth of human tumor cell lines. *Cancer Res.* **55**, 5302–5309.

Song, S. Y., Meszoely, I. M., Coffey, R. J., Pietenpol, J. A., and Leach, S. D. (2000). K-Ras-independent effects of the farnesyl transferase inhibitor L-744,832 on cyclin B1/cdc2 kinase activity, G2/M cell cycle progression and apoptosis in human pancreatic ductal adenocarcinoma cells. *Neoplasia* **2**, 261–272.

Sun, J., Qian, Y., Hamilton, A. D., and Sebti, S. M. (1995). Ras CAAX peptidomimetic FTI 276 selectively blocks tumor growth in nude mice of a human lung carcinoma with K-Ras mutation and p53 deletion. *Cancer Res.* **55**, 4243–4247.

Sun, J., Qian, Y., Hamilton, A. D., and Sebti, S. M. (1998). Both farnesyltransferase and geranylgeranyltransferase I inhibitors are required for inhibition of oncogenic K-Ras prenylation but each alone is sufficient to suppress human tumor growth in nude mouse xenografts. *Oncogene* **16**, 1467–1473.

Suzuki, N., Urano, J., and Tamanoi, F. (1998). Farnesyl transferase inhibitors induce cytochrome *c* release and caspase 3 activation preferentially in transformed cells. *Proc. Natl. Acad. Sci. U.S.A.* **95**, 15356–15361.

Symons, M. (1996). Rho family GTPases: The cytoskeleton and beyond. *Trends Biochem. Sci.* **21**, 178–181.

Takahashi, M., Mukai, H., Toshimori, M., Miyamoto, M., and Ono, Y. (1998). Proteolytic activation of PKN by caspase-3 or related protease during apoptosis. *Proc. Natl. Acad. Sci. U.S.A.* **95**, 11566–11571.

Tapon, N., and Hall, A. (1997). Rho, Rac and Cdc42 GTPases regulate the organization of the actin cytoskeleton. *Curr. Opin. Cell Biol.* **9**, 86–92.

Van Aelst, L., and D'Souza-Schorey, C. (1997). Rho GTPases and signaling networks. *Genes Dev.* **11**, 2295–2322.

van Golen, K. L., Davies, S., Wu, Z. F., Wnag, Y., Bucana, C. D., Root, H., Chandrasekharappa, S., Strawderman, M., Ethier, S. P., and Merajver, S. D. (1999). A novel putative low-affinity insulin-like growth factor-binding protein, LIBC (lost in inflammatory breast cancer), and RhoC GTPase correlate with the inflammatory breast cancer phenotype. *Clin. Cancer Res.* **5**, 2511–2519.

van Golen, K. L., Wu, Z.-F., Qiao, X. T., Bao, L. W., and Marajver, S. D. (2000). RhoC GTPase, a novel transforming oncogene for human mammary epithelial cells that partially recapitulates the inflammatory breast cancer phenotype. *Cancer Res.* **20**, in press.

Whyte, D. B., Kirschmeier, P., Hockenberry, T. N., Nunez-Olivia, I., James, L., Catino, J. J., Bishop, W. R., and Pai, J. K. (1997). K- and N-ras geranylgeranylated in cells treated with farnesyl protein transferase inhibitors. *J. Biol. Chem.* **272,** 14459–14464.

Yoshioka, K., Hakamori, S., and Itoh, K. (1999). Overexpression of small GTP-binding protein RhoA promotes invasion of tumor cells. *Cancer Res.* **59,** 2004–2010.

Zalcman, G., Closson, V., Linares-Cruz, G., Leregours, F., Honore, N., Tavitian, A., and Olofsson, B. (1995). Regulation of Ras-related RhoB protein expression during the cell cycle. *Oncogene* **10,** 1935–1945.

RAS Binding Compounds

Oliver Müller and Alfred Wittinghofer

Department of Structural Biology
Max-Planck-Institut für Molekulare Physiologie
44227 Dortmund, Germany

I. Introduction
II. Ras Cycle and Ras–Raf Signaling Pathway
III. The Structure of Ras Proteins
IV. Drug Target Sites of Ras
V. Conclusions and Outlook
References

I. INTRODUCTION

Because of its important role in human tumor development, the Ras protein has been the target of many different screening approaches for the identification of novel drugs against cancer. Although Ras is one of the most extensively studied proteins in cellular signal transduction cascades, it appears that we are still far from fully understanding its physiological role(s). In this review we will first introduce the Ras protein with the focus on its biochemical and structural features. Second, we will give an overview on potential anticancer drugs that directly interact with Ras and potentially inhibit its mitogenic functions. For more detailed aspects of Ras biology, other reviews might be consulted (e.g., Campbell *et al.*, 1998; Downward, 1998; Shields *et al.*, 2000; Wittinghofer and Waldmann, 2000; Kuhlmann and Herrmann, 2000).

Tumor-Suppressing Viruses, Genes, and Drugs
Innovative Cancer Therapy Approaches
Copyright © 2002 by Academic Press.
All rights of reproduction in any form reserved.

II. Ras CYCLE AND Ras–Raf SIGNALING PATHWAY

The Ras gene is a dominant oncogene which plays a key role in tumor development. With a mutation rate of approximately 30%, the Ras gene is the most
prevalently mutated proto-oncogene in human tumors. Tumor-relevant Ras gene
mutations, which all result in single amino acid replacements, are found at either
of the three codons 12, 13, and 61. The Ras protein (of 21 kDa) is the key in one
of the most important signaling pathways for the regulation of cell growth and
proliferation. The mutations lead to a constitutively active G protein with a continuous, unregulated signaling capacity. Thus, the Ras protein, in particular the
oncogenically mutated form, represents an ideal target for a drug against cancer.
The G protein Ras can be regarded as a molecular switch that is cycling between
a GDP bound inactive "switch-OFF" state and a GTP bound active "switch-ON"
state. Only in the active GTP bound state, is the Ras able to bind strongly to
effector molecules which become activated in order to further transmit the signal
in their pathway (Figure 1).

Since the switch-ON and switch-OFF reactions in the cycle of Ras are
intrinsically very slow, the regulatory input by Ras interacting proteins determines
the lifetime of the two states. The guanine nucleotide exchange factors (GEFs)
such as SOS catalyze the dissociation of GDP from the normal Ras, facilitating
the loading with GTP, which is the more prevalent guanine nucleotide in
eukaryotic cells. GEFs themselves are activated by recruitment to the inner side of

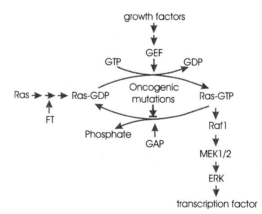

Figure 1 The GTP–GDP cycle of Ras. The GTP bound
Ras transduces the mitogenic signal by binding
its effectors such as Raf1. The GAP-dependent
GTP hydrolysis by Ras is blocked by oncogenic
mutations in Ras.

the plasma membrane by adaptor proteins such as Grb-2. Adaptor proteins bind the cytosolic part of tyrosine-phosphorylated growth factor receptors.

In its active form, the GTP bound Ras directly activates cytosolic effector proteins of which the best studied is the Ser–Thr-kinase c-Raf-1. Raf then activates the MEK–MAPK signaling cascade through phosphorylation of MEK1 and MEK2. The MEKs in turn phosphorylate the ERK family kinases, leading to an amplification of the signal. Activated ERK can interact with both cytosolic and nuclear transcription factors in order to stimulate progression into and through the mitotic cell cycle.

The Ras transmitted signal is terminated by hydrolysis of the bound GTP to GDP by the GTPase activity of normal Ras. The slow intrinsic GTPase activity is stimulated by Ras-specific GTPase activating proteins (GAPs) by several orders of magnitude. The dissociation of Ras and Raf was shown to be very fast, in contrast to the intuitive belief that the Ras–Raf complex should be long-lived (Gorman et al., 1996; Sydor et al., 1998). This highly dynamic equilibrium of Ras–Raf interaction ensures that GAPs, which cannot bind Ras simultaneously with Raf, are able to turn off the normal Ras after dissociation. GAPs directly interact with Ras and provide an amino acid residue which is essential for a rapid GTP hydrolysis (Mittal et al., 1996)

The oncogenic mutations in the Ras gene have been shown to confer to Ras the capacity to transform normal cells into malignant tumor cells (Scheffzek et al., 1997). Oncogenic mutations may have two different consequences on the biochemical level. The Gln residue at codon 61 is crucial for the intrinsic GTPase activity of Ras, that is, its ability to hydrolyze the bound GTP, or the interaction with the GAPs (Scheffzek et al., 1998). Mutations at codon 12 or 13 block the interaction with the GAPs by steric constraints (Scheffzek et al., 1998). Therefore, the major proportion of each oncogenic Ras mutant is GTP bound and thereby constitutively active. In contrast to the wild-type Ras the intrinsic GTPase activity of the oncogenic Ras is no longer stimulated by GAPs. The consequence of this constitutive activation is a continuous and more intense signal for growth and proliferation. Consistent with this model are the experimental findings that constitutively active Raf, MEK, or ERKs can partially mimic the effects of oncogenic Ras, and that specific inhibitors of Raf, MEK, and ERK kinases can block these effects (Block et al., 1996).

III. THE STRUCTURE OF Ras PROTEINS

The best characterized isoform of all Ras proteins is the Ha-Ras, although in human tumors Ki-Ras and N-Ras are more frequently mutated. The biochemical and structural properties are, however, very similar between these Ras isoforms.

Ha-Ras consists of 189 amino acids with five highly conserved sequence elements which are also found in other Ras related proteins. Two of these motifs are responsible for specific recognition of the guanine nucleotide, three are necessary for binding of the phosphate groups and complexation of a Mg^{2+} ion, and for hydrolysis of GTP. The first 166 amino acids are necessary and sufficient for the biochemical properties of Ras, whereas the C-terminal end does not seem to be involved in any direct interaction with other proteins. Instead, the last four amino acids CVLS constitute a motif which is called a CAAX box, where C stands for Cys, A for aliphatic, and X for any amino acid. The CAAX box is the recognition sequence for the enzyme farnesyltransferase (FT) which thioalkylates the Cys with a C_{15} prenyl group. This modification is the first in a series of posttranslational modifications which are responsible for anchoring Ras into the inner site of the plasma membrane (Willumsen et al., 1984; Cox and Der, 1997; Hancock et al., 1989, 1990). The membrane anchorage is essential for its biological signal transducing activity, even for constitutively active oncogenic Ras proteins that carry the mutations found in human tumors. FT is the key enzyme for this posttranslational modification.

The three-dimensional (3D) structure of Ras was solved both in the triphosphate and diphosphate forms by using analogs of GTP such as GppNHp and $Gpp(CH_2)p$, which are non- or slowly hydrolyzing analogs (Pai et al., 1989, 1990; Milburn et al., 1990; Privé et al., 1992; Schlichting et al., 1990). These data showed that the structural changes are confined to only two small areas which have been defined as switch I and switch II. The structures showed how the conformational change is triggered since both switch regions are bound to the γ-phosphate via two invariant amino acid residues, switch I by Thr-35 and switch II by Gly-60 (Wittinghofer and Pai, 1991). Removal of the γ-phosphate in the course of GTP hydrolysis allows these regions to relax and adopt a new conformation in what may be termed a loaded spring mechanism (Wittinghofer and Nassar, 1996).

A number of mutants of Ras were structurally investigated also, most notably oncogenic mutants of Ras such as G12V, G12R, G12D, and Q61L (Privé et al., 1992; Krengel et al., 1990; Franken et al., 1993). They showed a similar overall structure as compared to the wild-type Ras. Although small changes around the γ-phosphate group were observed in all of these structures, the analysis did not provide any consistent explanation for the inability of these mutants to hydrolyze GTP. GAP increases the k_{cat} of the intrinsic GTPase reaction by the factor of 10^5 compared to the unstimulated reaction (Wiesmüller and Wittinghofer, 1992; Gideon et al., 1992; Eccleston et al., 1993). Originally, it was suggested that a conformational change of the Ras induced by GAP determined the overall rate of the intrinsic reaction and that this rate is activated by GAP (Neal et al., 1990). This would have meant that Ras would be in principle a perfect

GTPase in the wrong conformation and just needed GAP to become activated. As an alternative model, it was discussed that Ras was an imperfect enzyme and needed the presence of extra chemical groups on GAP for fast hydrolysis (Schweins et al., 1995; Rensland et al., 1991). The latter model was supported in several biochemical studies (Mittal et al., 1996; Ahmadian et al., 1997). The weak binding of GAP to Ras in the GTP-state was enhanced considerably when GTP (or GppNHp) was replaced by GDP and fluoroaluminate. In the α subunits of heterotrimeric G-proteins, AlF_4^- and GDP were shown to mimic the transition state of GTP hydrolysis. For Ras, this state could only be shown in the presence of GAP, indicating an additional stabilization of the transition state by the interacting GAP. The issue was finally solved by the structure solution of GAP and the Ras–GAP complex (Scheffzek et al., 1996, 1997). The 3D structure indeed shows an Arg residue from GAP penetrating the active site of Ras and contacting the aluminium fluoride that is supposed to mimic the γ-phosphoryl group. This "Arg finger" is interpreted as the GAP residue hypothesized from biochemical experiments to contribute catalytic assistance to GTP hydrolysis on Ras. It appears that a negative charge developing on the γ-phosphate during GTP hydrolysis is stabilized by the positively charged guanidinium side of Arg, leading to faster hydrolysis. The structural analysis also explained why the GTPase reaction is slow in the oncogenic Ras and cannot be stimulated by GAP. A further catalytic effect of GAP binding is the stabilization of switch II, thereby fixing Ras residue Gln61 in a catalysis-competent position close to the attacking water molecule. For Gln61, one of the two most frequently mutated amino acids found in oncogenic Ras, the involvement in GTP hydrolysis was apparent from the Ras and the Ras–Ras GAP structure. Glutamine-61 is necessary to stabilize the hydroxyl ion or the polarized water molecule that attacks the phosphoryl group. The oncogenic effect of mutated Gly-12 could also be rationalized by the Ras–GAP complex structure. Any amino acid side chain at this position leads to a clash with the main chain of the GAP Arg finger, presumably leading to deranged conformations in the catalytic center of the complex.

IV. DRUG TARGET SITES OF Ras

Finding a compound that efficiently inhibits the mitogenic signals induced by the oncogenic Ras mutants would represent a breakthrough in the search for new anticancer therapeutics. Therefore, Ras has long been considered as an attractive target in the search for novel anticancer drugs. There are several major sites and biochemical functions of Ras which have been points of attack. For each of these target sites, promising candidate molecules with potential as anti-Ras drugs have been described.

Mainly, these have been as follows:

1. The C-terminal end of Ras which is posttranslationally modified. This modification is necessary for the membrane anchorage and the biological signaling function of the Ras protein.
2. The GDP–GTP exchange reaction which is the switch-ON reaction of the Ras signaling activity. The nucleotide exchange of Ras bound GDP to GTP leads to activation of the Ras protein.
3. The GTP hydrolysis as the switch-OFF of the Ras activity. The intrinsic GTPase of the Ras protein is very slow. To terminate the active state, the GTPase of the wild-type Ras is activated by a member of the GAP family.
4. The interaction of Ras with its major downstream effector proteins. The direct interaction with an effector protein is the prerequisite for the signaling activity of the Ras.

A. The C-terminal end of Ras

The Ras protein is posttranslationally modified at its carboxyl-terminal end. Since this modification is crucial for the membrane anchorage of the Ras protein and Ras function, the carboxyl-terminal region represents a potential target site for anti-Ras drugs. The key enzyme for the posttranslational modification of the COOH-end of Ras is farnesyltransferase (FT). FT is responsible for transferring a lipid anchor to the carboxyl-terminal end of the Ras. Ras without farnesylation remains inactive in the cytosol, and cells do not respond to mitogenic growth factors, that is, they do not proliferate. Hence the inhibition of anchor attachment has become an attractive pharmacological target (Cox and Der, 1997). Molecules that inhibit FT are currently being tested in several clinical studies as anticancer drugs. Prominent examples include competitive analogs of the FT lipid substrate, peptides or peptidomimetic molecules that compete with the Ras in the FT reaction. For details of FT inhibitors, see Chapter 14.

A new class of non-FT inhibitors has been identified to inhibit the membrane anchorage of Ras. Transfarnesylthiosalicylic acid (FTS) was shown to decrease the amount of active membrane-bound Ras (Marciano *et al.*, 1995; Marom *et al.*, 1995; Aharonson *et al.*, 1998; Haklai *et al.*, 1998). FTS is a synthetic small molecule that mimics the *S*-farnesyl moiety of Ras which serves as a recognition unit for membrane anchorage. FTS inhibits cell growth and transformation driven by oncogenic Ras (Jansen *et al.*, 1999; Elad *et al.*, 1999). FTS inhibits human tumor growth in animal models as a result of its ability to inhibit Ras signaling and reverses the transformed phenotype of tumor cells (Gana-Weisz *et al.*, 1997; Egozi *et al.*, 1999; Weisz *et al.*, 1999). FTS releases Ras from constrains on its lateral diffusion which is followed by dislodgement of Ras from its membrane anchorage domains (Niv *et al.*, 1999). A progressive

I

screening approach was chosen to select RNA aptamers which bind to the C-terminal end of Ras (Gilbert et al., 1997). The Cys-farnesylated peptide Boc-Cys-Lys-Ser-Lys-Thr-Lys-Cys-OH representing the C-terminal end of the Ras was immobilized, and interacting RNA aptamers were selected out of a random pool of RNA sequences. After 10 rounds of selection two sequence motifs were identified that showed high affinity to the farnesylated peptide, and a 10-fold lower affinity to the nonfarnesylated peptide. Next, the identified RNA aptamers could be shown in the cell culture experiments to inhibit Ras function by preventing its access to the cell membrane. In a related approach a library of branched peptidic molecules was generated and investigated for its ability to bind the nonfarnesylated C-terminal end of Ras (Dong et al., 1999). From the entire library which consisted of more than 150,000 compounds, individual members were identified and investigated for their ability to prevent the farnesylation of Ras-peptides and GFP-CAAX fusion proteins. One of the identified compounds (**I**) inhibited the farnesylation of a Ras peptide with an IC_{50} value of 100 μM and displayed sequence selectivity. These findings show that it is possible to prevent Ras processing not only by interfering with one of the involved enzymes but also by blocking the access of the enzyme to its substrate using synthetic molecules to mask the substrate.

B. The GDP–GTP exchange reaction

Guanine nucleotide exchange factors (GEFs) are the immediate activators of Ras. On interaction of a GEF with Ras, the binding to the nucleotide is weakened and its dissociation is accelerated. Because of the large molar excess of GTP over GDP in the cell, it is predominantly GTP that rebinds to Ras and transforms it into its active state. Independently of GEFs, oncogenically mutated Ras proteins are mostly bound to GTP because of the inhibition of the GTPase reaction. Therefore, the search for inhibitors of the guanine nucleotide exchange

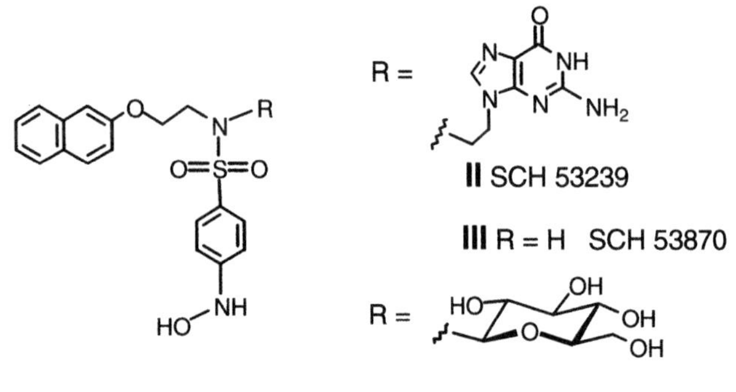

II SCH 53239

III R = H   SCH 53870

IV SCH 54292

reaction might not represent a suggestive strategy for the identification of drugs against Ras induced tumors. Nevertheless, drugs identified in these studies might still be useful in the therapy of tumors induced by components upstream of Ras. On the basis of molecular modeling, compounds SCH53239 (**II**), SCH53870 (**III**), and SCH54292 (**IV**) were designed to compete with GDP for binding in the nucleotide binding site. Indeed, these compounds inhibited nucleotide exchange on Ras with IC_{50} values of 0.5 to 0.7 μM, but they do not compete with GTP binding. NMR analysis of SCH54292 (**IV**) bound to Ras revealed that it binds to the switch II region of the Ras protein and not to the nucleotide binding site. The switch II region, which comprises residues 60 to 70, is involved in Ras interaction with nucleotide exchange factors. The treatment of NGF stimulated PC12 cells with SCH53870 (**III**) leads to inhibition of Ras signaling (Taveras *et al.*, 1997).

Several GTP analogs with alteration in the ribose part displayed drastically decreased binding affinity to both normal and mutated Ras. Pyrazolo[3,4-*b*]quinoline ribosides (**V, VI**) and the morpholino analogs (**VII**) were found to be moderate *in vitro* inhibitors of the nucleotide exchange process of oncogenic Ras. Surprisingly, fully protected triesters (**V**) and the corresponding

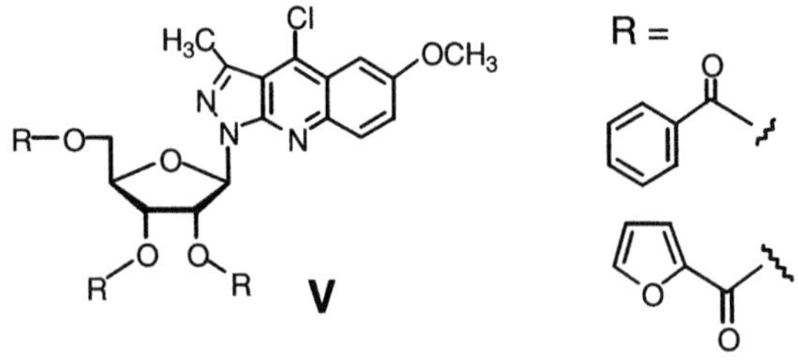

V

thionocarbonates **(VI)** were the most effective ribo-compounds displaying IC_{50} values in the low micromolar range (Noonan et al., 1991).

These examples show that one can, in principle, inhibit the exchange of GDP for GTP on Ras and thus prevent its activation. However, rather than trying to inhibit nucleotide exchange itself or inhibit the Ras–GEF interaction, it might be more reasonable to develop noncompetitive inhibitors of the GEF catalyzed nucleotide exchange, similar to Brefeldin A. This compound stabilized the abortive Arf GDP–Sec7 domain complex. A similar noncompetitive approach for the Ras–GEF complex would be inhibitory for the interaction with effectors that use an overlapping interface (Cherfils and Chardin, 1999).

C. The GTP hydrolysis

The major problem with oncogenic Ras is its inability to hydrolyze GTP. Thus, it was always considered a possibility that stimulating the GTPase activity of oncogenic Ras would be a valuable target for the development of anti-Ras drugs. The discovery that the introduction of an Arg into the active site of Ras by the GAPs stimulates the GTPase very efficiently has strengthened this idea considerably,

VIII   R = NH$_2$: DABP-GTP

IX   R = H: MABP-GTP

although GAP itself is unable to act on oncogenic Ras. Anti-Ras therapeutics on the basis of the GTP hydrolysis are supposed to accelerate the GTPase reaction of the oncogenic constitutively active Ras protein. The concept of GTPase inducers as Ras inhibitors has gained a further support by some recent data with the GTP analog diaminobenzophenone-phosphoroamidate-GTP (DABP-GTP) (**VIII**) (Ahmadian *et al.*, 1999). Normally, substitution of the β-phosphate by any ester or amide completely blocks hydrolysis of that compound. In the case of DABP-GTP (**VIII**), it was found (first for the G_α subunits of heterotrimeric G proteins and then in a more thorough study on Ras) that this analog is efficiently broken down to GDP on the G proteins. The amino group of the aromatic ring is mostly responsible for this effect, since MABP-GTP (**IX**) reacts only very slowly. Interestingly, the Gln at position 61 is not necessary for this effect. Actually, all oncogenic mutants of Ras including those of Gly-12 react DABP-GTP (**VIII**) even more efficiently than the wild-type Ras to yield GDP bound inactive Ras and DABP-Pi. From these data, one can conclude that oncogenic Ras can be induced to break the β,γ-phosphate bond efficiently just by supplying the proper chemistry into the active site. The GTP moiety fulfills the function to direct DABP right into the nucleotide binding pocket of the Ras protein where it activates nucleotide breakdown. Thus, the defective GTPase reaction of oncogenic Ras mutants can be rescued by using DABP-GTP (**VIII**) instead of GTP, arguing that the GTPase switch of mutant Ras is not irreversibly damaged.

Unfortunately, GTP analogs like DABP-GTP (**VIII**) are not good lead compounds for the development of anti-Ras drugs for two reasons. First, the affinity of Ras for GTP, which is the major nucleotide within living cells, is much higher than its affinity for GABP-GTP (**VIII**). This renders the nucleotide exchange of the bound GTP against the DABP-GTP (**VIII**) in the oncogenic mutant Ras protein ineffective. Second, treatment with nucleotide analogs might lead to strong side effects caused by affecting many other G proteins. Nevertheless, the successful identification of a small compound homing into the active site of Ras and carrying reactive chemical groups might open the way into a whole new strategy for anti-Ras drug development. In this way, the more recently solved structure of Ras bound to DABP-GppNp will be very helpful (Ahmadian *et al.*, 1999).

D. The interaction with effector proteins

The direct interaction of Ras with its effector proteins is a key event in Ras mediated signal transduction. The inhibition of the interaction between Ras and one or several of its downstream effectors could open up new alternatives to interfere with and regulate Ras-signaling. Ras function is mediated through interaction with various effectors. Although these functionally different effectors share a common fold, they lack sequence homology and use different residues for the interaction with Ras. Accordingly, it appears possible to develop selective inhibitors.

The best characterized interaction between Ras and an effector protein is its binding to c-Raf-1. It is believed that the main function of Ras is to recruit Raf to the plasma membrane where in a chain of events that includes allosteric interactions, phosphorylation, and interactions with other proteins such as the 14-3-3 family, the Raf somehow becomes activated (Daum *et al.*, 1994; McCormick and Wittinghofer, 1996; Morrison and Cutler, 1997).

Inhibiting protein–protein interactions with small molecules is generally considered to be difficult. This is due to the fact that the interface usually covers a large non-solvent accessible surface area between 1000 to 5000 Å^2 (in the case of the Ras–Raf binding domain 1300 Å^2). Protein–protein interfaces are additionally rather flat and do not contain deep water filled pockets as found in enzyme active sites. Furthermore, it is difficult to conclude from structural data of the interacting amino acids which are responsible for the binding. Mutational studies have shown that residues in the interface of a protein–protein complex are not equally contributing to the binding energy and that sometimes one to three residues are responsible for the major part of the binding energy, in a way that is not at all obvious from the 3D protein structure (Clackson and Wells, 1995; Schreiber and Fersht, 1993; Conte *et al.*, 1999). It might thus be necessary to construct new types of scaffolds covering a large surface area specifically designed for the protein in question. Nevertheless, the following examples show that, despite the high complexity of protein–protein interactions, it should be possible to develop small molecules as inhibitors of Ras-mediated protein–protein interactions. Apart from a small hydrophobic patch, the Ras–Raf interaction is mediated predominantly by oppositely charged or hydrophilic side chains forming specific salt bridges and hydrogen bonds, respectively. Peptides have been described that correspond to the Ras-binding domains of the downstream targets or to the region of Ras corresponding to the effector binding region.

The interaction between Raf and Ras-GTP was reportedly inhibited by penta- and decapeptides as well as by hepta- and octapeptides derived from the sequence of the Ras binding domain of c-Raf-1 (Clark *et al.*, 1996; Barnard *et al.*, 1995, 1998). Some of these compounds were reported to inhibit the Ras–Raf binding with IC$_{50}$ values as low as 7 μM (observed for one of the heptapeptides) and block the Ras-mediated activation of mitogen activated protein

(MAP) kinase. However, many other groups failed to reproduce these observations, in particular with the hepta- and octapeptides. Furthermore, in many cases, the reason for their inhibition is not obvious, as these peptides were not derived from the interface of the Ras–Raf RBD complex as determined from the 3D structure (Nassar et al., 1995, 1996).

A series of 13 to 28 amino acid peptides, derived from the effector binding region of Ras and bearing various amino acid substitutions, were investigated as inhibitors of the association of Ras with different effectors (Ohnishi et al., 1998). The peptide corresponding to the original Ras sequence inhibited Ras association with three different effectors at K_i values of 1 to 10 μM. Introducing amino acid substitutions corresponding to Ras mutations led, however, to peptides that selectively inhibited one or two Ras–effector interactions. Thus, the specificity observed with the whole Ras protein was retained in the effector region peptide. However, peptides are very large molecules and thus might not serve as lead compounds.

A promising candidate molecule to inhibit the Ras–Raf interaction is the physiological sulfide metabolite (X) of the nonsteroidal anti-inflammatory drug sulindac (XI) (Herrmann et al., 1998). Sulindac is widely used in the prevention and the therapy of adenomatous polyps in patients with inherited cancer predisposition. Rectal maintenance treatment resulted in complete reversion and prevention of rectal adenomas in patients with the inherited disease familial adenomatous polyposis coli (FAP) (Willumsen et al., 1984; Pasricha et al., 1995; Winde et al., 1995). Until recently, the potent antiproliferative effect of sulindac has only been ascribed to its inhibition of the cyclooxygenases, and the resulting decrease of the concentrations of mitogenic eicosanoids. Nevertheless, it was shown that sulindac and its physiological sulfide metabolite (X) can inhibit cell growth and carcinogenesis independent from the eicosanoid metabolism. The drug seems to target Ras mutated tumors more effectively than tumors without Ras mutations (Thompson et al., 1997). Indeed, sulindac sulfide strongly inhibits Ras induced malignant cell

X   Sulindac Sulfide   XI   Sulindac

transformation. Biochemical data suggest that this inhibition is due to a direct interaction of sulindac sulfide with the Ras (Herrmann *et al.*, 1998). Because of its low affinity, sulindac sulfide (**X**) itself cannot be regarded as a potent and specific anti-Ras drug. However, the molecule might serve as a lead structure in further studies.

V. CONCLUSIONS AND OUTLOOK

Oncogenic Ras plays a pivotal role in the initiation and the progression of human tumors. Hence the identification of a Ras-binding compound that selectively inhibits the transforming activity of Ras would represent a milestone in anticancer drug development. Although the Ras protein has been the subject of intense research scrutiny since 1982, structural and biochemical data gave just first insights into the basic principles of Ras function, its activation, inactivation, and signal transmission. This information forms the starting point for the development of, and the screening for, novel anti-Ras drugs. The examples described in this chapter, show that it is possible to inhibit Ras signaling by targeting various sites of the protein by small synthetic molecules of different chemical classes. Currently, the most promising target sites of Ras are its well-characterized biochemical properties, that is, the posttranslational modification, the effector protein binding, and the GTP hydrolysis. Of these, the most advanced approach is the development of drugs inhibiting the Ras farnesylation. Farnesyltransferase inhibitors are already being tested in various phases of clinical trials. In comparison, anti-Ras molecules that directly interact with Ras are just beginning to be characterized on the academic level. Nevertheless, the growing number of data about the complex function of Ras and the availability of new assay methods will push forward the search for new anti-Ras drugs. In the near future, two parallel developments may be expected. First, interactions of Ras with other proteins will be elaborated in the cellular context. The results will give new insights into complex biological patterns. Second, the obtained experimental data together with novel and established physicochemical techniques will be utilized in high throughput screening systems to investigate new substances in terms of their ability to interfere with Ras function.

References

Aharonson, Z., Gana-Weisz, M., Varsano, T., Haklai, R., Marciano, D., and Kloog, Y. (1998). Stringent structural requirements for anti-Ras activity of S-prenyl analogues. *Biochim. Biophys. A* **1406**, 40–50.

Ahmadian, M. R., Hoffmann, U., Goody, R. S., and Wittinghofer, A. (1997). Individual rate constants for the interaction of Ras proteins with GTPase-activating proteins determined by fluorescence spectroscopy. *Biochemistry* **36**, 4535–4541.

Ahmadian, M. R., Zor, T., Vogt, D., Kabsch, W., Selinger, Z., Wittinghofer, A., and Scheffzek, K. (1999). Guanosine triphosphatase stimulation of oncogenic Ras mutants. *Proc. Natl. Acad. Sci. U.S.A.* **96,** 7065–7070.

Barnard, D., Diaz, B., Hettich, L., Chuang, E., Zhang, X., Avruch, J., and Marshall, M. (1995). Identification of the sites of interaction between c-Raf-1 and Ras-GTP. *Oncogene* **10,** 1283–1290.

Barnard, D., Sun, H., Batur, L., and Marshall, M.S. (1998). In vitro inhibition of Ras–Raf association by short peptides. *Biochem. Biophys. Res. Commun.* **247,** 176–180.

Block, C., Janknecht, R., Herrmann, C., Nassar, N., and Wittinghofer, A. (1996). Quantitative structure–activity analysis correlating Ras/Raf interaction in vitro to Raf activation in vivo. *Nat. Struct. Biol.* **3,** 244–251.

Campbell, S. L., Khosravifar, R., Rossman, K. L., Clark, G. J., and Der, C. J. (1998). Increasing complexity of Ras signaling. *Oncogene* **17,** 1395–1413.

Cherfils, J., and Chardin, P. (1999). GEFs: Structural basis for their activation of small GTP-binding proteins. *Trends Biochem. Sci.* **24,** 306–311.

Clackson, T., and Wells, J. A. (1995). A hot spot of binding energy in a hormone–receptor interface. *Science* **267,** 383–386.

Clark, G. J., Drugan, J. K., Terrell, R. S., Bradham, C., Der, C. J., Bell, R. M., and Campbell, S. (1996). Peptides containing a consensus Ras binding sequence from Raf-1 and the GTPase activating protein NF1 inhibit Ras function. *Proc. Natl. Acad. Sci. U.S.A.* **93,** 1577–1581.

Conte, L. L., Chothia, C., and Janin, J. (1999). The atomic structure of protein–protein recognition sites. *J. Mol. Biol.* **285,** 2177–2198.

Cox, A. D., and Der, C. J. (1997). Farnesyltransferase inhibitors and cancer treatment: Targeting simply Ras? *Biochim. Biophys. Acta* **1333,** F51-F71.

Daum, G., Eisenmann-Tappe, I., Fries, H.-W., Troppmair, J., and Rapp, U. R. (1994). The ins and outs of Raf kinases. *Trends Biochem. Sci.* **19,** 474–480.

Dong, D. L., Liu, R., Sherlock, R., Wigler, M. H., and Nestler, H. P. (1999). Molecular forceps from combinatorial libraries prevent the farnesylation of Ras by binding to its carboxyl terminus. *Chem. Biol.* **6,** 133–141.

Downward, J. (1998). Ras signalling and apoptosis. *Curr. Opin. Genet. Dev.* **8,** 49–54.

Eccleston, J. F., Moore, K. J. M., Morgan, L., Skinner, R. H., and Lowe, P. N. (1993). Kinetics of interaction between normal and proline 12 Ras and the GTPase-activating proteins, p120-GAP and neurofibromin. The significance of the intrinsic GTPase rate in determining the transforming ability of ras. *J. Biol. Chem.* **268,** 27012–27019.

Egozi, Y., Weisz, B., Gana-Weisz, M., Ben-Baruch, G., and Kloog, Y. (1999). Growth inhibition of Ras-dependent tumors in nude mice by a potent Ras-dislodging antagonist. *Int. J. Cancer* **80,** 911–918.

Elad, G., Paz, A., Haklai, R., Marciano, D., Cox, A., and Kloog, Y. (1999). Targeting K-Ras 4B by S-trans, trans farnesyl thiosalicylic acid. *Biochim. Biophys. A* **1452,** 228–242.

Franken, S. M., Scheidig, A. J., Krengel, U., Rensland, H., Lautwein, A., Geyer, M., Scheffzek, K., Goody, R. S., Kalbitzer, H. R., Pai, E. F., and Wittinghofer, A. (1993). Three-dimensional structures and properties of a transforming and a nontransforming glycine-12 mutant of p21H-ras. *Biochemistry* **32,** 8411–8420.

Gana-Weisz, M., Haklai, R., Marciano, D., Egozi, Y., Ben-Baruch, G., and Kloog, Y. (1997). The Ras antagonist S-farnesylthiosalicylic acid induces inhibition of MAPK activation. *Biochem. Biophys. Res. Commun.* **239,** 900–904.

Gideon, P., John, J., Frech, M., Lautwein, A., Clark, R., Scheffler, J. E., and Wittinghofer, A. (1992). Mutational and kinetic analyses of the GTPase-activating protein (GAP)–p21 interaction: The C-terminal domain of GAP is not sufficient for full activity. *Mol. Cell. Biol.* **12,** 2050–2056.

Gilbert, B. A., Sha, M., Wathen, S. T., and Rando, R. R. (1997). RNA aptamers that specifically bind to a K Ras-derived farnesylated peptide. *Bioorg. Med. Chem.* **5,** 1115–1122.

Gorman, C., Skinner, R. H., Skelly, J. V., Neidle, S., and Lowe, P. N. (1996). Equilibrium and kinetic measurements reveal rapidly reversible binding of Ras to Raf. *J. Biol. Chem.* **271**, 6713–6719.

Haklai, R., Gana-Weisz, M., Elad, G., Paz, A., Marciano, D., Egozi, Y., Ben-Baruch, G., and Kloog, Y. (1998). Dislodgement and accelerated degradation of Ras. *Biochemistry* **37**, 1306–1314.

Hancock, J. F., Magee, A. I., Childs, J. E., and Marshall, C. J. (1989). All ras proteins are polyisoprenylated but only some are palmitoylated. *Cell* **57**, 1167–1177.

Hancock, J. F., Paterson, H., and Marshall, C. J. (1990). A polybasic domain or palmitoylation is required in addition to the CAAX motif to localize p21ras to the plasma membrane. *Cell* **63**, 133–139.

Herrmann, C., Block, C., Geisen, C., Haas, K., Weber, C., Winde, G., Möröy, T., and Müller, O. (1998). Sulindac sulfide inhibits Ras signaling. *Oncogene* **17**, 1769–1776.

Jansen, B., Schlabaur-Wadl, H., Kahr, H., Ress, E., Mayer, B. X., Eichler, H., Pehamberger, H., Gana-Weisz, M., Ben-David, E., Kloog, Y., and Wolff, K. (1999). Novel Ras antagonist blocks human melanoma growth. *Proc. Natl. Acad. Sci. U.S.A.* **96**, 14019–14024.

Krengel, U., Schlichting, I., Scherer, A., Schumann, R., Frech, M., John, J., Kabsch, W., Pai, E. F., and Wittinghofer, A. (1990). Three-dimensional structures of H-ras p21 mutants: Molecular basis for their inability to function as signal switch molecules. *Cell* **62**, 539–548.

Kuhlmann, J., and Herrmann, C. (2000). Biophysical characterization of the Ras protein. *Top. Curr. Chem.* **211**, 61–116.

McCormick, F., and Wittinghofer, A. (1996). Interactions between Ras proteins and their effectors. *Curr. Opin. Biotechnol.* **7**, 449–456.

Marciano, D., Ben-Baruch, G., Marom, M., Egozi, Y., Haklai, R., and Kloog, Y. (1995). Farnesyl derivatives of rigid carboxylic acids inhibitors of ras-dependent cell growth. *J. Med. Chem.* **38**, 1267–1272.

Marom, M., Haklai, R., Ben-Baruch, G., Marciano, D., Egozi, Y., and Kloog, Y. (1995). Selective inhibition of ras-dependent cell growth by farnesyl thiosalicylic acid. *J. Biol. Chem.* **270**, 22263–22270.

Milburn, M. V., Tong, L., DeVos, A. M., Brünger, A., Yamaizumi, Z., Nishimura, S., and Kim, S.-H. (1990). Molecular switch for signal transduction: Structural differences between active and inactive forms of protooncogenic ras proteins. *Science* **247**, 939–945.

Mittal, R., Ahamdian, M. R., Goody, R. S., and Wittinghofer, A. (1996). Formation of a transition-state analog of the Ras GTPase reaction by Ras-GDP, tetrafluoroaluminate, and GTPase-activating proteins. *Science* **273**, 115–117.

Morrison, D. K., and Cutler, R. E. (1997). The complexity of Raf-1 regulation. *Curr. Opin. Cell Biol.* **9**, 174–179.

Nassar, N., Horn, G., Herrmann, C., Scherer, A., McCormick, F., and Wittinghofer, A. (1995). The 2.2 Å crystal structure of the Ras-binding domain of the serine/threonine kinase c-Raf1 in complex with Rap1A and a GTP analogue. *Nature* **375**, 554–560.

Nassar, N., Horn, Herrmann, C., Block, C., Janknecht, R., and Wittinghofer, A. (1996). Ras/Rap effector specificity determined by charge reversal. *Nat. Struct. Biol.* **3**, 723–729.

Neal, S. E., Eccleston, J. F., and Webb, M. R. (1990). Hydrolysis of GTP by p21NRAS, the N-RAS protooncogene product, is accompanied by a conformational change in the wild-type protein: Use of a single fluorescent probe at the catalytic site. *Proc. Natl. Acad. Sci. U.S.A.* **87**, 3652–3565.

Niv, H., Gutman, O., Henis, Y., and Kloog, Y. (1999). Membrane interactions of a constitutively active GFP-K-Ras 4B and their role in signaling: Evidence from lateral mobility studies. *J. Biol. Chem.* **274**, 1606–1613.

Noonan, T., Brown, N., Dudycz, L., and Wright G. (1991). Interaction of GTP derivatives with cellular and oncogenic ras-p21 proteins. *J. Med. Chem.* **34**, 1302–1307.

Ohnishi, M., Yamawaki-Kataoka, Y., Kariya, K., Tamada, M., Hu, C.-D., and Kataoka, T. (1998). Selective inhibition of Ras interaction with its particular effector by synthetic peptides corresponding to the Ras effector region. *J. Biol. Chem.* **273**, 10210–10215.

Pai, E. F., Kabsch, W., Krengel, U., Holmes, K. C., John, J., and Wittinghofer, A. (1989). Structure of the guanine-nucleotide-binding domain of the Ha-ras oncogene product p21 in the triphosphate conformation. *Nature* **341**, 209–214.

Pai, E. F., Krengel, U., Petsko, G. A., Goody, R. S., Kabsch, W., and Wittinghofer, A. (1990). Refined crystal structure of the triphosphate conformation of H-ras p21 at 1.35 A resolution: Implications for the mechanism of GTP hydrolysis. *EMBO J.* **9**, 2351–2359.

Pasricha, P. J., Bedi, A., O'Connor, K., Rashid, A., Akhtar, A. J., Zahurak, M. L., Piantadosi, S., Hamilton, S. R., and Giardiello, F. M. (1995). The effects of sulindac on colorectal proliferation and apoptosis in familial adenomatous polyposis. *Gastroenterology* **109**, 994–998.

Privé, G. G., Milburn, M. V., Tong, L., DeVos, A. M., Yamaizumi, Z., Nishimura, S., and Kim, S.-H. (1992). X-ray crystal structures of transforming p21 ras mutants suggest a transition-state stabilization mechanism for GTP hydrolysis. *Proc. Natl. Acad. Sci. U.S.A.* **80**, 3649–3653.

Rensland, H., Lautwein, A., Wittinghofer, A., and Goody, R. S. (1991). Is there a rate-limiting step before GTP cleavage by H-ras p21? *Biochemistry* **30**, 11181–11185.

Scheffzek, K., Lautwein, A., Kabsch, W., Ahmadian, M. R., and Wittinghofer, A. (1996). Crystal structure of the GTPase-activating domain of human p120GAP and implications for the interaction with Ras. *Nature* **384**, 591–596.

Scheffzek, K., Ahmadian, M. R., Kabsch, W., Wiesmüller, L., Lautwein, A., Schmitz, F., and Wittinghofer, A. (1997). The Ras–RasGAP complex: Structural basis for GTPase activation and its loss in oncogenic Ras mutants. *Science* **277**, 333–338.

Scheffzek, K., Ahmadian, M. R., and Wittinghofer, A. (1998). GTPase-activating proteins: Helping hands to complement an active site. *Trends Biochem. Sci.* **23**, 257–262.

Schlichting, I., Almo, S. C., Rapp, G., Wilson, K., Petratos, K., Lentfer, A., Wittinghofer, A., Kabsch, W., Pai, E. F., and Petsko G. A. (1990). Time-resolved X-ray crystallographic study of the conformational change in Ha-Ras p21 protein on GTP hydrolysis. *Nature* **345**, 309–315.

Schreiber, G., and Fersht, A. R. (1993). The refolding of cis-and trans-peptidylprolyl isomers of barstar. *Biochemistry* **32**, 11195–11202.

Schweins, T., Geyer, M., Scheffzek, K., Warshel, A., Kalbitzer, H. R., and Wittinghofer, A. (1995). Substrate-assisted catalysis as a mechanism for GTP hydrolysis of p21ras and other GTP-binding proteins. *Nat. Struct. Biol.* **2**, 36–44.

Shields, J. M., Pruitt, K., McFall, A., Shaub, A., and Der, C. J. (2000). Understanding Ras: 'it ain't over 'til it's over'. *Trends Cell Biol.* **10**, 147–154.

Sydor, J. R., Engelhard, M., Wittinghofer, A., Goody, R. S., and Herrmann, C. (1998). Transient kinetic studies on the interaction of Ras and the Ras-binding domain of c-Raf-1 reveal rapid equilibration of the complex. *Biochemistry* **37**, 14292–14299.

Taveras, A. G., Remiszewski, S. W., Doll, R. J., *et al.* (1997). Ras oncoprotein inhibitors: The discovery of potent, ras nucleotide exchange inhibitors and the structural determination of a drug–protein complex. *Bioorg. Med. Chem.* **5**, 125–133.

Thompson, H. J., Jiang, C., Lu, J., Mehta, R. G., Piazza, G. A., Paranka, N. S., Paukcu, R., and Ahnen, D. J. (1997). Sulfone metabolite of sulindac inhibits mammary carcinogenesis. *Cancer Res.* **57**, 267–271.

Weisz, B., Giehl, K., Gana-Weisz, M., Egozi, Y., Ben-Baruch, G., Marciano, D., Gierschick, P., and Kloog, Y. (1999). A new functional Ras antagonist inhibits human pancreatic tumor growth in nude mice. *Oncogene* **18**, 2579–2588.

Wiesmüller, L., and Wittinghofer, A. (1992). Expression of the GTPase activating domain of the neurofibromatosis type 1 (NF1) gene in *Escherichia coli* and role of the conserved lysine residue. *J. Biol. Chem.* **267**, 10207–10210.

Willumsen, B. M., Christensen, A., Hubbert, N. L., Papageorge, A. G., and Lowy, D. R. (1984). The p21 ras C-terminus is required for transformation and membrane association. *Nature* **310**, 583–586.

Winde, G., Schmid, K. W., Schlegel, W., Fischer, R., Osswald, H., and Bunte, H. (1995). Complete reversion and prevention of rectal adenomas in colectomized patients with familial adenomatous polyposis by rectal low-dose sulindac maintenance treatment. Advantages of a low-dose nonsteroidal anti-inflammatory drug regimen in reversing adenomas exceeding 33 months. *Dis. Col. Rect.* **38**, 813–830.

Wittinghofer, A., and Nassar, N. (1996). How Ras-related proteins talk to their effectors. *Trends Biochem. Sci.* **21**, 488–491.

Wittinghofer, A., and Pai, E. F. (1991). The structure of Ras protein: A model for a universal molecular switch. *Trends Biochem. Sci.* **16**, 382–387.

Wittinghofer, A., and Waldmann, H. (2000). Ras-A molecular switch involved in tumor formation. *Angew. Chem. Int. Ed.* **39**, 4192–4214.

Actin-Binding Drugs: MKT-077 and Chaetoglobosin K (CK)

Hiroshi Maruta

Ludwig Institute for Cancer Research
Royal Melbourne Hospital
Parkville/Melbourne 3050, Australia

I. Introduction
II. MKT-077: F-Actin Bundler
III. Chaetoglobosin K: F-Actin Capper
References

I. INTRODUCTION

As discussed in Chapter 9, during malignant transformation of fibroblasts or epithelial cells caused by SV40 virus or oncogenes such as v-Src and v-Ha-RAS, actin stress fibers rapidly disassemble, and actin-based membrane ruffle and microspikes are induced (Weber *et al.*, 1974; Bar-Sagi and Feramisco, 1986). These changes in actin-cytoskeleton are in part due to the downregulation of several distinct genes encoding actin-binding proteins such as gelsolin and vinculin (Vandekerckhove *et al.*, 1990), and are also in part due to inactivation of several distinct F-actin cappers and bundlers, such as tensin and cortactin through PI-3 kinase and Rac/CDC42 GTPases (He *et al.*, 1998; Maruta *et al.*, 1999). Furthermore, overexpression of F-actin cappers such as gelsolin and tensin or F-actin bundlers such as vinculin and HS1 suppresses RAS or SV40-induced malignant transformation (Fernandez *et al.*, 1992; Muellauer *et al.*, 1993; He *et al.*, 1998; Tikoo *et al.*, 1999). These observations strongly suggest the anticancer potential of specific chemical compounds

Tumor-Suppressing Viruses, Genes, and Drugs
Innovative Cancer Therapy Approaches
Copyright © 2002 by Academic Press.
All rights of reproduction in any form reserved.

which cap or bundle actin filaments. We indeed found more recently that both the F-actin bundling compound MKT-077 and the F-actin capping antibiotic CK (chaetoglobosin K) are able to suppress RAS transformation *in vitro* and *in vivo* (Tikoo *et al.*, 1999, 2000). Thus, in this chapter I shall focus the discussion mainly on the anti-RAS cancer potential of these two unique actin-binding chemicals.

II. MKT-077: F-ACTIN BUNDLER

A. Background

The rhodacyanine dye MKT-077 (for the chemical structure, see Figure 1A), which was originally developed as a photosensitizer used for the processing of color films, has been shown to be highly toxic toward a variety of human cancer cells, but far less toxic toward some normal cells, and, therefore, has been in clinical trials as a novel anticancer drug (Koya *et al.*, 1996). However, so far no direct comparison has been made between the malignant cells with any known mutation of either a single oncogene or tumor suppressor gene and the parental

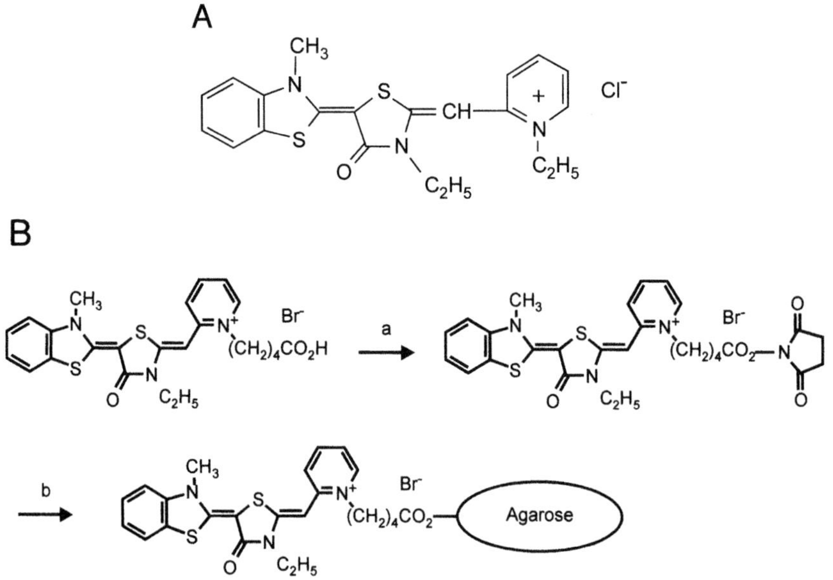

Figure 1 MKT-077 and its analog. (A) The chemical structure of MKT-007. (B) The MKT-077 analog (Compound 1, top left) conjugated to agarose beads.

normal cells of the exact same species and tissue origin. Therefore, it remained to be clarified which oncogene (or dysfunction of which tumor suppressor gene) is responsible for its selective toxicity, and which protein is its major target in malignant cells. Interestingly, one of the MKT-077 sensitive cancer cell lines is human bladder carcinoma EJ where c-Ha-RAS has been mutated at position 12 (Gly to Val), thereby being oncogenic (Koya et al., 1996; Tabin et al., 1982).

We more recently found that RAS oncogene is responsible for the selective toxicity of MKT-077, by comparing the effect of this drug between normal and v-Ha-RAS-transformed NIH 3T3 cells (Tikoo et al., 2000). Furthermore, by an MKT-077 sepharose affinity chromatography, we have identified at least two distinct proteins that selectively bind MKT-077 in RAS transformants, which turned out to be actin and HSC70, a constitutive member of the HSP70 heat-shock ATPase family (Tikoo et al., 2000).

B. RAS-dependent selective toxicity of MKT-077

MKT-077, at a concentration of $1.5\,\mu M$, inhibits the growth of RAS-transformants by more than 60%, but causes no effect on that of the parental normal cells. Furthermore, it inhibits almost completely (by more than 95%) the focus formation of RAS transformants by piling up on top of each other at their confluence. Moreover, MKT-077 even at concentrations below $1.5\,\mu M$ almost completely inhibits the anchorage-independent growth (colony formation in soft agar) of RAS-transformants. Since the parental normal cells never form any focus, or colony in soft agar, the MKT-077 treated RAS transformed cells behave like the normal cells. These results seem to be compatible with the previous observation that malignant cells such as human EJ bladder carcinoma cells, carrying an oncogenic c-Ha-RAS mutant, are around 100 times more sensitive to this drug than normal cells such as CV-1 (Koya et al., 1996). Interestingly, in nude mice, the growth of sarcomas derived from v-Ha-RAS transformed NIH 3T3 cells is strongly inhibited by MKT-077 at 5 mg/kg [intraperitoneally (i.p.) every other day for 2 weeks]. Fifty percent of the treated mice remain to be free of tumors, and the size of the growing tumors in the treated mice is reduced by more than 50% (Tikoo et al., 2000). These observations clearly indicate that (1) oncogenic RAS mutations are responsible, at least in part, for the selective cytostatic effect or toxicity of this drug toward malignant cells, and that (2) MKT-077 reverses the RAS-induced malignant phenotype both in vitro and in vivo.

How does MKT-077 inhibit selectively the growth of RAS-transformants? It has been claimed previously that MKT-077 is selectively accumulated in mitochondria of the target cancer cells such as the human pancreatic carcinoma cell line CRL-1420, but not in normal cells such as African green monkey kidney cell line CV-1 (Koya et al., 1996). However, in RAS-transformed NIH 3T3 cells, such a selective accumulation of MKT-077 in their mitochondria has never been observed (Tikoo *et al.*, 2000), suggesting that the major responsible target of MKT-077 is not a mitochondrial component.

C. Actin and Hsc70: The major targets of MKT-077

To identify and affinity-purify the major target proteins of MKT-077 in RAS-transformants, we have generated an immobilized derivative of MKT-077 (see Figure 1B) by conjugating the compound 1, a carboxyl derivative of MKT-077, with Sepharose beads (Kawakami *et al.*, 1998; Tikoo *et al.*, 2000). This conjugate is called ligand-beads here. First, using the ligand-beads, we have identified two major proteins of 45 kDa and 75 kDa that bind MKT-077 in the cytosol or cytoskeletal fractions of RAS-transformants, but not of the parental normal cells (Tikoo *et al.*, 2000). Subsequent microsequencing of these two proteins has revealed that the 45 kDa protein is actin isoforms, whereas the 75 kDa protein is HSC70, a constitutive member of the heat-shock HSP70 family (Giebel *et al.*, 1988), and not the HSP70, a heat-inducible member (Hunt and Calderwood, 1990).

D. MKT-077 bundles actin filaments and blocks membrane ruffling

We have further confirmed the direct binding of this drug to both monomeric (G) and filamentous (F) forms of actin, incubating both purified G- and F-actin solution with the ligand-beads, Moreover, we found that MKT-077 superprecipitates actin filaments with MKT-077 under the conditions where actin filaments alone are hardly precipitated (Tikoo *et al.*, 2000), suggesting that MKT-077 causes the formation of either actin meshworks or bundles by cross-linking actin filaments loosely or tightly, respectively. Using the standard electron microscope (EM) negative staining technique, we have confirmed that MKT-077 indeed bundles actin filaments (Tikoo *et al.*, 2000). Interestingly, 1.5 μM MKT-077 strongly inhibits RAS/Rac-induced membrane ruffling (Tikoo *et al.*, 2000), as do two other anti-RAS cancer drugs, SCH51344 and a cytochalasin called CK, by blocking a Rac-mediated pathway (Walsh *et al.*, 1997; Tikoo *et al.*, 1999). This observation suggests that MKT-077 interferes with the function of actin filaments in RAS transformants.

E. MKT-077 inhibits the actin-myosin interaction

The ATPase heads of each myosin II (double-headed myosin) such as smooth muscle myosin also cross-link actin filaments, and actin filaments activate Mg^{2+}-ATPase activity of smooth muscle myosin when the light chain of myosin II is fully phosphorylated (Ye *et al.*, 1997). We found that MKT-077 strongly inhibits the Mg^{2+}-ATPase activity of the actomyosin (actin–myosin II complex), but not that of the phosphorylated myosin II alone (Tikoo et al., 2000). These results clearly indicate that MKT-077 selectively binds actin filaments to block their interactions with myosin ATPase heads, without any effect on myosin heads alone.

Our finding that, like a few F-actin-bundling proteins such as HS1, α-actinin and vinculin, the F-actin bundling drug MKT-077 suppresses RAS-transformation by blocking membrane ruffling suggests that other selective F-actin bundling compounds are also potentially useful for the chemotherapy of RAS-associated cancers. Furthermore, our observations clearly indicate that the mitochondrial proteins or lipids are not the primary targets of MKT-077, at least in the RAS transformants. Perhaps, the previously observed accumulation of MKT-077 in mitochondria might be a phenomenon specific for a particular cell type(s), rather than for malignant transformation in general.

Actin is a housekeeping protein. It is constitutively present even in the normal cells. Why does MKT-077 fail to bind actin in the normal cells? In normal cells, and not in RAS-transformed cells, the major actin-cytoskeleton is present in actin stress fibers where actin filaments and myosin II form a tight complex. It is most likely that myosin II protects actin filaments from binding to MKT-077 in the stress fibers, as MKT-077 and myosin II apparently interfere with their binding to actin filaments. In fact no staining of actin stress fibers with the fluorescence MKT-077 was observed in normal cells.

F. The role of MKT-077 in the regulation of HSC70–Mot-2

We have more recently found that MKT-077 binds both HSC70 and its related protein called MOT-2, which inactivates the tumor suppressor p53 by blocking its nuclear localization, and that MKT-077 reactivates p53 by blocking the p53–MOT-2 interaction (Wadhwa et al., 1998, 2000). Since RAS activates cyclin-dependent kinases (CDKs) by overexpressing cyclin D1 (Maruta, 1996), whereas p53 inactivates CDKs by overexpressing a CDK inhibitor called WAF1/CIP1/p21 (Wadhwa et al., 1998), it is possible that MKT-077 might antagonize RAS indirectly by reactivating the tumor suppressor p53.

Interestingly, an MKT-077 analog called FJ-5774 was found to be much more potent than MKT-077 itself to reactivate p53 and subsequently induce WAF1 expression in RAS transformed cells, and its maximum effect was observed at $50\,\mu M$ (Wadhwa et al., 2000). However, FJ-5774 at this concentration or less was found to be equally toxic for both normal cells and RAS transformants (Nheu, T., Shishido, T., and Maruta, H., unpublished observation, 2000). Furthermore, MKT-077 at $1.5\,\mu M$ or less, which fully suppresses RAS transformation, is not sufficient for any significant reactivation of p53 or induction of WAF1 expression in the

RAS-transformants (Wadhwa *et al.*, 2000). These observations suggest it rather unlikely that the interaction of MKT-077 with HSC70 or MOT-2 significantly contributes to its selective toxicity toward RAS transformants.

In conclusion, although the anticancer drug MKT-077 binds two major target proteins in RAS transformants, actin and the chaperon ATPases HSC70 or MOT-2, it appears to be actin, and not the chaperone ATPases that is mainly responsible for MKT-077-induced selective toxicity toward RAS transformants. Like the F-actin capping drug CK (Tikoo *et al.*, 1999) and the drug SCH51344 (Walsh *et al.*, 1997), the F-actin bundling drug MKT-077 blocks RAS/Rac-induced membrane ruffling to suppress RAS transformation. These findings strongly suggest the anti-RAS potential of other specific F-actin capping or bundling drugs or inhibitors specific for the Rac–CDC42-dependent kinase PAK such as the drug CEP-1347 and the cell-permeable peptide WR-PAK18 that selectively block the membrane ruffling (Maruta *et al.*, 1999; He *et al.*, 2001). For detail of the PAK-specific inhibitors, see Chapter 18.

III. CHAETOGLOBOSIN K: F-ACTIN CAPPER

Cytochalasins and phalloidins are two distinct families of antibiotics that bind actin filaments and control the dynamics or equilibrium of actin polymerization or depolymerization (Cooper, 1987). Cytochalasins such as cytochalasin D (CD) form a 1:1 complex with actin monomer through Val-139 and Ala-295 (Ohmori *et al.*, 1992), and this complex shortens actin filaments by capping selectively the plus end of actin filaments. Cytochalasins, in particular CD, have been wildly used to study the role of actin cytoskeleton in a variety of cell functions such as cell motility, cell adhesion to a solid substratum, cytokinesis, and endocytosis. The function of phalloidins appears to be the opposite to that of cytochalsins. Phalloidins bind only actin polymers (filaments) and not actin monomers, and they block the depolymerization of actin filaments. Thus, their fluorescent derivatives have been mainly used to demonstrate the intracellular localization of actin filaments. In this section, the anticancer potential of a unique cytochalasin called chaetoglobosin K (CK) will be discussed.

A. Cytochalasins

Since cytochalasins were first reported to inhibit cell movement and cytokinesis (cytoplasmic cleavage), and also induce the extrusion of nucleus (Carter, 1967), more than two dozen distinct cytochalasins have been isolated from a variety of microorganisms (Tanenbaum, 1978; Yahara *et al.*, 1982). The majority of them are either 10-phenyl or 10-indolyl analogs of the cytochalasin ring. One common biological property of all these cytochalasins is to cap the plus end of actin filaments, block actin polymerization at that end, and eventually stop membrane ruffling of cells (Yahara *et al.*, 1982). However, some of them such as cytochalasin B (CB) show a side (additional biological) effect such as binding to a GLUT family of glucose transporters to block glucose uptake of cells, although dihydrocytochalasin (DCB) and CD do not bind GLUT family proteins (Kan *et al.*, 1994). In addition to blocking membrane ruffling, all cytochalasins, except for CK, cause rounding up of cells (Yahara *et al.*, 1982). Although the exact cause of this cell rounding up still remains to be determined, clearly it has nothing to do with their F-actin capping activity. Thus, CK appears to be so far the best suited member of the cytochalasin family for us to examine the effect of F-actin capping alone on RAS transformation.

B. Chaetoglobosin K inhibits anchorage-dependent growth of RAS transformants

Chaetoglobosin K (CK) was isolated from the plant fungus *Diplodia macrospora* (Cutler *et al.*, 1980). It belongs to 10-indolylcytochalasin family (for the chemical structure, see Figure 2). The specific reason why we have chosen CK is that, unlike any other cytochalasins, CK does not cause either rounding up of cells or contraction of actin cables, suggesting that CK has, so far, the least side effects among more than 24 distinct cytochalasins (Yahara *et al.*, 1982). CK at $2\,\mu M$ almost completely inhibits both actin polymerization *in vitro* and membrane ruffling of cells (Yahara *et al.*, 1982).

CK at $2\,\mu M$ almost completely inhibits the colony formation in soft agar (anchorage-independent growth) of RAS transformants, indicating that,

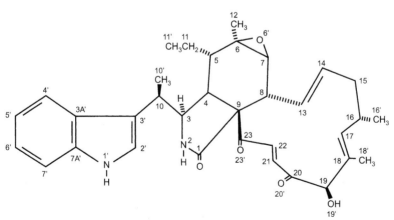

Figure 2 Chemical structure of chaetoglobosin K (CK).

like tensin, CK suppresses the malignant phenotype (Tikoo *et al.*, 1999). Our observations that two distinct F-actin capping molecules, that is, tensin and CK, suppress the RAS-induced malignant phenotype strongly suggest, if not prove, that capping of actin filaments at the plus ends alone is sufficient to block one of the RAS signaling pathways essential for its oncogenicity. This notion is compatible with the fact that RAS induces the uncapping of actin filaments at the plus ends through the Rac–PIP2 pathway.

However, unlike tensin, CK causes an apoptosis of RAS-transformed NIH 3T3 cells, and less effectively normal NIH 3T3 cells, indicating that CK has an F-actin capping-independent side effect(s). CK-induced apoptosis is at least in part due to CK-induced inhibition of a Ser–Thr kinase called AKT or PKB (Tikoo *et al.*, 1999). PKB is activated by RAS through an end product of PI-3 kinase, and is essential for cell survival, that is, the prevention from apoptosis, of the fibroblasts (Marte and Downward, 1997). Like PAKs, PKB phophorylates and inactivates an apoptosis agonist called BAD which forms a heterodimer with the apoptosis antagonist Bcl-2 (Schuermann *et al.*, 2000).

C. The ICE/caspase-1 inhibitor N1445 blocks the CK-induced apoptosis

In an attempt to rescue normal cells from the CK-induced apoptosis, several distinct inhibitors of apoptotic proteases were screened first for their ability to block the apoptosis of normal cells caused by serum withdrawal (SW). A specific ICE/caspase-1 inhibitor called N1445 (Z-Asp-CH$_2$-DCB) most effectively rescues the normal cells from the SW-induced apoptosis (Tikoo *et al.*, 1999). Furthermore, N1447 completely abolishes the CK-induced apoptosis of the normal cells by reactivating PKB, but without affecting the CK-induced suppression of RAS transformation (Tikoo *et al.*, 1999). Interestingly, other apoptosis inhibitors such as Z-VAD-FMK, which inhibits the conversion of the proCPP32 to the apoptotic CPP32/caspase-3 (Greidinger *et al.*, 1996), fail to abolish the CK-induced apoptosis (Tikoo *et al.*, 1999). Furthermore, N1445 fails to abolish the apoptosis caused by LY294002, a PI-3 kinase inhibitor (Tikoo *et al.*, 1999). Thus, N1447 appears to be a rather specific inhibitor for the CK-induced apoptosis.

These findings clearly indicate that CK activates a specific apoptotic pathway involving the ICE protease/caspase-1 which is also activated by serum withdrawal, but not by inhibition of PI-3 kinase. It is very unlikely, however, that CK exerts its apoptotic action simply through its F-actin capping activity, because the overexpression of the F-actin capping protein tensin causes no apoptosis, but suppresses RAS transformation. Both serum (growth factors) withdrawal and CB cause apoptosis by reducing glucose uptake through its binding to the glucose transporter family GLUT, whereas DCB blocks glucose uptake only weakly and does not cause any apoptosis (Kan *et al.*, 1994). We found that, like DCB, CK inhibits only weakly the basal glucose uptake of normal NIH 3T3 cells, whereas

CD has no effect on the glucose uptake, but still causes apoptosis that is almost completely abolished by N1445 (Tikoo et al., 1999). These observations suggest that the CK-induced apoptosis differs from the CB-induced apoptosis, and is probably caused by a mechanism(s) other than the glucose starvation.

To clarify whether the CK-induced apoptosis involves any inactivation of PKB, we treated the fibroblasts with either CK or the PI-3 kinase inhibitor LY294002 as a positive control, and then the PKB activity immunoprecipitated with a specific antibody against PKB from the resulting cell lysates was assayed. Like LY294002, CK significantly reduces the PKB activity. Interestingly, N1445 is a potent activator of PKB, and abolishes completely the CK-induced inhibition of PKB activity (Tikoo et al., 1999). These observations clearly indicate that (1) CK causes apoptosis at least in part by blocking the activation of PKB as does LY294002, and (2) CK-induced inactivation of PKB is prevented by this caspase-1 inhibitor which activates PKB. It still remains to be clarified, however, whether CK inactivates PKB directly or a upstream component essential for the activation of PKB, and how the caspase-1 inhibitor causes the activation of PKB.

More importantly in a practical sense, the N1445 does not reduce the ability of CK to suppress the malignancy (anchorage-independent growth) of RAS transformants at all. These observations suggest the possibility that the combination of this unique cytochalasin (CK), and the specific apoptosis inhibitor N1445, would provide a potentially useful anti-RAS tumor suppressing remedy.

D. *In vivo* effects of cytochalasins

CD was tested for its anticancer potential, using mice and rats carrying several distinct tumors almost 3 decades ago (Katagiri and Matsuura, 1971). However, its lethal doses (5 mg/kg/day) toward the whole animals turned out to be pretty close to its antitumor doses (around 2.5 mg/kg/day), its potential as an anticancer drug remains clearly questionable. More recently, although CB (100 mg/kg single dose) could significantly (by 10 days) delay the death of mice caused by either lung carcinomas or melanomas, which occur in 20 days, it failed to prevent their death beyond 30 days (Bousquet et al., 1990). Furthermore, it still remains unknown whether these tumors carry any oncogenic RAS mutants.

Thus, we are currently testing this CK-N1445 combination chemotherapy of RAS-associated cancers, using experimental animals such as nude or RAS-transgenic mice. The LD_{100} of CK alone toward chickens is around 50 mg/kg single dose, and no mortalities were observed at 25 mg/kg (Cutler et al., 1980). Our preliminary experiment with nude mice carrying sarcomas derived from v-Ha-RAS-transformed NIH 3T3 cells has revealed that the treatment with CK (2 mg/kg, i.p., every other day) in 30% dimethylsulfoxide (DMSO), in the presence or absence of N1445, has no detectable adverse effect on any health conditions of these mice, and significantly reduces the size of these sarcomas over 20 days. However, the inclusion of 1% Tween-20 in the solvent DMSO, in an attempt to increase the solubility of CK, causes an acute neurotoxicity most likely by breaking the blood–brain barrier against CK. Thus, further optimization of the treatment conditions would be required for both the safety and improvement of its therapeutic efficacy.

References

Bar-Sagi, D., and Feramisco, J. (1986). Induction of membrane ruffling and fluid-phase pinocytosis in quiescent fibroblasts by Ras proteins. Science 233, 1061–1065.

Bousquet, P., Paulsen, L., Fondy, C., Lipski, K., Loucy, K., and Fondy, T. (1990). Effects of cytochalasin B in culture and in vivo on murine Madison 109 lung carcinomas and B16 melanoma. Cancer Res. 50, 1431–1439.

Carter, S. (1967). Effects of cytochalasins on mammalian cells. Nature 213, 261–264.

Cooper, J. (1987). Effects of cytochalasin and phalloidin on actin. J. Cell Biol. 105, 1473–1478.

Cutler, H. G., Crumley, F., et al. (1980). Chaetoglobosin K: A new plant growth inhibitor and toxin from Diplodia macrospora. J. Agric. Food Chem. 28, 139–142.

Fernandez, J. L. R., Geiger, B., Salmon, D., Sabanay, I., Zoeller, M., and Ben-Ze'ev, A. (1992). Suppression of tumorigenesis in SV40-transformed cells after transfection with vinculin cDNA. J. Cell Biol. 119, 427–438.

Giebel, L., Dworniczak, B., and Bautz, E. (1988). Developmental regulation of a constitutively expressed mouse mRNA encoding a 72 kDa heat-shock family protein. Dev. Biol. 125, 200–207.

Glueck, U., Kwiatkowski, D., and Ben-Ze'ev, A. (1993). Suppression of tumorigenesis in SV-40-transformed 3T3 cells transfected with α-actinin cDNA. Proc. Natl. Acad. Sci. U.S.A. 90, 383–387.

Greidinger, E., Miller, D., Yamin, T. T., Casciola-Rosen, L., and Rosen, A. (1996). Sequential activation of these distinct ICE-like activities in Fas-ligated Jurkat cells. FEBS Lett. 390, 299–303.

He, H., Watanabe, T., Zhan, X., et al. (1998). Role of PIP2 in Ras/Rac-induced disruption of the cortactin–actomyosin II complex and malignant transformation. Mol. Cell Biol. 18, 3829–3837.

He, H., Hirokawa, Y., Manser, E., et al. (2001). Signal therapy for RAS-induced cancers in combination of AG 879 and PP1, specific inhibitors for ErbB2 and Src family kinases, that block PAK activation. Cancer J. 7, 191–202.

Hunt, C., and Calderwood, S. (1990). Characterization and sequence of a mouse Hsp70 gene and its expression in mouse cell lines. Gene 87, 199–204.

Kan, O., Baldwin, S., and Whetton, A. (1994). Apoptosis is regulated by the rate of glucose transport in an interleukin 3-dependent cell line. J. Exp. Med. 180, 917–923.

Katagiri, K., and Matsuura, S. (1971). Antitumor activity of cytochalasin D. J. Antibiot. 24, 722–723.

Kawakami, M., Suzuki, N., Sudo, Y., et al. (1998). Development of an enzyme-linked immunosorbent assay (ELISA) for antitumor agent MKT-077. Anal. Chim. Acta 362, 177–186.

Koya, K., Li, Y., Wang, H., et al. (1996). MKT-077, a novel rhodacyanine dye in clinical trials, exhibits anticarcinoma activity in preclinical studies based on selective mitochondrial accumulation. Cancer Res. 56, 538–543.

Marte, B. M., and Downward, J. (1997). PKB/AKT: Connecting PI-3 kinase to cell survival and beyond. Trends Biochem. Sci. 22, 355–358.

Maruta, H. (1996). Downstream of RAS and action-cytoskeleton. In "Regulation of the RAS Signaling Network" (H. Maruta and A. Burgess, eds.), pp. 139–180. Springer-Verlag/R. G. Landes, Heidelberg/Austin.

Maruta, H., He, H., Tikoo, A., *et al.* (1999). G proteins, PIP2, and actincytoskeleton in the control of cancer growth. *Microsc. Res. Tech.* **47**, 61–66.

Muellauer, L., Fujita, H., Shizaki, A., and Kuzumaki, N. (1993). Tumor-suppressive function of mutated gelsolin in Ras-transformed cells. *Oncogene* **8**, 2531–2536.

Ohmori, H., *et al.* (1992). Direct proof that the primary site of action of cytochalasin on cell motility is actin. *J. Cell Biol.* **116**, 933–941.

Schuermann, A., *et al.* (2000). PAK1 phosphorylates the death agonist BAD and protects cells from apoptosis. *Mol. Cell. Biol.* **20**, 453–461.

Tabin, C. J., Bradley, S. M., Bargmann, C. I., *et al.* (1982). Mechanism of activation of a human oncogene. *Nature* **300**, 143–149.

Tanenbaum, S. (1978). Cytochalasins. Biochemical and cell biological aspects. *Front. Biol.* **46**, 1–14.

Tikoo, A., Cutler, H., Lo, S. H., *et al.* (1999). Treatment of RAS-induced cancers by the F-actin cappers tensin and chaetoglobosin K, in combination with the caspase-1 inhibitor N1445. *Cancer J. Sci. Am.* **5**, 293–300.

Tikoo, A., Shakri, R., Connolly, L., *et al.* (2000). Treatment of RAS-induced cancers by the F-actin-bundling drug MKT-077. *Cancer J. Sci. Am.* **6**, 162–168.

Vandekerckhove, J., Bauw, G., Vancompernolle, K. Honore, B., and Celis, J. (1990). Comparative two-dimensional gel analysis and microsequencing identifies gelsolin as one of the most prominent down-re gulated markers of transformed human fibroblast and epithelial cells. *J. Cell Biol.* **111**, 95–102.

Wadhwa, R., Takano, S., Robert, M., *et al.* (1998). Inactivation of tumor suppressor p53 by Mot-2, a hsp70 family member. *J. Biol. Chem.* **273**, 29586–29591.

Wadhwa, R., Sugihara, T., Yoshida, A., *et al.* (2000). Selective toxicity of MKT-077 to cancer cells is mediated by its binding to HSP70 family protein, mortalin, and reactivation of p53 function. *Cancer Res.* **60**, 6818–6821.

Walsh, A., Dhanasekaran, M., Bar-Sagi, D., *et al.* (1997). SCH51344-induced reversal of RAS-transformation is acompanied by the specific inhibition of the RAS and Rac-dependent cell morphology pathway. *Oncogene* **15**, 2553–2560.

Weber, K., Lazarides, E., Goldman, R., Vogel, A., and Pollack, R. (1974). Localization and distribution of actin fibers in normal, transformed and revertant cells. *Cold Spring Harbor Symp. Quant. Biol.* **39**, 363–369.

Yahara, I., Harada, F., Sekita, S., Yoshihira, K., and Natori, S. (1982). Correlation between effects of 24 different cytochalasins on cellular structures and events and those on actin *in vitro*. *J. Cell Biol.* **92**, 69–78.

Ye, L. H., Hayakawa, K., Kishi, H., *et al.* (1997). The structure and function of the actin-binding domain of myosin light chain kinase of smooth muscle. *J. Biol. Chem.* **272**, 32182–32189.

Tyr Kinase Inhibitors as Potential Anticancer Agents: EGF Receptor and ABL Kinases

Antony W. Burgess

Ludwig Institute for Cancer Research/CRC for Cellular Growth Factors
Parkville/Melbourne 3050, Australia, and
Institute for Advanced Studies
Hebrew University
Jerusalem 91904, Israel

I. INTRODUCTION

When it was discovered that the oncogenic component of the Rous sarcoma virus (v-Src) was a mutant form of a host gene, tumor biology was changed forever (Fujita *et al.*, 1978). Within a few short years it was apparent that v-Src was a constitutively active form of a cellular tyrosine kinase (Collett *et al.*, 1980).

Tumor-Suppressing Viruses, Genes, and Drugs
Innovative Cancer Therapy Approaches

Although tyrosine (Tyr) kinases were unusual (rare) cellular enzymes, it was clear that a constitutively active form of c-Src played an important role in the processes associated with oncogenic transformation. The search for other Tyr kinases quickly led to the identification of the enzymic activity of the epidermal growth factor (EGF) receptor (EGFR), whose intracellular kinase domain is activated on the ligation of the extracellular domain (ECD) by the EGF family ligands (Ushiro and Cohen, 1980).

During the 1980s and 1990s dozens of Tyr kinases were identified and characterized (Hunter, 1998). Many of these kinases are known to be involved with the signaling from growth factor or cytokine receptors. Tyr kinases are known to regulate many cellular processes from cell–cell interactions and membrane signaling to cytoskeletal, cytoplasmic, and nuclear protein associations. Activated forms of the kinases have already been associated with many cancers and appear to be a key element in the maintenance of the oncogenic state. Indeed by inhibiting tumor associated kinases it is possible to either kill tumor cells directly (Dorsey et al., 2000) or to sensitize them to the killing action of cytotoxic drugs or other biological effectors (e.g., antibodies) (Ciardiello et al., 1999; Fang et al., 2000).

In many normal tissues receptor Tyr kinases are dormant, consequently, inhibitors will not have a major effect on the tissue physiology (Donaldson and Cohen, 1992). In contrast, the growth factor receptor systems appear to be activated in most cancer cells, so as soon as a component of the signaling pathways is inhibited, the tumor cells stop proliferating and some cells even undergo apoptosis. Initial clinical trials in the area of signaling therapeutics have started (Goldman, 2000; O'Dwyer and Druker, 2000), and it is already clear that we have reached a new era in the development of anticancer agents which target the molecular lesion(s) associated with specific cancers. Targeting the oncogenic lesions should increase the effectiveness of cancer cell killing as well as reduce the general toxicity associated with more conventional cancer therapeutics.

Although receptor kinases can be targeted by extracellular agents (Baselga et al., 2000), many oncogenic Tyr kinases are intracellular enzymes which rely on the binding and hydrolysis of ATP to affect their phosphoryl transfer function. Consequently, several strategies have been developed in an attempt to inhibit the activation of receptor Tyr kinases (see Figure 1). Apart from extracellular agents such as peptide or protein antagonists, antibodies, ligand binding molecules, and dominant negative receptors, intracellular agents aimed at blocking autocrine growth factor secretion (Mercola et al., 1990), receptor expression (Witters et al., 1999), translocation or activation, and Tyr kinase inhibitors have already been used to stop the proliferation of cancer cells.

Many of these agents are both difficult to design and synthesize. Clearly, a major action of the ligands for EGFR is to stimulate its dimerization (Yarden and Schlessinger, 1985; Zidovetzki et al., 1981). However, our knowledge of the

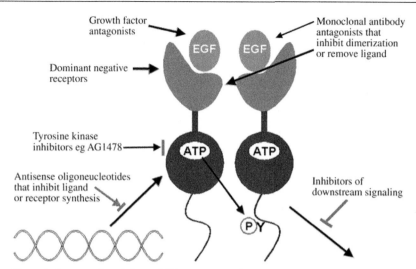

Figure 1 Strategies for inhibiting EGF receptor signaling.

dimerization interface is minimal (Groenen *et al.*, 1997), making the synthesis of dimerization antagonists difficult. Furthermore, our ability to inhibit protein–protein interactions is limited by our understanding of the forces that stabilize protein interactions. Protein–protein interaction sites tend to be large flat surfaces, at present it is still difficult to identify and/or design small molecules capable of competing effectively with these types of interactions. Initial studies that concentrated on inhibition of the kinase were thwarted by the dual requirements for crossing the cell membrane and competing with a highly charged molecule such as ATP. Furthermore, because of the fact that so many enzymes used ATP it was presumed that ATP antagonists would tend to be so nonspecific, that the biological effects of these types of inhibitors would be difficult to control. Indeed several natural product inhibitors of ATP binding had been known for many years [e.g., isoflavones such as quercetin (Glossmann *et al.*, 1981) and genestein (Akiyama *et al.*, 1987)]; these molecules are neither potent nor specific. More recently modified natural products have been synthesized, and the potency of these kinase inhibitors has increased [e.g., staurosporine analogs (Lazarovici *et al.*, 1996)], however, the specificity of these agents still leaves much to be desired.

In this short review of the current status and clinical potential of Tyr kinase inhibitors there will be a focus on two systems: the EGFR family (Mendelsohn, 2000) and the Abl kinase (Tsatsanis and Spandidos, 2000). These targets are being evaluated in the clinic (Baselga *et al.*, 2000; Druker *et al.*, 1996; Goldman, 2000), and their development typifies the challenges for many signaling inhibitors. The comments and examples used in this chapter are designed to introduce the challenges of signaling therapeutics, rather than to attempt an exhaustive review of all targets or inhibitor strategies.

II. Tyr KINASE INHIBITORS

In the late 1980s Gazit and Levitzki initiated a systematic study to synthesize and test small molecules which would inhibit Tyr kinases (Yaish *et al.*, 1988). They based their search on the structure of erbstatin, which inhibited the EGFR with a K_i of ~14 μM (Umezawa *et al.*, 1986). They were searching for substrate competitive inhibitors capable of inhibiting the EGFR kinase more potently than the insulin receptor kinase. The name tyrphostins was proposed for compounds that inhibit Tyr kinases. Their initial success encouraged themselves and others to search for more potent and more specific inhibitors. Levitzki, Gazit, and their colleagues discovered a number of new tyrphostins with specificity for the EGFR, the insulin receptor, the platelet-derived growth factor (PDGF) receptor (PDGFR), and an oncogenic derivative of c-Abl (Bcr-Abl) (Anafi *et al.*, 1992). Several of these inhibitors had K_i values in the submicromolar range, and against a limited panel of kinases showed remarkable specifity. Indeed some of their compounds could clearly discriminate between closely related receptor kinases such as the EGFR/ErbB1 and HER2/ErbB2 (Osherov *et al.*, 1993). One surprise from their studies was that these inhibitors seemed to compete for both substrate and ATP binding, indicating that there must be some overlap between these sites.

The next major advance in the development of the tyrphostins came from a systematic study of quinazoline derivatives. A small molecule generally referred to as AG1478 was discovered to inhibit the EGFR at low nanomolar concentrations (see Figure 2A), and this molecule could penetrate the cell membrane to inhibit the signaling functions of the EGFR (Osherov and Levitzki, 1994; Ward *et al.*, 1994). Despite the search for water soluble inhibitors, almost all of

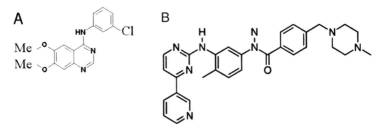

Figure 2 (A) AG-1478, an EGFR-specific inhibitor. (B) STI-571, an inhibitor specific for the Abl, PDGFR, and Kit Tyr-kinases.

the quinazoline compounds were poorly soluble in water, but highly soluble in dimethylsulfoxide (DMSO). Quite unexpectedly a simple substitution of F for Cl in AG1478 led to an inhibitor with an apparent K_i of <10 pM and with similar specificity for the EGFR [i.e., there was no inhibition of the PDGFR, the FGFR, the CSF-1R, the insulin receptor, or the Src kinase even at 50 μM (Fry et al., 1994)]. This specificity was obtained despite the fact that the quinazoline analogs in their present form displace ATP binding rather than substrate binding. The development of the EGFR as a clinical agent is described in more detail in the EGFR section of this chapter.

III. CHRONIC MYELOGENOUS LEUKEMIA

Chronic myelogenous leukemia (CML) was identified in the early part of the nineteenth century (Donne, 1844; Geary, 2000). In its final stages, the blood from CML patients resembled pus more than blood. Until the last quarter of the twentieth century, CML was a fatal disease (Goldman et al., 1986). With the introduction of allogeneic bone marrow transplantation, survival times began to increase. However, the treatments were empirical and often involved traumatic purging of the bone marrow.

Murine leukemia viruses have led to many advances in our understanding of the processes that lead to cancer. The Abelson murine leukemia virus (Abelson and Rabstein, 1970a,b) was no exception and has proved to be a key factor in identifying the molecular lesion responsible for CML. The analysis of the viral oncogene v-Abl (Prywes et al., 1983; Reddy et al., 1983) quickly led to the identification of a human homolog (Heisterkamp et al., 1983; Prywes et al., 1983). Experimental data indicated that v-Abl might be an activated tyrosine kinase (Ponticelli et al., 1982; Witte et al., 1981).

An extraordinary series of papers between 1982 and 1985 associated an oncogenic form of the Abl gene with the 9;22 translocation which occurs in chronic myeloid leukemia (Canaani et al., 1984; Cooper and Hunter, 1981; Gale and Canaani, 1984; Ozanne et al., 1982; Shtivelman et al., 1985). This translocation produces a unique mRNA encoding the Bcr-Abl fusion protein, which is a major determining feature of the leukemia (Collins and Groudine, 1983; Groffen et al., 1984; Konopka et al., 1984; Lewis et al., 1983; Wang and Baltimore, 1983). This enzyme was one of the first targets for the screening of the tyrphostins, and indeed a submicromolar inhibitor (AG957) for Bcr-Abl was reported in 1992 (Anafi et al., 1992). Interestingly, this tyrphostin inhibitor appears to discriminate between c-Abl and Bcr-Abl (or v-Abl). Several years later the screening of a large chemical library based around the 2-phenyl-aminopyrimidines yielded a potent Bcr-Abl inhibitor STI-571 (see Figure 2B) which has an IC_{50} of ~25 nM (Buchdunger et al., 1996). While the specificity of this inhibitor was rather broad, namely, it also inhibits the PDGFR and c-Kit, its potency for Bcr-Abl and its ability to kill chronic myelogenous leukemia cells has led to the attempts to develop STI-571 as a signaling therapeutic.

IV. EPIDERMAL GROWTH FACTOR RECEPTOR

During the last 20 years scientists have sought to unravel the mysteries of ligand binding, receptor activation, receptor cross talk, receptor regulation, and biological signaling for the EGF receptor family (Schlessinger, 2000). Despite the thousands of studies there are many details of these processes which are poorly understood, and the biological significance of either receptor expression or signaling in tissues is largely unknown. Neither the extracellular domain (ECD), nor the intracellular kinase have yielded to direct structural analysis; although the availability of related structures such as the IGF-1R-ECD (Adams et al., 2000; Garrett et al., 1998), the fibroblast growth factor receptor (FGFR) kinase (Plotnikov et al., 1999), and Src kinase structures (Sicheri and Kuriyan, 1997; Waksman et al., 1993) have facilitated the development of preliminary three-dimensional (3D) models for the structures of both the EGFR-ECD (Jorissen et al., 2000) and the EGFR kinase (Groenen et al., 1997).

In this brief review, some of the most recent studies relevant to the activation of the EGFR and the initiation of its biochemical signaling program are outlined. Where possible the various signaling events have been related to particular biological processes. Of necessity many of the studies are from experiments on cell lines or even cell free systems. Approaching the function of the EGFR/ErbB1 or its ligands in animal biology or pathology has proved to be an extraordinarily difficult challenge. The complexity of EGFR signaling is exacerbated by the presence of three other EGFR family members, ErbB2 (HER2), ErbB3 (HER3), and ErbB4 (HER4) which are often displayed on the surface of the same cells as the EGFR (Olayioye et al., 1998; Pinkas-Kramarski et al., 1998). The four EGFR family members can interact with each other to form heteromeric, higher order complexes. Clearly, the different receptor complexes have different ligand specificity and regulatory mechanisms. The heteromeric kinases will have different signaling (substrate) preferences. Since there are more than 10 distinct ligands interacting with the members of the EGFR, the outcomes of receptor activation are difficult to identify in vitro, let alone in animals.

Every week more than 100 papers are published on the EGF–EGFR system. Any attempt to review even particular aspects of the field will necessarily become selective and far from comprehensive. Consequently, a few examples of the specificity, use, and clinical application of several inhibitors or antagonists are used to represent similar results from many other studies with EGFR inhibitors or antagonists.

V. ANTAGONISTS OF THE EPIDERMAL GROWTH FACTOR RECEPTOR EXTRACELLULAR DOMAIN

Although there are no ligand analogs known to antagonize the human EGFR, several antibodies have been isolated which compete with EGF or prevent receptor activation (Baselga and Mendelsohn, 1997). There are also antibodies that recognize the mutant form of the EGFR or oncogenic forms of the receptor (Sampson et al., 2000) (http://www.licr.org/05_pro/antibody.htm). One antibody against HER2 (ErbB2) called herceptin has already been registered for use as a drug for the treatment of breast cancer (Baselga, 2000; Pegram et al., 2000). Although there are some indications that herceptin shows some efficacy in advanced breast cancer, the results still leave much to be desired (Feldman et al., 2000), and it is now critical that we learn how to combine herceptin with other biological or cytotoxic agents. The EGFR is often overexpressed in cancers such as head and neck, breast, and prostate cancers. Indeed cancer patients that overexpress the EGFR usually have a poor prognosis. Initial studies with C225 from John Mendelsohn's laboratories have demonstrated the antitumor activity of the C225 antibody. Nude mice bearing an established tumor (A431) were cleared of the tumor after treatment with C225 (Mendelsohn, 2000). Phase I studies with this antibody have been completed, and there have been minimal problems with the initial doses (Baselga et al., 2000). Promising synergy with the other chemotherapeutics was also observed (see later section).

In some tumors with highly amplified expression of the EGFR there is also a mutant form of the EGFR known as EGFRvIII (or $\Delta2-7$ EGFR) (Lax et al., 1990). This receptor is missing a major part of the EGF binding domain. Consequently, it is not influenced by EGF or its homologs; however, it is constitutively active and is obviously involved in the transformed phenotype of advanced gliomas (Collins, 1993; Ekstrand et al., 1991; Weichselbaum et al., 1989) as well as head and neck cancers (Rikimaru et al., 1992; Weichselbaum et al., 1989). An antibody raised against the unique peptide at the junctional splice site of the mutant form can bind to the mutant receptor and induce the killing of tumor cells in nude mice (Sampson et al., 2000). Indeed, when mice inoculated with tumor cells expressing the mutant EGFR are treated with this antibody, no tumors are observed. In brain tumors and head and neck tumors this mutation is quite common, so this antibody could well form the basis of an effective treatment for these tumors.

Another antibody (MAb 806) that has been raised against the truncated, oncogenic form of the EGFR (http://www.licr.org/05_pro/antibody.htm)

has even more interesting properties in that it recognizes overexpressed EGFR as well as the truncated form of the EGFR. This antibody only recognizes the normal EGFR when it has been denatured. The nature of this epitope is still unclear, but this antibody is also a strong candidate for development as an anticancer drug.

Although unarmed antibodies can kill tumor cells effectively, there have been many attempts to load antibodies with cytotoxic agents such as radioactivity, inhibitors of DNA synthesis, toxins, or toxic enzymes. Radioactivity (e.g., ^{125}I or ^{131}I) is probably the easiest of these toxic regimes to control, but it is often difficult to deliver a sufficient dose of irradiation to achieve a satisfactory antitumor response. Small molecule toxins conjugated to the antibody can produce an increased therapeutic index, but often the conjugation is difficult, and the toxic molecule can leak from the antibody leading to nonspecific killing. Fusion constructs between antibody single chains and bacterial toxins have worked well in mice (Schmidt et al., 1999), however, most of these toxins are quite immunogenic and a course of therapy becomes difficult. Several groups have attempted to overcome the immunogenicity and leakage by using human proteins fused to the EGF molecule (Psarras et al., 1998). These fusion proteins kill cells that overexpress the EGFR most effectively, but in animals (including humans) there will be binding to the EGFR on liver cells, so unacceptable side effects might be expected to be associated with this strategy.

Other strategies have also been used to antagonize the EGFR in vitro: dominant negative forms of the receptor can block signaling and reduce cell production (Xie et al., 1997), similarly, inhibiting the production of autocrine transforming growth factor-α using antisense oligonucleotides essentially stops cell proliferation (Sizeland and Burgess, 1992), and blocking the expression of the EGFR using antisense oligonucleotides or mRNA inhibits tumorigenicity (Pu et al., 2000). All of these strategies require extraordinarily efficient gene transfer technology if they are to serve as anticancer treatments. This technology is still quite a long way from the clinic, and while it may become a valuable anticancer tool in the future, other reagents are likely to dominate the development of cancer therapeutics for the next few years.

VI. CHEMICAL INHIBITORS OF THE KINASE DOMAIN OF THE EPIDERMAL GROWTH FACTOR RECEPTOR

There are now many potent small molecule inhibitors of the EGFR kinase. The most potent of these molecules are either quinazolines (Bridges et al., 1996; Boschelli, 1999) or pyrrolo-pyrimidines (Boschelli, 1999). After the realization that the quinazolines bound to the ATP binding site, other ATP analogs extensive structure activity relationship (SAR) studies yielded potent inhibitors of both classes of compounds. Many of these studies were designed to increase the potency, solubility, and pharmacological properties of the inhibitors. In particular, substituents such as a morpholino group were added to improve solubility (Ciardiello *et al.*, 2000), and the halogens on the anilino ring were varied or replaced by ethenyl substituents. Although the reported IC_{50} values vary from low picomolar to low micromolar, the methods of analysis vary so much that it is actually difficult to determine whether there has been any improvement in the actual potency of the inhibitors. However, it is clear that the pharmacological properties have changed and that some of the inhibitors are suitable for oral delivery.

Determining the specificity of these inhibitors is problematic. Almost all cell lines and immunoprecipitate (IP) kinase assays contain mixtures of the EGFR family heterodimers. In most cases the inhibitors show some cross-reactivity with ErbB2, but the difficulties associated with receptor isolation have interfered with the specificity analyses. Interestingly, although the experimental structure of the EGFR kinase domain has not yet been determined, the availability of structures for homologous proteins has allowed reasonable models to be constructed for the EGFR kinase.

Docking studies with the different inhibitors have helped in the understanding of the mode of binding, and several new inhibitors designed to bind irreversibly via Cys773 have been made successfully (Discafani *et al.*, 1999). One of the irreversible quinzalone inhibitors was shown to link covalently to the expected EGFR residue, but the potency was not significantly different to the reversible high potency inhibitors. One interesting observation with the irreversible, rationally designed EGFR inhibitors was that these molecules inhibited cell lines overexpressing ErbB2 more potently than the cell lines overexpressing the EGFR (Discafani *et al.*, 1999). Perhaps the modifications not only created an irreversible inhibitor, but also changed the specificity in favor of ErbB2.

In many ways our theoretical and experimental tools for inhibitor design are still very primitive. We recognize that accurate predictions about binding configurations and affinities require detailed free energy calculations, but the current design algorithms use crude estimates of the binding "energy." Even if we could extend the calculations toward free energy, the parameterization of molecular mechanical force fields were not designed to calculate free energies for ligand binding systems. Our present algorithms would not predict why a chloroquinazolone was almost 1000-fold less potent than the bromo derivative of the same molecule. Similarly, the differences at the ATP binding sites for EGFR and ErbB2 are so small that it is difficult to imagine that the current design algorithms could lead to inhibitors which are selective for one or the other receptor.

The small molecule inhibitors are clearly potent anticancer drugs against human tumor xenografts that overexpress the EGFR. In nude mouse experiments, EGFR inhibitors can decrease tumor growth rates dramatically without toxic effects on the animals. The inhibitors can be administered intravenously (i.v.), intraperitoneally (i.p.), or orally with similar effects. Although some tumors appear to die when treated with EGFR kinase inhibitors, many seem to simply stop growing in the presence of the inhibitor, but on the removal of the inhibitor the tumors start to grow again.

Interestingly, by combining either the EGFR receptor antagonists or the EGFR inhibitors with cytotoxic drugs leads to increased killing of the tumor cells (Ciardiello et al., 2000). This synergistic action of the anti-EGFR drugs and the cytotoxic drugs has led to studies that have improved our understanding of tumor biology and the design of new therapeutic strategies to improve the therapeutic effects of the EGFR antagonists or inhibitors.

VII. EPIDERMAL GROWTH FACTOR RECEPTOR ANTAGONISTS OR INHIBITORS ACT SYNERGISTICALLY TO KILL TUMOR CELLS

It has been known for more than 10 years that anti-EGFR antibodies could act in synergy with cytotoxic drugs such as doxorubicin or cis-platinum to inhibit the growth of human tumor xenografts in nude mice (Aboud-Pirak et al., 1988). In vitro, the EGFR kinase inhibitors are capable of inducing apoptosis in cell lines that require exogenous or autocrine EGFR ligands for their continuous proliferation. However, clearly there are cell lines where antagonist antibodies or kinase inhibitors simply block cell cycling, thus inducing a state of cytostasis, rather than cell killing. Furthermore, cancer cells often acquire a resistance to the cytotoxic effects of ionizing irradiation or DNA damaging drugs (Nagane et al., 1996, 1998).

When human tumor xenografts are exposed simultaneously to anti-EGFR antibodies and a cytotoxic drug there is profoundly more killing (Aboud-Pirak et al., 1988). Indeed the synergistic actions of the antibodies and the drug not only reduced the rate of tumor growth, but in many animals the tumor is eliminated completely. This synergy has been extended to a number of antibodies and many different cytotoxic agents such as topotecan, doxorubicin, cis-platinum, etoposide, and ionizing radiation.

Similar results have been observed when the EGFR kinase inhibitors are combined with cytotoxic drugs (Han et al., 1996; Nagane et al., 1998). Again tumors that appear to be driven by EGFR associated events are killed more effectively in the presence of both the inhibitor and the cytotoxic drug. However, it must be said that the timing and dosing for the treatment schedules either in culture or in animals still needs to be optimized. While the agents act synergistically when given at the same time, there is some indication that with the sequential application of the agents (e.g., when topotecan is given before the EGFR inhibitor) the level of apoptosis induced can double. The scheduling might be particularly important when attempting to combine antiangiogenic agents with the inhibitors as drug accessibility, the induction of apoptosis, and the biological interactions between the agents may be expected to exhibit quite different kinetics when compared to cultures in vitro.

Advanced brain tumors are invariably associated with increased signaling from the EGFR (Ekstrand et al., 1991). In many cases this is due to amplification of the EGFR gene and an associated deletion in some of the amplicons, which leads to a defective extracellular domain and constitutive activation of the EGFR kinase. These tumors are sensitive to EGFR kinase inhibitors and are killed most effectively by combinations of cytotoxic drugs such as cis-platinum and EGFR inhibitors. The synergy between the agents has been attributed to the stimulation of an antiapoptotic signaling protein bcl-X_L by the activated EGFR. The increase in bcl-X_L prevents the induction of apoptosis by the cis-platinum; when the mutant form of the EGFR is inhibited by AG1478 the levels of bcl-X_L and the cis-platinum induces apoptosis (Nagane et al., 1998).

Interestingly, the effects of the EGFR inhibitors and antibody antagonists can also be augmented by interfering with other signaling pathways. In an established tumor cell model of renal carcinoma, when animals were treated with the monoclonal anti-EGFR antibody C225 and/or an antisense oligonucleotide that interfered with the production of protein kinase A (Tortora et al., 1999), the single agents delayed tumor growth. However, the combination of both agents caused regression of the tumors.

Although the EGFR antibody antagonists and the EGFR kinase inhibitors target the same molecule there is evidence that when used together the two agents induce more potent killing of EGFR dependent tumors. The reasons for this synergy have not been identified as yet. The antibodies in vivo might induce immunological killing as well as interference with EGFR signaling, and these killing pathways may be more effective when the kinase is inhibited by small molecules. There also appears to be increased killing when antibodies against both the EGFR and HER-2 are combined (Ye et al., 1999). Some of the small molecular weight inhibitors of the EGF receptors appear also to interfere with HER-2 signaling. It is not clear whether the broader specificity will improve the antitumor potency of these inhibitors or whether there will be increased side effects associated with the shutting down of two EGF receptor family members.

A combination of cis-platinum and the EGFR antibody C225 has already completed phase I trials in head and neck and lung cancer patients. Not only was the combination of drugs well tolerated (70% of the patients completed 12 weeks of therapy), but several patients experienced disease stabilization and two partial responses were noted (Baselga et al., 2000).

There is much to be learned about the use of EGFR antagonists and inhibitors before their optimal use in cancer therapy is achieved, but clinical trials have started and even as single agents the antibody antagonists and kinase inhibitors appear to offer the promise of a new therapeutic option for the treatment of advanced cancers.

VIII. THE EFFECTS OF Abl INHIBITORS ON LEUKEMIA

The Bcr-Abl translocation of CML patients is present in most of the blood cell lineages (Fialkow *et al.*, 1977; Kabarowski and Witte, 2000). Thus, CML is a disease of the hemopoietic stem cell, but its biological effects are usually manifest in the development of myelogenous leukemia. In many patients the leukemic (Bcr-Abl) clones coexist with normal hemopoietic stem cells, and conventional chemotherapy can restore, albeit temporarily, normal hemopoiesis (Deininger and Goldman, 1998). Indeed it is possible that methods might be developed which can separate and expand normal hemopoietic stem cells from the leukemic bone marrow so that after bone marrow purging, patients can be rescued using the normal, autologous bone marrow stem cells (Goldman *et al.*, 1986). The expression of Bcr-Abl in several hemopoietic cell lines can release the need for exogenous growth factors. Until more recently the mechanism by which Bcr-Abl achieved autonomy from exogenous cytokine stimulation was unclear. It may be that an activated form of abl in the cytosol alters signaling downstream of c-RAS (thus circumventing the normal requirements for cytokine stimulation); however, Bcr-Abl also induces the expression of granulocyte-colony stimulating factor (G-CSF) and interleukin-3 (IL-3) (Jiang *et al.*, 1999). There is evidence that the autocrine G-CSF and IL-3 contribute to the leukemic phenotype of the Bcr-Abl expressing cells. Although it has been associated with CML, the autocrine production of granulocyte–macrophage colony stimulating factor does not appear to be important in CML (Bi *et al.*, 2000).

The development of CML clearly depends on the expression and activity of Bcr-Abl. From the early 1980s, when the involvement of Abl was discovered, scientists have been attempting to devise methods for removing the effects of Bcr-Abl on leukemic cell survival and production. Antisense oligonucleotides have proved effective in culture (Rowley *et al.*, 1999), and there is even some evidence for inhibition of leukemic cell production in animal models. But therapeutic delivery systems for medical applications are still some time away and are likely to encounter considerable difficulties before they prove useful as therapeutic agents for CML. Similarly, vectors expressing catalytic RNAs, which can destroy the Bcr-Abl mRNA, have been shown to be effective agents for killing cells stimulated by Bcr-Abl, but it will be some time before this technology can be used in the treatment of human leukemia.

Novartis had developed a series of 2-phenylaminopyrimidine analogs for the inhibition of tyrosine kinases. They showed that one of these compounds CGP-53716 was capable of inhibiting the platelet-derived growth factor receptor

(PDGFR) in a kinase assay with an IC_{50} of 100 nM (Zimmermann *et al.*, 1996). CGP-53716 seemed to be quite specific for the PDGFR, and initially Novartis reported that the only other kinase inhibited by this class of molecules was the abl kinase ($IC_{50} = 600$ nM). Other reports indicated that CGP-53716 also inhibits EGF and bFGF stimulated cell division (Major and Keiser, 1997). SAR studies led to the synthesis of more potent PDGFR inhibitors, and one of these compounds, CGP-57148, was found to inhibit the abl kinase with an IC_{50} of 40 nM and the PDGFR with an IC_{50} of 50 nM. Again CGP-57148 was initially reported to be quite specific for these two kinases, but subsequent work has shown that CGP-57148 also inhibits the c-Kit kinase (Heinrich *et al.*, 2000).

CGP-57148 is an effective drug for stopping the proliferation of Bcr-Abl positive human leukemic cells in soft agar cultures and in nude mice (Druker *et al.*, 1996). The drug was identified by Novartis as a therapeutic candidate and renamed signaling therapeutic inhibitor-571 (STI-571).

Brian Druker, a key scientist–clinician in the STI-571 development program, described the initial results to a meeting of the American Society of Hematology in December 1999 (O'Dwyer and Druker, 2000). At doses of 300 mg daily, all CML patients responded to the drug. Indeed all of the 31 CML patients who received 300 mg or more of STI-571 daily for at least 4 weeks achieved a complete hematological remission. The responses were seen within 3 weeks of commencing therapy and were maintained for more than 8 months. Many patients had cytogenetic responses, and the side effects were minimal. There was some myelotoxicity (this might be due to the activity of STI-571 on c-Kit). This drug is still in short supply, and patients have only been on treatment for 2 years, but the initial responses are most encouraging (Goldman, 2000; O'Dwyer and Druker, 2000). Patients who are too old to receive bone marrow transplantation or lack suitable donors may well have an effective alternative treatment. Although Bcr-Abl[+] acute lymphoblastic leukemia (ALL) patients have also responded to STI-571, the duration of these responses is considerably shorter than in the CML patients.

The development of STI-571 is in its early phases; however, for a single agent the initial successes have been most encouraging. It is quite possible that resistance to STI-571 will develop (e.g., through amplification and/or mutation of the *Bcr-Abl* gene or downstream effectors), so more effective killing strategies need to be developed. The use of STI-571 in combination with cytotoxic drugs or biological effectors such as α-interferon offer new possibilities, and a recent report of synergistic effects of STI-571 and an inhibitor of the JAK2 kinase offers another approach to improving the use of this drug (Sun X *et al.*, 2001). We are entering a new era for the development of anticancer drugs. Many critical targets have been identified, and chemists are designing and synthesizing molecules with sufficient specificity for these for therapeutic development. As the results from the three-dimensional structural analyses of targets and inhibitor complexes accumulate it should be possible to design anticancer drugs with even higher potency and selectivity. By combining these design features with modifications to facilitate formulation and delivery, chemists will develop many more effective signaling therapeutics during this decade. It is unlikely that a single drug will be therapeutic during this decade.

It is unlikely that a single agent will be sufficient to reverse the effects of the multiple genetic aberrations associated with most cancer cells. Clearly, if we are to achieve effective anticancer treatments with signaling therapeutics, we will need to block more than one signaling pathway and learn how to stimulate the different apoptotic mechanisms in the cells targeted by the signaling inhibitors.

References

Abelson, H. T., and Rabstein, L. S. (1970a). Influence of prednisolone on Moloney leukemogenic virus in BALB-c mice. *Cancer Res.* **30**, 2208–2212.

Abelson, H. T., and Rabstein, L. S. (1970b). Lymphosarcoma: Virus-induced thymic-independent disease in mice. *Cancer Res.* **30**, 2213–2222.

Aboud-Pirak, E., Hurwitz, E., Pirak, M. E., Bellot, F., Schlessinger, J., and Sela, M. (1988). Efficacy of antibodies to epidermal growth factor receptor against KB carcinoma *in vitro* and in nude mice. *J. Natl. Cancer Inst.* **80**, 1605–1611.

Adams, T. E., Epa, V. C., Garrett, T. P., and Ward, C. W. (2000). Structure and function of the type 1 insulin-like growth factor receptor. *Cell. Mol. Life Sci.* **57**, 1050–1093.

Akiyama, T., Ishida, J., Nakagawa, S., Ogawara, H., Watanabe, S., Itoh, N., Shibuya, M., and Fukami, Y. (1987). Genistein, a specific inhibitor of tyrosine-specific protein kinases. *J. Biol. Chem.* **262**, 5592–5595.

Anafi, M., Gazit, A., Gilon, C., Ben-Neriah, Y., and Levitzki, A. (1992). Selective interactions of transforming and normal abl proteins with ATP, tyrosine-copolymer substrates, and tyrphostins. *J. Biol. Chem.* **267**, 4518–4523.

Baselga, J. (2000). Current and planned clinical trials with trastuzumab (Herceptin) [in process citation]. *Semin. Oncol.* **27**, 27–32.

Baselga, J., and Mendelsohn, J. (1997). Type I receptor tyrosine kinases as targets for therapy in breast cancer. *J. Mammary Gland Biol. Neoplasia* **2**, 165–174.

Baselga, J., Pfister, D., Cooper, M. R., Cohen, R., Burtness, B., Bos, M., D'Andrea, G., Seidman, A., Norton, L., Gunnett, K., Falcey, J., Anderson, V., Waksal, H., and Mendelsohn, J. (2000). Phase I studies of anti-epidermal growth factor receptor chimeric antibody C225 alone and in combination with cisplatin. *J. Clin. Oncol.* **18**, 904–914.

Bi, S., Gao, X., Devemy, E., Chopra, H., Venugopal, P., Raza, A., and Preisler, H. D. (2000). Cytokine production by *in vitro* processed and unprocessed haematopoietic cells. *Cytokine* **12**, 1124–1128.

Boschelli, D. (1999). Small molecule inhibitors of receptor tyrosine kinases. *Drugs of the Future* **24**, 1–35.

Bridges, A. J., Zhou, H., Cody, D. R., Rewcastle, G. W., McMichael, A., Showalter, H. D., Fry, D. W., Kraker, A. J., and Denny, W. A. (1996). Tyrosine kinase inhibitors. 8. An unusually steep structure–activity relationship for analogues of 4-(3-bromoanilino)-6,7-dimethoxyquinazoline (PD 153035), a potent inhibitor of the epidermal growth factor receptor. *J. Med. Chem.* **39**, 267–276.

Buchdunger, E., Zimmermann, J., Mett, H., Meyer, T., Muller, M., Druker, B. J., and Lydon, N. B. (1996). Inhibition of the Abl protein-tyrosine kinase in vitro and in vivo by a 2-phenylaminopyrimidine derivative. *Cancer Res.* **56,** 100–104.

Canaani, E., Gale, R. P., Steiner-Saltz, D., Berrebi, A., Aghai, E., and Januszewicz, E. (1984). Altered transcription of an oncogene in chronic myeloid leukaemia. *Lancet* **1,** 593–595.

Ciardiello, F., Bianco, R., Damiano, V., De Lorenzo, S., Pepe, S., De Placido, S., Fan, Z., Mendelsohn, J., Bianco, A. R., and Tortora, G. (1999). Antitumor activity of sequential treatment with topotecan and anti-epidermal growth factor receptor monoclonal antibody C225. *Clin. Cancer Res.* **5,** 909–916.

Ciardiello, F., Caputo, R., Bianco, R., Damiano, V., Pomatico, G., De Placido, S., Bianco, A. R., and Tortora, G. (2000). Antitumor effect and potentiation of cytotoxic drug activity in human cancer cells by ZD-1839 (Iressa), an epidermal growth factor receptor-selective tyrosine kinase inhibitor. *Clin. Cancer Res.* **6,** 2053–2063.

Collett, M. S., Purchio, A. F., and Erikson, R. L. (1980). Avian sarcoma virus-transforming protein, pp60src shows protein kinase activity specific for tyrosine. *Nature* **285,** 167–169.

Collins, S. J., and Groudine, M. T. (1983). Rearrangement and amplification of c-abl sequences in the human chronic myelogenous leukemia cell line K-562. *Proc. Natl. Acad. Sci. U.S.A.* **80,** 4813–4817.

Collins, V. P. (1993). Amplified genes in human gliomas. *Semin. Cancer Biol.* **4,** 27–32.

Cooper, J. A., and Hunter, T. (1981). Changes in protein phosphorylation in Rous sarcoma virus-transformed chicken embryo cells. *Mol. Cell. Biol.* **1,** 165–178.

Deininger, M. W., and Goldman, J. M. (1998). Chronic myeloid leukemia. *Curr. Opin. Hematol.* **5,** 302–308.

Discafani, C. M., Carroll, M. L., Floyd, M. B., Jr., Hollander, I. J., Husain, Z., Johnson, B. D., Kitchen, D., May, M. K., Malo, M. S., Minnick, A. A., Jr., Nilakantan, R., Shen, R., Wang, Y. F., Wissner, A., and Greenberger, L. M. (1999). Irreversible inhibition of epidermal growth factor receptor tyrosine kinase with in vivo activity by N-[4-[(3-bromophenyl)amino]-6-quinazolinyl]-2-butynamide (CL-387,785). *Biochem. Pharmacol.* **57,** 917–925.

Donaldson, R. W., and Cohen, S. (1992). Epidermal growth factor stimulates tyrosine phosphorylation in the neonatal mouse: association of a M(r) 55,000 substrate with the receptor. *Proc. Natl. Acad. Sci. U.S.A.* **89,** 8477–8481.

Donne, A. (1844). Cours de microscopie complementaire des etudes medicales. *Balliere, Paris* **135,** 196.

Dorsey, J. F., Jove, R., Kraker, A. J., and Wu, J. (2000). The pyrido[2,3-d]pyrimidine derivative PD180970 inhibits p210Bcr-Abl tyrosine kinase and induces apoptosis of K562 leukemic cells. *Cancer Res.* **60,** 3127–3131.

Druker, B. J., Tamura, S., Buchdunger, E., Ohno, S., Segal, G. M., Fanning, S., Zimmermann, J., and Lydon, N. B. (1996). Effects of a selective inhibitor of the Abl tyrosine kinase on the growth of Bcr-Abl positive cells. *Nat. Med.* **2,** 561–566.

Ekstrand, A. J., James, C. D., Cavenee, W. K., Seliger, B., Pettersson, R. F., and Collins, V. P. (1991). Genes for epidermal growth factor receptor, transforming growth factor alpha, and epidermal growth factor and their expression in human gliomas in vivo. *Cancer Res.* **51,** 2164–2172.

Fang, G., Kim, C. N., Perkins, C. L., Ramadevi, N., Winton, E., Wittmann, S., and Bhalla, K. N. (2000). CGP57148B (STI-571) induces differentiation and apoptosis and sensitizes Bcr-Abl-positive human leukemia cells to apoptosis due to antileukemic drugs. *Blood* **96,** 2246–2253.

Feldman, A. M., Lorell, B. H., and Reis, S. E. (2000). Trastuzumab in the treatment of metastatic breast cancer: Anticancer therapy versus cardiotoxicity. *Circulation* **102,** 272–274.

Fialkow, P. J., Jacobson, R. J., and Papayannopoulou, T. (1977). Chronic myelocytic leukemia: Clonal origin in a stem cell common to the granulocyte, erythrocyte, platelet and monocyte/macrophage. *Am. J. Med.* **63,** 125–130.

Fry, D. W., Kraker, A. J., McMichael, A., Ambroso, L. A., Nelson, J. M., Leopold, W. R., Connors, R. W., and Bridges, A. J. (1994). A specific inhibitor of the epidermal growth factor receptor tyrosine kinase. *Science* **265,** 1093–1095.

Fujita, D. J., Tal, J., Varmus, H. E., and Bishop, J. M. (1978). *env* Gene of chicken RNA tumor viruses: Extent of conservation in cellular and viral genomes. *J. Virol.* **27,** 465–474.

Gale, R. P., and Canaani, E. (1984). An 8-kilobase abl RNA transcript in chronic myelogenous leukemia. *Proc. Natl. Acad. Sci. U.S.A.* **81,** 5648–5652.

Garrett, T. P., McKern, N. M., Lou, M., Frenkel, M. J., Bentley, J. D., Lovrecz, G. O., Elleman, T. C., Cosgrove, L. J., and Ward, C. W. (1998). Crystal structure of the first three domains of the type-1 insulin-like growth factor receptor. *Nature* **394,** 395–399.

Geary, C. G. (2000). The story of chronic myeloid leukaemia. *Br. J. Haematol.* **110,** 2–11.

Glossmann, H., Presek, P., and Eigenbrodt, E. (1981). Quercetin inhibits tyrosine phosphorylation by the cyclic nucleotide-independent, transforming protein kinase, pp60src. *Naunyn-Schmiedeberg's Arch. Pharmacol.* **317,** 100–102.

Goldman, J. M. (2000). Tyrosine-kinase inhibition in treatment of chronic myeloid leukaemia. *Lancet* **355,** 1031–1032.

Goldman, J. M., Apperley, J. F., Jones, L., Marcus, R., Goolden, A. W., Batchelor, R., Hale, G., Waldmann, H., Reid, C. D., Hows, J., Gordon-Smith, E., Catovsky, D., and Galton, D.A.G. (1986). Bone marrow transplantation for patients with chronic myeloid leukemia. *N. Engl. J. Med.* **314,** 202–207.

Groenen, L. C., Walker, F., Burgess, A. W., and Treutlein, H. R. (1997). A model for the activation of the epidermal growth factor receptor kinase involvement of an asymmetric dimer? *Biochemistry* **36,** 3826–3836.

Groffen, J., Stephenson, J. R., Heisterkamp, N., Bartram, C., de Klein, A., and Grosveld, G. (1984). The human c-abl oncogene in the Philadelphia translocation. *J. Cell. Physiol. Suppl.* **3,** 179–191.

Han, Y., Caday, C. G., Nanda, A., Cavenee, W. K., and Huang, H. J. (1996). Tyrphostin AG 1478 preferentially inhibits human glioma cells expressing truncated rather than wild-type epidermal growth factor receptors. *Cancer Res.* **56,** 3859–3861.

Heinrich, M. C., Griffith, D. J., Druker, B. J., Wait, C. L., Ott, K. A., and Zigler, A. J. (2000). Inhibition of c-kit receptor tyrosine kinase activity by STI 571, a selective tyrosine kinase inhibitor. *Blood* **96,** 925–932.

Heisterkamp, N., Groffen, J., and Stephenson, J. R. (1983). The human v-abl cellular homologue. *J. Mol. Appl. Genet.* **2,** 57–68.

Hunter, T. (1998). The Croonian Lecture 1997. The phosphorylation of proteins on tyrosine: Its role in cell growth and disease. *Philos. Trans. R. Soc. London B. Biol. Sci.* **353,** 583–605.

Jiang, X., Lopez, A., Holyoake, T., Eaves, A., and Eaves, C. (1999). Autocrine production and action of IL-3 and granulocyte colony-stimulating factor in chronic myeloid leukemia. *Proc. Natl. Acad. Sci. U.S.A.* **96,** 12804–12809.

Jorissen, R. N., Epa, V. C., Treutlein, H. R., Garrett, T. P., Ward, C. W., and Burgess, A. W. (2000). Characterization of a comparative model of the extracellular domain of the epidermal growth factor receptor. *Protein Sci.* **9,** 310–324.

Kabarowski, J. H., and Witte, O. N. (2000). Consequences of BCR-ABL expression within the hematopoietic stem cell in chronic myeloid leukemia [in process citation]. *Stem Cells* **18,** 399–408.

Konopka, J. B., Watanabe, S. M., and Witte, O. N. (1984). An alteration of the human c-abl protein in K562 leukemia cells unmasks associated tyrosine kinase activity. *Cell* **37,** 1035–1042.

Lax, I., Bellot, F., Honegger, A. M., Schmidt, A., Ullrich, A., Givol, D., and Schlessinger, J. (1990). Domain deletion in the extracellular portion of the EGF-receptor reduces ligand binding and impairs cell surface expression. *Cell Regul.* **1,** 173–188.

Lazarovici, P., Rasouly, D., Friedman, L., Tabekman, R., Ovadia, H., and Matsuda, Y. (1996). K252a and staurosporine microbial alkaloid toxins as prototype of neurotropic drugs. *Adv. Exp. Med. Biol.* **391,** 367–377.

Lewis, J. P., Watson-Williams, E. J., Lazerson, J., and Jenks, H. M. (1983). Chronic myelogenous leukemia and genetic events at 9q34. *Hematol. Oncol.* **1,** 269–274.

Major, T. C., and Keiser, J. A. (1997). Inhibition of cell growth: Effects of the tyrosine kinase inhibitor CGP 53716. *J. Pharmacol. Exp. Ther.* **283,** 402–410.

Mendelsohn, J. (2000). Blockade of receptors for growth factors: An anticancer therapy—the fourth annual Joseph H. Burchenal American Association of Cancer Research Clinical Research Award Lecture. *Clin. Cancer Res.* **6,** 747–753.

Mercola, M., Deininger, P. L., Shamah, S. M., Porter, J., Wang, C. Y., and Stiles, C. D. (1990). Dominant-negative mutants of a platelet-derived growth factor gene. *Genes Dev.* **4,** 2333–2341.

Nagane, M., Coufal, F., Lin, H., Bogler, O., Cavenee, W. K., and Huang, H. J. (1996). A common mutant epidermal growth factor receptor confers enhanced tumorigenicity on human glioblastoma cells by increasing proliferation and reducing apoptosis. *Cancer Res.* **56,** 5079–5086.

Nagane, M., Levitzki, A., Gazit, A., Cavenee, W. K., and Huang, H. J. (1998). Drug resistance of human glioblastoma cells conferred by a tumor-specific mutant epidermal growth factor receptor through modulation of Bcl-XL and caspase-3-like proteases. *Proc. Natl. Acad. Sci. U.S.A.* **95,** 5724–5729.

O'Dwyer, M. E., and Druker, B. J. (2000). Status of bcr–abl tyrosine kinase inhibitors in chronic myelogenous leukemia [in process citation]. *Curr. Opin. Oncol.* **12,** 594–597.

Olayioye, M. A., Graus-Porta, D., Beerli, R. R., Rohrer, J., Gay, B., and Hynes, N. E. (1998). ErbB-1 and ErbB-2 acquire distinct signaling properties dependent upon their dimerization partner. *Mol. Cell. Biol.* **18,** 5042–5051.

Osherov, N., and Levitzki, A. (1994). Epidermal-growth-factor-dependent activation of the src-family kinases. *Eur. J. Biochem.* **225,** 1047–1053.

Osherov, N., Gazit, A., Gilon, C., and Levitzki, A. (1993). Selective inhibition of the epidermal growth factor and HER2/neu receptors by tyrphostins. *J. Biol. Chem.* **268,** 11134–11142.

Ozanne, B., Wheeler, T., Zack, J., Smith, G., and Dale, B. (1982). Transforming gene of a human leukaemia cell is unrelated to the expressed tumour virus related gene of the cell. *Nature* **299,** 744–747.

Pegram, M. D., Konecny, G., and Slamon, D. J. (2000). The molecular and cellular biology of HER2/neu gene amplification/overexpression and the clinical development of herceptin (trastuzumab) therapy for breast cancer [in process citation]. *Cancer Treat. Res.* **103,** 57–75.

Pinkas-Kramarski, R., Shelly, M., Guarino, B. C., Wang, L. M., Lyass, L., Alroy, I., Alimandi, M., Kuo, A., Moyer, J. D., Lavi, S., Eisenstein, M., Ratzkin, B. J., Seger, R., Bacus, S. S., Pierce, J. H., Andrews, G. C., Yarden, Y., and Alamandi, M. (1998). ErbB tyrosine kinases and the two neuregulin families constitute a ligand-receptor network [published errata appear in *Mol. Cell. Biol.* 1998 Dec;18(12):7602 and 1999 Dec;19(12):8695]. *Mol. Cell. Biol.* **18,** 6090–6101.

Plotnikov, A. N., Schlessinger, J., Hubbard, S. R., and Mohammadi, M. (1999). Structural basis for FGF receptor dimerization and activation. *Cell* **98,** 641–650.

Ponticelli, A. S., Whitlock, C. A., Rosenberg, N., and Witte, O. N. (1982). *In vivo* tyrosine phosphorylations of the Abelson virus transforming protein are absent in its normal cellular homolog. *Cell* **29,** 953–960.

Prywes, R., Foulkes, J. G., Rosenberg, N., and Baltimore, D. (1983). Sequences of the A-MuLV protein needed for fibroblast and lymphoid cell transformation. *Cell* **34,** 569–579.

Psarras, K., Ueda, M., Yamamura, T., Ozawa, S., Kitajima, M., Aiso, S., Komatsu, S., and Seno, M. (1998). Human pancreatic RNase1–human epidermal growth factor fusion: An entirely human 'immunotoxin analog' with cytotoxic properties against squamous cell carcinomas. *Protein Eng.* **11,** 1285–1292.

Pu, P., Liu, X., Liu, A., Cui, J., and Zhang, Y. (2000). Inhibitory effect of antisense epidermal growth factor receptor RNA on the proliferation of rat C6 glioma cells *in vitro* and *in vivo*. *J. Neurosurg.* **92**, 132–139.

Reddy, E. P., Smith, M. J., and Srinivasan, A. (1983). Nucleotide sequence of Abelson murine leukemia virus genome: Structural similarity of its transforming gene product to other onc gene products with tyrosine-specific kinase activity. *Proc. Natl. Acad. Sci. U.S.A.* **80**, 3623–3627.

Rikimaru, K., Tadokoro, K., Yamamoto, T., Enomoto, S., and Tsuchida, N. (1992). Gene amplification and overexpression of epidermal growth factor receptor in squamous cell carcinoma of the head and neck. *Head Neck* **14**, 8–13.

Rowley, P. T., Kosciolek, B. A., and Kool, E. T. (1999). Circular antisense oligonucleotides inhibit growth of chronic myeloid leukemia cells. *Mol. Med.* **5**, 693–700.

Sampson, J. H., Crotty, L. E., Lee, S., Archer, G. E., Ashley, D. M., Wikstrand, C. J., Hale, L. P., Small, C., Dranoff, G., Friedman, A. H., Friedman, H. S., and Bigner, D. D. (2000). Unarmed, tumor-specific monoclonal antibody effectively treats brain tumors. *Proc. Natl. Acad. Sci. U.S.A.* **97**, 7503–7508.

Schlessinger, J. (2000). Cell signaling by receptor tyrosine kinases [in process citation]. *Cell* **103**, 211–225.

Schmidt, M., Maurer-Gebhard, M., Groner, B., Kohler, G., Brochmann-Santos, G., and Wels, W. (1999). Suppression of metastasis formation by a recombinant single chain antibody-toxin targeted to full-length and oncogenic variant EGF receptors. *Oncogene* **18**, 1711–1721.

Shtivelman, E., Lifshitz, B., Gale, R. P., and Canaani, E. (1985). Fused transcript of *abl* and *bcr* genes in chronic myelogenous leukaemia. *Nature* **315**, 550–554.

Sicheri, F., and Kuriyan, J. (1997). Structures of Src-family tyrosine kinases. *Curr. Opin. Struct. Biol.* **7**, 777–785.

Sizeland, A. M., and Burgess, A. W. (1992). Anti-sense transforming growth factor α oligonucleotides inhibit autocrine stimulated proliferation of a colon carcinoma cell line. *Mol. Biol. Cell.* **3**, 1235–1243.

Sun X, L. J., Layton, J. E., Elefanty, A., and Lieschke, G. J. (2001). Comparison of effects of the tyrosine kinase inhibitors AG957, AG490 and STI-571 on bcr–abl-expressing cells, demonstrating synergy between AG490 and STI-571. *Blood* **97**, 2008–2015.

Tortora, G., Caputo, R., Pomatico, G., Pepe, S., Bianco, A. R., Agrawal, S., Mendelsohn, J., and Ciardiello, F. (1999). Cooperative inhibitory effect of novel mixed backbone oligonucleotide targeting protein kinase A in combination with docetaxel and anti-epidermal growth factor-receptor antibody on human breast cancer cell growth. *Clin. Cancer Res.* **5**, 875–881.

Tsatsanis, C., and Spandidos, D. A. (2000). The role of oncogenic kinases in human cancer (Review). *Int. J. Mol. Med.* **5**, 583–590.

Umezawa, H., Imoto, M., Sawa, T., Isshiki, K., Matsuda, N., Uchida, T., Iinuma, H., Hamada, M., and Takeuchi, T. (1986). Studies on a new epidermal growth factor-receptor kinase inhibitor, erbstatin, produced by MH435-hF3. *J. Antibiot. (Tokyo)* **39**, 170–173.

Ushiro, H., and Cohen, S. (1980). Identification of phosphotyrosine as a product of epidermal growth factor-activated protein kinase in A-431 cell membranes. *J. Biol. Chem.* **255**, 8363–8365.

Waksman, G., Shoelson, S. E., Pant, N., Cowburn, D., and Kuriyan, J. (1993). Binding of a high affinity phosphotyrosyl peptide to the Src SH2 domain: Crystal structures of the complexed and peptide-free forms. *Cell* **72**, 779–790.

Wang, J. Y., and Baltimore, D. (1983). Cellular RNA homologous to the Abelson murine leukemia virus transforming gene: Expression and relationship to the viral sequence. *Mol. Cell. Biol.* **3**, 773–779.

Ward, W. H., Cook, P. N., Slater, A. M., Davies, D. H., Holdgate, G. A., and Green, L. R. (1994). Epidermal growth factor receptor tyrosine kinase. Investigation of catalytic mechanism, structure-based searching and discovery of a potent inhibitor. *Biochem. Pharmacol.* **48**, 659–666.

Weichselbaum, R. R., Dunphy, E. J., Beckett, M. A., Tybor, A. G., Moran, W. J., Goldman, M. E., Vokes, E. E., and Panje, W. R. (1989). Epidermal growth factor receptor gene amplification and expression in head and neck cancer cell lines. *Head Neck* **11,** 437–442.

Witte, O. N., Ponticelli, A., Gifford, A., Baltimore, D., Rosenberg, N., and Elder, J. (1981). Phosphorylation of the Abelson murine leukemia virus transforming protein. *J. Virol.* **39,** 870–878.

Witters, L., Kumar, R., Mandal, M., Bennett, C. F., Miraglia, L., and Lipton, A. (1999). Antisense oligonucleotides to the epidermal growth factor receptor. *Breast Cancer Res. Treat.* **53,** 41–50.

Xie, W., Paterson, A. J., Chin, E., Nabell, L. M., and Kudlow, J. E. (1997). Targeted expression of a dominant negative epidermal growth factor receptor in the mammary gland of transgenic mice inhibits pubertal mammary duct development. *Mol. Endocrinol.* **11,** 1766–1781.

Yaish, P., Gazit, A., Gilon, C., and Levitzki, A. (1988). Blocking of EGF-dependent cell proliferation by EGF receptor kinase inhibitors. *Science* **242,** 933–935.

Yarden, Y., and Schlessinger, J. (1985). The EGF receptor kinase: Evidence for allosteric activation and intramolecular self-phosphorylation. *Ciba Found. Symp.* **116,** 23–45.

Ye, D., Mendelsohn, J., and Fan, Z. (1999). Augmentation of a humanized anti-HER2 mAb 4D5 induced growth inhibition by a human–mouse chimeric anti-EGF receptor mAb C225. *Oncogene* **18,** 731–738.

Zidovetzki, R., Yarden, Y., Schlessinger, J., and Jovin, T. M. (1981). Rotational diffusion of epidermal growth factor complexed to cell surface receptors reflects rapid microaggregation and endocytosis of occupied receptors. *Proc. Natl. Acad. Sci. U.S.A.* **78,** 6981–6985.

Zimmermann, J., Caravatti, G., Mett, H., Meyer, T., Muller, M., Lydon, N. B., and Fabbro, D. (1996). Phenylamino-pyrimidine (PAP) derivatives: A new class of potent and selective inhibitors of protein kinase C (PKC). *Arch. Pharm. (Weinheim, Ger.)* **329,** 371–376.

Antagonists of Rho Family GTPases: Blocking PAKs, ACKs, and Rock

Hiroshi Maruta,[1] **Hong He, and Thao Nheu**
Ludwig Institute for Cancer Research
Royal Melbourne Hospital
Parkville/Melbourne 3050, Australia

I. Rho FAMILY GTPases (Rho, Rac, AND CDC42)

Rho family GTPases (Rho, Rac, and CDC42) share around 30% sequence homology with RAS family GTPases, and are mainly responsible for the regulation of actomyosin-based cytoskeleton organization (Hall, 1998; Ridley, 1998). In general, Rho GTPases are responsible for actin stress fiber formation, and Rac and CDC42 are essential for induction of membrane ruffling and microspike formation, respectively. Oncogenic RAS mutants such as v-Ha-RAS affect these cytoskeletal structures, namely, disrupting actin stress fibers and inducing both membrane ruffle and microspikes. Interestingly, we and others showed that both Rac and CDC42 are activated by RAS and are

[1] To whom correspondence should be addressed.

Tumor-Suppressing Viruses, Genes, and Drugs
Innovative Cancer Therapy Approaches

essential for RAS transformation (Qiu *et al.*, 1995, 1997; Nur-E-Kamal *et al.*, 1999; Liliental *et al.*, 2000). However, although Rho is also essential for RAS transformation (Lebowitz *et al.*, 1995), there is so far no evidence that RAS activates Rho. In this chapter, we will introduce several distinct molecules, peptides, or drugs, that block the oncogenic signaling mediated by Rho family GTPases.

II. BLOCKING PAKs

As discussed in Chapter 16, a few drugs such as MKT-077, CK, and SCH51344 that block Rac-induced membrane ruffling suppress RAS transformation (Walsh *et al.*, 1997; Tikoo *et al.*, 1999, 2000). These observations suggest that membrane ruffling may be essential for RAS transformation. However, none of these drugs blocks the function of Rac directly. MKT-077 bundles actin filaments, and CK caps the plus end of actin filaments. The target of SCH51344 has not been identified as yet. Thus, it would be of great interest to develop a Rac-specific inhibitor that blocks only the function of Rac, and does not affect the function of either CDC42 or Rho. However, no such inhibitor has been identified or created as yet.

One of Rac effectors belongs to a Ser/Thr kinase family called PAKs. The first member of this family is a myosin I heavy chain kinase that was isolated from the soil amoeba *Acanthamoeba*, and it is essential for actin activation of this unique myosin ATPase (Maruta and Korn, 1977). PAKs are activated not only by Rac, but also by CDC42 (Manser *et al.*, 1994). More recently, PAKs were shown to require another protein called PIX for their full activation (Manser *et al.*, 1998). The N-terminal SH3 domain of PIX binds a Pro-rich motif of PAK (residues 186–203, PAK18), and the fragment PAK18 selectively blocks the PAK–PIX interaction (see Figure 1). Microinjection of the PAK18 blocks membrane ruffling of PC12 cells (Obermeier *et al.*, 1998). These observations suggest that PAKs are essential for membrane ruffling. Furthermore, a so-called dominant negative mutant of PAK was reported to inhibit RAS transformation of Rat-1 fibroblasts, but not NIH 3T3 fibroblasts (Tang *et al.*, 1997). The latter report alone is confusing, and even contradicts previous observations that NIH 3T3 fibroblasts require both Rac and CDC42 for RAS transformation (Qui *et al.*, 1995; 1997). This mutant still binds (and sequesters) both Rac and CDC42 as well as PIX and PAK substrates when (or if) it is overexpressed. Therefore, in theory, it should have inhibited RAS transformation of both Rat-1 and NIH 3T3 fibroblasts. Furthermore, because this mutant is not specific for PAKs, it is not clear whether PAKs are essential for RAS transformation.

A. WR-PAK18

To clarify this point, using a more specific inhibitor for PAKs, we have created a cell-permeable peptide called WR-PAKs (see Figure 1), a conjugate of the penetratin WR and PAK18 (Maruta *et al.*, 1999a; He *et al.*, 2001). The penetratin is a peptide vector of 16 amino acids that delivers any small peptides of less than 100 amino acids into target cells (Derossi *et al.*, 1998). 10 μM WR-PAK18 selectively inhibits the growth of v-Ha-RAS transformed NIH 3T3 fibroblasts, but has little effect on the growth of the parental normal NIH 3T3 cells (He *et al.*, 2001). The vector WR alone or a mutant of WR-PAK18 called R192A where Arg corresponding to residue 192 of PAK1 is replaced by Ala (see Figure 1), which no longer binds the SH3 domain of PIX, have no effect on the growth of RAS transformants, suggesting that the binding of WR-PAK18 to PIX is essential for its anti-RAS action. Furthermore, we confirmed that 10 μM WR-PAK18 is sufficient to almost completely block PAK activation in RAS transformants. Again, neither WR alone nor the mutant R192A affects the PAK activation. Membrane ruffling of RAS transformants is inhibited by WR-PAK18, but not by

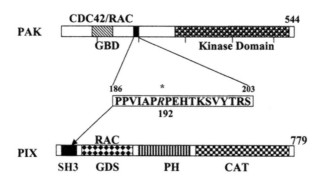

WR-PAK18
(WR)   **(PAK18)**
RRWRRWWRRWWRRWRR-PPVIAPRPEHTKSVYTRS
A (R192A)

Figure 1 Schematic presentation of the PAK–PIX interaction. GBD, GTPase (CDC42/Rac)-binding domain; GDS, GDP dissociation stimulating (GDP/GTP exchange) domain for Rac; WR-PAK18, the conjugate of a penetratin (WR) to the PAK18 corresponding to residues 186–203 of human PAK1; R192A, a mutant of WR–PAK18 carrying a mutation (Arg to Ala) at position 192 of human PAK1.

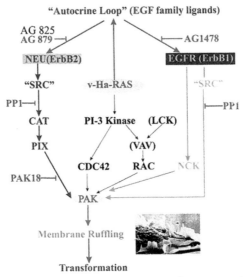

Figure 2 Two independent autocrine pathways essential
for v-Ha-RAS-induced PAK activation.

either WR alone or the mutant R192A. These observations clearly indicate that
PAKs are required for RAS-induced transformation as well as membrane ruffling
of NIH 3T3 cells.

Interestingly, however, we found that oncogenic RAS mutants absolutely require two independent autocrine pathways for the activation of PAKs
which involve ErbB1 and ErbB2 (for a summary, see Figure 2), in addition to
Rac, CDC42, and PIX (He *et al.*, 2001). It has been known that oncogenic
RAS mutants activate several distinct genes that encode epidermal growth factor (EGF) family ligands such as transforming growth factor (TGF)-α, HB-EGF,
amphiregulin, and heregulin as well as vascular EGF (VEGF) through the
Raf–MEK–Erk kinase cascade (Glick *et al.*, 1991; Higashiyama *et al.*, 1991;
Normanno *et al.*, 1994; Grugel *et al.*, 1995; Mincione *et al.*, 1996). Although
VEGF is a paracrine factor which activates FLK-1 receptor kinase of vascular
endothelial cells that is essential for angiogenesis of RAS transformants and
other carcinomas, the autocrine role of these EGF family ligands in RAS transformation still remains to be clarified. Since oncogenic RAS mutants are constitutively activated, they no longer require any of these autocrine factors for their
activation per se. However, RAS transformants definitely require autocrine
factors for PAK activation, which are secreted from the cells. Within 10 min
after removal of these autocrine factors from the culture medium, PAK activity

of RAS transformants sharply declines, and then is gradually restored within 60 min by biosynthesis or secretion of these autocrine factors (He *et al.*, 2001).

B. ErbB1-specific inhibitor (AG 1478)

AG 1478 inhibits selectively ErbB1 (IC_{50} is 3 nM), and does not inhibit any other kinases such as ErbB2 and platelet-derived growth factor (PDGF) receptor (even at $100 \mu M$) *in vitro* (Levitzki and Gazit, 1995). We found that AG 1478 (200 nM) inhibits both PAK activation in RAS transformants, and their anchorage-independent growth in soft agar, without any effect on their anchorage-dependent growth (He *et al.*, 2001). Because AG 1478 does not inhibit PAKs directly *in vitro*, it is clear that ErbB1 is essential for RAS-induced PAK activation. Interestingly, however, the inhibition by AG 1478 can reach only around 50% even at higher concentrations, suggesting that another receptor for EGF family ligands is also involved in RAS-induced PAK activation. Heregulin, for instance, activates both ErbB3 and ErbB4, and as a consequence the kinase-free ErbB3 forms a heterodimer with ErbB2 and activates the latter. Using nude mice, which are transplanted with v-Ha-RAS transformants, we have examined the effect of AG 1478 [20 mg/kg, intraperitonealy (i.p.), every other day] on the growth of RAS-induced sarcomas. The AG 1478 treatment reduces the growth of these sarcomas by 50% (He *et al.*, 2001).

C. ErbB2-specific inhibitors (AG 825 and AG 879)

AG 825 inhibits selectively ErbB2 (IC_{50} is around $0.4 \mu M$), and its IC_{50} for ErbB1 is around $10 \mu M$ (Levitzki and Gazit, 1995). We found that $0.8 \mu M$ AG 825 inhibits both PAK activation in RAS transformants and their anchorage-independent growth, without any effect on their anchorage-dependent growth (He *et al.*, 2001). These observations indicate that ErbB2 is also essential for RAS-induced PAK activation. Again the inhibition by AG 825 can reach only 50%, even at higher concentrations, supporting the notion that both ErbB1 and ErbB2 almost equally contribute to RAS-induced PAK activation and RAS transformation. Indeed the combination of AG 1478 ($0.2 \mu M$) and AG 825 ($0.8 \mu M$) almost completely inhibits both PAK activation and RAS transformation *in vitro* (He *et al.*, 2000a), confirming that ErbB1 and ErbB2 are the two major receptors essential for the RAS-induced autocrine-dependent PAK activation and RAS transformation. Because AG 825 is rather metabolically unstable in experimental animals, another ErbB2-specific inhibitor called AG 879 (Levitzki and Gazit, 1995) was used for *in vivo* experiments to test the therapeutic efficacy or potency of the ErbB2 inhibitor against RAS-associated cancers. The treatment of mice by AG 879 (20 mg/kg, i.p., every other day) reduces the growth of RAS-induced sarcomas by at least 85%, and keeps 50% of the treated mice absolutely free of tumors (He *et al.*, 2001).

D. SRC family kinase-specific inhibitors (PP1 and PP2)

What mediates the ErbB1- or ErB2-dependent activation of PAKs? ErbB1 can be linked to PAKs through an adapter protein called NCK, which contains both SH2 and SH3 domains (Galisteo *et al.*, 1996). The SH3 domain of NCK binds the N-terminal Pro-rich motif (residue 11–18) of PAK, and when ErbB1 is autophosphorylated on its binding to its ligands such as EGF and TGF-α, the SH2 domain of NCK binds the phosphorylated Tyr residues of ErbB1 to form a tertiary complex of ErbB1–NCK–PAKs. Such a complex formation could bring PAKs to the plasma membranes, where their activators such as CDC42 and Rac are localized.

It is also known that both ErbB1 and ErbB2 activates Src family kinases (Osherov and Levitzki, 1994; Sheffield, 1998). The direct activator of these Src family kinases is a Tyr phosphatase called PTP-α which dephosphorylates the critical Tyr residues near the C-terminus of Src family kinases (Zheng *et al.*, 2000). However, the molecular mechanism underlying ErbB1- or ErbB2-induced activation of this phosphatase remains to be clarified. Furthermore, the PAK-binding partner or activator PIX binds another protein called CAT which is a substrate of Src family kinases (Bagrodia *et al.*, 1999). Although the biological role of the Tyr phosphorylation of CAT still remains to be determined, it is clear that the CAT–PIX interaction is essential for PAK activation. CAT binds the C-terminal half (residues 400–646) of PIX, and a variant of PIX called p50 that lacks the CAT-binding domain acts as a dominant negative mutant of PIX, blocking PAK activation (Bagrodia and Cerione, 1999). Thus, it is conceivable that a member of Src family kinases is involved in ErbB1- or ErbB2-induced activation of PAKs.

To test this hypothesis, we used PP1, an inhibitor specific for Src family kinases (Hanke *et al.*, 1996; Schindler *et al.*, 1999). In fibroblasts such as NIH 3T3 cells, three distinct members of Src family kinases are expressed: c-Src, c-Fyn, and c-Yes. The IC_{50} of PP1 for c-Src is around 170 nM, whereas that for c-Fyn is around 6 nM in test tubes (Schindler *et al.*, 1999). 10 nM PP1 is sufficient to inhibit both PAK activation in RAS transformants and their anchorage-independent growth by 50%, indicating that a Src family kinase(s) other than c-Src itself is essential for RAS-induced PAK activation and malignant transformation *in vitro* (He *et al.*, 2000). To determine which receptor, ErbB1 or ErbB2, requires this Src family kinase to activate PAKs, we examined whether AG 1478 or AG 825 potentiates the PPI action to inhibit either RAS-induced PAK activation or malignant transformation *in vitro*. Both AG 1478 and AG 825 failed to affect the action of PP1 toward these two phenotypes (He *et al.*, 2000), suggesting that the PP1-sensitive kinase is involved in both ErbB1 and ErbB2 pathways leading to PAK activation.

Treatment of nude mice by PP1 (20 mg/kg, i.p., every other day) alone keeps 50% of the treated mice absolutely free of RAS-induced sarcomas, and reduces the size of the growing tumors by 50% (He *et al.*, 2000). To improve further the therapeutic efficacy of PP1, we combined PP1 with a few other anti-cancer drugs such as AG 879, AG 1478, and MKT-077 each of which alone reduces the size of RAS sarcomas or the number of tumor-bearing mice or both by around 50% (Tikoo *et al.*, 2000; He *et al.*, 2001). We found that a combination of AG 879 and PP1 (20 mg/kg of each) almost completely suppresses the growth of RAS sarcomas (see Figure 3), and causes no detectable adverse effect on any health conditions of the treated mice (He *et al.*, 2001). These observations strongly suggest that the usage of drugs such as PPI and AG 879, which indirectly block PAK activation in RAS transformants, would be an alternative approach toward the treatment of RAS-associated cancers which represent more than 30% of all human cancers, notably more than 90% of pancreatic carcinomas and 50% of colon carcinomas (Bos, 1989). We are currently determining, again using the nude mouse model, how efficiently this combination therapy suppresses the growth of several distinct human pancreatic carcinoma cell lines, all of which carry oncogenic Ki-RAS mutants (Kato *et al.*, 1999).

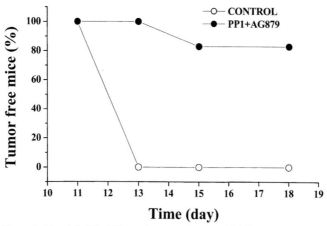

Figure 3 The AG 879–PP1 combination therapy of RAS tumors in mouse model. The growth of RAS sarcomas in nude mice treated with both AG 879 and PP1 (20 mg/kg of each) was examined. No sarcoma was grown in the treated mice, except one in which only a tiny tumor grew, while all control mice developed large sarcomas.

E. PAKs-specific inhibitors

If one could find a potent PAK-specific inhibitor (PAKI), a single chemical compound that alone is able to inhibit directly PAKs in a highly specific manner at a very low concentration (with the IC_{50} around 1 nM or less), it would be the ideal therapeutic for the treatment of RAS-associated cancers. So far, such a PAKI has not been identified among the preexisting chemical compounds (or newly generated from scratch as yet). However, there are a few clues to a rational designing of such a PAKI. An antibiotics called staurosporine (ST), which belongs to indolo-carbazole family compounds, is so far the most potent inhibitor for PAKs, although it is not specific for PAKs (Zeng *et al.*, 2000). It also inhibits cyclin-dependent kinases (CDKs) by binding the ATP-binding pocket of their kinase domain. The three-dimensional (3D) structure of the ST–CDK complex was determined (Lawrie *et al.*, 1997). This structure indicates that the N-methyl group of the hexose ring of ST, which appears to contact directly Asp-86 of CDK by forming a salt-bridge, plays the key role in inhibiting CDKs. Very recently, the 3D structure of PAK kinase domain was also determined (Lei *et al.*, 2000). The structural comparison between CDK2 and PAKs in their ATP-binding pocket has revealed that Asp-86 of CDK2 appears to be replaced by neutral amino acids such as Ser and Cys in PAK family kinases (see Figure 4, R. Jorissen and H. Treutlein, unpublished data, 2000). Thus, the N-methyl group of ST may not be essential for the ST–PAK interaction. In fact, our structure–function

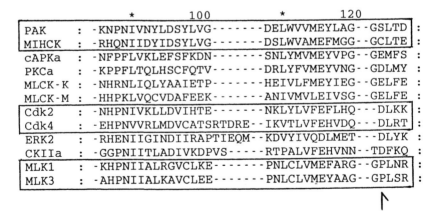

Figure 4   Amino acid sequence alignment of PAKs, MLK3, and CDKs. The sequence alignment of the ATP-binding pockets of PAKs, CDKs, and MLKs (boxed) and several other Ser/Thr kinases where the sugar ring of ST (and ATP) binds is highlighted. The arrowed residues correspond to Asp-86 of CDK2 and Ser-350 of PAK1.

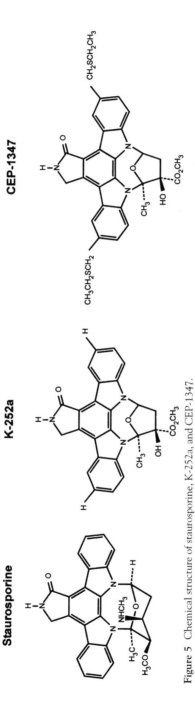

Figure 5 Chemical structure of staurosporine, K-252a, and CEP-1347.

relation analysis of more than a dozen distinct ST analogs toward PAKs supports this notion (T. Nheu, *et al.*, unpublished data).

More importantly, another antibiotic called K-252a, in which the hexose ring of ST is replaced by a pentose ring, and the N-methyl group is replaced by a carboxymethyl group (see Figure 5), appears to be more specific for PAKs than ST, although the former is less potent than the latter (T. Nheu and H. Maruta, unpublished data, 2000). Furthermore, we recently found that a K-252a analog called KT-7515 or CEP-1347 (see Figure 5) directly inhibit PAKs in a quite selective manner (T. Nheu, *et al.*, unpublished data). KT-7515 was previously reported to inhibit activation of JNKs (c-Jun N-terminal kinases), but indirectly, suggesting that a kinase(s) upstream of JNKs is a direct target of KT-7515 (Maroney *et al.*, 1998). Unlike K252a, KT-7515 does not inhibit the majority of common kinases such as Trk (NGF receptor), PKA (cAMP-dependent kinase), PKC, myosin light chain kinase (MLCK), and PI-3 kinase (Maroney *et al.*, 1998). Interestingly, KT-7515 was more recently suggested to inhibit another CDC42/Rac-dependent Ser/Thr kinase called MLK3 (Maroney *et al.*, 2001). Like PAKs, MLK3 activates JNK family kinases (Vacratsis and Gallo, 2000). Interestingly, in MLK3, Asp 86 of CDK2 is replaced by another neutral amino acid (Pro). Such a functional–structural similarity between PAKs and MLK3 suggests the possibility that KT-7515 is a specific inhibitor for the Rac/CDC42-dependent kinases (PAKs and MLK3). It was found that PAKs and MLK3 form a complex within cells under certain conditions (M. L. Schmitz, personal communication, 2000). Based on both the predicted 3D structure of the PAK–K252a complex, and our biological data on the structure–function relationship of K252a analogs, we are currently replacing a critical side chain of KT-7515 to develop a much more potent PAK-specific inhibitor which would be useful for the clinical use.

III. BLOCKING CDC42 PATHWAYS (ACKs AND N-WASP)

A few years ago it was reported that overexpression of a so-called dominant negative (N-17) mutant of CDC42 suppresses RAS-transformation (Qiu *et al.*, 1997). Strictly speaking, this observation only supports, but has not proven as yet, the notion that CDC42 is essential for RAS transformation, simply because this N-17 mutant is not a specific inhibitor for CDC42. It can sequester not only CDC42-specific GDP-dissociation stimulators (GDSs) such as FGD1, but also many other nonspecific GDSs for Rho family GTPases such as DBL, OST, and Tiam-1 which activate Rho or Rac, in addition to CDC42 (Maruta, 1998; Feig, 1999; Nur-E-Kamal *et al.*, 1999).

Thus, we have created a CDC42-specific inhibitor from ACK-1, a Tyr kinase that binds only CDC42 in the GTP-bound form (Manser *et al.*, 1993).

This inhibitor corresponds to the 42-amino acid fragment of ACK-1 (residues 504–545), called ACK42, that binds the CDC42–GTP complex (Nur-E-Kamal et al., 1999). ACK42 inhibits the interactions of CDC42 with its effectors such as ACK family kinases, PAK family kinases, and WASP family proteins as well as CDC42 GAPs such as p190-A and p190-B (Maruta, 1998). We found that overexpression of ACK42 in RAS-transformants suppresses malignant transformation (Nur-E-Kamal et al., 1999). It also inhibits NGF-induced neurite outgrowth in PC12 cells (Nur-E-Kamal et al., 1999; Maruta et al., 1999b). In addition, GST–ACK42 fusion protein serves as a ligand specific for the activated form of CDC42 (GTP–CDC42 complex) which is quite useful for monitoring the activation of CDC42 in cells. Using GST–ACK42, we have shown for the first time that oncogenic RAS activates CDC42 in NIH 3T3 fibroblasts, and both NGF and EGF promote the activation of CDC42 by activating normal RAS in PC12 cells (Nur-E-Kamal et al., 1999). Thus, it is now clear that, like Rac, CDC42 is activated by RAS, and it is essential for RAS transformation. It was also shown that PI-3 kinase is involved in RAS-induced activation of CDC42, in addition to Rac (Liliental et al., 2000).

A cell-permeable ACK42 derivative called WR-ACK42 was also generated by conjugating ACK42 with the peptide vector called WR or penetratin which can deliver any small peptides of less than 100 amino acids into target cells (Nur-E-Kamal et al., 1999). WR-ACK42 inhibits CDC42-induced microspike formation and RAS transformation of NIH 3T3 fibroblasts without affecting the growth of normal fibroblasts (Nur-E-Kamal et al., 1999). The efficiency of this peptide to block the CDC42 pathways depends on its affinity for the CDC42–GTP complex.

Thus, we have screened for a high affinity mutant of ACK42 by generating a series of ACK42 mutants. A mutant called K34-ACK42, in which Arg 34 is replaced by Lys (see Figure 6), binds 15 times more efficiently than the

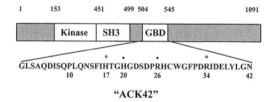

1. Binds only CDC42 in the GTP-Bound Form.
2. Suppresses both NOG and RAS Transformation.

Figure 6 Primary structure of ACK42, the CDK42–GTP binding domain of ACK-1. Replacement of Arg-34 by Lys markedly increases the affinity of ACK42 for CDC42.

wild-type ACK42 (Maruta *et al.*, 1999b). This mutant would be a much better ligand for monitoring CDC42 activation in cells, and should serve as a more potent anti-RAS cancer reagent.

IV. BLOCKING Rho PATHWAYS

A. Bacterial exotoxin C3

There are several bacterial toxins that modify (activate or inactivate) some members of Rho family GTPases. One of them is a *Clostridium botulinum* exotoxin, or exoenzyme, called C3 that inactivates selectively Rho GTPases by ADP-ribosylating Asn at position 41 within their effector domain (Sekine *et al.*, 1989). This modification interferes with the interactions of Rho GTPases with their effectors, although it does not affect their interactions with GAPs (Sekine *et al.*, 1989). The matured form of C3 is a protein of 204 amino acids. It is poorly cell-permeable, and to test its biological (physiological or pharmacological) effect on cells, it is necessary to microinject it into target cells or deliver it in a complex with a basic liposome such as lipofectamine across the plasma membranes (Maruta *et al.*, 1999b). Among the several distinct members of Rho GTPases, at least RhoA, RhoB, RhoC, and RhoG appear to be the substrates for C3, whereas Rho E resists the action of C3 (Nobes *et al.*, 1998).

A RAS farnesyltransferase inhibitor (FTI) called L-739, 749 suppresses RAS transformation primarily by blocking farnesylation of Rho B, instead of RAS (Lebowitz *et al.*, 1995), suggesting that at least Rho B is essential for RAS transformation. This FTI induces apoptosis of RAS transformants in suspension (anoikis), in addition to activation of the WAF-1 gene which encodes a CDK inhibitor (Lebowitz *et al.*, 1995; Du *et al.*, 1999). Interestingly, in the presence of this FTI, Rho B is geranylgeranylated (GG), instead of farnesylated (F), and Rho B-GG activates the WAF-1 gene, but does not promote anoikis of RAS transformants (Du *et al.*, 1999), suggesting that Rho B-F might act as an anoikis antagonist.

We found that C3 ($50\,\mu g/ml$) inhibits selectively the growth of RAS transformants, but not that of the parental normal cell growth (Maruta *et al.*, 1999b). Since RhoB is inactivated by ADP-ribosylation in the presence of C3, the function of Rho B-F should be lost. Thus, C3-treated RAS transformants must become susceptible to anoikis. Interestingly, in the presence of $10\,\mu M$ WR-PAK18, fibroblasts in suspension suffer from a rapid death (anoikis) (Kim, Y., He, H., Maruta, H., and Bertics, P., unpublished observation, 2000). PAKs inactivate the death agonist Bad by phosphorylation at positions 112 and 136 (Schuermann *et al.*, 2000). How does the cell adhesion to solid substratum block the anoikis? Does it activates PAKs, or inactivate the *Bad* gene?

The cell–substratum interaction definitely activates PAKs, at least in part by translocating Rac to the plasma membranes (del Pozo et al., 2000). As a consequence, Bad is inactivated by PAK-mediated phosphorylation. Similarly, in the presence of $5-10\,\mu M$ WR-PAK18 or WR-BIN13, serum-starved RAS transformants, even attached to solid substratum, are subjected to a rapid death (H. He and H. Maruta, unpublished observation, 2000). WR-BIN13 is a conjugate of WR with the cytoplasmic FAK-binding domain of β-integrin (residues 752–764) that selectively blocks the interaction of FAK with β-integrin (Hungerford et al., 1996). Like WR-PAK18, WR-BIN13 selectively inhibits the growth of RAS transformants, but not the parental normal fibroblasts (H. He and H. Maruta, unpublished observation, 2000). These observations suggest that either serum or the cell–substratum interaction (anchorage) is required for RAS transformants to keep PAK activity above a certain level that promotes both cell survival and growth. Once the PAK level goes down below that threshold, the cells would start to die. For normal cells to survive, both serum and anchorage are essential, simply because serum cannot activate PAK without anchorage.

B. Rock inhibitor Y-27632

Rho subfamily GTPases activate several distinct effectors. One of them is a Ser/Thr kinase called Rock (Narumiya et al., 1998). Rock is a Rho-specific effector, and is not activated by any other GTPases such as Rac and CDC42. Rock phosphorylates several distinct proteins such as myosin II light chain (MLC), myosin-binding regulatory subunit (MBS) of myosin phosphatase, and LIM-kinase (Narumiya et al., 1998; Maekawa et al., 1999). Phosphorylation of MLC stimulates the actin-activation of myosin II ATPase and formation of bipolar myosin filaments (Amano et al., 1996). Phosphorylation of MBS inactivates the myosin phosphatase, and therefore indirectly stimulates the MLC phosphorylation (Kimura et al., 1996). Phosphorylation of LIM-kinase at Thr-508 causes the activation of this kinase that phosphorylates cofilin at Ser-3 (Maekawa et al., 1999; Ohashi et al., 2000). As mentioned previously, PAK also phosphorylates and activates LIM-kinase at Thr-508 (Edwards et al., 1999). Phosphorylation of cofilin abolishes its ability to severe actin filaments (Moriyama et al., 1996).

Interestingly, a pyridine derivative called Y-27632, which selectively inhibits smooth muscle contraction by inhibiting Ca-sensitization, turned out to be a Rock-specific inhibitor (Uehata et al., 1997). This inhibitor does not inhibit either PAK or MCL kinase. Accordingly, this inhibitor abolishes both actin stress fibers and focal adhesion, which are induced by Rho and Rock in normal fibroblasts, but not membrane ruffling, which is induced by Rac and

PAK. Furthermore, $10\,\mu M$ Y-27632 suppresses RAS transformation *in vitro* (anchorage-independent growth) at least partially (Sahai *et al.*, 1999). However, a possible usage of this inhibitor as an anticancer drug would be rather limited, as it causes a few side effects including the relaxation of blood vessels and stomach muscle, leading to a drop of blood pressure and insufficient digestion.

V. Rac-SPECIFIC INHIBITORS?

Rac induces actin-based membrane ruffling and activates several distinct effectors such as Por1 and Sra-1 (van Aelst *et al.*, 1996; Kobayashi *et al.*, 1998) as well as Ser/Thr kinases such as PAKs and MLKs that in turn activate JNK pathways (Manser *et al.*, 1994; Maruta *et al.*, 1999a; Vacratsis and Gallo, 2000). Among these effectors, Por1 and Sra-1 were reported to bind only Rac, whereas PAKs and MLKs bind both Rac and CDC42. A few distinct drugs such as SH51344, MKT-077, and chaetoglobosin K (CK) that block Rac-induced membrane ruffling suppress RAS transformation (Walsh *et al.*, 1997; Tikoo *et al.*, 1999, 2000). Thus, if one could develop a Rac-specific inhibitor, it would be quite useful for the treatment of RAS-associated cancers. However, so far no Rac-specific molecule has been developed. Although the so-called dominant negative mutant (N17) of Rac was wildly used to block Rac in the past, its action is not really specific for Rac, since it binds (and sequesters) many nonspecific GDSs (GDP-dissociation stimulators) such as DBL, OST, and Tiam-1 that activate CDC42 or Rho, in addition to Rac (Maruta, 1998; Feig, 1999).

So we have tried to develop a small fragment derived from either POR1 or Sra-1 which binds only the Rac–GTP complex, but none of them showed any high-affinity for Rac. Then we have identified the minimal Rac–CDC42-binding fragment of 40 amino acids (PAK40, residues 69–108) from human PAK1 (Maruta *et al.*, 2001). Our attempt to generate by a series of PAK40 mutations of a Rac-specific mutant that binds only Rac, and not CDC42, has been so far unsuccessful. Thus, such a Rac-specific inhibitor might remain to be a Holy Grail in this field for some years. Interestingly, a point mutation in human PAK3, R67C, was more recently reported to be associated with X-linked nonspecific mental retardation (Bienvenu *et al.*, 2000). In this PAK3 mutant, the highly conserved Arg-67 (corresponding to Arg-72 of human PAK1) is replaced by Cys. Although this mutation occurs in the minimal CDC42–Rac-binding domain (PAK40), it does not affect its binding to either CDC42 or Rac at least *in vitro* (T. Nheu and H. Maruta, unpublished observation, 2000). Thus, the possible contribution of this mutation to an alteration in the neuronal function of PAK3 still remains to be clarified.

References

Amano, M., Ito, M., Kimmura, K., *et al.* (1996). Phosphorylation and activation of myosin by Rho-associated kinase (Rho-kinase). *J. Biol. Chem.* **271,** 20246–20249.

Bagrodia, S., and Cerione, R. (1999). PAK to the future. *Trends Cell Biol.* **9,** 350–355.

Bagrodia, S., Bailey, D., Lenard, Z., *et al.* (1999). A Tyr-phosphorylated protein that binds an important regulatory region on the Cool family of PAK-binding proteins. *J. Biol. Chem.* **274,** 22393–22400.

Bienvenu, T., *et al.* (2000). Missense mutation in PAK3, R67C, causes X-linked nonspecific mental retardation. *Am. J. Med. Genet.* **93,** 294–298.

Bos, J. (1989). Ras oncogenes in human cancer. A review. *Cancer Res.* **49,** 4682–4689.

del Pozo, M., Price, L., Alderson, N., *et al.* (2000). Adhesion to the extracellular matrix regulates the coupling of the small GTPase Rac to its effector PAK. *EMBO J.* **19,** 2008–2014.

Derossi, D., Chassaing, G., and Prochiantz, A. (1998). Trojan peptides: The penetratin system for intracellular delivery. *Trends Cell Biol.* **8,** 84–87.

Du, W., Lebowitz, P., and Prendergast, G. (1999). Cell growth inhibition by farnesyltransferase inhibitors is mediated by gain of geranylgeranylated Rho B. *Mol. Cell Biol.* **19,** 1831–1840.

Edwards, D., Sanders, L., Bokoch, G., *et al.* (1999). Activation of LIM-kinase by PAK couples Rac/CDC42 GTPase signaling to actin cytoskeletal dynamics. *Nat. Cell Biol.* **1,** 253–259.

Feig, L. (1999). Tools of the trade: Use of dominant-inhibitory mutants of RAS family GTPases. *Nat. Cell Biol.* **1,** E25–E27.

Galisteo, M., Chernoff, J. Su, Y. C., *et al.* (1996). The adaptor protein NCK links receptor Tyr kinase with the Ser/Thr kinase PAK1. *J. Biol. Chem.* **271,** 20997–21000.

Glick, A., Sporn, M., and Yuspa, S. (1991). Altered regulation of TGFβ-1 and TGFα in primary keratinocytes and papillomas expressing v-Ha-RAS. *Mol. Carcinog.* **4,** 210–219.

Grugel, S., *et al.* (1995). Both v-Ha-Ras and v-Raf stimulate expression of the VEGF in NIH 3T3 cells. *J. Biol. Chem.* **270,** 25915–25919.

Hall, A. (1998). Rho GTPases and the actin cytoskeleton. *Science* **279,** 509–514.

Hanke, J., Gardner, J., Dow, R., *et al.* (1996). Discovery of a novel, potent, and Src family-selective Tyr kinase inhibitor. *J. Biol. Chem.* **271,** 695–701.

He, H., Hirokawa, Y., Manser, E., *et al.* (2001). Signal therapy of RAS-induced cancers in combination of AG879 and PP1, specific inhibitors for ErbB2 and Src family kinases, that block PAK activation. *Cancer J.* **7,** 191–202.

He H., Hirokawa Y., Levitzki A., *et al.* (2000). An anti-RAS cancer potential of PP1, an inhibitor specific for SRC family kinases: *In vitro* and *in vivo* studies. *Cancer J. Sci. Am.* **6,** 243–248.

Higashiyama, S., Abraham, J., Miller, J., *et al.* (1991). A heparin-binding growth factor secreted by macrophage-like cells is related to EGF. *Science* **251,** 936–939.

Hungerford, J., Compton, M., Matter, M., *et al.* (1996). Inhibition of FAK in cultured fibroblasts results in apoptosis. *J. Cell Biol.* **135,** 1383–1390.

Kato, M., Shimada, Y., Tanaka, H., *et al.* (1999). Characterization of six cell lines established from human pancreatic adenocarcinomas. *Cancer* **85,** 832–840.

Kimura, K., Ito, M., Amano, M., *et al.* (1996). Regulation of myosin phosphatase by Rho and Rho-associated kinase (Rho-kinase). *Science* **273,** 245–248.

Kobayashi, K., Kuroda, S., Fukata, M., *et al.* (1998). p140Sra-1 (specifically Rac-associated protein) is a novel specific target for Rac GTPase. *J. Biol. Chem.* **273,** 291–295.

Lawrie, A., Noble, M., Tunnah, P., *et al.* (1997). Protein kinase inhibition by staurosporine revealed in details of the molecular interaction with CDK2. *Nat. Struct. Biol.* **4,** 796–800.

Lebowitz, P., Davide, J., and Prendergast, G. (1995). Evidence that farnesyltransferase inhibitors suppress Ras transformation by interfering with Rho activity. *Mol. Cell Biol.* **15,** 6613–6622.

Lei, M., Lu, W., Meng, W., *et al.* (2000). Structure of PAK1 in an autoinhibited conformation reveals a multistage activation switch. *Cell* **102,** 387–397.

Levitzki, A., and Gazit, A. (1995). Tyrosine kinase inhibition: An approach to drug development. *Science* **267**, 1782–1788.

Liliental, J., Moon, S. Y., Lesche, R., *et al.* (2000). Genetic deletion of the PTEN tumor suppressor gene promotes cell motility by activation of Rac1 and CDC42 GTPases. *Curr. Biol.* **10**, 401–404.

Maekawa, M., Ishizaki, T., Boku, S., *et al.* (1999). Signaling from Rho to the actin cytoskeleton through protein kinases Rock and LIM-kinase. *Science* **285**, 895–989.

Manser, E., Leung, T., Salihuddin, H., *et al.* (1993). A non-receptor Tyr kinase that inhibits the GTPase activity of CDC42. *Nature* **363**, 364–367.

Manser, E., Leung, T., Salihuddin, H., *et al.* (1994). A brain Ser/Thr kinase activated by CDC42 and Rac. *Nature* **367**, 40–46.

Manser, E., Loo, T. H., Koh, C. G., *et al.* (1998). PAK kinases are directly coupled to the PIX family of nucleotide exchange factors. *Mol. Cell* **1**, 183–192.

Maroney, A., Glicksman, M., Basma, A., *et al.* (1998). Motoneuron apoptosis is blocked by CEP-1347 (KT-7515), a novel inhibitor of the JNK signaling pathway. *J. Neurosci.* **18**, 104–111.

Maroney, A., Finn, J., Connors, T., *et al.* (2001). CEP-1347 (KT7515), a semisynthetic inhibitor of the mixed lineage kinase family. *J. Biol. Chem.* **276**, 25302–25308.

Maruta, H. (1998). Regulators of Ras/Rho family GTPase: GAPs, GDSs and GDIs. *In* "G Proteins, Cytoskeleton and Cancer" (H. Maruta and K. Kohama, eds.), pp. 151–170. Landes Biosciences, Austin, Texas.

Maruta, H., and Korn, E. D. (1977). *Acanthamoeba* cofactor protein is a heavy chain kinase required for actin-activation of the Mg^{2+}-ATPase activity of *Acanthamoeba* myosin I. *J. Biol. Chem.* **252**, 8329–8332.

Maruta, H., He, H., Tikoo, A., *et al.* (1999a). G proteins, PIP2, and actin-cytoskeleton in the control of cancer growth. *Microsc. Res. Tech.* **47**, 61–66.

Maruta, H., He, H., Tikoo, A., *et al.* (1999b). Cytoskeletal tumor suppressors that block oncogenic RAS signaling. *Ann. N.Y. Acad. Sci.* **886**, 48–57.

Maruta, H., He, H., and Nheu, T. (2001). Interfering with RAS signaling using membrane permeable peptides or drugs. *In* "GTPase Protocols: The RAS Superfamily" (E. Manser and T. Leung, eds.). Humana Press, Totowa, NJ. In press.

Mincione, G., Bianco, C., Kannan, S., *et al.* (1996). Enhanced expression of heregulin in ErbB2 and c-Ha-RAS transformed mouse and human mammary epithelial cells. *J. Cell Biochem.* **60**, 437–446.

Moriyama, K., Iida, K., and Yahara, I. (1996). Phosphorylation of Ser 3 of cofilin regulates its essential function on actin. *Gene to Cells* **1**, 73–86.

Narumiya, S., Watanabe, N., and Ishizaki, T. (1998). Rho effectors: Structure and function. *In* "G Proteins, Cytoskeleton and Cancer (H. Maruta and K. Kohama, eds.), pp. 225–237. Landes Biosciences, Austin, Texas.

Nobes, C., Lauritzen, I., Mattei, M., *et al.* (1998). A new member of the Rho family, Rnd1, promotes disassembly of actin filament structures and loss of cell adhesion. *J. Cell Biol.* **14**, 187–197.

Normanno, N., Selvan, M., Qi, C., *et al.* (1994). Amphiregulin as an autocrine growth factor for c-Ha-Ras/c-ErbB2-transformed human mammary epithelial cells. *Proc. Natl. Acad. Sci. U.S.A.* **91**, 2790–2794.

Nur-E-Kamal, M. S. A., Kamal, J. M., Qureshi, M. M., *et al.* (1999). The CDC42-specific inhibitor derived from ACK-1 blocks v-Ha-RAS-induced transformation. *Oncogene* **18**, 7787–7793.

Obermeier, A., Ahmed, S., Manser, E., *et al.* (1998). PAK promotes morphological changes by acting upstream of Rac. *EMBO J.* **17**, 4328–4339.

Ohashi, K., Nagata, K., Maekawa, M., *et al.* (2000). Rock activates LIM-kinase by phosphorylation at Thr 508 within the activation loop. *J. Biol. Chem.* **275**, 3577–3582.

Osherov, N., and Levitzki, A. (1994). EGF-dependent activation of the Src family kinases. *Eur. J. Biochem.* **225**, 1047–1053.

Qiu, R. G., Che, J., Kim, D., *et al.* (1995). An essential role for Rac in RAS transformation. *Nature* **374,** 457–459.

Qiu, R. G., Abo, A., McCormick, F., *et al.* (1997). CDC42 regulates anchorage-independent growth and is essential for RAS transformation. *Mol. Cell. Biol.* **17,** 3449–3458.

Ridley, A. (1998). Rho family GTPases and actin cytoskeleton. *In* "G Proteins, Cytoskeleton and Cancer" (H. Maruta and K. Kohama, eds.), pp. 211–223. Landes Biosciences, Austin, Texas.

Sahai, E., Ishizaki, T., Narumiya, S., *et al.* (1999). Transformation mediated by RhoA requires activity of Rock kinases. *Curr. Biol.* **9,** 136–145.

Schindler, T., Sicheri, F., Pico, A., *et al.* (1999). Crystal structure of HCK in complex with a Src family-selective Tyr kinase inhibitor. *Mol. Cell* **3,** 639–648.

Schuermann, A., Mooney, A., Sanders, L., *et al.* (2000). PAK1 phosphorylates the death agonist Bad and protects cells from apoptosis. *Mol. Cell. Biol.* **20,** 453–461.

Sekine, A., Fujiwara, M., and Narumiya, S. (1989). Asp residue in the Rho gene product is the modification site for botulinum ADP-ribosyltransferase (C3). *J. Biol. Chem.* **264,** 8602–8605.

Sheffield, L. (1998). c-Src activation by ErbB2 leads to attachment-independent growth of human breast epithelial cells. *Biochem. Biophys. Res. Commun.* **250,** 27–31.

Tang, Y., Chen, Z., Ambrose, D., *et al.* (1997). Kinase-deficient PAK1 mutants inhibit RAS transformation of Rat-1 fibroblasts. *Mol. Cell. Biol.* **17,** 4454–4464.

Tikoo, A., Cutler, H., Lo, S. H., *et al.* (1999). Treatment of RAS-induced cancer by the F-actin cappers tensin and chaetoglobosin K, in combination with the caspase-1 inhibitor N1445. *Cancer J. Sci. Am.* **5,** 293–300.

Tikoo, A., Shakri, R., Connolly, L., *et al.* (2000). Treatment of RAS-induced cancers by the F-actin-bundling drug MKT-077. *Cancer J. Sci. Am.* **6,** 162–168.

Uehata, M., Ishizaki, T., Satoh, H., *et al.* (1997). Calcium sensitization of smooth muscle mediated by a Rho-associated protein kinase in hypertension. *Nature* **389,** 990–994.

Vacratsis, O., and Gallo, K. (2000). Zipper-mediated oligomerization of the MLK3 is not required for its activation by CDC42, but is necessary for its activation of the JNK pathway. *J. Biol. Chem.* **275,** 27893–27900.

van Aelst, L., Joneson., T., and Bar-Sagi, D. (1996). Identification of a novel Rac-interacting protein involved in membrane ruffling. *EMBO J.* **15,** 3778–3786.

Walsh, A., Dhanasekaran, M., Bar-Sagi, D., *et al.* (1997). SCH51344-induced reversal of RAS-transformation is accompanied by the specific inhibition of the RAS and RAC-dependent cell morphology pathway. *Oncogene* **15,** 2553–2560.

Zeng, Q., Lagunoff, D., Masaracchia, R., *et al.* (2000). Endothelial cell retraction is induced by PAK2 monophosphorylation of myosin II. *J. Cell Sci.* **113,** 471–482.

Zheng, X. M., Resnick, R., and Shalloway, D. (2000). A phosphotyrosine displacement mechanism for activation of Src by PTP-alpha. *EMBO J.* **19,** 964–978.

Integrin Antagonists as Cancer Therapeutics

Chandra C. Kumar[1] and Lydia Armstrong
Department of Tumor Biology
Schering-Plough Research Institute
Kenilworth, New Jersey 07033

I. INTRODUCTION

Integrins are a family of cell surface receptors that mediate the attachment of cells to extracellular matrix (ECM) proteins or to other cells (Ruoslahti, 1996; Hynes, 1999). Cells secrete ECM, a network of proteins into the intercellular space. ECM exerts profound influence over cells. It is composed of structural and regulatory molecules, some of which include fibronectin, laminin, collagen, vitronectin, and von Willebrand factor. Integrins serve to link the ECM to the cytoskeleton inside the cell. Although integrins were initially viewed as relatively simple adhesion molecules, it has become clear that attachment of cells to ECM is a crucial event. Most normal cells will not survive or proliferate unless they are adhering to a substrate, a phenomenon referred to as anchorage dependence. In addition to

[1] To whom correspondence should be addressed.

Tumor-Suppressing Viruses, Genes, and Drugs
Innovative Cancer Therapy Approaches
Copyright © 2002 by Academic Press.
All rights of reproduction in any form reserved.

linking ECM or adjacent cells to the intracellular cytoskeleton, integrins serve as signal transducers (Hynes, 1994; Giancotti and Rouslahti, 1999; Kumar, 1998). The signaling pathways affected by integrins include: (1) activation of Rho family of GTPases leading to organization of actin-cytoskeletaleton; (2) activation of mitogen-activated protein (MAP) kinases, and (3) activation of other protein and lipid kinases. These signaling pathways allow integrins to regulate cell-cycle progression, cell survival, and gene expression in addition to their effects on cell adhesion and morphology. Binding of growth factors such as epidermal growth factor (EGF), or platelet derived growth factor (PDGF) to their receptors is not sufficient for optimal activation of the signaling cascades; input from the integrin family of receptors is also necessary. There is considerable cross talk and coopera-tion between growth factor receptors and integrins. Thus integrins are known to physically interact with growth factor receptors and influence each other's activ-ity, in addition to multiple inputs into common signaling pathways (Vuori and Ruoslahti, 1994; Miyamoto et al., 1996; Moro et al., 1998; Schneller et al., 1997; Soldi et al., 1999). A variety of physiologically important processes depend on the ability of cells to recognize and in turn respond to the immediate environment. These include angiogenesis, wound healing, bone resorption, and inflammation.

Integrins are composed of noncovalently associated α and β chains that form heterodimeric receptor complexes (Hynes, 1999; Ruoslahti, 1996). The α and β subunits contain a large extracellular domain, a short transmem-brane domain, and a cytoplasmic carboxyl-terminal domain of variable length. The extracellular domains of both the α and β chains form the ligand binding domain. In mammals, 18 α subunits and 8 β subunits are known. Many $\alpha-\beta$ combinations fail to occur, but at least two dozen are well defined. Each $\alpha\beta$ combination has its own binding specificity and signaling properties. Integrins not only bind ligands present in the extracellular matrix such as fibronectin, collagen, and vitronectin, certain integrins can also bind to soluble ligands such as fibrinogen or to counterreceptors such as intracellular adhesion molecules (ICAMs) on adjacent cells (Hynes, 1994).

As integrins bind to ECM, they become clustered at the membrane and associate with actin-binding and signaling proteins promoting the assembly of actin stress fibers. As a result, ECM proteins, integrins, and cytoskeletal proteins assemble into aggregates known as focal adhesion plaques (Miyamoto et al., 1995; Burridge et al., 1998) (Figure 1). Hence, these receptors serve to integrate the ECM with the cytoskeleton, a property for which they are named as integrins. Many of the integrins recognize the RGD (Arg-Gly-Asp) sequence in their matrix proteins (Ruoslahti, 1996). Nevertheless, they are capable of dis-tinguishing different RGD-containing proteins such that some bind primarily to fibronectin and others to vitronectin. Integrins are also expressed in a cell-type specific manner; Thus one group of integrins such as $\alpha_5\beta_1$, $\alpha_v\beta_3$, and $\alpha_v\beta_6$ is associated with migration and proliferation in various cell types. Many other

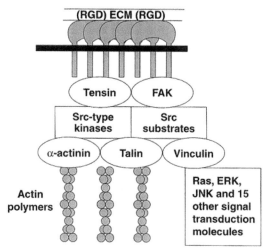

Figure 1 Model for the arrangements of different proteins in a focal adhesion complex. Arrangements of cytoskeletal proteins that colocalize to focal adhesion complexes *in vivo* and bind to integrins *in vitro* are shown. Ligand occupancy, integrin aggregation, and tyrosine phosphorylation leads to the accumulation of F-actin and associated cytoskeletal proteins in a massive adhesive and signaling complex. ECM, Extracellular matrix; RGD, Arg-Gly-Asp; JNK, Jun kinase. Reproduced from Kumar, C. C. (1998). Signaling by integrin receptors. *Oncogene* **17,** 1365–1373.

integrins are expressed in selective cell types such as $\alpha_{IIb}\beta_3$ in platelets and $\alpha_6\beta_4$ in epithelial cells (Hynes, 1994; Ruoslahti, 1996).

II. SIGNALING PATHWAYS ACTIVATED BY INTEGRINS

In addition to growth factors and nutrients, many normal cells require adhesion to the ECM to proliferate. A number of signaling proteins have been found to be associated with the focal adhesion complexes, suggesting the involvement of integrin engagement in the activation of signaling pathways. In addition, the receptors for insulin, PDGF, EGF, and vascular endothelial growth factor (VEGF) are optimally activated by their ligands only when cells are attached to ECM (Vuori and Ruoslahti, 1994; Miyamoto *et al.*, 1996; Moro *et al.*, 1998; Schneller *et al.*, 1997;

Soldi *et al.*, 1999). Certain integrins are also associated with specific growth factor receptors. For example, integrin $\alpha_v\beta_3$ can be immunoprecipitated in complex with insulin, PDGF, and VEGF receptors (Vuori and Ruoslahti, 1994; Soldi *et al.*, 1999), whereas $\alpha_5\beta_1$ and perhaps other β_1 integrin receptors associate with EGF receptors (Miyamoto *et al.*, 1995; Schneller *et al.*, 1997).

A. Role of integrins in mitogenic signaling pathways

Integrin-mediated signaling is necessary for the optimal activation of growth factor receptors (Figure 2). Two such pathways are activated by most integrins. First, integrins facilitate the activation of the extracellular-signal regulated kinase (ERK) family of MAP kinases (Schlaepfer *et al.*, 1994; Wary *et al.*, 1996; Kumar, 1998; Giancotti and Ruoslahti, 1999). In some cells, signaling along the RAS–ERK cascade is blocked at the level of Raf or MEK in the absence of attachment (Renshaw *et al.*, 1997; Lin *et al.*, 1997). Integrins remove this block, perhaps by activating Rac or phosphatidylinositol (PI) 3-kinase (King *et al.*, 1997). Second, integrins activate the c-jun NH_2-terminal kinases (JNKs) (Miyamoto *et al.*, 1995; Oktay *et al.*, 1999). JNKs activation leads to the phosphorylation of JUN which in combination with c-fos forms the AP1 transcription factor complex. The AP1 complex regulates genes that are important for cell proliferation (Davis, 2000). Since, most growth factors are poor activators of the JNK pathway, the ability of integrins to activate this kinase may explain why cell proliferation requires integrin-mediated adhesion.

Integrin-mediated signals are necessary for cell cycle progression. Studies have indicated that cell adhesion is specifically required for the induction of cyclin D1 and for the activation of the cyclin E–cdk2 complex in early–mid G1 phase (Fang *et al.*, 1996). Although cyclin D1 is regulated by cell adhesion at both the transcriptional and translational levels, the effect of cell adhesion on cyclin E–cdk2 activity appears to be indirect and mediated by downregulation of the cdk inhibitors p21 and p27 (Zhu *et al.*, 1996). In addition, there is evidence that integrin linked kinase (ILK) and activation of the PI-3 kinase–Akt pathway promotes translocation of β-catenin to the nucleus, and β-catenin can regulate the cyclin D1 promoter (Tetsu and McCormick, 1999; Schuttman *et al.*, 1999).

B. Role of integrins in cell survival signaling mechanisms

Many mammalian cells depend on adhesion to the ECM for their continued survival (Ruoslahti and Reed, 1994). Loss of their attachment to the matrix causes apoptosis in many cell types (Frisch and Francis, 1994; Meredith *et al.*, 1993; Frisch and Ruoslahti, 1997). This phenomenon is called anoikis (meaning "homelessness" in Greek). Anoikis is a mechanism to insure that cells, which

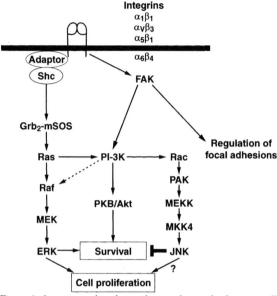

Figure 2 Integrin-mediated signaling pathways leading to cell proliferation and cell survival. Activation of the Ras–ERK signaling pathway by integrins involves the recruitment of Shc via a transmembrane adaptor protein. Although all β_1 and α_v integrins can regulate focal adhesions via FAK, only some of them can recruit Shc, activate Ras–ERK signaling, and promote cell-cycle progression. Cell survival is promoted by integrins via Ras–ERK and Akt. Jnk may inhibit cell survival if ERK is not activated. There is also evidence for the involvement of PI-3K in integrin-mediated activation of the Raf–ERK pathway. See text for details. Reproduced from Kumar, C. C. (1998). Signaling by integrin receptors. *Oncogene* **17,** 1365–1373.

are displaced from their natural environment, are eliminated. Anoikis is controlled by ECM in an integrin-specific manner. The $\alpha_5\beta_1$ integrin, which binds to fibronectin, induces expression of Bcl-2, protecting cells from apoptosis resulting from growth factor withdrawal (Zhang *et al.*, 1995). Other integrins such as $\alpha_v\beta_1$ or $\alpha_v\beta_3$ and various β_1 integrins fail to rescue cells from apoptosis under these conditions. It is possible that under other conditions cells depend on other integrins for survival (Brooks *et al.*, 1994a; Boudreau *et al.*, 1995). Thus vascular endothelial cells seem to depend on $\alpha_v\beta_3$ receptor for their survival (Brooks *et al.*, 1994a). Inhibition of $\alpha_v\beta_3$ function during angiogenesis with anti-$\alpha_v\beta_3$

antibodies both inhibited their proliferation and induced apoptosis, whereas treatment with antibodies that block β_1 integrin function did not (Brooks et al., 1994a). Thus integrin-mediated attachment to ECM is a general requirement for cell survival, but survival under special circumstances may require a particular integrin.

Integrin-mediated cell adhesion leads to activation of the PI-3 kinase – Akt pathway (King et al., 1997; Khwaja et al., 1997). As with other signaling mechanisms, the PI-3 kinase and PKB – Akt pathway is also activated on growth factor stimulation (Franke et al., 1997a,b). The mechanism by which integrin engagement leads to PI-3 kinase activation in normal cells is not known. One mechanism may be through focal adhesion kinase (FAK), as this kinase has been found to associate with the SH2 domains of p85 of PI-3 kinase, so it may contribute to the constitutive activation of PI-3 kinase. Akt promotes cell survival in part by phosphorylating and inactivating proapoptotic proteins Bad, Procaspase-9, Forkhead, and glycogen synthase kinase-3 (GSK-3) (Datta et al., 1999).

III. ROLE OF INTEGRINS IN NEOPLASTIC TRANSFORMATION

Normal cells require attachment to ECM and stimulation by serum or growth factors for proliferation. This anchorage requirement and serum-dependency is lost in neoplastic cells, although anchorage independence is the feature that correlates best with tumorigenicity in vivo (Freedman and Shin, 1974). The exact mechanism by which tumor cells acquire anchorage-independence is not known. Many oncogene-encoded proteins activate the integrin-mediated signaling pathways mainly through the Rho family proteins (Kumar, 1998). Specifically, oncogenic RAS mutants have been shown to activate the cell morphology pathway through RAC (Joneson et al., 1996), and more recent studies have shown that activation of both the cell morphology pathway and the ERK pathway are essential for transformation (Joneson et al., 1996). Constitutive activation of both the integrin-mediated pathway and the growth factor pathways is necessary to elicit serum and anchorage-independent status to the transformed cells. Activation of a step on the growth factor portion only should lead to enhanced proliferation but not anchorage-independent growth. Activation of a step on an integrin pathway prior to convergence would be expected to lead to anchorage-independent but not serum-dependent growth. Thus oncogenes such as dbl and lbc which function as constitutively active guanine nucleotide exchange factors for the Rho family GTPases, induce anchorage-independent but not serum-dependent growth (Schwartz et al., 1996). lbc and dbl were identified by their ability to induce focus formation and tumorigenicity in vivo when expressed in 3T3 cells (Toksoz and Williams, 1994; Eva and Aaronson, 1985). Thus both lbc and dbl transformants show the same serum dependence as the control normal cells. These studies show that anchorage-independence and serum independence can be separated.

Changes in integrin gene expression, both qualitative and quantitative, have been documented in a number of epithelial malignancies (Serini et al., 1996), including carcinoma of the lung (Damjanovich et al., 1992), breast (D'Ardenne et al., 1991), colon (Kouloulis et al., 1993), gastric (Tani et al., 1996), and pancreas (Shimoyama et al., 1995) to name a few. The degree of activation of integrins in adherent cells can also vary, and general lowering of adhesiveness is thought to be important in allowing tumor cells to dislodge from their original site and invade (Ruoslahti, 1999). The general approach of blocking integrin function in tumors cells by disrupting their interaction with ECM using small peptides from various ligands has been pursued by various laboratories and companies (e.g., laminin peptides, thrombospondin peptides, collagen fragments, fibronectin fragments). To date the most promising anticancer approach by inhibiting integrin function is directed toward newly synthesized normal endothelial cells formed during tumour-induced angiogenesis. A number of pharmaceutical companies are developing small molecular weight integrin antagonists for the treatment of neoplastic disease by targeting the tumor vasculature.

IV. ROLE OF INTEGRINS IN TUMOR-INDUCED ANGIOGENESIS

Angiogenesis is the process by which new blood vessels are formed from pre-existing vessels (Folkman, 1992, 1995). The process of angiogenesis is very selective and is controlled by the local balance between factors that stimulate new vessel growth and factors that inhibit it (Folkman, 1995). Tumor cells elicit a complex angiogenic response as a result of increased secretion of angiogenic factors and decreased production of angiogenic inhibitors (Liotta et al., 1991). The process of angiogenesis involves the following steps (Figure 3): (1) tumor cells begin to secrete cytokines and growth factors that activate the vascular endothelial cells of a nearby blood vessel to proliferate and migrate toward the tumor; (2) the migration of vascular cells is facilitated by the secretion of proteolytic enzymes that degrade the ECM; (3) cell adhesion receptors facilitate endothelial cell migration by interacting with the ECM proteins; and (4) the vascular endothelial cells differentiate and form a mature vessel. A crucial role for ECM in the development of vasculature in physiological and pathological conditions has been demonstrated. The invasion, migration, and proliferation of vascular endothelial cells during angiogenesis are regulated by integrins.

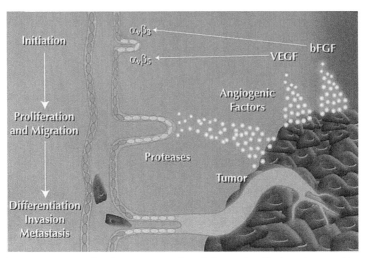

Figure 3 Tumor-induced angiogenesis. Growth factors (small spheres) secreted by tumor cells (large spheres) stimulate vascular endothelial cells to proliferate (initiation) and invade into the tumor tissue (migration). Integrins $\alpha_v\beta_3$ and $\alpha_v\beta_5$ mediate the attachment of endothelial cells to matrix proteins and facilitate survival and migration. Finally, endothelial cells differentiate to form lumens, making new blood vessel offshoots from a preexisting vessel. (See color plates.)

A. Role of integrins $\alpha_v\beta_3$ and $\alpha_v\beta_5$ in tumor-induced angiogenesis

Integrin $\alpha_v\beta_3$ has been found to play a very significant role in the process of angiogenesis (Brooks et al., 1994a,b). $\alpha_v\beta_3$ is a promiscuous receptor as it is capable of interacting with a number of ECM proteins including vitronectin, fibrinogen, fibronectin, and thrombospondin (Varner et al., 1995; Varner and Cheresh, 1996), as well as other proteins with different biological functions, including fibroblast growth factor-2 (FGF2) (Rusnati et al., 1997) and metalloproteinase MMP-2 (Brooks et al., 1996). In addition, $\alpha_v\beta_3$ has been shown to associate with activated PDGF, insulin, and VEGF receptors, to facilitate optimal activation of cell proliferative signaling pathways (Giancotti and Ruoslahti, 1999; Kumar, 1998) and to prevent apoptosis (Stromblad et al., 1996). This integrin is minimally expressed in normal blood vessels but is significantly upregulated in new blood vessels formed in response to a variety of tumors. A characteristic feature of $\alpha_v\beta_3$ that makes it an attractive target for therapeutic intervention is its relatively limited cellular distribution. It is not generally expressed on epithelial cells and is expressed only at low levels on a subset of B cells, some cells of macrophage lineage, smooth muscle cells, and activated endothelial cells (Shattil, 1995; Cheresh, 1991). Integrin $\alpha_v\beta_3$

is also expressed on certain invasive tumors including metastatic melanoma (Albelda *et al.*, 1990; Petitclerc *et al.*, 1999) and late stage glioblastoma (Gladson and Cheresh, 1991), where it contributes to their malignant phenotype.

Brooks *et al.* (1994a,b) have shown that antibody and peptide antagonists of integrin $\alpha_v\beta_3$ inhibit angiogenesis on the chick chorioallantoic membrane when introduced intravenously into the chick embryo. Remarkably, $\alpha_v\beta_3$ antagonists had very little effect on preexisting blood vessels, indicating the usefulness of targeting this receptor for therapeutic benefit without adverse side effects. Evidence has been presented indicating that antagonists of integrin $\alpha_v\beta_3$ inhibit tumor-induced angiogenesis by selectively promoting apoptosis of vascular endothelial cells (Brooks *et al.*, 1994a). More recently, we have demonstrated that antagonists of $\alpha_v\beta_3$ induce apoptosis by a direct signaling event (Brassard *et al.*, 1999). Altogether, these findings indicate a key role for integrin $\alpha_v\beta_3$ in a signaling event critical for the survival and ultimately the differentiation of vascular cells undergoing angiogenesis *in vivo*.

Studies have implicated a related integrin, $\alpha_v\beta_5$, in angiogenesis under certain conditions. For example, Friedlander *et al.* (1995) have shown that antibody antagonists of $\alpha_v\beta_3$ inhibit FGF2 stimulated angiogenesis, and antagonists of integrin $\alpha_v\beta_5$ inhibit VEGF-stimulated angiogenesis in the corneal and chorioallantoic membrane models. More recent studies have shown that kinase-deleted mutants of Src blocked VEGF-induced but not FGF2-induced angiogenesis (Elicieri *et al.*, 1999). These results suggest that FGF2 and VEGF may activate different angiogenic pathways that require $\alpha_v\beta_3$ and $\alpha_v\beta_5$, respectively. Therefore, dual antagonists of $\alpha_v\beta_3$ and $\alpha_v\beta_5$ may be useful in blocking tumor-induced angiogenesis.

B. Role of integrin $\alpha_5\beta_1$ in angiogenesis

The integrin $\alpha_5\beta_1$ represents a specific fibronectin receptor that plays a central role in anchorage-dependent survival and apoptosis and is frequently lost or downregulated in transformed cells (Giancotti and Ruoslahti, 1990). The $\alpha_5\beta_1$ integrin, when not bound to fibronectin, transmits a negative growth signal to the cell, which can be reverted by attachment to fibronectin (Juliano and Varner, 1993). Conversely, ectopic expression of integrin $\alpha_5\beta_1$ has been demonstrated to regulate tumor cell proliferation *in vitro* as well as tumor growth *in vivo* (Giancotti and Ruoslahti, 1990). Although α_v integrins have been shown to play critical roles in angiogenesis, studies using α_v $(-/-)$ knockout mice suggest that other receptors and their ligands may also regulate this process (Bader *et al.*, 1998). More recent studies indicate that $\alpha_5\beta_1$ and its ligand fibronectin are coordinately upregulated on blood vessels in human tumor biopsies and play a critical role in angiogenesis (McDonald *et al.*, 2000). Importantly, antibody, peptide, and novel nonpeptide antagonists of integrin $\alpha_5\beta_1$

blocked angiogenesis induced by several growth factors but had very little effect on angiogenesis induced by VEGF in several animal models (Kim *et al.*, 2000). For details on angiogenesis and its inhibitors in general, see the Chapter 13 by Achen and Stacker.

V. INTEGRIN ANTAGONISTS AS ANTIANGIOGENESIS AGENTS

The realization that integrins conform to classic descriptions of signaling receptors suggests that it should be possible to design synthetic agents that antagonize integrins. A number of potent, RGD-derived peptides and peptidomimetics ($\alpha_v\beta_3$ antagonists) have been identified, several of which are active in preclinical animal models (Table 1) (Miller *et al.*, 2000). About 80 patent applications claiming nonpeptide antagonists have been published (and several have been issued). SmithKline Beecham (King of Prussia, PA) initially extended their efforts on platelet integrin antagonists and identified SB223245, a potent antagonist of $\alpha_v\beta_3$ ($K_i = 2$ nM in a receptor binding assay), with good selectivity over $\alpha_{IIb}\beta_3$ receptor ($K_i = 30,000$ nM). Merck (West Point, PA) described L74815, ($K_i = 0.3$ nM in receptor binding assay); however, this compound also inhibited ADP-stimulated platelet aggregation, suggesting high affinity for the platelet integrin, $\alpha_{IIb}\beta_3$. Hoechst (Frankfurt, Germany) identified several series of $\alpha_v\beta_3$ antagonists with good affinity (IC$_{50}$ 20 nM) in inhibiting ligand binding to $\alpha_v\beta_3$. Dupont (Wilmington, DE) described a series of β_3 antagonists derived from a structural template originally used for $\alpha_{IIb}\beta_3$ antagonists. Dupont evaluated several compounds in mouse models of angiogenesis and tumorigenesis (Kerr *et al.*, 1999). In a mouse matrigel model of angiogenesis, subcutaneous administration of SM256 (IC$_{50}$ 4 nM) decreased blood vessel formation with an ED$_{50}$ of 0.055 ug/kg/day.

More recently, we have described the pharmacological and biological characterization of a nonpeptide small molecule (SCH 221153) that is a potent inhibitor of both $\alpha_v\beta_3$ and $\alpha_v\beta_5$ integrin receptors (Kumar *et al.*, 2001). We have shown that SCH 221153 inhibits vascular endothelial cell adhesion and proliferation mediated by FGF2 and VEGF. SCH 221153 inhibited FGF2-induced angiogenesis in chick chorioallantoic membrane and inhibited the growth of human tumor xenografts in SCID mice.

A. Integrin antagonists in clinical trials

Vitaxin, developed by Ixsys (San Diego, CA), is a humanized version of the mouse monoclonal antibody, LM609, that functionally blocks $\alpha_v\beta_3$ integrin receptor (Table 2). This antibody has been shown to target angiogenic blood vessels and cause suppression of tumor growth in various animal models (Eliceiri

Table 1 Examples of Integrin Antagonists

Structure of $\alpha_v\beta_3$ antagonists	Company and product
	SmithKline Beecham WO-09905107 (2/99) $IC_{50} \sim 23$ nM Vn Binding
	Merck & Co., Inc WO-09808840 (3/98) IC_{50} range 0.1 to 100 nM Cell Adhesion $IC_{50} = 1 \mu M$
	Hoechst AG EP 00820988 (1/93) IC_{50} (kistan binding) = 30 nM
	Hoechst
	DuPont Merck *Bioorg. Med. Chem. Lett.* 1997, 1371 $IC_{50} = 1.1$ nM (ELISA)
	Merck L748415
	Scripps
	CP4632 Meiji Seika Kaisha
	Schering-Plough SCH 221153

Table 2 Antiangiogenic Integrin Antagonists in Clinical Trials

Company	Product	Structure	Status	Comments
Ixsys/MedImmune	Vitaxin	Antibody	Phase II	Humanized momoconal antibody selective for $\alpha_v\beta_3$
E. Merck/Scripps	EMD-121974		Phase II/III	Cyclic RGDfV peptide with activity for $\alpha_v\beta_3$ and $\alpha_v\beta_5$
Monsanto-Searle	SC-68448		Phase I	Small synthetic molecule antagonist for $\alpha_v\beta_3$

and Cheresh, 2000). Vitaxin, represents a potentially novel cancer therapy, and it may be an effective and safe treatment for a wide spectrum of tumors types. Ixsys had progressed Vitaxin through phase I in the first quarter of 1999. Of the 14 cancer patients evaluated, 8 showed disease stabilization and/or some objective tumor reduction. Importantly, there was no evidence of toxicity at all doses tested, even when administered for 22 consecutive months (Gutheil et al., 2000). Phase II cancer trials are underway to evaluate Vitaxin during long-term treatments.

Merck KGaA and the Scripps Research Institute (La Jolla, CA) are codeveloping a cyclic peptide RGDfV antagonist of $\alpha_v\beta_3$ and $\alpha_v\beta_5$ (EMD-121974). Results of a phase I trial in 25 patients with solid tumors demonstrated no dose-limiting toxicities up to 850 mg/m²/infusion, and stable disease was observed in two patients. EMD-121974 is currently in phase II clinical trials (E. Merck) in patients with Kaposi's sarcoma and brain tumors. Phase II/III trials in combination with other treatments are planned for patients with colorectal, breast, and non-small cell lung cancer (Table 2).

As mentioned in the previous section, many small molecule antagonists of $\alpha_v\beta_3$ are currently under development by many pharmaceutical companies. However, Monsanto-Searle has progressed an orally bioavailable compound into phase I clinical trials. In preclinical models this series of compounds inhibited Rice Leydig tumor growth in a dose-dependent manner by 85%, decreased the density of tumor microvessels, and inhibited lung metastasis by >90% (Carron et al., 1998).

VI. CONCLUSIONS AND FUTURE PERSPECTIVES

Research on integrin family of receptors has led to a number of therapeutic applications and identification of targets for tumor therapy. One integrin-directed drug, an anti-β_3 integrin antibody for the prevention of arterial restenosis, has reached the marketplace. Several other integrin based drugs such as the peptides containing the integrin-binding RGD sequence (Ruoslahti, 1996), and mimics of such peptides that specifically block individual integrins are also under development. These drugs target thrombosis [$\alpha_{IIb}\beta_3$ in platelets (Cheng et al., 1994)], osteoporosis [$\alpha_v\beta_3$ in osteoclasts (Flores et al., 1992)], and tumor-induced angiogenesis [$\alpha_v\beta_3$ in neovascular endothelial cells (Carron et al., 1998)]. In a more recent exciting study, Ruoslahti's group used the in vivo selection of phage display libraries to isolate peptides that home specifically to tumor blood vessels (Pasqualini et al., 1997; Ruoslahti, 2000). Two of these peptides, one containing α_v integrin-binding Arg-Gly-Asp motif and the other an Asn-Gly-Arg motif, were conjugated to the anticancer drug doxorubicin to demonstrate enhanced efficacy of the drug against human breast cancer xenografts in nude mice (Arap et al.,

1998a,b). These peptide–drug conjugates exhibited significantly less toxicity, suggesting that it may be possible to develop targeted chemotherapeutic strategies based on the selective expression of integrins in tumor vasculature.

In summary, it is clear that signaling pathways activated by growth factor receptors and integrin receptors are extensively intertwined, most importantly at the level of RAS, PI-3K, and FAK. In addition, certain integrins physically form complexes with growth factor receptors. Understanding the mechanisms by which tumor cells gain the anchorage-independence status is an important area of investigation and may lead to the identification of novel targets for anticancer drug discovery.

Acknowledgment

We thank Dr. Elizabeth Smith for help with the preparation of the tables.

References

Albelda, S. M., Mette, S. A., Elder, D. E., Steward, R., Danjanovich, L., Herlyn, M., and Buck, C. A. (1990). Integrin distribution in malignant melanoma: Association of the β_3 subunit with tumor progression. *Cancer Res.* **50**, 6757–6764.

Arap, W., Pasqualini, R., and Ruoslahti, E. (1998a). Cancer treatment by targeted drug delivery to tumor vasculature in a mouse model. *Science* **279**, 377–380.

Arap, W., Pasqualini, R., and Ruoslahti, E. (1998b). Chemotherapy targeted to tumor vasculature. *Curr. Opin. Oncol.* **10**, 560–565.

Bader, B. L., Rayburn, H., Crowley, D., and Hynes, R. O. (1998). Extensive vasculogenesis, angiogenesis, and organogenesis precede lethality in mice lacking all alpha v integrins. *Cell* **95**, 507–519.

Boudreau, N., Sympson, C. J., Werb, Z., and Bissel, M. J. (1995). Suppression of ICE and apoptosis in mammary epithelial cells by extracellular matrix. *Science* **267**, 891–893.

Brassard, D. L., Maxwell, E., Malkowski, M., Nagabhushan, T. L., Kumar, C. C., and Armstrong, L. (1999). Integrin $\alpha_v\beta_3$ mediated activation of apoptosis. *Exp. Cell Res.* **251**, 33–45.

Brooks, P. C., Clark, R. A. F., and Cheresh, D. A. (1994a). Requirement of vascular integrin $\alpha_v\beta_3$ for angiogenesis. *Science* **264**, 569–571.

Brooks, P. C., Montgomery, A. M. P., Rosenfeld, M., Reisfeld, R. A., Hiu, T., Klier, G., and Cheresh, D. A. (1994b). Integrin $\alpha_v\beta_3$ antagonists promote tumor regression by inducing apoptosis of angiogenic blood vessels. *Cell* **79**, 1157–1164.

Brooks, P. C., Stromblad, S., Sanders, L. C., von Schalscha, T. L., Aimes, R. T., Stetlet-Stevenson, W. G., Quigley, J. P., and Cheresh, D. A. (1996). Localization of matrix metalloproteinase MMP-2 to the surface of invasive cells by interaction with integrin $\alpha_v\beta_3$. *Cell* **85**, 683–693.

Burridge, K., Fath, K., Kelly, G., Nuckolls, G., and Turner, C. (1988). Focal adhesions: Transmembrane junctions between the extracellular matrix and the cytoskeleton. *Annu. Rev. Cell Biol.* **4**, 487–525.

Carron, C. P., Meyer, D. M., Pegg, J. A., Engleman, W., Nickols, M. A., Settle, S. L., Westlin, W. F., Ruminski, P. G., and Nickols, G. A. (1998). A peptidomimetic antagonist of the integrin $\alpha_v\beta_3$ inhibits leydig cell tumor growth and the development of hypercalcemia of malignancy. *Cancer Res.* **58**, 1930–1935.

Cheng, S., Craig, W. S., Mullen, D., Tschopp, J. F., Dixon, D., and Pierschbacher, M. D. (1994). Design and synthesis of novel cyclic RGD-containing peptides as highly potent and selective integrin alpha 11b beta 3 antagonists. *J. Med. Chem.* **37**, 1–8.

Cheresh, D. A. (1991). Structure, function and biological properties of integrin $\alpha_v\beta_3$ on human melanoma cells. Cancer Metastasis rev. 10, 3–10.

Damjanovich, L., Albelda, S. M., and Mette, S. A. (1992). The distribution of intergrin cell adhesion receptors in normal and malignant lung tissue. Am. J. Resp. Cell Mol. Biol. 6, 197–206.

D'Ardenne, A. J., Richmon, P. I., Horton, M. A., Mcaulay, A. E., and Jordan, S. (1991). Co-ordinate expression of the alpha 6 integrin laminin receptor subunit and laminin in breast cancer. J. Pathol. 165, 213–220.

Datta, S. R., Brunet, A., and Greenberg, M. E. (1999). Cellular survival: A play in three Akts. Genes Dev. 13, 2905–2927.

Davis, R. J. (2000). Signal transduction by the JNK group of MAP kinases. Cell 103, 239–252.

Eliceiri, B. P., and Cheresh, D. A. (2000). Role of α_v integrins during angiogenesis. Cancer J. Sci. Am. Suppl. 3, S245–S249.

Elicieri, B. P., Paul, R., Schwartzberg, P. L., Hood, J. D., Leng, J., and Cheresh, D. A. (1999). Selective requirement for Src kinases during VEGF-induced angiogenesis and vascular permeability. Mol. Cell Biol. 4, 915–924.

Eva, A., and Aaronson, S. A. (1985). Isolation of a new human oncogene from a diffuse B-cell lymphoma. Nature 316, 273–275.

Fang, F., Orend, G., Watanabe, N., Hunter, T., and Ruoslahti, E. (1996). Dependence of cyclin E–CDK2 kinase activity on cell anchorage. Science 271, 499–502.

Flores, M. E., Norgard, M., Heinegard, D., Reinholt, F., and Andersson, G. (1992). RGD-directed attachment of isolated rat osteoclasts to osteopontin, bone sialoprotein, and fibronectin. Exp. Cell Res. 201, 526–530.

Folkman, J. (1992). The role of angiogenesis in tumor growth. Semin. Cancer Biol. 3, 65–71.

Folkman, J. (1995). Angiogenesis in cancer, vascular, rheumatoid and other diseases. Nat. Med. 1, 27–30.

Franke, T. F., Kaplan, D. R., and Cantley, L. C. (1997a). PI3K: Downstream AKTion blocks apoptosis. Cell 88, 435–437.

Franke, T. F., Kaplan, D. R., Cantley, L. C., and Toker, A. (1997b). Direct regulation of the Akt protooncogene product by phospahatidyl-inositol-3,4-biphosphate. Science 275, 665–668.

Freedman, V. H., and Shin, S. (1974). Cellular tumorigenicity in nude mice: Correlation with cell growth in semi-solid medium. Cell 3, 355–359.

Friedlander, M., Brooks, P. C., Shaffer, R. W., Kincaid, C. M., Varner, J. A., and Cheresh, D. A. (1995). Definition of two angiogenic pathways by distinct alpha v integrins. Science, 270, 1500–1502.

Frisch, S. M., and Francis, H. J. (1994). Disruption of epithelial cell–matrix interactions induces apoptosis. J. Cell Biol. 124, 619–626.

Frisch, S. M., and Ruoslahti, E. (1997). Integrins and anoikis. Curr. Opin. Cell Biol. 9, 701–706.

Giancotti, F. G., and Ruoslahti, E. (1990). Elevated levels of the $\alpha_5\beta_1$ fibronectin receptor suppress the transformed phenotype of Chinese hamster ovary cells. Cell 60, 849–859.

Giancotti, F. G., and Ruoslahti, E. (1999). Integrin signaling. Science 285, 1028–1032.

Gladson, C., and Cheresh, D. A. (1991). Glioblastoma expression of vitronectin and $\alpha_v\beta_3$ integrin. J. Clin. Invest. 88, 1924–1932.

Gutheil, J. C., Campbell, T. N., Pierce, P. R., Watkins, J. D., Huse, W. D., Bodkin, D. J., and Cheresh, D. A. (2000). Targeted antiangiogenic therapy for cancer using Vitaxin: A humanized monoclonal antibody to the integrin $\alpha_v\beta_3$. Clin. Cancer Res. 6, 3056–3061.

Hynes, R. O. (1994). The impact of molecular biology on models for cell adhesion. BioEssays 16, 663–669.

Hynes, R. (1999). Cell adhesion: Old and new questions. Trends Cell Biol. 12, 33–37.

Joneson, T., White, M., Wigler, M., and Bar-Sagi, D. (1996). Stimulation of membrane ruffling and MAP kinase activation by distinct effectors of RAS. Science 271, 810–812.

Juliano, R. L., and Varner, J. A. (1993). Adhesion molecules in cancer: The role of integrins. *Curr. Opin. Cell Biol.* **5**, 812–818.

Kerr, J. S., Wexler, R. S., Mousa, S. A., Robinson, C. S., Wexler, E. J., Mohamed, S., Voss, M. E., Devenny, J. J., Czerniak, P. M., Gudzelak, A., and Slee, A. M. (1999). Novel small molecule alpha v integrin antagonists: Comparative anti-cancer efficacy with known angiogenesis inhibitors. *Anticancer Res.* **19**, 959–968.

Khwaja, A., Rodriguez-Viciana, P., Wennstrom, S., Warne, P. H. A., and Downward, J. (1997). Matrix adhesion and Ras transformation both activate a phosphoinositide 3-OH kinase and protein kinase B/Akt cellular survival pathway. *EMBO J.* **16**, 2783–2793.

Kim, S., Bell, K., Mousa, S. A., and Varner, J. A. (2000). Regulation of angiogenesis *in vivo* by ligation of alpha 5 beta 1 with the central cell-binding domain of fibronectin. *Am. J. Pathol.* **156**, 1345–1362.

King, W. G., Mattaliano, M. D., Chan, T. O., Tsichilis, P. N., and Brugge, J. S. (1997). Phosphatidylinositol 3-kinase is required for integrin-stimulated AKT and Raf-1/mitogen-activated protein kinase pathway activation. *Mol. Cell. Biol.* **17**, 4406–4418.

Kouloulis, G. K., Virtanen, I., Moll, R., Quaranta, V., and Gould, V. E. (1993). Immunolocalization of integrins in the normal and neoplastic colonic epithelium. *Virchows Arch. B.* **63**, 373–383.

Kumar, C. C. (1998). Signaling by integrin receptors. *Oncogene* **17**, 1365–1373.

Kumar, C. C., Malkowski, M., Yin, Z., Tanghetti, E., Yaremko, B., Nechuta, T., Varner, J., Liu, M., Smith, E. M., Neustadt, B., Presta, M., and Armstrong, L. (2001). Inhibition of angiogenesis and tumor growth by SCH 221153, a dual $\alpha_v\beta_3$ and $\alpha_v\beta_5$ integrin receptor antagonist. *Cancer Res.* **61**, 2232–2238.

Lin, T. H., Chen, Q., Howe, A., and Juliano, R. L. (1997). Cell anchorage permits efficient signal transduction between Ras and its downstream kinases. *J. Biol. Chem.* **272**, 8849–8852.

Liotta, L. A., Steeg, P. S., and Steller-Stevenson, G. S. (1991). Cancer metastasis and angiogenesis: An imbalance of positive and negative regulation. *Cell* **64**, 327–336.

McDonald, D. M., Munn, L., and Jain, R. K. (2000). Vasculogenic mimicry: How convincing, how novel, and how significant? *Am. J. Pathol.* **156**, 383–388.

Meredith, J. E., Fazeli, B., and Schwartz, M. A. (1993). The extracellular matrix as a cell survival factor. *Mol. Biol. Cell* **4**, 953–961.

Miller, W. H., Deenan, R. M., Willette, R. N., and Lark, M. W. (2000). Identification and *in vivo* efficacy of small-molecule antagonists of integrin $\alpha_v\beta_3$ (the vitronectin receptor). *Drug Discovery Today* **5**, 397–408.

Miyamoto, S., Teramoto, H., Cosco, O. A., Gutkind, J. S., Burbelo, P. D., Akiyama, S. K., and Yamada, K. M. (1995). Integrin function: Molecular hierarchies of cytoskeletal and signaling molecules. *J. Cell Biol.* **131**, 791–805.

Miyamoto, S., Teramoto, H., Gutkind, J. S., and Yamada, K. M. (1996). Integrins can collaborate with growth factors for phosphorylation of receptor tyrosine kinases and MAP kinase activation: Roles of integrin aggregation and occupancy of receptors. *J. Cell Biol.* **135**, 1633–1640.

Moro, L., Venturino, M., Bozzo, C., Silengo, L., Altruda, F., Beguinot, L., Tarone, G., and Defilippi, P. (1998). Integrins induce activation of EGF receptor: Role in MAP kinase induction and adhesion-dependent cell survival. *EMBO J.* **17**, 6622–6632.

Oktay, M., Wary, K. K., Dans, M., Birge, R. B., and Giancotti, F. G. (1999). Integrin-mediated activation of focal adhesion kinase is required for signaling to Jun NH2-terminal kinase and progression through the G1 phase of the cell cycle. *J. Cell Biol.* **145**, 1461–1470.

Pasqualini, R., Koivnem, E., and Ruoslahti, E. (1997). Alpha v integrins as receptors for tumor targeting by circulating ligands. *Nat. Biotechnol.* **15**, 542–546.

Petitclerc, E., Stromblad, S., von Schalsck, T. L., Mitjans, F., Piulats, J., Montgomery, A. M. P., Cheresh, D. A., and Brooks, P. A. (1999). Integrin $\alpha_v\beta_3$ promotes M21 melanoma growth in human skin by regulating tumor cell survival. *Cancer Res.* **59**, 2724–2730.

Renshaw, M. W., Ren, X., and Schwartz, M. A. (1997). Growth factor activation of MAP kinase requires cell adhesion. *EMBO J.* **16**, 5592–5599.

Ruoslahti, E. (1996). RGD and other recognition sequences for integrins. *Annu. Rev. Cell Dev. Biol.* **12**, 697–715.

Ruoslahti, E. (1999). Fibronectin and its integrin receptors in cancer. *Adv. Cancer Res.* **76**, 1–20.

Ruoslahti, E. R. D. (2000). An address system in the vasculature of normal tissues and tumors. *Annu. Rev. Immunol.* **18**, 813–827.

Ruoslahti, E., and Reed, J. (1994). Anchorage dependence, integrins, and apoptosis. *Cell* **77**, 477–478.

Rusnati, M. T. E., Dell'Era, P., Gualandris, A., and Presta, M. (1997). $\alpha_v\beta_3$ Integrin mediates the cell-adhesive capacity and biological activity of basic fibroblast growth factor (FGF2) in cultured endothelial cells. *Mol. Biol. Cell* **8**, 2449–2461.

Schlaepfer, D. D., Hanks, S. K., Hunter, T., and Van der Geer, P. (1994). Integrin-mediated signal transduction linked to Ras pathway by GRB2 binding to focal adhesion kinase. *Nature* **372**, 786–791.

Schneller, M., Vuori, K., and Ruoslahti, E. (1997). $\alpha_v\beta_3$ integrin associates with activated insulin and PDGFβ receptors and potentiates the biological activity of PDGF. *EMBO J.* **16**, 5600–5607.

Schuttman, M., Zhurinsky, J., Simcha, I., Albanese, C., D'Amico, C., Pestell, R., and Ben-Ze'ev, A. (1999). The cyclin D_1 gene is a target of the β-catenin/LEF-1 pathway. *Proc. Natl. Acad. Sci. U.S.A.* **96**, 5522–5527.

Schwartz, M. A., Toksoz, D., and Khosravi-Far, R. (1996). Transformation by Rho exchange factor oncogenes is mediated by activation of an integrin-dependent pathway. *EMBO J.* **15**, 6525–6530.

Serini, G., Trusolino, L., Saggiorato, E., Cremona, O. L., Derossi, M. L., Angeli, A., Orlandi, F., and Marchisio, P. C. (1996). Changes in integrin and E-cadherin expression in neoplastic versus normal thyroid tissue. *J. Natl. Cancer Inst.* **88**, 442–449.

Shattil, S. J. (1995). Function and regulation of the β_3 integrins in hemostasis and vascular biology. *Thromb. Haemostat.* **10**, 3–10.

Shimoyama, S., Gansauge, F., Gansauge, S., Oohara, T., and Beger, H. G. (1995). Altered expression of extracellular matrix molecules and their receptors in chronic pancreatitis and pancreatic adenocarcinoma in comparison with normal pancreas. *Int. J. Pancreatol.* **18**, 227–234.

Soldi, R., Mitola, S., Strasly, M., Defilippi, P., Tarone, G., and Bussolino, F. (1999). Role of $\alpha_v\beta_3$ integrin in the activation of vascular endothelial growth factor receptor-2. *EMBO J.* **18**, 882–892.

Stromblad, S., Becker, J. C., Yebra, M., Brooks, P. C., and Cheresh, D. A. (1996). Suppression of p53 activity and p21WAF1/CIP1 expression by vascular cell integrin $\alpha_v\beta_3$ during angiogenesis. *J. Clin. Invest.* **98**, 426–433.

Tani, T., Karttunen, T., Kiviluoto, T., Kivilaakso, E., Burgeson, R. E., Sipponen, P., and Virtanen, I. (1996). $\alpha_6\beta_4$ integrin and newly deposited laminin-1 and laminin-5 form the adhesion mechanism of gastric carcinoma. Continuous expression of laminins but not that of collagen VII is preserved in invasive parts of the carcinomas: Implication for acquisition of the invading phenotype. *Am. J. Pathol.* **149**, 781–793.

Tetsu, O., and McCormick, F. (1999). β-catenin regulates expression of cyclin D_1 in colon carcinoma cells. *Nature* **398**, 422–426.

Toksoz, D., and Williams, D. A. (1994). Novel human oncogene *lbc* detected by transfection with distinct homology regions to signal transduction products. *Oncogene* **9**, 621–628.

Varner, J. A., and Cheresh, D. A. (1996). Tumor angiogenesis and the role of vascular cell integrin $\alpha_v\beta_3$. *In* "Important Advances in Oncology" (V. T. DeVita, S. Hellman, and S. A. Rosenberg, eds.), pp. 69–87. Lippincott-Raven, Philadelphia.

Varner, J. A., Brooks, P. C., and Cheresh, D. A. (1995). REVIEW: The integrin $\alpha_v\beta_3$: Angiogenesis and apoptosis. *Cell Adhes. Commun.* **3**, 367–374.

Vuori, K., and Ruoslahti, E. (1994). Association of insulin receptor substrate-1 with integrins. *Science* **266**, 1576–1578.

Wary, K. K., Mainiero, F., Isakoff, S. J., Marcanonio, E. E., and Giancotti, F. G. (1996). The adaptor protein Shc couples a class of integrins to the control of cell cycle progression. *Cell* **87**, 733–743.

Zhang, Z., Vuori, K., Reed, J. C., and Ruoslahti, E. (1995). The $\alpha_5\beta_1$ integrin supports survival of cells on fibronectin and up-regulates Bcl-2 expression. *Proc. Natl. Acad. Sci. U.S.A.* **92**, 6161–6165.

Zhu, X., Ohtsubo, M., Bohmer, R., Roberts, J. M., and Assoian, R. K. (1996). Adhesion-dependent cell cycle progression linked to the expression of cyclin D_1, activation of cyclin E-cdk2, and phosphorylation of the retinoblastoma protein. *J. Cell Biol.* **133**, 391–403.

Functional Rescue of Mutant p53 as a Strategy to Combat Cancer

Galina Selivanova and Klas G. Wiman[1]

Department of Oncology–Pathology
Karolinska Institute
Cancer Center Karolinska
Karolinska Hospital
SE-171 76 Stockholm, Sweden

I. Introduction
II. Multiple Pathways of p53-Induced Apoptosis
III. Regulation of p53 Activity
IV. Approaches toward Reactivation of Mutant p53
V. Implications for Tumor Therapy and Future Perspectives
References

I. INTRODUCTION

p53 responds to a variety of stress conditions, for instance DNA damage and hypoxia (for a review, see Prives and Hall, 1999). Moreover, expression of several proliferation-promoting genes and proto-oncogenes such as cellular c-myc, E2F-1, and ras, and viral E1A and SV40 large T antigen, induce p53 accumulation and a p53-dependent biological response (see Sharpless and DePinho, 1999). Induced p53 triggers cell cycle arrest and/or cell death by apoptosis. p53 is a transcription factor that binds specifically to DNA and regulates expression of a set of target genes. The p21, GADD45, and $14\text{-}3\text{-}3\text{-}\sigma$ genes are effectors for

---

[1]To whom correspondence should be addressed.

Tumor-Suppressing Viruses, Genes, and Drugs
Innovative Cancer Therapy Approaches
Copyright © 2002 by Academic Press.
All rights of reproduction in any form reserved.

p53-induced cell cycle arrest whereas bax, Fas, and the PIG genes are involved in p53-dependent apoptosis. p53 can also transrepress certain genes, including bcl-2, MAP4, presenelin 1, and the insulin-like growth factor 1 (IGF1) receptor gene (reviewed by Gottlieb and Oren, 1998). Both transactivation and transrepression of specific genes are important for p53-dependent apoptosis.

Inactivation of p53 by point mutation or deletion occurs in at least 50% of all human tumors. Tumor-derived p53 mutations cluster in the core domain that harbors p53:s specific DNA binding activity, indicating that sequence-specific transcriptional transactivation of target genes is crucial for the execution of p53:s tumor suppressor function (reviewed in Ko and Prives, 1996). If one takes into account alterations of upstream and downstream components of the p53 pathway and overexpression of p53-inhibiting cellular and viral proteins (see later), loss of p53 function appears to be an almost universal event in tumor development.

II. MULTIPLE PATHWAYS OF p53-INDUCED APOPTOSIS

The importance of transcriptional regulation by p53 for the induction of apoptosis has been demonstrated in a number of studies, and was elegantly confirmed in experiments using engineered mice and embryonic stem cells expressing transcriptionally defective p53 (Jimenez et al., 2000; Chao et al., 2000). p53 upregulates expression of Fas and TRAIL/KILLER/DR5, two cell surface death receptors from the tumor necrosis factor receptor (TNFR) family, both of which are very potent transmitters of apoptotic signals (reviewed by El-Deiry, 1998). In addition, p53 can induce depolimerization of microtubules through upregulation of microtubule-severing factor E1α (Kato et al., 1997), and downregulate the microtubule-stabilizing factor MAP4 (Duttaroy et al., 1998). Destabilization of microtubules leads to abrogation of the antiapoptotic activity of survivin and/or Bcl-2, resulting in apoptosis (Kato et al., 1997; Duttaroy et al., 1998; Murphy et al., 1996; Zhang et al., 1998).

Mitochondrial factors involved in p53-triggered apoptotic response include Bcl-2 family members Bcl-2 (Miyashita et al., 1994), Bax (McCurrach et al., 1997; Xiang et al., 1998; Yin et al., 1997), and Noxa (Oda et al., 2000). Simultaneous upregulation of the proapoptotic Bax and Noxa proteins and downregulation of the antiapoptotic Bcl-2 protein can presumably shift the equilibrium toward apoptosis. The p53-induced PIG genes that are involved in the generation of reactive oxygen species may also play an important role in the process of apoptosis (Polyak et al., 1997). The ability of p53 to induce cell death by blocking the IGF1 survival signaling through induction of IGF-BP3, a secreted protein that binds IGF1 (Buckbinder et al., 1995) and to repress the expression of the IGF1 receptor (Prisco et al., 1997) further illustrates how p53 may regulate the same cellular process at several levels.

Nuclear proteinsupregulated by p53 include cyclin G, recently shown to enhance apoptosis induced by TNF-α, retinoic acid, and serum starvation (Okamoto and Beach, 1994; Okamoto and Prives, 1999), and PAG608/wig-1, a nucleolar protein that induces apoptosis when ectopically overexpressed (Israeli *et al.*, 1997; Varmeh-Ziaie *et al.*, 1997).

Several studies have shown that p53 can trigger apoptosis in the absence of RNA and/or protein synthesis, that is, independently of transcription, at least in some cells (Wagner *et al.*, 1994; Caelles *et al.*, 1994; Bissonnette *et al.*, 1997). p53-mediated increase of Fas transport from cytoplasmic stores

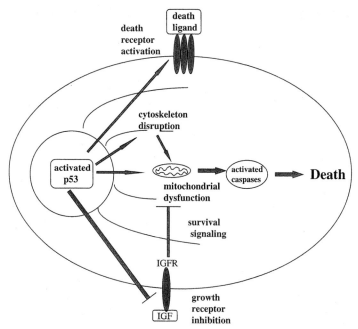

Figure 1   Multiple p53-induced apoptotic pathways. Upregulation of the Bax and PIG genes, along with downregulation of Bcl-2, induces aberrant mitochondrial function and release of cytochrome *c*, leading to caspase activation. Cytoskeleton disruption due to induction of the E1α microtubule-severing protein and downregulation of the microtubule-stabilizing MAP4 protein leads to abrogation of the apoptosis-protection function of Bcl-2 and survivin. Activation of expression of cell surface death receptors (e.g., Fas, DR5) transmits death signals to caspases. Block of survival signaling by downregulation of IGF1 receptor expression and upregulation of the secreted IGF1-binding IGF-BP3 protein sensitizes cells to apoptosis. p53 may also be involved in caspase activation thorough unknown protein–protein interactions.

to the cell surface is an example of a potential transcription-independent proapoptotic pathway (Bennett et al., 1998). The observation that p53 can induce caspase activation in cell-free extracts (Ding et al., 1998) supports the notion that protein–protein interactions could be important and even sufficient for p53-dependent apoptosis.

Available evidence supports the view that p53 induces apoptosis in a tissue- and signal-specific manner through several mechanisms, including both upregulation of apoptosis-promoting genes, downregulation of apoptosis-antagonizing genes, and transcription-independent mechanisms. Thus, as shown in Figure 1, p53 can transmit a death signal at many levels by regulating the expression of extracellular and cell surface proteins, cytoskeleton components, and mitochondrial and nuclear proteins.

III. REGULATION OF p53 ACTIVITY

Because p53 can trigger cell death, its activity must be rigorously controlled. This is achieved through two major pathways. First, the p53 protein is expressed at low levels due to rapid turnover mediated by ubiquitin-dependent degradation (Haupt et al., 1997; Kubbutat et al., 1997; Midgley and Lane, 1997). Second, newly synthesized p53 requires specific modifications to be able to bind specifically to DNA and transactivate target genes (Hupp and Lane, 1995). Oncogene activation, hypoxia, viral infection, ribonucleotide depletion, and DNA damage all cause a rapid increase in p53 levels due to protein stabilization. In addition, posttranslational modifications of the protein induced by stress conditions trigger a switch from a latent to an active form that binds DNA. Phosphorylation and dephosphorylation, acetylation, and/or protein–protein interactions may play a role for one or both levels of regulation (reviewed in Prives and Hall, 1999).

Various upstream stress signals seem to converge on the interaction between p53 and its negative regulator MDM-2. The levels of p53 and MDM-2 are balanced via two autoregulatory feedback loops. p53 transactivates the Mdm-2 gene that encodes a RING-finger protein that binds to the p53 transactivation domain and inhibits its transactivation function (Chen et al., 1993). Moreover, the MDM-2 protein functions as an E3 ubiquitin ligase that transports p53 from the nucleus to the cytoplasm, targeting it for proteasomal degradation (Haupt et al., 1997; Kubbutat et al., 1997; Tao and Levine, 1999; Honda and Yasuda, 1999). Under normal conditions, p53 is subjected to constitutive proteasomal degradation via MDM-2 and only becomes protected from destruction upon specific signals. Because the Mdm-2 gene is transcriptionally activated by p53, any signal that triggers p53 accumulation and activation will also induce MDM-2 and ultimately p53 degradation. MDM-2 thus acts to maintain low levels of p53 in the absence of stress signaling and terminates the p53 response to such signaling. In agreement with this notion, Mdm-2 null mice show an embryonic lethal phenotype, which is rescued in a p53 null background (Montes de Oca Luna *et al.*, 1995).

The association between MDM-2 and p53, as well as the E3 ubiquitin ligase activity of MDM-2, is regulated by phosphorylation events and binding to other factors such as p19/p14^{ARF} (see Prives, 1998). It is now clear that two major stress conditions that induce p53, that is, DNA damage and oncogene activation, both prevent MDM-2-mediated destruction of p53 but by different mechanisms. DNA damage induces phosphorylation of both p53 and MDM-2 which prevents their association (Shieh *et al.*, 1997, 1999, 2000; Unger *et al.*, 1999; Chehab *et al.*, 1999, 2000; Khosravi *et al.*, 1999). Oncogenes such as c-Myc, E2F1, activated ras, and E1A inhibit MDM-2 function by inducing the p19/p14^{ARF} protein that binds MDM-2 and inhibits ubiquitination of p53 (for a review, see Sherr and Weber, 2000). In addition, p53 represses transcription of the MDM-2 inhibitor p19/p14^{ARF}, creating another autoregulatory loop. Tumor cells with transcriptionally inactive mutant p53 express low levels of MDM-2, whereas p19/p14^{ARF} is often overexpressed. As a result, mutant p53 is usually expressed at high levels in tumor cells (Figure 2).

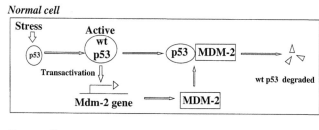

Figure 2   Accumulation of mutant p53 in tumor cells. In normal cells (top), p53 levels are low due to MDM-2-mediated proteasomal degradation. In tumor cells (bottom) mutant p53 is unable to activate the expression of its own destructor MDM-2. As a result, transcriptionally inactive p53 accumulates.

IV. APPROACHES TOWARD REACTIVATION OF MUTANT p53

At least 50% of human tumors carry mutant p53. The IARC p53 mutation database (http://www.iarc.fr/p53/index.html) now contains 15,000 entries—the largest number of reported mutations in one single gene accumulated so far. How can the frequent overexpression of nonfunctional mutant p53 in tumor cells be exploited for cancer therapy? We will discuss several possible approaches here.

More than 90% of p53 mutations in human tumors are clustered in the core domain. The great majority are point mutations resulting in substitution of only one amino acid residue (see Béroud and Soussi, 1998). The mutations are distributed throughout the p53 core domain, but they have one feature in common: they partially or completely abolish the specific DNA binding and transactivation function of p53, and consequently, the ability of p53 to suppress cell growth and induce apoptosis. Several residues, that is, R175, G245, R248, R249, R273, and R282, are mutated in tumors with an unusually high frequency, suggesting that they are particularly important for the tumor suppressor function of p53 (Béroud and Soussi, 1998). The resolution of the crystal structure of the p53 core domain bound to DNA has provided a framework for the understanding of the role of tumor-derived mutations for p53 activity (Cho et al., 1994). One of the most notable features revealed by the three-dimensional structure is that mutations are most frequent in the key regions of the core domain that are crucial for the specific DNA binding and maintenance of the proper folding of the core domain. So-called DNA contact mutants carry substitutions of residues that are directly involved in protein–DNA interactions (e.g., R248 and R273). On the other hand, so-called structural mutants have substitutions of residues that are important for the proper orientation of the two loops and loop–sheet–helix motifs that form the DNA binding surface. This destabilizes the secondary structure of the protein and thus abrogates DNA binding. Examples of structural mutants are R175, G245, R249, and R282 (Cho et al., 1994).

Is it possible to restore wild-type p53 function to mutant p53 proteins, in spite of their structural defects that abolish specific DNA binding? Yes, studies in several laboratories during recent years have demonstrated mutant p53 rescue, both with respect to specific DNA binding or transcriptional transactivation and biological response. First, it is clear that at least some mutant p53 proteins retain residual specific DNA binding. This is demonstrated by the fact that certain mutant p53 proteins that fail to bind DNA at 37°C can bind specifically to DNA at 32°C (Bargonetti et al., 1993; Zhang et al., 1994). Moreover, several common mutants can be reactivated for specific DNA binding by binding of PAb421, a monoclonal antibody that recognizes an epitope located in the p53 C-terminus. Similarly, C-terminal truncation and short synthetic peptides derived from the p53 C-terminus can reactivate the specific DNA binding of mutant p53 (Table 1). All these manipulations with the p53

Table 1 Reactivation of Tumor-Derived Mutant p53 Proteins

Location within the core domain	Mutation	Reactivation (+) or no reactivation (−) by		
		PAb421	Peptide	Low molecular weight compounds
β sandwich	K132E	$+^a$	ND^b	ND
	K132Q	$-^a$	$-^a$	ND
	C135Y	$-^a$	$-^a$	ND
	A138V	$-^a$	$+^a$	ND
	V143A	$-^a$	$+^{a,c}$	ND
	R156P	$-^a$	ND	ND
	L194F	$+^a$	ND	ND
Zn region	V173A	ND	ND	$+^{c,d}$
	R175H	$-^a$	$+^a$	ND
	R175S	ND	ND	$+^d$
	P177S	$+^a$	$+^a$	ND
	H179D	$-^a$	$-^a$	ND
DNA contact	S241F	ND	ND	$+^e$
	R248W	$-^a$	$+^a$	ND
	R248Q	$+^a$	$+^{f,d}$	ND
	R273H	$+^{a,c}$	$+^{a,c,f}$	$+^d$
	R273C	$+^a$	ND	ND
DNA region	G244C	$+^a$	$+^a$	ND
	G245S	$+^a$	$+^a$	ND
	R249S	$-^a$	$+^a$	$+^{c,d,e}$
	V272L	$+^a$	$+^a$	ND
	R280K	$-^a$	ND	ND
	D281G	$-^a$	$+^a$	ND
	R282P	$-^a$	$-^a$	ND
	R282W	$+^a$	ND	ND
	E285K	$+^a$	ND	ND

[a] Specific DNA binding.
[b] ND, Not determined.
[c] Sequence-specific transactivation function.
[d] Wild-type conformation.
[e] Tumor suppression function *in vivo*.
[f] p53-Dependent apoptosis.

C-terminus can also activate the specific DNA binding of latent wild-type p53 (Hupp et al., 1995), indicating a critical role for the p53 C-terminus in regulation of wild-type p53. Disruption of this negative regulation by PAb421 binding, competitive inhibition by an excess of C-terminal peptides, or C-terminal deletion, would thus allow specific DNA binding (reviewed by Selivanova et al., 1998). This may account for the ability of PAb421, C-terminal peptides, and C-terminal truncation to activate latent wild-type p53.

PAb421 has been shown to restore specific DNA binding to several p53 mutants, including both DNA contact mutants (e.g., R248Q and R273H) and structural mutants (e.g., P177S and G245S) (Table 1). Almost half of the mutants tested were reactivated by PAb421, including mutants that retain residual DNA binding as well as those that do not are susceptible to PAb421 reactivation. Furthermore, PAb421 can restore the transcriptional transactivation function to the R273H mutant p53 in living cells (Abarzúa et al., 1995). Also, single chain Fv fragments derived from the anti-p53 monoclonal antibodies PAb421 and 11D3 (similar to PAb421) restored the transactivation function of R273H mutant p53 in tumor cells (Caron de Fromentel et al., 1999). The specific DNA binding of both contact and structural p53 mutants was at least partially rescued by the deletion of the C-terminal regulatory domain (Wieczorek et al., 1996).

The finding that latent wild-type p53 could be activated for specific DNA binding in vitro by short peptides corresponding to the C-terminal inhibitory domain encouraged further studies aimed at reactivation of mutant p53 (Hupp et al., 1995). Could such peptides also reactivate mutant p53? Indeed, both contact and structural mutants were reactivated by a C-terminal peptide as shown by an in vitro DNA binding assay (Abarzúa et al., 1996). C-terminal peptides appear even more potent than antibody PAb421 (see Table 1). The potency of the reactivating peptide could be further improved by the addition of a 2000 molecular weight polyethylene glycol (PEG-2000) moiety to K381 (Abarzúa et al., 1996). Interestingly, glycerol was found to act as a chemical chaperone able to rescue some mutant p53 proteins with respect to transcriptional transactivation in cells (Ohnishi et al., 1999).

A C-terminal peptide was also able to rescue mutant p53 when microinjected or cotransfected in living cells (Table 1). Moreover, we showed that a C-terminal peptide corresponding to residues 361–382 in p53 could reactivate the two common mutant p53 proteins R248Q and R273H for induction of apoptosis. Efficient internalization of this C-terminal peptide in cells was achieved by fusion with a membrane-penetrating peptide derived from the Drosophila Antennapedia homeodomain protein (Derossi et al., 1994). This fusion peptide triggered p53-dependent growth suppression and/or death in different human tumor cell lines (Selivanova et al., 1997).

As discussed earlier, C-terminal truncation, binding of the monoclonal antibody PAb421, and synthetic peptides derived from the p53 C-terminus can probably eliminate negative regulation by the C-terminal tail of p53. However, the spectrum of p53 mutants that are rescued by C-terminal peptides is broader than that of mutants rescued by PAb421, suggesting that other mechanisms than disruption of C-terminal regulation also play a role for peptide-mediated mutant p53 reactivation. Consistent with this idea, C-terminal deletion in itself was not sufficient for full rescue of mutant p53, whereas the combination of C-terminal deletion and the introduction of a positively charged residue (T284R) restored potent growth suppression to mutant p53 (Wieczorek et al., 1996). Of note, the second-site mutations introduced in this study did not restore the wild-type conformation of p53.

We have studied the interaction of our C-terminal peptide with p53 and found that it binds to isolated wild-type as well as mutant p53 core domains (Selivanova et al., 1999). Peptide binding does not interfere with the specific DNA binding of the wild-type core domain, but in fact stimulates it. Moreover, a longer fragment of the p53 C-terminal domain protein added in trans restored specific DNA binding to mutant core domain proteins (Selivanova et al., 1999). There are at least two plausible mechanisms for the observed reactivation of mutant p53 by C-terminal peptides. First, since the peptides derived from the p53 C-terminus have a high content of basic Lys and Arg residues, their binding to the p53 core domain would introduce positive charges in the vicinity of the DNA contacting residues. This may create novel contacts with DNA, which may be particularly important for so-called DNA contact mutants. Second, the binding of C-terminal peptides to mutant p53 core domains might stabilize their folding and thus switch the equilibrium toward the wild-type conformation. This mechanism may be important for structural mutant p53 proteins.

An alternative approach for mutant p53 rescue is to generate second-site mutations that compensate for tumor-associated p53 mutations. The introduction of a positively charged residue in the vicinity of the DNA binding interface will serve to establish novel core domain–DNA contacts in DNA contact mutants. The deficient specific DNA binding and sequence-specific transactivation function of three hot spot mutants, R248Q, R273H, and R273C, could be compensated for by the substitution of Arg for Thr at position 284 (T284R). Moreover, the growth suppression function of the mutant p53 proteins was reconstituted by this substitution. Interestingly, combination of the T284R mutation and C-terminal truncation resulted in the most efficient growth suppression by mutant p53 (Wieczorek et al., 1996). This suggests that both abolishment of the negative regulatory function of the C-terminus and establishment of novel contacts with DNA are required for the complete rescue of mutant p53.

Another class of suppressor mutations was identified using a yeast-based selection of PCR-generated second-site mutations that reversed the effects of common p53 mutations (Brachmann *et al.*, 1998). The tumor-derived p53 mutants chosen for this study were known to have structural defects that result in partial unfolding of the core domain (Cho *et al.*, 1994). Some of them retained residual sequence-specific transactivation activity but were unable to induce apoptosis. Suppressor mutations were identified that restored the growth suppressor function of several mutants, including G244D, G244S, G245S, M246I, and R249S (partially), while some suppressor mutations only restored one specific mutant. For instance, N268D suppressed the V143A mutant only.

Structural studies using nuclear magnetic resonace (NMR) have shed more light on the molecular mechanism of second-site suppression. The N239Y and N268D suppressor mutations increase the global stability of the tumor-derived p53 mutants G245S and V143A (Nikolova *et al.*, 2000). The second-site mutation H168R is specific for the R249S mutant; it does not affect stability but restores DNA binding affinity, reverting some of the structural changes induced by the R249S mutation. With respect to implications for therapy, this study concluded that it is theoretically possible to restore the function of mutants such as V143A and G245S by small molecules that simply bind to and hence stabilize their native structure. According to the law of mass action, the unfolded mutant state will be converted to the active wild-type state. In contrast, some mutants, for example, R249S, will require alteration of the mutant structure and restoration of wild-type conformation (Nikolova *et al.*, 2000). It is interesting to note that the R249S mutant was susceptible to restoration by C-terminal peptide and low molecular weight compounds (see Table 1).

Measurements of the thermodynamic stability of tumor-derived mutant core domains combined with characterization of their residual DNA binding activity led to a more detailed classification of p53 mutants into distinct groups that correlate with four structural regions in the p53 core domain: DNA contact mutants, DNA-binding region mutants, Zn region mutants, and β sandwich mutants (Bullock *et al.*, 2000). Mutations in DNA-contacting residues (e.g., R273H) do not significantly destabilize the structure but affect DNA binding directly. DNA-binding region mutations result either in distorted folding with minimal DNA binding (as in R249S), partial destabilization with residual DNA binding at 20°C (G245S), or global denaturation with residual DNA binding (R282W). The V143A mutation in the β sandwich region also belongs to this class. The Zn region mutants R175H and C242S are globally denatured with abolished DNA binding. According to this classification, the G245S mutant is the most promising with regard to pharmacological rescue since its stability and DNA-binding affinity are not severely impaired. On the other hand, globally denatured β sandwich and Zn region mutants should be refractory to reactivation. However, as evident from Table 1, mutants from both the β sandwich and

Zn regions were functionally restored either by C-terminal peptide or low molecular weight compounds (see later). Thus, it appears difficult to predict from structural analyses how a specific p53 mutant protein will respond to functional reactivation.

The successful attempts to rescue mutant p53 discussed earlier have stimulated further work aimed at the identification of low molecular weight compounds that can reactivate mutant p53. Foster et al. (1999) screened a chemical library and found two small compounds, CP-31398 and CP-257042, that were able to preserve the wild-type-specific PAb1620 epitope upon heating of mutant p53 (mutants tested: R249S, R175H, R273H, V173A). CP-31398 also stabilized the active conformation of p53 (PAb1620⁺) in human tumor cells and in xenograft tumors in mice (mutant tested: V173A). p53 in a wild-type conformation accumulated in CP-31398-treated cells and was able to activate transcription (mutants V173A and R249S). Moreover, CP-31398 suppressed tumor growth in mice (mutants R249S and S241F) by up to 75%. CP-31398 and CP-257042 share some structural features, such as the presence of a hydrophobic group (polycyclic), a positively charged (amine) group, and a linker of certain length. The length of the linker was critical, possibly reflecting the necessity of precisely distancing or orienting two functional groups on the protein while providing a tether that enhances the stability of the active conformation.

A hydrophobic end present in both compounds most likely fits into a hydrophobic pocket of p53, whereas a positively charged group either binds to a negatively charged spot on p53 or facilitates stable DNA binding via additional DNA contacts. The exact molecular mechanism of p53 reactivation by compounds CP-31398 and CP-257042 remains to be resolved. However, the doses required for p53 reactivation *in vivo* are most probably toxic, so new p53-reactivating compounds with lower toxicity should be identified. Nonetheless, these studies have clearly demonstrated that mutant p53 can be rescued by small molecules in living cells.

In summary, more than 60% of all mutant p53 proteins tested so far were amenable to reactivation of specific DNA binding and/or sequence-specific transactivation, and in selected cases, induction of a p53-dependent biological response. Further structural studies, preferably of the cocrystal structure of full-length p53 and a reactivating molecule, will hopefully provide a more complete understanding of the mechanism behind mutant p53 reactivation. Taken together, the results obtained thus far suggest that pharmacological rescue of mutant p53 as a strategy for cancer therapy should be applicable to a large fraction of human tumors that carry mutant p53.

A principally different approach is based on restoration of mutant p53 mRNA rather than protein. Trans-splicing ribozymes were designed that were able to convert mutant p53 transcripts into wild-type in living cells (Watanabe and Sullenger, 2000). This resulted in reduction of mutant p53 expression along with the induction of wild-type p53 protein. Wild-type p53 generated in this fashion was functionally active, since it was able to induce the transcription of a p53 responsive reporter gene. This approach is promising, particularly if more efficient ribozyme delivery systems will be developed.

V. IMPLICATIONS FOR TUMOR THERAPY AND FUTURE PERSPECTIVES

Both chemotherapy and radiotherapy of cancer is aimed at triggering tumor cell death by apoptosis. Unfortunately, tumors often acquire resistance to such therapy, leading to relapse and progressive disease. Because both chemotherapeutic drugs and irradiation cause DNA damage and because p53 is a key component in the cellular response to DNA damage, inactivation of p53 in tumor cells has been viewed as a major obstacle for successful conventional anticancer therapy. Using a mouse tumor model, Lowe and colleagues (1994) showed that the efficacy of various anticancer drugs correlates with their ability to induce a p53-dependent apoptotic response. Furthermore, human tumors that are usually resistant to chemotherapy (e.g., lung, colon, and esophagus cancers) more often carry p53 mutations than chemosensitive tumors (e.g., testis tumors, certain lymphomas, and breast cancers), although exceptions exist. Many clinically used anticancer agents appear more active in the presence of wild-type p53 (Weinstein et al., 1997), whereas some are not (Cote et al., 1997). Nonetheless, clinical correlative studies of the relationship between the presence of p53 mutation and sensitivity to anticancer drugs and radiotherapy have generally indicated that p53 mutation is associated with therapy resistance and poor prognosis. For instance, p53 mutations were associated with a poor response to systemic adjuvant therapy along with radiotherapy in node-positive breast tumors (Bergh et al., 1995). Moreover, resistance to therapy at relapse has been shown to correlate with the acquisition of p53 mutations (Fisher, 1994). All this information strongly supports the need for efficient therapeutic methods to restore p53 function in tumor cells. Functional rescue of mutant p53 should restore p53-dependent apoptosis and, furthermore, greatly increase the sensitivity of tumor cells to chemotherapy and radiotherapy.

p53 activation in one tumor cell may inhibit growth and/or induce apoptosis in neighboring cells as well (Nishizaki et al., 1999). This so-called bystander effect may increase the efficiency of p53-activating cancer therapy, and also significantly improve the therapeutic effect of p53 gene therapy where only a fraction of the tumor cells can be reconstituted with wild-type p53. The bystander effect could be due to several mechanisms. First, the list of p53-responsive genes includes several secreted growth suppressors, such as IGF-BP3, transforming growth factor (TGF)-$\beta2$, inhibin-β, and serine protease inhibitors (Komarova et al., 1998). The secretion of such growth inhibitors from cells in which p53 is activated could suppress growth of adjacent cells in a tumor. Second, p53-mediated angiogenesis inhibition may contribute to the bystander effect. p53 can promote apoptosis indirectly by creating hypoxic conditions through inhibition of blood vessel formation. This effect involves transcriptional activation of angiogenesis inhibitors such as thrombospondin I and brain-specific angiogenesis inhibitor GD-AiF/BAL1, as well as downregulation of VEGF (vascular endothelial growth factor) expression (Dameron et al., 1994; Van Meir et al., 1994; Nishimori et al., 1997; Bouvet et al., 1998). Furthermore, a new class of molecular targets of p53 that have the potential to negatively regulate tumor invasion and metastasis has been identified. It includes tumor invasion and metastasis suppressors plasminogen activator inhibitor type 1 (PAI-1), KAI1, and maspin (Kunz et al., 1995; Mashimo et al., 1998; Zou et al., 2000).

Although p53 gene therapy has proven effective as a means to curb tumor growth or even completely eliminate tumors in some patients (see Wiman, 1999), this approach clearly has limitations. In particular, the p53 virus must be injected directly into the tumor lesion, since there is as yet no method for specific tumor cell targeting. This precludes efficient treatment of patients with disseminated disease. On the other hand, a therapeutic strategy based on a p53-reactivating drug should not require selective targeting of tumor cells. First, while reconstitution of p53 function would suppress growth and trigger apoptosis in tumor cells, there are reasons to assume that normal cells would respond mainly by transient growth arrest (Baker et al., 1990; Kim et al., 1999; D'Orazi et al., 2000). This could be due to differences between normal and tumor cells with regard to the control of p53 stability and/or the fact that p53-induced G1 cell cycle arrest is disrupted in many tumor cells due to inactivation of pRb. Moreover, the p16–pRb pathway is inactivated in most tumor cells (reviewed by Bartek et al., 1999). Since pRb normally suppresses apoptosis (Tsai et al., 1998), inactivation of this pathway presumably increases the sensitivity of tumor cells to diverse proapoptotic signals. Similarly, activation of c-myc and other oncogenes also sensitizes tumor cells to apoptosis (for a review, see Evan and Littlewood, 1998). p53 mutation may prevent oncogene-induced apoptotic signaling during tumor development. Reactivation of mutant p53 should trigger tumor cell death by restoring proapoptotic signaling to the death execution machinery. Consequently, normal cells that lack such proapoptotic signals should be unaffected.

Second, since mutant p53 proteins fail to induce expression of the MDM-2 protein that targets p53 for degradation, they usually accumulate at high levels in tumor cells (see Figure 2). In addition, it is possible that mutant p53 in tumor cells is modified by, for instance, phosphorylation and/or acetylation (see earlier) in response to oncogenic signaling and therefore in a sense is

Tumor cell

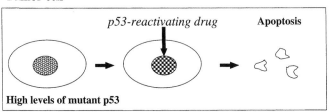

Normal cell

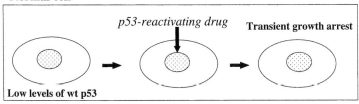

Figure 3 Selective induction of apoptosis by p53-reactivating drugs in tumor cells.
In normal cells with low p53 levels, activation of p53 will most likely
trigger transient growth arrest (top). In contrast, reactivation of mutant
p53 in tumor cells will trigger apoptosis due to high levels of mutant p53
and oncogene-induced sensitization for apoptosis (bottom).

already "activated". Thus, pharmacological rescue of its specific DNA binding
function should trigger a robust apoptotic response in tumor cells (Figure 3).

Some mutant p53 proteins may acquire new functions, for instance,
illegitimate activation of genes such as the multi-drug resistance (MDR) gene
(Dittmer et al., 1993). The observed resistance to therapy in tumors carrying
mutant p53 may therefore at least in part be due to a direct protective effect
against apoptosis induced by chemotherapy or radiotherapy (Blandino et al.,
1999). Moreover, it is conceivable that mutant p53 can inhibit the therapeutic
response by blocking p53-independent apoptotic pathways (Li et al., 1998).
Thus, functional rescue of mutant p53 appears advanageous for several reasons.

In summary, ample experimental evidence showing that the growth
suppressor activity of mutant p53 proteins representing both the DNA contact
and structural classes of mutants can be restored is now available. Mutant p53
reactivation as an anticancer strategy should thus be both feasible and applicable to many tumors. However, many questions remain, for instance whether
mutant p53 rescue will be sufficiently potent to induce tumor cell death *in vivo*
without affecting normal cells and tissues. Also, none of the molecules now
known to rescue mutant p53 will be useful as anticancer drugs themselves.

Detailed knowledge of the structure of a p53-reactivating small molecule when bound to p53 may help the design of more efficient and less toxic reactivators of p53. As an alternative approach, further screening of chemical libraries may lead to the identification of even more potent low molecular weight compounds that can rescue mutant p53 in human tumor cells. Work along these lines should provide a better understanding of how small molecules can rescue mutant p53 and should ultimately lead to efficient anticancer drugs. The first steps toward pharmacological rescue of mutant p53 in human tumors have already been taken.

Acknowledgments

We thank the Swedish Cancer Society (Cancerfonden), the Gustaf V Jubilee Fund, the Cancer Society of Stockholm, the Swedish Medical Research Council, and Åke Wibergs Stiftelse for generous support.

References

Abarzúa, P., LoSardo, J. E., Gubler, M. L., and Neri, A. (1995). Microinjection of monoclonal antibody PAb421 into human SW480 colorectal carcinoma cells restores the transcription activation function to mutant p53. Cancer Res. 55, 3490–3494.

Abarzúa, P., LoSardo, J. E., Gubler, M. L., Spathis, R., Lu, Y. A., Felix, A., and Neri, A. (1996). Restoration of the transcription activation function to mutant p53 in human cancer cells. Oncogene 13, 2477–2482.

Baker, S. J., Markowitz, S., Fearon, E. R., Willson, J. K., and Vogelstein, B. (1990). Suppression of human colorectal carcinoma cell growth by wild-type p53. Science 249, 912–915.

Bargonetti, J., Manfredi, J. J., Chen, X., Marshak, D. R., and Prives, C. (1993). A proteolytic fragment from the central region of p53 has marked sequence-specific DNA-binding activity when generated from wild-type but not from oncogenic mutant p53 protein. Genes Dev. 7, 2565–2574.

Bartek, J., Lukas, J., and Bartkova, J. (1999). Perspective: Defects in cell cycle control and cancer. J. Pathol. 187, 95–99.

Bennett, M., Macdonald, K., Chan, S., Luzio, J. P., Simari, R., and Weissberg, P. (1998). Cell surface trafficking of Fas: A rapid mechanism of p53-mediated apoptosis. Science 282, 290–293.

Bergh, J., Norberg, T., Sjogren, S., Lindgren, A., and Holmberg, L. (1995). Complete sequencing of the p53 gene provides prognostic information in breast cancer patients, particularly in relation to adjuvant systemic therapy and radiotherapy. Nat. Med. 1, 1029–1034.

Béroud, C., and Soussi, T. (1998). p53 gene mutation: Software and database. Nucleic Acids Res. 26, 200–204.

Bissonnette, N., Wasylyk, B., and Hunting, D. J. (1997). The apoptotic and transcriptional transactivation activities of p53 can be dissociated. Biochem. Cell Biol. 75, 351–358.

Blandino, G., Levine, A. J., and Oren, M. (1999). Mutant p53 gain of function: Differential effects of different p53 mutants on resistance of cultured cells to chemotherapy. Oncogene 18, 477–485.

Bouvet, M., Ellis, L. M., Nishizaki, M., Fujiwara, T., Liu, W., Bucana, C. D., Fang, B., Lee, J. J., and Roth, J. A. (1998). Adenovirus-mediated wild-type p53 gene transfer downregulates vascular endothelial growth factor expression and inhibits angiogenesis in human colon cancer. Cancer Res. 58, 2288–2292.

Brachmann, R. K., Yu, K., Eby, Y., Pavletich, N. P., and Boeke, J. D. (1998). Genetic selection of intragenic suppressor mutations that reverse the effect of common p53 cancer mutations. *EMBO J.* **17,** 1847–1859.

Buckbinder, L., Talbott, R., Velasco-Miguel, S., Takenaka, I., Faha, B., Seizinger, B. R., and Kley, N. (1995). Induction of the growth inhibitor IGF-binding protein 3 by p53. *Nature* **377,** 646–649.

Bullock, A. N., Henckel, J., and Fersht, A. R. (2000). Quantitative analysis of residual folding and DNA binding in mutant p53 core domain: Definition of mutant states for rescue in cancer therapy. *Oncogene* **19,** 1245–1256.

Caelles, C., Helmberg, A., and Karin, M. (1994). p53-dependent apoptosis in the absence of transcriptional activation of p53-target genes. *Nature* **370,** 220–223.

Caron de Fromentel, C., Gruel, N., Venot, C., Debussche, L., Conseiller, E., Dureuil, C., Teillaud, J. L., Tocque, B., and Bracco, L. (1999). Restoration of transcriptional activity of p53 mutants in human tumour cells by intracellular expression of anti-p53 single chain Fv fragments. *Oncogene* **18,** 551–557.

Chao, C., Saito, S., Kang, J., Anderson, C. W., Appella, E., and Xu, Y. (2000). p53 transcriptional activity is essential for p53-dependent apoptosis following DNA damage. *EMBO J.* **19,** 4967–4975.

Chehab, N. H., Malikzay, A., Stavridi, E. S., and Halazonetis, T. D. (1999). Phosphorylation of Ser-20 mediates stabilization of human p53 in response to DNA damage. *Proc. Natl. Acad. Sci. U.S.A.* **96,** 13777–13782.

Chehab, N. H., Malikzay, A., Appel, M., and Halazonetis, T. D. (2000). Chk2/hCds1 functions as a DNA damage checkpoint in G(1) by stabilizing p53. *Genes Dev.* **14,** 278–288.

Chen, J., Marechal, V., and Levine, A. J. (1993). Mapping of the p53 and mdm-2 interaction domains. *Mol. Cell. Biol.* **13,** 4107–4114.

Cho, Y., Gorina, S., Jeffrey, P. D., and Pavletich, N. P. (1994). Crystal structure of a p53 tumor suppressor–DNA complex: Understanding tumorigenic mutations. *Science* **265,** 346–355.

Cote, R. J., Esrig, D., Groshen, S., Jones, P. A., and Skinner, D. G. (1997). p53 and treatment of bladder cancer. *Nature* **385,** 123–125.

Dameron, K. M., Volpert, O. V., Tainsky, M. A., and Bouck, N. (1994). Control of angiogenesis in fibroblasts by p53 regulation of thrombospondin-1. *Science* **265,** 1582–1584.

D'Orazi, G., Marchetti, A., Crescenzi, M., Coen, S., Sacchi, A., and Soddu, S. (2000). Exogenous wt-p53 protein is active in transformed cells but not in their non-transformed counterparts: Implications for cancer gene therapy. *J. Gene Med.* **2,** 11–21.

Derossi, D., Joliot, A. H., Chassaing, G., and Prochiantz, A. (1994). The third helix of the Antennapedia homeodomain translocates through biological membranes. *J. Biol. Chem.* **269,** 10444–10450.

Ding, H. F., McGill, G., Rowan, S., Schmaltz, C., Shimamura, A., and Fisher, D. E. (1998). Oncogene-dependent regulation of caspase activation by p53 protein in a cell-free system. *J. Biol. Chem.* **273,** 28378–28383.

Dittmer, D., Pati, S., Zambetti, G., Chu, S., Teresky, A. K., Moore, M., Finlay, C., and Levine, A. J. (1993). Gain of function mutations in p53. *Nat. Med.* **4,** 42–46.

Duttaroy, A., Bourbeau, D., Wang, X. L., and Wang, E. (1998). Apoptosis rate can be accelerated or decelerated by overexpression or reduction of the level of elongation factor-1 α. *Exp. Cell Res.* **238,** 168–176.

El-Deiry, W. S. (1998). Regulation of p53 downstream genes. *Semin. Cancer Biol.* **8,** 345–357.

Evan, G., and Littlewood, T. (1998). A matter of life and cell death. *Science* **281,** 1317–1322.

Fisher, D. E. (1994). Apoptosis in cancer therapy: Crossing the threshold. *Cell* **78,** 539–542.

Foster, B. A., Coffey, H. A., Morin, M. J., and Rastinejad, F. (1999). Pharmacological rescue of mutant p53 conformation and function. *Science* **286,** 2507–2510.

Gottlieb, T. M., and Oren, M. (1998). p53 and apoptosis. *Semin. Cancer Biol.* **8,** 359–368.

Haupt, Y., Maya, R., Kazaz, A., and Oren, M. (1997). Mdm2 promotes the rapid degradation of p53. *Nature* **387,** 296–299.

Honda, R., and Yasuda, H. (1999). Association of p19(ARF) with Mdm2 inhibits ubiquitin ligase activity of Mdm2 for tumor suppressor p53. *EMBO J.* **18,** 22–27.

Hupp, T. R., and Lane, D. P. (1995). Two distinct signaling pathways activate the latent DNA binding function of p53 in a casein kinase II-independent manner. *J. Biol. Chem.* **270,** 18165–18174.

Hupp, T. R., Sparks, A., and Lane, D. P. (1995). Small peptides activate the latent sequence-specific DNA binding function of p53. *Cell* **83,** 237–245.

Israeli, D., Tessler, E., Haupt, Y., Elkeles, A., Wilder, S., Amson, R., Telerman, A., and Oren, M. (1997). A novel p53-inducible gene, PAG608, encodes a nuclear zinc finger protein whose over-expression promotes apoptosis. *EMBO J.* **16,** 4384–4392.

Jimenez, G. S., Nister, M., Stommel, J. M., Beeche, M., Barcarse, E. A., Zhang, X. Q., O'Gorman, S., and Wahl, G. M. (2000). A transactivation-deficient mouse model provides insights into trp53 regulation and function. *Nat. Genet.* **26,** 37–43.

Kato, M. V., Sato, H., Nagayoshi, M., and Ikawa, Y. (1997). Upregulation of the elongation factor-1α gene by p53 in association with death of an erythroleukemic cell line. *Blood* **90,** 1373–1378.

Khosravi, R., Maya, R., Gottlieb, T., Oren, M., Shiloh, Y., and Shkedy, D. (1999). Rapid ATM-dependent phosphorylation of MDM2 precedes p53 accumulation in response to DNA damage. *Proc. Natl. Acad. Sci. U.S.A.* **96,** 14973–14977.

Kim, A. L., Raffo, A. J., Brandt-Rauf, P. W., Pincus, M. R., Monaco, R., Abarzua, P., and Fine, R. L. (1999). Conformational and molecular basis for induction of apoptosis by a p53 C-terminal peptide in human cancer cells. *J. Biol. Chem.* **274,** 34924–34931.

Ko, L. J., and Prives, C. (1996). p53: Puzzle and paradigm. *Genes Dev.* **10,** 1054–1072.

Komarova, E. A., Diatchenko, L., Rokhlin, O. W., Hill, J. E., Wang, Z. J., Krivokrysenko, V. I., Feinstein, E., and Gudkov, A. V. (1998). Stress-induced secretion of growth inhibitors: A novel tumor suppressor function of p53. *Oncogene* **17,** 1089–1096.

Kubbutat, M. H., Jones, S. N., and Vousden, K. H. (1997). Regulation of p53 stability by Mdm2. *Nature* **387,** 299–303.

Kunz, C., Pebler, S., Otte, J., and von der Ahe, D. (1995). Differential regulation of plasminogen activator and inhibitor gene transcription by the tumor suppressor p53. *Nucleic Acids Res.* **23,** 3710–3717.

Li, R., Sutphin, P. D., Schwartz, D., Matas, D., Almog, N., Wolkowicz, R., Goldfinger, N., Pei, H., Prokocimer, M., and Rotter, V. (1998). Mutant p53 protein expression interferes with p53-independent apoptotic pathways. *Oncogene* **16,** 3269–3277.

Lowe, S. W., Bodis, S., McClatchey, A., Remington, L., Ruley, H. E., Fisher, D. E., Housman, D. E., and Jacks, T. (1994). p53 status and the efficacy of cancer therapy *in vivo*. *Science* **266,** 807–810.

McCurrach, M. E., Connor, T. M., Knudson, C. M., Korsmeyer, S. J., and Lowe, S. W. (1997). Bax-deficiency promotes drug resistance and oncogenic transformation by attenuating p53-dependent apoptosis. *Proc. Natl. Acad. Sci. U.S.A.* **94,** 2345–2349.

Mashimo, T., Watabe, M., Hirota, S., Hosobe, S., Miura, K., Tegtmeyer, P. J., Rinker-Shaeffer, C. W., and Watabe, K. (1998). The expression of the KAI1 gene, a tumor metastasis suppressor, is directly activated by p53. *Proc. Natl. Acad. Sci. U.S.A.* **95,** 11307–11311.

Midgley, C. A., and Lane, D. P. (1997). p53 protein stability in tumour cells is not determined by mutation but is dependent on Mdm2 binding. *Oncogene* **15,** 1179–1189.

Miyashita, T., Harigai, M., Hanada, M., and Reed, J. C. (1994). Identification of a p53-dependent negative response element in the bcl-2 gene. *Cancer Res.* **54,** 3131–3135.

Montes de Oca Luna, R., Wagner, D. S., and Lozano, G. (1995). Rescue of early embryonic lethality in mdm2-deficient mice by deletion of p53. *Nature* **378,** 203–206.

Murphy, M., Hinman, A., and Levine, A. J. (1996). Wild-type p53 negatively regulates the expression of a microtubule-associated protein. *Genes Dev.* **10,** 2971–2980.

Nikolova, P. V., Wong, K. B., DeDecker, B., Henckel, J., and Fersht, A. R. (2000). Mechanism of rescue of common p53 cancer mutations by second-site suppressor mutations. *EMBO J.* **19,** 370–378.

Nishimori, H., Shiratsuchi, T., Urano, T., Kimura, Y., Kiyono, K., Tatsumi, K., Yoshida, S., Ono, M., Kuwano, M., Nakamura, Y., and Tokino, T. (1997). A novel brain-specific p53-target gene, BAI1, containing thrombospondin type 1 repeats inhibits experimental angiogenesis. *Oncogene* **15,** 2145–2150.

Nishizaki, M., Fujiwara, T., Tanida, T., Hizuta, A., Nishimori, H., Tokino, T., Nakamura, Y., Bouvet, M., Roth, J. A., and Tanaka, N. (1999). Recombinant adenovirus expressing wild-type p53 is antiangiogenic: A proposed mechanism for bystander effect. *Clin. Cancer Res.* **5,** 1015–1023.

Oda, E., Ohki, R., Murasawa, H., Nemoto, J., Shibue, T., Yamashita, T., Tokino, T., Taniguchi, T., and Tanaka, N. (2000). Noxa, a BH3-only member of the Bcl-2 family and candidate mediator of p53-induced apoptosis. *Science* **288,** 1053–1058.

Ohnishi, T., Ohnishi, K., Wang, X., Takahashi, A., and Okaichi, K. (1999). Restoration of mutant TP53 to normal TP53 function by glycerol as a chemical chaperone. *Radiat. Res.* **151,** 498–500.

Okamoto, K., and Beach, D. (1994). Cyclin G is a transcriptional target of the p53 tumor suppressor protein. *EMBO J.* **13,** 4816–4822.

Okamoto, K., and Prives, C. (1999). A role of cyclin G in the process of apoptosis. *Oncogene* **18,** 4606–4615.

Polyak, K., Xia, Y., Zweier, J. L., Kinzler, K. W., and Vogelstein, B. (1997). A model for p53-induced apoptosis. *Nature* **389,** 300–305.

Prisco, M., Hongo, A., Rizzo, M. G., Sacchi, A., and Baserga, R. (1997). The insulin-like growth factor I receptor as a physiologically relevant target of p53 in apoptosis caused by interleukin-3 withdrawal. *Mol. Cell. Biol.* **17,** 1084–1092.

Prives, C. (1998). Signaling to p53: Breaking the MDM2–p53 circuit. *Cell* **95,** 5–8.

Prives, C., and Hall, P. A. (1999). The p53 pathway. *J. Pathol.* **187,** 112–126.

Selivanova, G., Iotsova, V., Okan, I., Fritsche, M., Strom, M., Groner, B., Grafstrom, R. C., and Wiman, K. G. (1997). Restoration of the growth suppression function of mutant p53 by a synthetic peptide derived from the p53 C-terminal domain. *Nat. Med.* **3,** 632–638.

Selivanova, G., Kawasaki, T., Ryabchenko, L., and Wiman, K. G. (1998). Reactivation of mutant p53: A new strategy for cancer therapy. *Semin. Cancer Biol.* **8,** 369–378.

Selivanova, G., Ryabchenko, L., Jansson, E., Iotsova, V., and Wiman, K. G. (1999). Reactivation of mutant p53 through interaction of a C-terminal peptide with the core domain. *Mol. Cell. Biol.* **19,** 3395–3402.

Sharpless, N. E., and DePinho, R. A. (1999). The INK4A/ARF locus and its two gene products. *Curr. Opin. Genet. Dev.* **9,** 22–30.

Sherr, C. J., and Weber, J. D. (2000). The ARF/p53 pathway. *Curr. Opin. Genet. Dev.* **10,** 94–99.

Shieh, S. Y., Ikeda, M., Taya, Y., and Prives, C. (1997). DNA damage-induced phosphorylation of p53 alleviates inhibition by MDM2. *Cell* **91,** 325–334.

Shieh, S. Y., Taya, Y., and Prives, C. (1999). DNA damage-inducible phosphorylation of p53 at N-terminal sites including a novel site, Ser20, requires tetramerization. *EMBO J.* **18,** 1815–1823.

Shieh, S. Y., Ahn, J., Tamai, K., Taya, Y., and Prives, C. (2000). The human homologs of checkpoint kinases Chk1 and Cds1 (Chk2) phosphorylate p53 at multiple DNA damage-inducible sites [published erratum appears in *Genes Dev.* 2000 Mar 15;14(6):750]. *Genes Dev.* **14,** 289–300.

Tao, W., and Levine, A. J. (1999). Nucleocytoplasmic shuttling of oncoprotein Hdm2 is required for Hdm2-mediated degradation of p53. *Proc. Natl. Acad. Sci. U.S.A.* **96,** 3077–3080.

Tsai, K. Y., Hu, Y., Macleod, K. F., Crowley, D., Yamasaki, L., and Jacks, T. (1998). Mutation of E2f-1 suppresses apoptosis and inappropriate S phase entry and extends survival of Rb-deficient mouse embryos. *Mol. Cell* **2,** 293–304.

Unger, T., Juven-Gershon, T., Moallem, E., Berger, M., Vogt Sionov, R., Lozano, G., Oren, M., and Haupt, Y. (1999). Critical role for Ser20 of human p53 in the negative regulation of p53 by Mdm2. *EMBO J.* **18,** 1805–1814.

Van Meir, E. G., Polverini, P. J., Chazin, V. R., Su Huang, H. J., de Tribolet, N., and Cavenee, W. K. (1994). Release of an inhibitor of angiogenesis upon induction of wild type p53 expression in glioblastoma cells. *Nat. Genet.* **8,** 171–176.

Varmeh-Ziaie, S., Okan, I., Wang, Y., Magnusson, K. P., Warthoe, P., Strauss, M., and Wiman, K. G. (1997). *Wig-1*, a new p53-induced gene encoding a zinc finger protein. *Oncogene* **15,** 2699–2704.

Wagner, A. J., Kokontis, J. M., and Hay, N. (1994). Myc-mediated apoptosis requires wild-type p53 in a manner independent of cell cycle arrest and the ability of p53 to induce p21waf1/cip1. *Genes Dev.* **8,** 2817–2830.

Watanabe, T., and Sullenger, B. A. (2000). Induction of wild-type p53 activity in human cancer cells by ribozymes that repair mutant p53 transcripts [in process citation]. *Proc. Natl. Acad. Sci. U.S.A.* **97,** 8490–8494.

Weinstein, J. N., Myers, T. G., O'Connor, P. M., Friend, S. H., Fornace, A. J., Jr., Kohn, K. W., Fojo, T., Bates, S. E., Rubinstein, L. V., Anderson, N. L., Buolamwini, J. K., van Osdol, W. W., Monks, A. P., Scudiero, D. A., Sausville, E. A., Zaharevitz, D. W., Bunow, B., Viswanadhan, V. N., Johnson, G. S., Wittes, R. E., and Paull, K. D. (1997). An information-intensive approach to the molecular pharmacology of cancer. *Science* **275,** 343–349.

Wieczorek, A. M., Waterman, J. L., Waterman, M. J., and Halazonetis, T. D. (1996). Structure-based rescue of common tumor-derived p53 mutants. *Nat. Med.* **2,** 1143–1146.

Wiman, K. G. (1999). p53 as a target for improved cancer therapy. *Emerging Therapeutic Targets* **3,** 347–353.

Xiang, H., Kinoshita, Y., Knudson, C. M., Korsmeyer, S. J., Schwartzkroin, P. A., and Morrison, R. S. (1998). Bax involvement in p53-mediated neuronal cell death. *J. Neurosci.* **18,** 1363–1373.

Yin, C., Knudson, C. M., Korsmeyer, S. J., and Van Dyke, T. (1997). Bax suppresses tumorigenesis and stimulates apoptosis *in vivo*. *Nature* **385,** 637–640.

Zhang, C. C., Yang, J. M., White, E., Murphy, M., Levine, A., and Hait, W. N. (1998). The role of MAP4 expression in the sensitivity to paclitaxel and resistance to vinca alkaloids in p53 mutant cells. *Oncogene* **16,** 1617–1624.

Zhang, W., Guo, X. Y., Hu, G. Y., Liu, W. B., Shay, J. W., and Deisseroth, A. B. (1994). A temperature-sensitive mutant of human p53. *EMBO J.* **13,** 2535–2544.

Zou, Z., Gao, C., Nagaich, A. K., Connell, T., Saito, S., Moul, J. W., Seth, P., Appella, E., and Srivastava, S. (2000). p53 regulates the expression of the tumor suppressor gene maspin. *J. Biol. Chem.* **275,** 6051–6054.

Index

Printed and bound by CPI Group (UK) Ltd, Croydon, CR0 4YY

08/05/2025

01864989-0001